AF698447

Fluids, Materials and Microgravity:
Numerical Techniques and Insights into Physics

Elsevier Internet Homepage - http://www.elsevier.com
Consult the Elsevier homepage for full catalogue information on all books, major reference works, journals, electronic products and services.

Elsevier Titles of Related Interest:

CRYSTAL GROWTH - FROM FUNDAMENTALS TO TECHNOLOGY
G. Müller
ISBN: 0-444-51386-8: 2004

CONCISE ENCYCLOPEDIA OF MATERIALS CHARACTERISATION
R. Cahn
ISBN: 0-08-044547-0: 2004

INTRODUCTION TO MATERIALS SCIENCE
J. Mercier
ISBN: 2-84-299286-5: 2003

ADVANCES IN SPACE BIOLOGY AND MEDICINE, VOLUME 5
S.L. Bonting
ISBN: 1-55938-296-1: 1996

Related Journals:
Elsevier publishes a wide-ranging portfolio of high quality research journals encompassing the topic of microgravity, and the behaviour of materials under specific conditions. A sample journal issue is available online by visiting the Elsevier web site (details at the top of this page). Leading titles include:

Acta Astronautica
Advances in Space Research (includes Cospar Information Bulletin)
Combustion and Flame
Journal of Biotechnology
Journal of Crystal Growth
Journal of Electrostatics
Tectonophysics
Materials Characterization
Materials Science and Engineering: A

All journals are available online via ScienceDirect: www.sciencedirect.com

To contact the Publisher
Elsevier welcomes enquiries concerning publishing proposals: books, journal special issues, conference proceedings, etc. All formats and media can be considered. Should you have a publishing proposal you wish to discuss, please contact, without obligation, the publisher responsible for Elsevier's Material Science programme:

Amanda Weaver
Publisher
Elsevier Ltd
The Boulevard, Langford Lane Phone: +44 1865 84 3634
Kidlington, Oxford Fax: +44 1865 843920
OX5 1GB, UK E.mail: a.weaver@elsevier.com

General enquiries, including placing orders, should be directed to Elsevier's Regional Sales Offices – please access the Elsevier homepage for full contact details (homepage details at the top of this page).

Fluids, Materials and Microgravity:
Numerical Techniques and Insights into Physics

Marcello Lappa

Napoli, Italy

2004

ELSEVIER

AMSTERDAM BOSTON HEIDELBERG LONDON NEW YORK OXFORD
PARIS SAN DIEGO SAN FRANCISCO SINGAPORE SYDNEY TOKYO

ELSEVIER B.V.
Sara Burgerhartstraat 25
P.O. Box 211
1000 AE, Amsterdam
The Netherlands

ELSEVIER Inc.
525 B Street
Suite 1900, San Diego
CA 92101-4495
USA

ELSEVIER Ltd.
The Boulevard
Langford Lane, Kidlington
Oxford OX5 1GB
UK

ELSEVIER Ltd.
84 Theobalds Road
London WC1X 8RR
UK

© 2004 Elsevier Ltd. All rights reserved.

This work is protected under copyright by Elsevier Ltd, and the following terms and conditions apply to its use:

Photocopying
Single photocopies of single chapters may be made for personal use as allowed by national copyright laws. Permission of the Publisher and payment of a fee is required for all other photocopying, including multiple or systematic copying, copying for advertising or promotional purposes, resale, and all forms of document delivery. Special rates are available for educational institutions that wish to make photocopies for non-profit educational classroom use.

Permissions may be sought directly from Elseviers Rights Department in Oxford, UK: phone (+44) 1865 843830, fax (+44) 1865 853333, e-mail: permissions@elsevier.com. Requests may also be completed on-line via the Elsevier homepage (http://www.elsevier.com/locate/permissions).

In the USA, users may clear permissions and make payments through the Copyright Clearance Center, Inc., 222 Rosewood Drive, Danvers, MA 01923, USA; phone: (+1) (978) 7508400, fax: (+1) (978) 7504744, and in the UK through the Copyright Licensing Agency Rapid Clearance Service (CLARCS), 90 Tottenham Court Road, London W1P 0LP, UK; phone: (+44) 20 7631 5555; fax: (+44) 20 7631 5500. Other countries may have a local reprographic rights agency for payments.

Derivative Works
Tables of contents may be reproduced for internal circulation, but permission of the Publisher is required for external resale or distribution of such material. Permission of the Publisher is required for all other derivative works, including compilations and translations.

Electronic Storage or Usage
Permission of the Publisher is required to store or use electronically any material contained in this work, including any chapter or part of a chapter.

Except as outlined above, no part of this work may be reproduced, stored in a retrieval system or transmitted in any form or by any means, electronic, mechanical, photocopying, recording or otherwise, without prior written permission of the Publisher. Address permissions requests to: Elsevier's Rights Department, at the fax and e-mail addresses noted above.

Notice
No responsibility is assumed by the Publisher for any injury and/or damage to persons or property as a matter of products liability, negligence or otherwise, or from any use or operation of any methods, products, instructions or ideas contained in the material herein. Because of rapid advances in the medical sciences, in particular, independent verification of diagnoses and drug dosages should be made.

First edition 2004

Library of Congress Cataloging in Publication Data
A catalog record is available from the Library of Congress.

British Library Cataloguing in Publication Data
A catalogue record is available from the British Library

ISBN: 0 08 044508 X

The paper used in this publication meets the requirements of ANSI/NISO Z39.48-1992 (Permanence of Paper).
Printed in The Netherlands.

Working together to grow
libraries in developing countries

www.elsevier.com | www.bookaid.org | www.sabre.org

ELSEVIER BOOK AID International Sabre Foundation

To my sons Francesco Romano and Arianna,
To my wife Maria Lucia, to my Parents
and to the memory of Natale Lappa

Preface

Gravity dominates everything on Earth, from the way life has developed to the way materials interact. But aboard a spacecraft orbiting the Earth, the effects of gravity are barely felt. In this "microgravity environment", scientists can conduct experiments that are all but impossible to perform on Earth. In this virtual absence of gravity as we know it, space flight gives scientists a unique opportunity to study the states of matter (solids, liquids and gases), and the forces and processes that affect them.

In practice, microgravity or near-weightlessness corresponds to a free-fall situation and in this condition various phenomena are significantly altered, in particular convection, buoyancy, hydrostatic pressure and sedimentation. When microgravity conditions are attained, fluids do "incredible" things.

Scientific disciplines affected include fluid physics and transport phenomena, combustion, crystal growth and solidification, biological processes and biotechnology.

Microgravity is instrumental in unraveling processes that are interwoven or overshadowed in normal gravity. It can therefore be regarded as an important tool for improving models of complex phenomena and hence manufacturing processes on Earth. Critical knowledge gained from microgravity experiments, in fact, is validating new, more complex models, accelerating the current trend towards predictable and reproducible phenomena, and enabling the development of new industrial processes (i.e. a commercial return from space research activities based on the application on the ground of the knowledge obtained in space).

Within this context the present book develops working engineering models that can be easily employed in applications, while providing a rigorous mathematical and numerical framework for deeper understanding and effective treatment of phenomena encountered in microgravity and/or unmasked by this environment.

Mathematical modeling is the art and craft of building a system of equations that is both sufficiently complex to do justice to physical reality and sufficiently simple to focus on the most significant aspects of the given situation. Numerical simulation is the art of solving these equations. In this book, a comprehensive study of fundamental concepts and simulation methods is presented. Partial differential equations are used as the basis for the methods. However, the analysis is not limited to these aspects. A number of prototype applications, in fact, is presented.

This means that the reader is taken beyond the theoretical to demonstrate how multiphase flow equations can be solved to provide applied, practical, predictive solutions to a variety of technological problems both on Earth and in space. Along these lines, a significant effort is provided throughout the volume to show how the prototype numerical examples fit the corresponding experimental processes. The consistency of numerical predictions with experimental data suggests, in fact, that rate-controlling steps are taken into account, that simplifications do not distort actual behavior and finally provides validation for the theoretical models and techniques.

It is also worthwhile to stress how the proposed simulations, exhibiting significant capabilities to predict and elucidate experimental observations, in turn lead to identify cause-and-effect relationships, i.e. they are propaedeutical to discerning heretofore unknown physical mechanisms. Thus the book can also be regarded as an additional step in the evolutionary progress towards the full understanding of the "physics of microgravity" and related processes.

New material is included along with a compilation of published material. The text gains information from fifty of the author's relevant and recent papers to illustrate the philosophy of modeling, the practical applications and the insights into the physics. It is conceived in order to be a useful reference guide for other specialists in these disciplines as well as an advanced level text for students taking part in courses on CFD (computational fluid dynamics), or on numerical methods for materials engineering and similar techniques. It is directed at readers already engaged or starting to be engaged in these topics. Engineers, designers and students will find the necessary numerical techniques and the revealing insights into the behavior of many phenomena usually overlooked and/or obscured by gravity (including historical developments and very recent contributions). Often a deductive approach is followed with systems of growing complexity being treated as the book progresses.

Numerical methods are important since they are a decisive tool to reduce the number of expensive space experiments. They are propaedeutical to plan and improve the experimental setups and to optimize the new production techniques suggested by the amount of knowledge obtained in space.

Superimposed on this is the fact that most of the scientists carrying out research on Earth (as well as the undergraduate and the Ph.D. students) cannot directly access microgravity platforms. For this reason the use of numerical computations is of paramount importance (often it is the only way) for the investigation of the properties of materials and fluids in the zero-g (simulated) environment, for the understanding of forces and processes affecting them (hidden or undervalued in normal gravity) and finally for the design of the aforementioned new manufacturing methods to be used on Earth.

The volume provides a rich learning environment covering inorganic, organic and living (tissues) materials and therefore could be used, in principle, by different groups of readers:

(1) Professionals and students in the metallurgy and foundry field; the text, in fact, covers the latest developments in the understanding of some still poorly known phenomena controlling the properties of cast materials, and indeed the properties of final solidified alloys;
(2) Researchers and scientists looking for new high-quality semiconductor crystals and related new production techniques needed to help advance progress in computer science and technology;
(3) Organic chemists and materials scientists who are now coordinating their efforts in designing macromolecular crystals for a variety of physical and chemical applications; organic crystal growers will also find in this book the detailed analysis of the various effects governing the morphological evolution of the crystals.
(4) Scientists, clinicians and engineers working in the new field of tissue engineering; the text offers many novel mathematical approaches including a detailed coverage of cell–tissue interactions at cellular and molecular levels; tissue surface kinetics, biochemical, and fluid-mechanical environments, etc.

The book is wide-ranging since the coverage reflects the multidisciplinary nature of space research.

Some unexpected theoretical kinships existing among the different subjects are elucidated and emphasized (for instance, those dealing with the presence of moving and/or interacting interfaces). Despite the very different genesis (inorganic, organic, biological, etc.), many problems are treated within the common framework of Volume of Fraction and Level-Set numerical methods and other similar Eulerian or Lagrangian techniques. This is an example of the fact that herein a large amount of information is transmitted from one field to the others in terms of models and numerical strategies. This philosophy is also used in the attempt to build a common source made available for the scientific community under the optimistic idea that the contacts established among the different fields will develop into an ongoing, mutually beneficial dialogue.

Napoli, 03 March 2004 Marcello Lappa

Acknowledgements

I would like to thank Professor N. Imaishi and Dr. S. Yasushiro (University of Kyushu, Institute for Materials, Chemistry and Engineering, Division of Advanced Device Materials, formerly Department of Advanced Material Study, Fukuoka) for some helpful discussions about the behavior of liquid metals under microgravity, Professor T. Hibiya (Fundamental Research Laboratories, NEC Corporation and University of Tokyo) for making possible the visit to the astonishing NEC private laboratories in Japan (Tsukuba) devoted to research on semiconductors, Dr. B.C. Sim and Professor A. Zebib (Rutgers University – New Jersey) and Dr. V. Shevtsova (University of Bruxelles) and Dr. H.C. Kuhlmann (University of Bremen) for some helpful discussions about the liquid-bridge problem, Professor A.A. Chernov (Cooperative Research in Biotechnology and Materials Science, Universities Space Research Association (USRA) at NASA Marshall Space Flight Center), Dr. L. Carotenuto (Microgravity Advanced Research and Support Center – MARS, Napoli) and Dr. F. Otàlora (Laboratorio de Estudios Cristalográficos, IACT, Campus Fuentenueva (Facultad de Ciencias)) for some helpful information about the kinetics of macromolecular crystals, Dr. T.G. Hammond (Tulane University Medical Center, Tulane Astrobiology Center, Center for BioEnvironmental Research), Professor A. Taber (Department of Biomedical Engineering, Washington University), and Dr. G. Vunjak-Novakovic (Harvard – Massachusetts Institute of Technology – Division of Health Sciences and Technology) for the very helpful data provided about the rotating bioreactors and the sensitivity of tissue to fluid-dynamic shear forces.

I am also indebted to Professor R. Monti (University of Naples, Italy) and Dr. Dainotti (Alenia Aerospace Industries) for the guidance provided in recent years, Dr. D. Sleeman (Elsevier Publishing Editor – Materials Science Group) for the necessary spark at the root of the book, Dr. C. Piccolo (MARS Center) for providing many experimental results and Dr. G. De Chiara (MARS Center) for preparing many figures and sketches. All the figures dealing with numerical simulations come from my own results, with the exception of Figs. 2.12 and 4.43 which were contributed by Professor Imaishi and Dr. Yasushiro.

I also acknowledge the Italian Aerospace Center (CIRA), the Italian Interuniversity Center for Supercomputing (CINECA) that allowed some calculations on the Silicon Graphics Power Challenge Supercomputer and on the Cray T3E massively parallel computer to be made and the Microgravity Advanced Research and Support Center (MARS) whose laboratories were used for conducting many of the experiments.

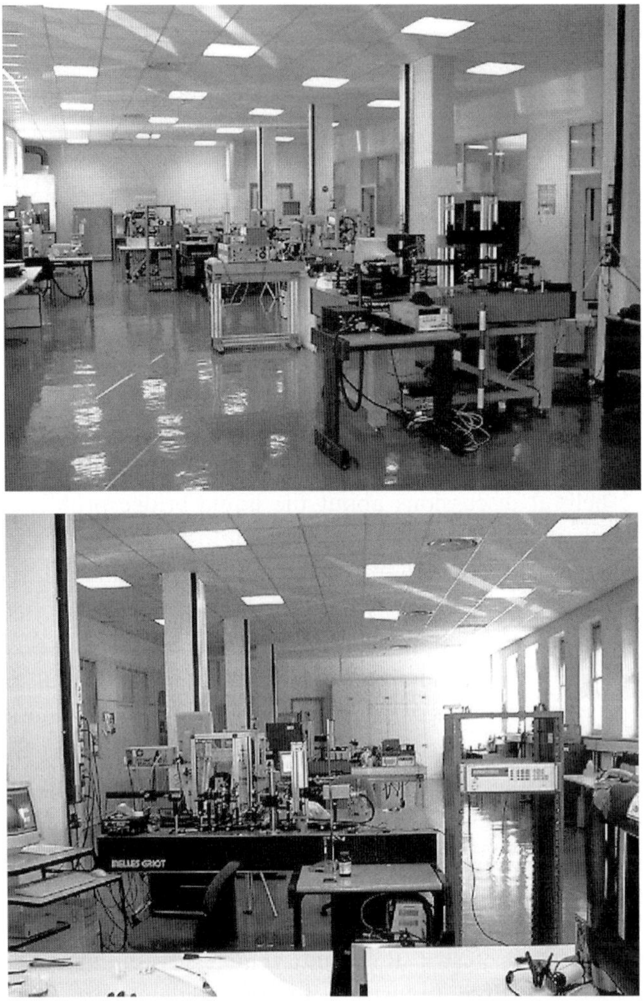

Microgravity Advanced Research and Support Center – Laboratories

Finally, I must thank my sons for their patience, and my wife for the support provided with the final stages of the book revision that I carried out during summer 2004 in the pleasant atmosphere of Villanmare-Vibonati (Sapri, Italy).

List of contents

CHAPTER 4

Growth of semiconductors: the floating zone technique

CHAPTER 6
Macromolecular crystal growth at macroscopic length scales

CHAPTER 7
The growth of biological tissues

CHAPTER 1

Space research

1.1. What is microgravity?

Gravitational attraction is a fundamental property of matter that exists throughout the known universe. The presence of Earth creates a gravitational field that acts to attract objects with a force inversely proportional to the square of the distance between the center of the object and the center of Earth. When the acceleration of an object acted upon only by Earth's gravity at the Earth's surface is measured, it is commonly referred to as one-g (1g) or one Earth gravity. This acceleration is approximately 9.8 meters per second squared (m/s^2). The weight of an object is the gravitational force exerted on it by Earth.

While the mass of an object is constant and the weight of an object is constant (ignoring differences in g at different locations on the Earth's surface), the environment of an object may be changed in such a way that its apparent weight changes.

Although gravity is a universal force, there are times, in fact, when it is not desirable to conduct scientific research under its full influence. In these cases, scientists perform their experiments in "microgravity," a condition in which the effects of gravity are greatly reduced, sometimes described as "weightlessness." This description brings to mind images of astronauts and objects floating around inside an orbiting spacecraft, seemingly free of Earth's gravitational field, but these images are misleading. The pull of Earth's gravity actually extends far into space. To reach a point where the Earth's gravity is reduced to one-millionth of that on Earth's surface, one would have to be 6.37 million kilometers away from Earth (almost 17 times farther away than the Moon). Since spacecraft usually orbit only 200–450 km above Earth's surface, there must be another explanation for the microgravity environment found aboard these vehicles (Rogers et al., 1997).

Astronauts floating in the Shuttle appear weightless, not because they have escaped Earth's gravity, but because they are in a state of free-fall. Any object falling due only to the force of gravity (free-falling) is experiencing microgravity and appears to be weightless. For example, a scale taped to an apple falling from a tree would register zero because it is falling with the apple. A spacecraft in a circular orbit is actually in a continuous state of free-fall at an altitude and speed that cause its fall to match the curvature of the Earth (a circular orbit results when the centripetal acceleration of uniform circular motion, v^2/r; v = velocity of the object, r = distance from the center of the object to the center of the Earth, is the same as that due to gravity). All objects carried by an orbiting spacecraft are also in a state of free-fall and so appear to be weightless.

As anticipated in the foregoing, the name given to such an environment is microgravity. The prefix micro- (μ) derives from the original Greek "mikros" meaning small. By this definition, a microgravity environment is one in which the apparent weight of a system is small compared to its actual weight due to gravity.

Quantitative systems of measurement, such as the metric system, commonly use micro to mean one part in a million. Using that definition, the acceleration experienced by an object in a microgravity environment would be one-millionth (10^{-6}) of that experienced at the Earth's surface. In practice, the microgravity environments used by scientific researchers range from about one per cent of Earth's gravitational acceleration (aboard aircraft in parabolic flight) to better than one part in a million (for example, onboard Earth-orbiting research satellites).

1.2. Microgravity facilities and platforms

A microgravity environment provides a unique laboratory in which scientists can observe and explore physical events, phenomena, and processes that are normally masked by the effects of Earth's gravity. For example, crystal growth from melt and solutions, capillary effects, multiphase flow, diffusive transport processes, can under normal gravity conditions be dominated by effects such as buoyancy, thermal and solutal convection and sedimentation.

Before embarking on the detailed coverage of these aspects, the following pages provide quite a broad overview of the different microgravity platforms and facilities to familiarize the reader with the microgravity environment and its peculiarities.

As discussed in Section 1.1, from a physical point of view, any object in free-fall experiences microgravity conditions, which occur when the object falls toward the Earth with an acceleration equal to that due to gravity alone.

Many factors, however, contribute to the effective experienced accelerations, i.e. the quality of the microgravity environment depends on the mechanism used to create it.

Very brief periods of microgravity can be achieved on Earth by dropping objects from tall structures. Longer periods are created through the use of airplanes, rockets, and spacecraft (see, e.g., Chen and Xiang, 1990). Research aircraft can expose experiments to a maximum of 30 s of low gravity while the aircraft approaches the top of a steep climb and begins a sharp descent. This parabolic curve is generally repeated many times during each flight, which is used primarily to perform experiments requiring short times for experimental equipment tests.

Low-cost sounding rockets, such as Space Processing Applications Rockets, also have parabolic flight paths – they ascend and then descend, rather than proceeding into orbit around the Earth. Sounding-rocket flights provide some minutes of low gravity. Although these periods of microgravity are brief, the test facilities are beneficial both for space-flight preparation and for some actual microgravity research.

The different facilities that can provide a microgravity environment for scientific research can be classified according to the time provided to perform the experiment (see Fig. 1.1).

1.2.1 *Drop towers and tubes*

These facilities are long shafts used for dropping experiments (see, e.g., Fig. 1.2). They essentially consist of a special tube, or pit, where vacuum can be realized and where a capsule with the payload is dropped. While the experiments drop, they are in free-fall, or microgravity conditions. Scientists make measurements and use video cameras to observe the experiments as they fall. The drop tubes are capable of accommodating small, uncontained material samples. Drop towers differ from drop tubes in their ability to accommodate large experimental packages, as opposed to the small uncontained material samples. A drop tower can use a drop shield, which contains the experimental package and isolates the experiment from aerodynamic

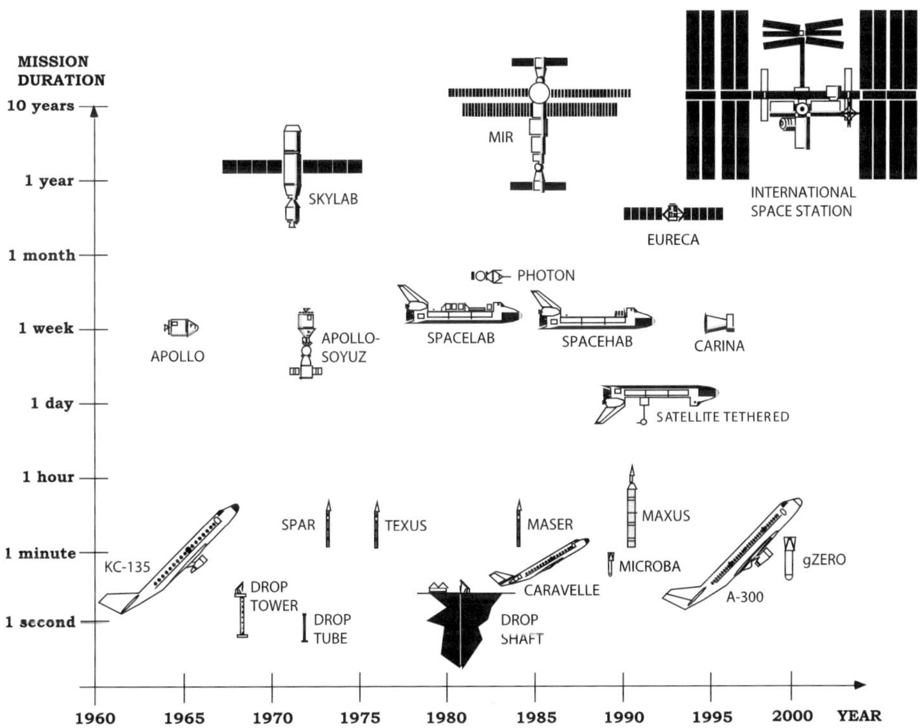

Fig. 1.1 Microgravity facilities and related microgravity time.

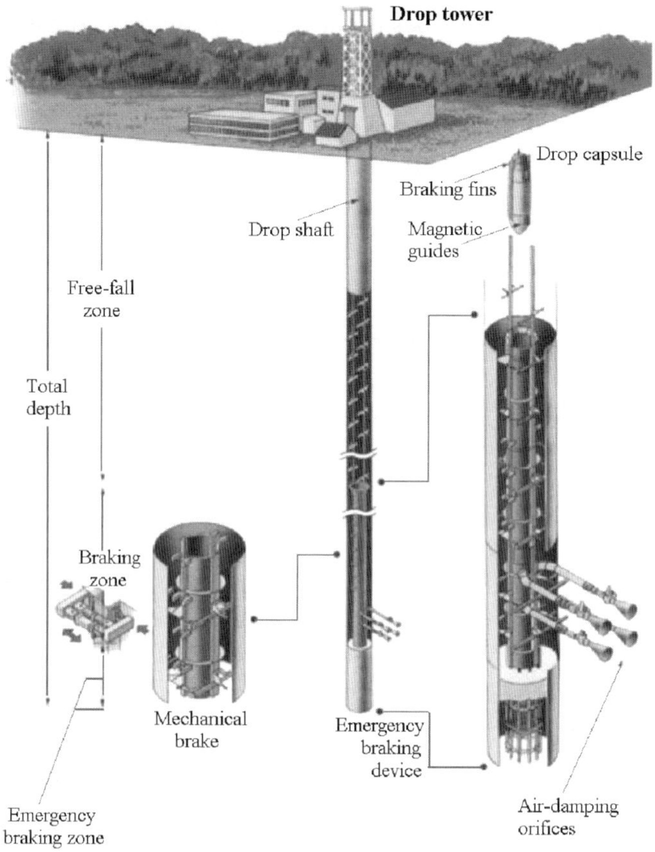

Fig. 1.2 Sketch of a drop tower.

drag during free-fall if air is present. In this way, it is possible to achieve few (3 or 4) seconds of microgravity by free-falling.

1.2.2 *Parabolic flights*

In a parabolic flight, the aircraft is put into a suborbital trajectory that provides free-fall, or weightlessness. Each cycle begins by having the aircraft perform an aerobatic maneuver that starts from level flight, and pitches up to approximately 45° nose-high and wings level subjecting the passengers to a 2-g pull up lasting about 20 s. After that, the aircraft engines are powered back and the airplane is launched into the same parabolic trajectory that a ball would follow, providing everyone inside the airplane with around 20 s of total weightlessness for experimentation purposes (Fig. 1.3). At the bottom of the parabola, the aircraft slowly pulls out of its

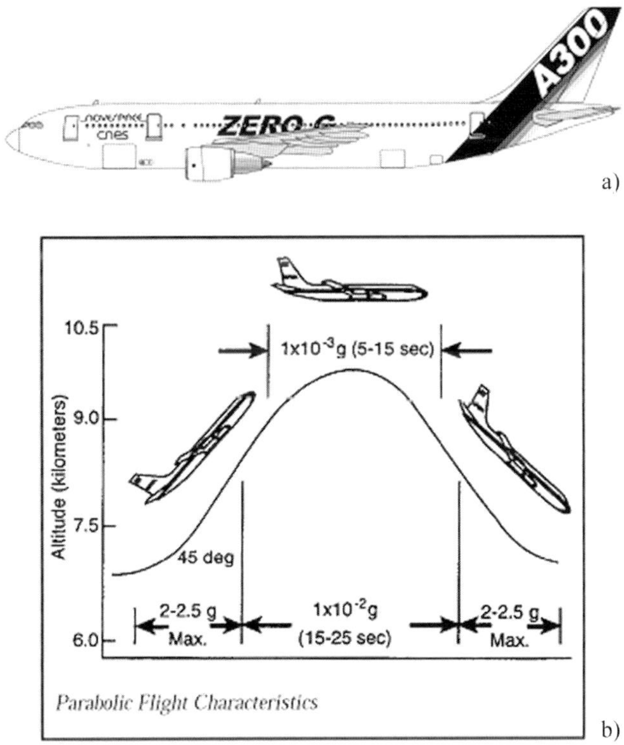

Figs. 1.3 (a) Aircraft used for parabolic flights, (b) sketch of suborbital parabolic trajectory.

dive and levels off for the next arc, restoring weight to the cabin. As many as 40 arcs can be flown on a typical flight so that scientists can conduct several experiments or can repeat short runs of one experiment many times. During the parabola, net accelerations drop as low as 1.5×10^{-2} g for about 15–25 s.

1.2.3 *Sounding rockets*

Sounding rockets take their name from the nautical term "to sound," which means to take measurements. These rockets are basically divided into two parts: a solid–fuel rocket motor and a payload. Many of the motors used in sounding-rocket programs are surplus military motors, which keep down the cost of the rocket. The payload is the section that carries the instruments to conduct the experiment and sends the data back to Earth (see, e.g., Monti et al., 1998a).

These rockets produce higher-quality microgravity conditions for longer periods than airplanes, or drop towers, and tubes. An experiment is placed on the rocket, which is launched and then allowed to free-fall back to Earth. A sounding rocket follows a parabolic arc, like the aircraft, but goes above the Earth's atmosphere, where air drag does not disturb microgravity conditions. The typical flight profile (see Fig. 1.4a) of a sounding rocket is the following: subsequent to a launch and as the rocket motor uses up its propellants it separates from the vehicle; the payload continues into space after separation from the motor(s) and begins conducting the experiments; when the experiments are completed, the payload re-enters the atmosphere and a parachute is deployed, bringing the payload gently back to Earth; the payload is then retrieved (by retrieving the payload a considerable saving can be achieved because the payload or parts of the payload and experiments can be refurbished and flown again).

The main difference between a sounding rocket and an orbital launch vehicle is the velocity reached. In fact, a sounding rocket does not reach the velocity (in terms of (km/s)) needed to go into orbit, and after achieving the maximum altitude comes back to Earth.

The experiments experience several minutes of microgravity before the rocket re-enters the atmosphere. Acceleration levels are usually around 10^{-5} g.

Therefore, sounding rockets provide a reasonably economical means of conducting engineering tests for instruments and devices used on satellites and other spacecraft, prior to their use in more expensive activities. Also, because of their low

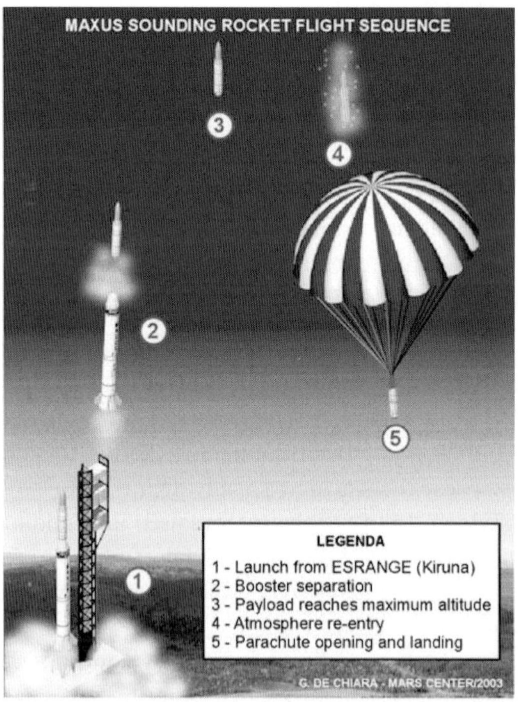

a)

Figs. 1.4 (a) Typical flight profile of a sounding rocket, (b) NASA sounding-rocket family, (c) Texus, Maser and Maxus sounding rockets.

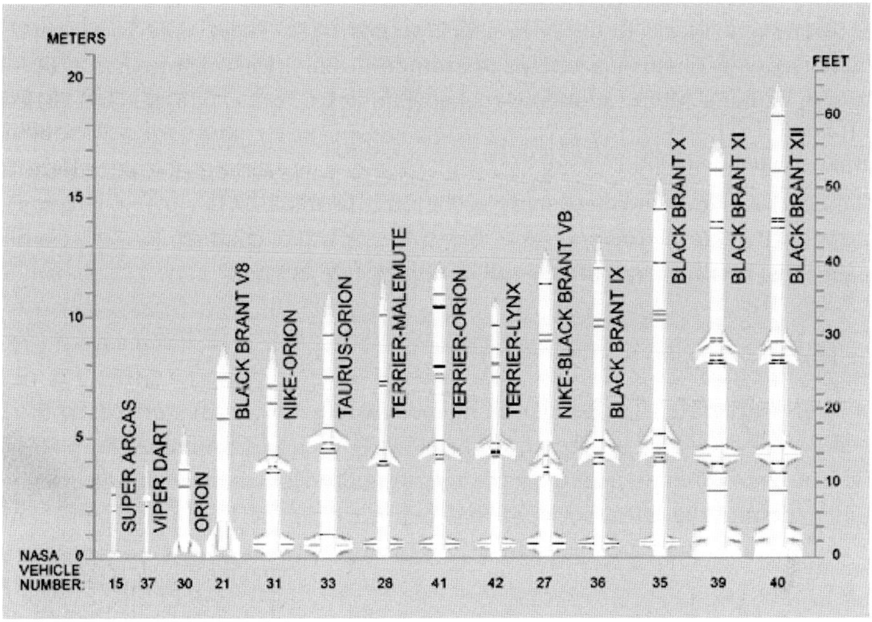

b)

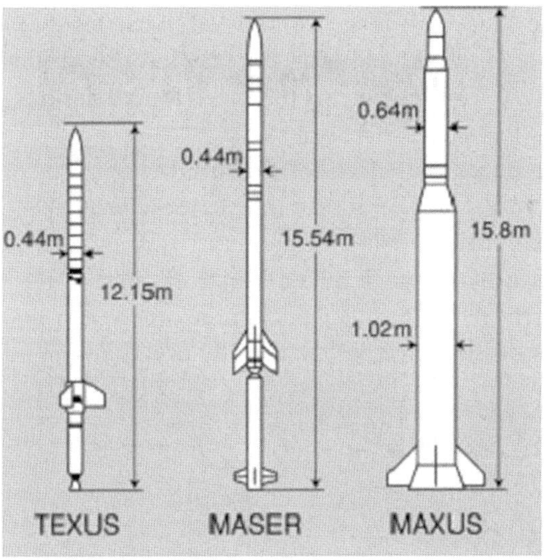

c)

Figs. 1.4 Continued.

cost and short mission lead time, they are valuable tools for undergraduate and graduate students conducting research in the microgravity environment.

NASA currently uses 15 different sounding rockets (Fig. 1.4b). The rockets come in a variety of sizes from the single-stage Super Arcas, which stands 7 ft (3 m) tall, to the four-stage Black Brant XII, which stands 65 ft (20 m) tall. These rockets can carry payloads of various weights to altitudes from 48 km to more than 1200 km.

Europe exploits three sounding-rocket families: Texus, Maser, and Maxus (Fig. 1.4c).

Although drop facilities, airplanes, and rockets can establish a reduced-gravity environment, all these facilities share a common problem. After a few seconds or minutes, Earth gets in the way and free-fall stops. To conduct longer scientific investigations, another type of free-fall is needed (see below).

1.2.4 *Foton recoverable capsule*

The Foton spacecraft is a Russian-built recoverable spacecraft derived from the manned Vostok/Voskhod spacecraft of the 1960s (Fig. 1.5). ESA chose this platform to carry out several experiments in the microgravity environment due to its good reliability (despite the latest accident that happened to the Foton M1 launcher in 2002) and the long period of permanence in orbit (2 weeks).

1.2.5 *The Space Shuttle*

The Space Shuttle is a spacecraft (Fig. 1.6) that often carries special laboratories for conducting scientific research. The Shuttle can stay in orbit around the Earth for up to 17 days, which is ideal for experiments that need longer periods in microgravity to be successful.

While a free-fall reduces the effects of gravity, being in an orbiting laboratory introduces, however, other accelerations that cause effects that are indistinguishable from those due to gravity. When a spacecraft is in orbit around the Earth, the orbit is actually defined by the path of the center of mass of the spacecraft around the center of the Earth. Any object in a location other than on the line traversed by the center of mass of the spacecraft is actually in a different orbit around Earth.

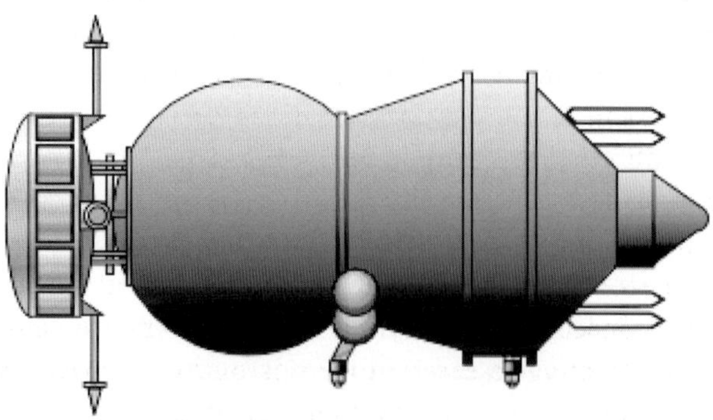

Fig. 1.5 The Foton recoverable capsule.

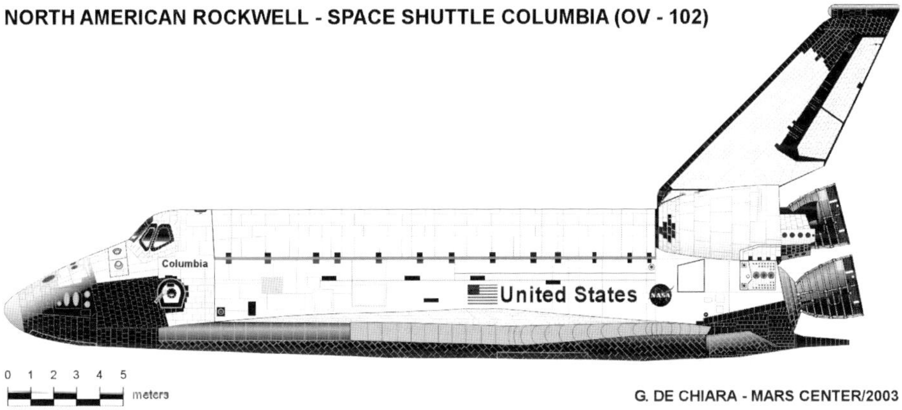

Fig. 1.6 The Space Shuttle.

Owing to this, all objects not attached to the spacecraft move relative to the orbiter center of mass. Other relative motions of unattached objects are related to aerodynamic drag on the vehicle and spacecraft rotations. A spacecraft in a low-Earth orbit experiences some amount of drag due to interactions with the atmosphere. An object within the vehicle, however, is protected from the atmosphere by the spacecraft itself and does not experience the same deceleration that the vehicle does. The floating object and the spacecraft therefore are moving relative to each other. Similarly, rotation of the spacecraft due to orbital motion causes a force to act on objects fixed to the vehicle but not on objects freely floating within it. On average for the Space Shuttles, the quasisteady accelerations resulting from the sources discussed above (position in the spacecraft, aerodynamic drag, and vehicle rotation) are on the order of 1×10^{-6} g, but vary with time due to variations in the atmospheric density around the Earth and due to changes in the Shuttle orientation. In addition to these quasisteady accelerations, many operations on spacecraft cause vibrations of the vehicle and the payloads (experiment apparatus). These vibrations are often referred to as "g-jitters," because their effects are similar to those that would be caused by a time-varying gravitational field. Typical sources for vibrations are experiment and spacecraft fans and pumps, motion of centrifuges, and thruster firings. With a crew onboard to conduct experiments, additional vibrations can result from crew activities. The combined acceleration levels that result from the quasisteady and vibratory contributions are generally referred to as the microgravity environment of the spacecraft. On the Space Shuttles, the types of vibration-causing operations discussed earlier tend to create a cumulative background microgravity environment of about 1×10^{-4} g, considering contributions for all frequencies below 250 Hz.

1.2.6 *The International Space Station*

A space station is a spacecraft that can remain in the Earth's orbit for as long as several decades. Using this facility, a scientist can conduct an experiment over several months and leave the experimental equipment on the station permanently, rather than exchanging it with each mission, as is done with the Space Shuttle.

Figure 1.7 shows some sketches and different views of the International Space Station (ISS). The ISS is the most important international cooperative program ever undertaken in the scientific and technological fields. The assembly of the ISS, which began in November 1998, will be completed in 2007/2008 after approximately 60 assembly missions carried out by the Shuttle and Russian launchers. It is a joint program between the USA, Russia, Japan, Europe (ESA), and Canada and is the greatest work ever realized in time of peace, engaging 16 nations all over the World. Being permanently on orbit, the ISS provides for the first time the opportunity to perform systematic, long-lasting investigations in many scientific disciplines.

Fig. 1.7 Sketches and different views of the International Space Station.

ISS components:

- 6 laboratory modules (2 USA – a laboratory and a Centrifuge Accommodation Module (CAM), 1 European Space Agency (ESA) – Columbus Orbital Facility, 1 Japanese – Experiment Module (JEM), 2 Russian Research Modules);
- 2 habitation modules (1 American, 1 Russian);
- 4 logistic modules (3 Russian, 1 American);
- 2 airlocks (1 Russian, 1 American);
- 2 trusses (1 American, 1 Russian);
- 24 solar panels (16 American, 8 Russian);
- 11 radiator panels (10 American, 1 Russian);
- 3 external robotic arms (1 Canadian, 1 Japanese, 1 European).

The ISS in numbers:

- flies at a height between 350 and 450 km above the Earth;
- completes an Earth orbit revolution every hour and a half;
- is as big as a football field (108 × 79 meters);
- weight = 450 tons;
- hosts 6/7 crew members;
- disposes of 110 kW of power;
- to be completed will require more than 60 space launches (Shuttle, Proton, Soyuz);
- will be completed in 2007/2008 and will be operative for at least 10 years.

For further details see: "International Space Station: A guide for European users," ESA SP-1202, ESA publication Division ISBN 92-9092-407-1.

Science modules The ISS features three main science modules (Fig. 1.8): the US-built (Destiny) Lab, Europe's Columbus Laboratory (ESA-COF) and the Japanese Kibo experiment module.

The Destiny Lab was launched on 7 February 2001 and is an 8.5-m long, cylindrical, pressurized module. It has a capacity of 24 rack locations and can accommodate 13 laboratory racks.

The Columbus Laboratory (see Fig. 1.9) is a 6.8-m long cylinder that contains 75 m^3 of internal space. This space accommodates an entire suite of science laboratories. The module, in fact, has room for 10 International Standard Payload Racks, each hosting an entire laboratory in miniature – complete with power and cooling systems, and video and data links to researchers back on Earth. Four external mounting points allow experiments to be conducted in the vacuum of space.

Kibo is Japan's first manned space facility that allows up to four astronauts to perform scientific experiments. It consists of four components: two experimental

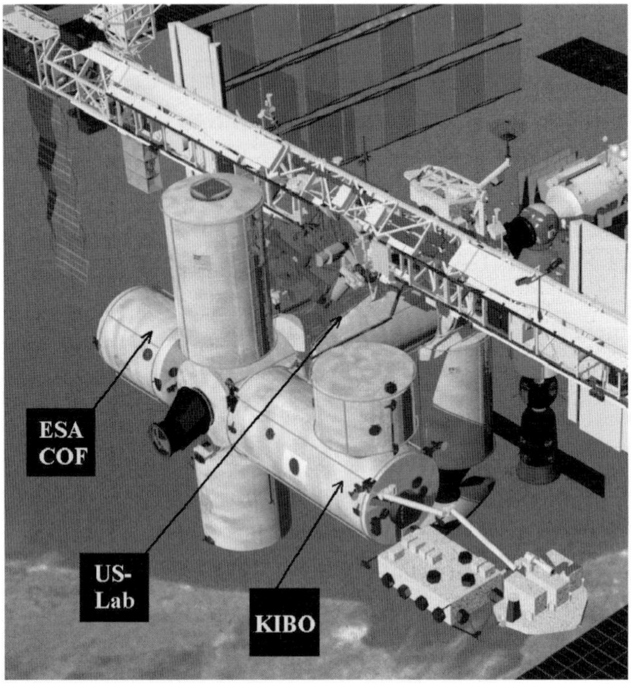

Fig. 1.8 Main science modules on the ISS (courtesy of NASA).

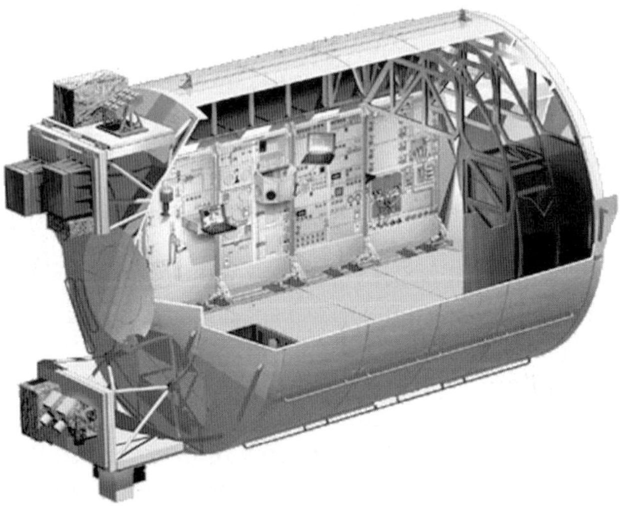

Fig. 1.9 The Columbus (ESA-COF) is a general-purpose laboratory developed by ESA, accommodating the following multiuser facilities: the Fluid Science Laboratory, the Materials Science Laboratory, the Biolab, the European Drawer Rack, the European Physiology Modules (the image is courtesy of ESA).

facilities (one pressurized and one exposed), logistics modules attached to each of them and a manipulator arm to be used for experiments.

Laboratories and facilities The list of acronyms and abbreviations below provides disjoint glimpses of the rich variety of multiuser laboratories and facilities accommodated in the different ISS science modules for studies in the field of fluids (e.g., the American Fluids and Combustion Facility, FCF; the European Fluid Science Laboratory, FSL; the Japanese Fluid Physics Experimental Facility, FPEF), inorganic material science (e.g., the Materials Sciences Laboratory, MSL with the Low Gradient Furnace LGF, the Solidification and Quenching Furnace SQF and the Float Zone Furnace FMF; the High Gradient Directional Solidification Furnace Experiment Module, HGDS; the Directional Solidification and Vapor Transport Experiment Module, DSVT; the Advanced Tubular Furnace with Integrated Thermal analysis Under Space conditions, TITUS; the Quench Module Insert, QMI; the Diffusion Module Insert, DMI; the Advanced Thermal Environment Furnace ATEN; the Advanced Furnace for Microgravity Experiment with X-ray radiography, AFEX; the Gradient Heating Furnace, GHF, etc.) organic material science (the Protein Crystallization Diagnostics Facility, PCDF; the Solution Protein Crystal Growth Facility, SPCF; etc.) and biotechnology (the BIOLAB; etc.).

Figure 1.10a shows the FSL facility. The FSL is a multiuser laboratory for experiments in the strange behavior of weightless liquids. It has been conceived as the most advanced generation of facilities in the field of fluid physics to be conducted in a pressurized environment. The MSL (shown in Fig. 1.10c) is a multiuser facility for the melting and solidification of conductive metals, alloys, or semiconductors, in ultrahigh vacuum, or in high-purity gaseous atmospheres. This is especially

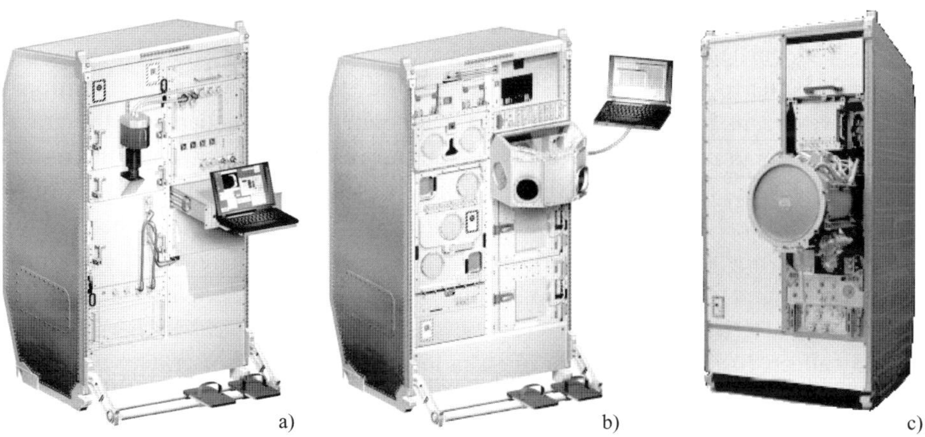

Fig. 1.10 (a) Fluid Science Laboratory (FSL), (b) Biolab, (c) Material Science Laboratory (MSL).

important for reactive materials, whose properties can be very sensitive to contamination. The Biolab facility (shown in Fig. 1.10b) is a laboratory designed to support biological experiments on microorganisms, cells, tissue cultures, small plants and small invertebrates. Finally, the European Drawer Rack provides a flexible experiment carrier for a large variety of scientific disciplines, and provides the accommodation and resources to experimental modules housed within standardized drawers and lockers (for instance, the aforementioned Protein Crystallization Diagnostics Facility, PCDF).

The microgravity environment for the ISS, its science modules and therefore for the different facilities and laboratories therein housed will be discussed in depth in Chapter 2.

1.3. From basic research to industrial applications

Space research needs to be presented to the public as what it really is: a natural, human attraction for the unknown without forgetting, however, that a serious scientific effort always bears a commitment to be beneficial for humankind.

Currently, the motivation for microgravity-based studies often begins with a purely scientific interest to question, observe and understand certain fundamental laws and basic phenomena. But, as the short history of the development of semiconductors clearly illustrates, the answers to purely scientific questioning can quickly be taken up and applied to practical problems in the development of technical systems and, from time to time, in the discovery and introduction of new processes with a major industrial and economic impact.

From an historical point of view, microgravity was initially regarded as an ideal "site" for the production of many materials. The dreams of manufacturing "perfect" crystals using zero gravity began in the 1960s during the Apollo era. Talk of manufacturing other materials continued through the 1970s and into the 1980s. Then reality set in, particularly with the Challenger disaster. Often the materials were not sufficiently better to warrant the costs. The value added did not exceed the additional cost. However, an immense and significant amount was learned about the gravitational and nongravitational effects on materials process, both through the results of space experiments and through related ground-based research. In fact, many results were unanticipated and some still await full explanation. There have been many surprises, and not all experimental results have yet been satisfactorily explained. Gravity was found to influence processes that were thought to be gravity independent. On the contrary, by means of microgravity experimentation, researchers focused on many effects (usually masked or overshadowed by the effect of gravity on Earth) that were also proven to influence the processes carried out on the ground.

Along these lines, there is no doubt that at the present stage the results obtained in microgravity are mostly of fundamental nature, quantifying, as explained before, the theoretical models of gravity influences on fluid phenomena or leading to better insights into the significance of forces and interactions which, during experiments on Earth are masked by gravity-induced flows; however, as anticipated, it has been demonstrated that such effects can be relevant in a number of technical terrestrial areas.

Materials processing on Earth has often been improved and it is difficult to imagine how the materials-processing industries could have flourished without the engineers and scientists who received their training by working on microgravity materials processing. This knowledge has proven to be extremely useful in the improvement and innovation of materials processing on Earth (see also the excellent discussion in Wilcox and Regel, 2001).

In the light of these arguments, now it is widely recognized that microgravity is an interesting and useful condition to carry out research in, rather than a really convenient means for industrial production (by this definition microgravity is no longer considered as a special, expensive "technology", or production site but rather as a significant parameter for the investigation of phenomena which play a certain role in industrially relevant processes on Earth).

Using both space-flight and ground-based experiments, researchers throughout the world are working to benefit industrial, economic, social and "health" aspects of life on Earth. The approach implemented to pursue these objectives is based on the integration of different competencies at the international level and intends to promote the direct involvement of industry (see, e.g., Spenger and Peters, 1997).

The domains of microgravity science research are fundamental physics, fluid physics, materials science, and biotechnology (see, for example, the glimpses of the microgravity facilities provided in Fig.1.10). Hereafter, selected examples of research will be introduced where microgravity as a tool is relevant for application-oriented and industrial research. Each area demonstrates a potential for applications and commercial spinoffs.

The objective of this discussion is to identify the promising application-oriented areas, and to provide the researchers from the industry and universities with a basis on which they could estimate possible benefits from the utilization of space as an additional tool. The approach is based on the evaluation of results of scientific microgravity experiments, mostly covering areas of fundamental research, and an estimation of their potential to solve questions of applied or application-oriented research. These questions have been identified by the analysis of actual research projects with participation of industry or of consortia including contract research organizations with industrial and/or biotechnological research objectives.

The review discussed in this chapter, however, is necessarily limited in scope and depth, both because of the page limit and because of the amount of published literature on these subjects that is now enormous. Consequently, the emphasized topics have been selected according to the opinion and the experience of the author (see also Section 1.7). Owing to the elaborate nature of some of these studies, it has not always been possible to fit an adequate account of them into the framework of this book, but attempts are made to give some indication of the most important results, and of the methods employed. The author apologizes to those whose work or publications are not described or cited. The following sections in particular follow the classification introduced by Rogers et al. (1997).

1.4. Research in fluid physics under microgravity

Everyone has practical experience with fluids (liquids and gases) and knows how a fluid behaves under normal circumstances. Gravity drives much of the fluid behavior we are accustomed to on Earth. When microgravity conditions are attained, however, fluids can do incredible things. Many of our intuitive expectations do not hold up in microgravity, because other forces such as surface tension (the tendency of molecules at a fluid's surface to attract each other) control fluid behavior.

For instance, surface tension causes droplets of any liquid to form almost perfect spheres when the influence of gravity is absent. On Earth, gravity distorts the shape when a liquid is resting on or attached to a surface. While these differences in fluid behavior often present engineers and astronauts with practical problems, they also offer scientists unique opportunities to explore different aspects of the physics of fluids.

A very common example of gravity's impact on fluids is the creation of flows due to density differences (buoyancy-induced convection). Consider, for instance, what happens when a container of water is heated from below. As the water in the bottom is heated by conduction through the container, it becomes less dense than the unheated cooler water above. Because of gravity, the cooler, more dense water sinks to the bottom of the container and the heated water rises to the top due to buoyancy. A circulation pattern is produced that mixes the hot water with the colder water. This is an example of buoyancy-driven (or gravity-driven) convection. The convection causes the water to be heated more quickly and uniformly than if it were heated by conduction (diffusion) alone. This is the same density-driven convection process to which we refer to when we state matter-of-factly that "hot air rises." In addition to mixing, density differences can also cause things to differentially settle through a process called "sedimentation." In this process, the more dense components of mixtures of immiscible fluids or solid particles in fluids settle to the bottom of a container due to gravity.

Because gravity's effect on fluids is so strong, scientists cannot determine what effect other (still poorly known) forces have on fluid behavior. Many of these forces become dominant in microgravity, allowing scientists to observe them without the competing influence of gravity. From a practical point of view, research conducted in microgravity is increasing our understanding of fluid physics (otherwise hidden in normal-gravity conditions) to provide a foundation for predicting, controlling, and improving a vast range of technological processes. The behavior of fluids, in fact, is the core of many phenomena in materials science, and biotechnology science.

1.4.1 *Dynamics and stability*

The goal of this branch of fluid-dynamic research is to describe and predict the motions of fluids in response to applied surface and/or body forces.

In many circumstances, these forces scale with a nondimensional (properly defined) number. When this number (hereafter referred to as a "characteristic" number) is small, the problem is not so difficult because there is a unique correspondence between the given boundary and internal forcing data and the predicted motions. But when the characteristic number is larger, the dynamics of the system can be complicated; there are many solutions; nonuniqueness is the rule; sets of solutions must be described, and the stable and observable subsets must be separated from the others.

A mathematical basis for the study of these difficult problems is the theory of stability and the theory of bifurcation, even if, as will be pointed out in subsequent chapters, also direct numerical discretization and solution of the system model equations (the Navier–Stokes equations in their complete form) can provide significant information about these topics. In this section some typical cases dealing with the microgravity environment are introduced and discussed.

Hereafter, following Joseph (1976) the boundary conditions (for instance, surface-tension forces acting on the interface separating two liquids or a liquid and a gas) and the body forces (gravitational buoyancy) that drive the motion are referred to as "data" of the problem.

When the characteristic number is small, all solutions of the Navier–Stokes equations tend to a single one, determined by the data after the initial conditions have died away. So if the data are steady (for instance, a steady body force), the solution is steady; if the data are time periodic (g-jitters), so is the solution. In other words, the flow that evolves after a time is uniquely determined by the data, independent of initial conditions. This unique flow has the maximum symmetry consistent with the data.

At higher values of the aforementioned number, however, this uniqueness is lost. Other flows with different, usually more complicated patterns of symmetry are then observed after transients have decayed away. For instance, a convection that is spatially uniform in certain directions can be replaced by motions that are spatially periodic, quasiperiodic, or aperiodic. And motions that are steady can be taken over by flows that are time periodic, quasiperiodic, or aperiodic. So at larger values of the characteristic number, flows that do not follow the symmetry of the boundary conditions can appear.

The loss of symmetry implies the existence of a new solution that bifurcates from the pre-existent one. If the characteristic number is increased many times the bifurcations become faster (bifurcations sequence) and faster until the system is chaotic and turbulent.

This breakdown in the symmetry of solutions is especially dramatic in the turbulent condition. In such flows, the connection between the data and the flow is very elusive; even with steady data, flows arise whose behavior after a long time is aperiodic with no regularity. The spatial structure of turbulent flow is also very complicated with many small eddies and fluctuations that are sometimes called random because nobody knows how to characterize them in a precise way. These flows are extremely sensitive to initial conditions. Two flows with the same data but slightly different initial conditions evolve into two very different flows; the connections between the data and the flow are subtle and elusive. It may be useful to think of turbulence as the least symmetric state of motion consistent with the given data.

Instability of flows and their transition to turbulence are widespread phenomena in engineering and nature, and are also important in many applied sciences. Even if limited to the first or the second transition of the flow along the sequence that leads to complete turbulence (hereafter referred to as "route to chaos"), the loss of spatial and/or temporal symmetry can have, in fact, detrimental effects in a number of technological processes where the spatial symmetry of the thermo-fluid-dynamic field and/or steady conditions are crucial factors to guarantee a good quality of the final product (the growth process of many inorganic and organic crystals and materials).

Within the context of these studies, a microgravity experimentation is very relevant and useful as it makes it possible to observe how some forces, interwoven or overshadowed in normal gravity conditions (for instance, surface-tension forces), or peculiar of the microgravity environment (the aforementioned g-jitters and similar oscillatory body forces), can give rise to the behaviors (flow transitions and bifurcations) described above.

Without the complications of gravity-driven convection flows on Earth, it becomes possible to test in space fundamental theories of three-dimensional laminar, oscillatory, and turbulent flow generated by these forces. As mentioned before, all are of substantial practical as well as theoretical interest on Earth. By improving the basics for predicting and controlling the behavior of fluids, in fact, new possibilities open up for improving a whole range of industrial processes.

For the case of pure surface-tension-driven flows (the so-called Marangoni or thermocapillary convection), for instance, a special configuration (the liquid bridge, a drop of liquid with cylindrical or quasicylindrical free liquid/gas interface held between two disks at different temperatures) has been conceived for the microgravity environment and has become, over the years, a paradigm configuration for the study of these flows, their stability, and their bifurcations.

Due to the complexity of Marangoni flow and of the related sequence of bifurcations (up to the onset of full turbulence), the liquid bridge has also been used by many investigators simply as a reference configuration for testing their ability to predict the behavior of nonlinear systems regardless of any possible technological application.

Similarly to the case of Marangoni convection, the nonlinear response of fluid systems to g-jitter disturbances has become a topic of great interest in the last decade. Also in this case, in fact, the system can exhibit a transition to subsequent complex patterns of flow (thermovibrational convection) according to the amplitude, oscillation frequency, and orientation of the external forces (see, e.g., Lizée and Alexander, 1997) and deviate in structure and magnitude from the diffusive conditions that should be ideally maintained by the microgravity condition (in space, these flows usually represent an undesired effect when carrying out experimentation that would benefit from the establishment of quiescent regimes; on Earth, fluid motion induced by high-frequency vibrations can be used as a useful means to "counteract" possible instabilities of buoyancy and/or Marangoni convection).

The numerical simulation of such instabilities using various time-dependent three-dimensional solvers is exceptionally CPU-time consuming. It is a very challenging problem requiring a tremendous effort in terms of simulations and computational time. A very large number of computational grid points (and consequently a large number of unknown variables), in fact, is necessary. Coarse grids cannot be used because a low resolution does not permit capture of the small-scale structure with increasing degrees of complexity that are formed when the characteristic number increases.

The alternative numerical approach, based on the so-called "linear stability analysis," allows one to avoid the straightforward integration in time.

The related mathematical treatment of the problem generally proceeds along the following lines. An initial flow that represents a stationary state of the system is considered (the so-called "basic state"). By supposing that the various physical variables describing the flow suffer small (infinitesimal) increments, the equations governing these increments are obtained.

In obtaining these equations from the relevant equation of motion, all products and powers (higher than the first) of the increments are neglected, while retaining only those terms that are linear in them. The theory derived on the basis of such linearized equations is called the linear stability theory (Chandrasekhar, 1981).

Since stability means stability with respect to all possible (infinitesimal) disturbances, for an investigation on stability to be complete, it is necessary that the reaction of the system to all possible disturbances be examined. This is accomplished by expressing an arbitrary disturbance as a superposition of certain basic possible modes and examining the stability of the system with respect to each of these modes.

In practice, this approach considers all possible three-dimensional infinitesimal perturbations and usually provides better understanding of the instability phenomena, as well as more precise critical values of the governing parameters. At the same time it requires the solution of a series of eigenproblems of very high order (the order is equal to the number of scalar unknowns of a numerical method). The transition from a steady to periodic behavior as the characteristic parameter is varied corresponds to a Hopf bifurcation where an eigenvalue of the Jacobian matrix of the governing equations crosses the imaginary axis (i.e. the real part of the eigenvalue becomes positive with a corresponding value of imaginary part $\neq 0$). If a single real eigenvalue crosses through zero to the positive half plane (i.e. the real part of the eigenvalue becomes positive, but the corresponding imaginary part is zero), the bifurcation is steady, i.e. some breakdown in the initial symmetry of the flow occurs but such a phenomenon is not associated with the onset of oscillatory convection.

Despite the aforementioned advantages offered to the investigators by the linear stability analyses, it is the opinion of the author (and the present book provides a solid basis to such a conviction) that the recent development of numerical methods and high-performance computing is enabling the possibility to approach these problems in terms of numerical solution of the Navier–Stokes equations in their nonlinear and complete form.

This philosophy provides a wealth of additional aspects that cannot be obtained on the basis of linear stability computations, such as: the effective amplitude of the disturbances when they are saturated; the behavior of the system far from the threshold point; a quite exhaustive picture of the subsequent stages of temporal evolution, etc.

A more complete understanding of how fluids can undergo transition to different stages of evolution will lead to the ability to better control (reduce or enhance) fluid flow in systems used in space and on Earth.

Further details on the intrinsic nature and genesis of Marangoni, buoyancy, and thermovibrational convection as well as on their instabilities and bifurcations will be discussed in Chapter 2.

1.4.2 *Interfacial phenomena*

This argument refers to the behaviors a fluid displays at its interface (boundary) with another fluid or a solid. One example of an interface is the boundary between a liquid surface and air; another is the boundary between a liquid and its container. Many forces affect the behavior of fluids at the interfaces.

Interface dynamics relate to the interaction of surfaces in response to gravity, heating, cooling, and chemical influences. A better understanding of this topic will contribute to improved material processing and other applications.

Interfacial phenomena, such as the wetting and spreading of two immiscible liquids or the spreading of a fluid across a solid surface, are, in fact, ubiquitous in nature and technology.

Wetting is a dominant factor in material-processing techniques, including soldering processes, film and spray coating, liquid injection from an orifice, and the production of metal alloys. Interfaces dominate the properties and behavior of advanced composite materials, where wetting of the constituent materials dictates the processing of such materials. Understanding and controlling wetting and spreading pose both scientific and technological challenges.

In reduced gravity, wetting determines the configuration and location of fluid interfaces, thus greatly influencing, if not dominating, the behavior of multiphase fluid systems. This environment provides scientists with an excellent opportunity to study the wetting and surface-tension forces that are normally masked by the force of Earth's gravity.

1.4.3 *Multiphase flows*

The motion of drops and bubbles in a fluid medium is of fundamental importance in many natural physical processes as well as a host of industrial activities. Drops or bubbles floating in a fluid can be moved by a number of forces, including gravity, which is normally the dominant mechanism for their motion on Earth, and interfacial tension gradients induced at the surface of the drop or bubble through variations of temperature and/or concentration of a solute. The latter mechanisms (Marangoni effects) can play a major role under microgravity conditions where sedimentation and gravity-driven convection are largely eliminated.

For example, in solidification of metals and processing of glass under reduced-gravity conditions, thermally driven (Marangoni) motion of gas bubbles provides an alternate mechanism for their removal from the melt.

This phenomenon can also be important in two-phase systems involving liquid drops dispersed in the continuous phase. This is the case, for instance, in immiscible liquid–liquid systems and in liquid mixtures where a thermal quench of the initially homogeneous mixture exhibiting a miscibility gap from temperatures above the gap to lower temperatures inside the gap actually causes phase separation of the homogeneous liquid to occur (formation of drops, see also Section 1.5.1).

It is known that the thermocapillary (Marangoni) motion of these drops in a temperature gradient often leads to an undesired spatial separation, both on Earth and in low-gravity environments. If properly understood, however, thermocapillary effects may be used in a positive way to control multiphase fluid transport and phase separation on Earth.

1.4.4 *Complex fluids*

Complex fluids are gases or liquids that contain particles of other substances dispersed within them. This causes them to have unusual properties. For instance, granular systems are composed of a series of similar solid objects (such as grains of sand or pebbles) tightly packed in a liquid. Granular systems can flow under certain conditions and are considered a complex fluid.

A relevant example for one of the aims of the present book falling into this category of problems deals with the nucleation and growth of macromolecular crystals from solution. A liquid with dispersed solid crystals, in fact, behaves as a complex fluid where the dynamics of both phases can be affected by normal or residual gravity (via buoyancy-induced convection in the liquid, via gravity-induced sedimentation for the grains and related complex interplay of both effects).

In microgravity, where the effects of gravity-driven convection and sedimentation are reduced, scientists can observe other influences on the properties of complex fluids and learn to better control and predict these properties on Earth.

1.5. Research in materials science

Materials science is an extremely broad field that encompasses the study of all materials. Materials scientists seek to understand the formation, structure, and properties of materials on various scales, ranging from the atomic to the microscopic and to the macroscopic (large enough to be visible).

The properties a material has (such as how strong it is or whether it conducts electricity) are determined by its structure. Hence, establishing quantitative and predictive relationships between the way a material is produced (processing), its structure (how atoms or larger inclusions are arranged), and its properties is of paramount importance.

Gravity is a major obstacle to gaining this understanding also in this case. Materials production in industry inevitably includes steps that are strongly influenced by gravity-driven processes. One of the steps commonly involved in processing metals, in fact, is solidification, in which the material begins as a liquid and is cooled to form a solid.

Both convection and sedimentation can influence the structure of the material (additional details about these aspects are provided in further sections).

To understand the relationship of processing to a material's structure better, scientists are conducting materials science experiments in microgravity, where the aforementioned undesired effects are reduced. The opportunity to investigate these steps under microgravity will help to develop a more fundamental understanding of the production process and hence improve product quality and yield, and, in some cases, lead to introduction of new products.

1.5.1 *Metals and alloys*

Metal alloys constitute an important category of engineered materials.

Currently, in particular, the principal area of interest deals with a special group of metals (and their organic and transparent surrogates) known as "immiscible alloys."

When an immiscible alloy is cooled, it demixes into two melts of different compositions and densities (the so-called minority and majority phases). This is very similar to a mixture of water and oil at room temperature in which the two components separate from each other after having been shaken into a near single solution. The difference between the immiscible alloy and the water/oil system, besides the materials, is that the latter is already below the temperature at which the demixing process occurs spontaneously. All the shaking performed after phase decomposition will only break the droplets into smaller droplets, but will not break up the droplets to the size of an atom, which is required to have a homogeneous solution (hence the appropriate denomination of "immiscible alloys").

As a consequence, the materials exhibiting such a behavior in the liquid state, transform from a single-phase liquid to substances of more than one phase when they are solidified. A well-known example is the alloy manganese–bismuth. Solidifying liquid Mn–Bi results in two different solids, each of which has a chemical composition that differs from the liquid. The solidified minority phase is distributed as rods, particles, or layers through the other solid (continuous matrix, or majority phase).

The current interest in these special alloys is motivated by the fact that they present significant potential applications. These include uses as catalysts, superconductors, magnetic materials, and high-performance electrical contacts.

When these materials are processed on Earth, the resultant substances have a structure that is influenced by gravity. On the ground, in fact, when the melt is cooled it will separate from a single solution into a two-phase liquid with the lighter component lying on top of the heavier one (if the mixture is in a container, the droplets will form at the container walls whereupon gravity will pull the denser droplets to the bottom or allow the less dense drops to float to the top). For these reasons casting of liquid immiscible alloys on Earth by classical casting routines has often been tried, but has never led to the desired results.

On the contrary, when processed in microgravity, it is expected that the end product should consist of an evenly dispersed, multiphase structure. The lack of gravity should, in principle, cause a more uniform distribution of the droplets and thus provide unique properties to the alloy after it solidifies.

It is worthwhile to stress how experiments dealing with these topics carried out in space have confirmed on the one hand the possibility to improve the final solidified alloys, and have led on the other hand to the identification of a number of gravity-independent phenomena that can also significantly affect the corresponding processes on Earth. It has been shown, in fact, that molten, immiscible metals still undergo spatial separation of the phases in microgravity when cooled, but they do so in a very unusual way. Instead of the denser metal settling at the bottom, one component (and not necessarily the denser) migrates towards the hot side of the container. These phenomena exhibit an outstanding theoretical kinship with those already discussed in Sections 1.4.2 and 1.4.3 dealing with fundamental research in the field of fluids and multiphase systems.

1.5.2 *Electronic materials*

These materials are essential to many modern devices, such as computers and communication systems. A material's ability to conduct electricity is determined by its chemical and crystalline structure. The most perfect crystalline arrangement of atoms does not always produce the properties that are needed for specific electronic devices. Intentionally adding small amounts of impurities (dopants) to a material can drastically change its ability to conduct electricity.

One aspect of the solidification of semiconductors that influences their microstructures is the shape of the boundary, or interface, that exists between the melt and the solidifying material. It is known that during the solidification process, the shape of the solidifying interface can go through a series of transitions. This instability is influenced by many factors. Gravity plays an important role in a number of them. In particular, buoyancy-driven convection can deeply influence the shape of the solidifying interface.

In addition, on Earth, convection disturbs the orderly arrangement of the crystal and can cause dopants to be distributed unevenly. In microgravity, researchers can control better the arrangement of atoms and find ways to produce better electronic materials on Earth.

Also in this case the virtual absence of gravity-dependent phenomena in microgravity allows researchers to study underlying events that are normally obscured by the effects of gravity and that are therefore difficult or impossible to study quantitatively on Earth.

Microgravity also supports an alternative approach to studying and/or producing materials called containerless processing. Containerless processing has an advantage over normal processing in that containers can contaminate the materials being processed inside them. Melt containment leads to serious adverse side effects, including the establishment of mechanical stresses arising from the interaction between the forming solid and the rigid containment. In addition, there are some cases in which there are no containers that will withstand the very high temperatures and corrosive environments needed to work with certain materials.

Within this context it should be pointed out that the rapid development that has taken place over the past thirty years in the field of electronic and optoelectronic devices has been fostered on the one hand by the increasing basic knowledge of solid-state physics and on the other hand by the amount of theoretical knowledge about the behavior of fluids provided by experimentation in space. It is now firmly established that the diligent experimentation in a reduced-gravity environment, based on related meaningful ground-based research, can significantly enhance the scientific framework for crystal growth and thus provide for improved capabilities of crystal growth technology (Lenski, 1990). As a result of this new and sophisticated material engineering, methods have opened up the way for the production of compounds with characteristics particularly suitable for new applications.

A vast majority of semiconductor devices is currently based on wafers cut from single crystals of either elemental or compound semiconductors. In the following section, a brief description of the different growth techniques currently used to obtain high-quality single crystals from the melt (of initially polycrystalline materials) is provided. The prominent features of each method are discussed showing the advantages and disadvantages, intrinsic limitations and possible future developments. The existing terrestrial technology is reviewed briefly. Emphasis is given to the potential of microgravity and to a critical and focused comparison of the classical terrestrial methods with the aforementioned containerless processing (the Floating-Zone technique).

Floating-Zone (FZ) method During the Floating-Zone (FZ) process, a melt zone is established between the lower seed material and upper feed material by applying localized heating (see e.g., Fig. 1.11). This floating zone is moved along the rod (by means of relative motion of the heating device) in such a way that the crystal grows on the seed (which is below the melt) and simultaneously melting the feed material above the floating zone. The seed material, as well as the feed rod, is supported but no container is in contact with the growing crystal or the melt, which is held in place only by surface tension.

Of course, containerless processing on massive samples can only be done in the microgravity environment of space where the forces used for suspending and manipulating the specimens are not overwhelmed by gravity. Microgravity requires much smaller forces to control the position of containerless samples, so the materials being studied are not disturbed as much as they would be if they were levitated on Earth.

Under Earth conditions, the zone height is limited because the liquid will run down if the molten zone gets too big. This fact limits the possible diameter of crystals that are grown in Earth's gravity. In space, the maximum zone height is given by the circumference of the crystal. Therefore floating-zone experiments with higher zone heights and larger diameters are feasible under microgravity (see Fig. 1.12).

Along these lines, microgravity experimentation is enabling the production of limited quantities of high-quality large-size specimens and of materials that exhibit unique properties for use as benchmarks. This pioneering research is leading to

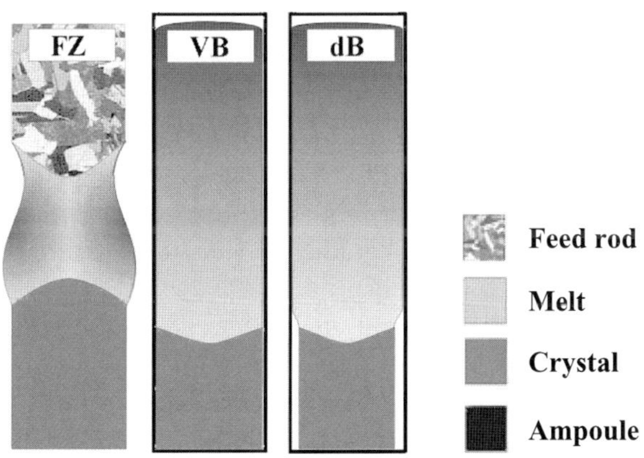

Fig. 1.11 Sketch of the Floating-Zone (FZ), The Vertical Bridgman (VB) and the detached Bridgman (dB) methods.

μg: FR15 μg: FR9 1g

Fig. 1.12 GaSb FZ single crystals grown during the Spacehab-4 mission (left and center). For comparison, a 1 g crystal, of the maximum diameter that can be grown by the FZ method, is shown on the right, after Cröll et al. (1998a). The crystals are single crystalline, exemplified by the appearance of facets at the crystal periphery.

next-generation commercial crystal products (for instance, most of the very pure silicon produced today is processed by the Floating-Zone technique).

Given our current expertise in growing semiconductor crystals as model systems for controlled studies of their enhanced properties and the certain future developments in creating proper conditions for the growth process, the possibility of using the Floating-Zone technique as a widely diffused industrial process is nearing reality. In future, the single crystals obtained by these techniques will be directly used for materials science and computer technology.

Bridgman method (directional solidification) In the Bridgman growth, directional solidification occurs within a vertical ampoule (VB) or a horizontal open boat (HB). The ampoule or the boat is inside a furnace with a temperature that varies from above the melting point at one end to below the melting point at the other end. Freezing is caused by moving the ampoule through a fixed furnace or vice versa, or by slowly lowering the furnace temperature with both ampoule and furnace immobile. This latter method is often called the "gradient-freeze technique" or "power-down method."

For semiconductors such as GaAs, HB is preferred to VB because of the observability of the growth process and the possibility of volume expansion. However, HB has some disadvantages related to the fact that the obtained crystals are characterized by semi-cylindrical cross-sections and Marangoni flow can arise on the free melt surface. In addition, buoyancy convection tends to be stronger in HB than in VB. Superimposed on this is also the fact that the numerical modeling of HB is more difficult owing to the free surface and the lack of axial symmetry.

In VB, the ampoule, which is, for example, a quartz glass or a graphite container, is necessary to support the melt, because, in contrast to the FZ method, the whole feed material (above the crystal) is molten.

The advantages of FZ over Bridgman include the absence of a container in contact with the melt (which normally is the source of impurities in the liquid) and the absence of a container in contact with the processed crystal, which frequently causes stress-induced defects in the crystal because of the different thermal contraction of the crystal and ampoule as they cool.

Some of these defects can also be reduced within the framework of directional-solidification-based methods if the crystal is not in contact with the ampoule material. Such a situation occurs in a very special case where the melt at the liquid/solid interface is detached from the wall and the crystal grows without contact. Growth under such conditions is known as "detached" Bridgman growth (dB) (see dB, Fig. 1.11). In the case of normal Bridgman growth, the melt and the crystal are in contact with the crucible wall, whereas in the case of detached Bridgman growth, there is a gap between the crystal and the ampoule wall (while the melt is in intimate contact with the container boundary).

Detachment (still a poorly understood phenomenon) can be influenced by parameters like the gravitational force (in fact, its mechanism cannot be reproducibly obtained under Earth conditions where it has occasionally occurred serendipitously), ampoule material (determines the contact angle between the container and the melt), and pressure forces above and below the melt.

Czochralski (CZ) method The Czochralski (CZ) method is relatively simple in concept but rather difficult to control. The method starts with chunks of a base material from which the desired crystalline structure can be grown (polysilicon for silicon crystals).

The melting of the base material takes place in a circular crucible, which can be heated or cooled arbitrarily at its bottom and sidewall. The crucible can rotate around its axis.

When the base material has melted and is at the proper temperature, a seed is brought into contact with the free melt surface to begin the growth process. The melt will freeze at the seed. At this point the crystal growth equipment begins to slowly pull the seed up from the melt as the material solidifies forming the high-quality crystals (Fig. 1.13). Solid crystals are afterwards cut to form thin semiconductor wafers from which, e.g., integrated circuits are produced.

To "counteract" flow arising in the melt owing to the buoyancy and surface-tension forces, the crucible where the crystal is being grown is rotated while the mechanism pulling it out of the melt is rotated in the opposite direction

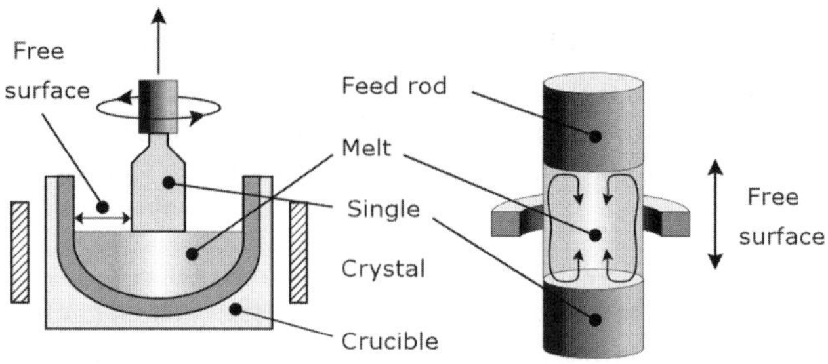

Fig. 1.13 Comparison between the Czochralski (CZ) and the Floating-Zone (FZ) Methods.

(in practice a relative rotatory motion is established between the crystal and the crucible with the intent to produce a cylindrical crystal). This, in turn, causes gas convection to occur in the crystal growth furnace, further complicating the process. A melt-encapsulant material is needed in the furnace to help control the process.

Convective-transport mechanisms affect the melt, the crystal/melt interface, the crystal itself, the encapsulant and the gas around the crystal and melt.

It is worthwhile to stress how, with such a technique, the crystal grows without wall contact but the melt can be subject to possible contamination from the crucible material. Like the case of HB and VB, this is a disadvantage and affects the quality of grown crystals adversely.

As previously emphasized, the growth by the FZ techniques is achieved in a containerless environment (without a crucible), and the grown crystals are free, in principle, from such contamination. However, the specimens grown by FZ on the ground are relatively small compared with those of CZ. This is mainly because, as previously explained, the melt must float freely (with a free surface between the melt and the ambient) between the feed at the top and the seed (or grown crystal) below and this is possible only for small size of the liquid zone in normal-gravity conditions.

From the foregoing, and, in particular, from all the arguments illustrated for the VB, FZ and CZ cases among other things, it is evident how most technological processes for semiconductor crystal growth from the melt are carried out in axisymmetric configurations. They require the application of axial and/or radial thermal gradients across the phase boundary; these gradients provide driving forces for free convection in all fluid phases involved. The melt is therefore subject to varying heat- and mass-transfer conditions, which in recent years have been found to be directly or indirectly responsible for a vast majority of bulk deficiencies in single crystalline electronic materials. In particular, instabilities of the melt flow usually lead to three-dimensional azimuthal effects, which strongly affect the quality of the growing crystals, and therefore are highly undesirable. It is necessary, therefore, to predict the appearance of these instabilities, to understand the related physical mechanisms, and to find the flow-control means capable of stabilizing the flow.

In many cases, stabilizing the primary instability (the first bifurcation of the flow) would mean stabilizing the process as a whole, which is extremely desirable for the various crystal-growth technologies illustrated earlier. This aspect provides the necessary theoretical link with the arguments illustrated in Section 1.4.1.

1.6. Basic questions in life sciences and organic materials

Biotechnology is the application of engineering and technology to life-sciences research. Biotechnology research in space focuses on protein crystal growth (growing organic crystals with thousands of atoms, i.e. macromolecular crystals) and on cell/tissue culturing (the study of how cells interact in a low-gravity or low-shear environment).

The primary obstacle to growing large crystals and tissue samples on Earth has been gravity. Both crystals and cells (or tissue samples) are typically grown in a liquid mixture (called mother liquor for protein crystals, culture medium for biological tissues). Gravity creates flows in the mixture due to density differences in the liquid (the aforementioned buoyancy-induced convection). It also causes heavier particles to settle down to the bottom of the container (due to sedimentation). Because solutal fluid flows and sedimentation disturb the growth of crystals and cells, one answer has been to conduct these growth processes in a microgravity environment, where such disturbances are virtually eliminated.

1.6.1 *Protein crystallization*

Proteins are the elementary building units of all living creatures and essential components for information and energy processing within the living systems.

The mother liquor is usually contained in a proper environment, specifically designed to favor growth, called the protein reactor. The protein reactor is a chamber where the strategy employed to bring about crystallization is to guide the system in a very gentle manner and very slowly toward a state of reduced solubility by modifying the properties of the solvent (growth from supersaturated solutions).

The main purpose in growing protein crystals in space is to advance our knowledge of the biological molecular structures. Researchers are using microgravity to help overcome a significant stumbling block in the determination of molecular structures: the difficulty of growing crystals suitable for structural analysis.

Scientists use X-ray diffraction to determine the three-dimensional molecular structure of a protein. They can calculate the location of the atoms that make up the protein based on the intensity and position of the spots formed by the diffracted X-rays. From high-resolution diffraction data, scientists can describe a protein's structure on a molecular scale and determine the parts of the protein that are important to its functions. Using computer analysis, they can then create and manipulate three-dimensional models of the protein and examine the intricacies of its structure to create a drug that "fits" into a protein's active site, like inserting a key into a lock to "turn off" the protein's function. But X-ray diffraction requires large, homogeneous crystals (about the size of a grain of table salt) for analysis. Unfortunately, crystals grown in Earth's gravity often have internal defects that make analysis by X-ray diffraction difficult or impossible. Space-Shuttle missions have shown that crystals of some proteins (and other complex biological molecules such as viruses) grown in orbit are larger and have fewer defects than those grown on Earth.

In practice, on Earth, buoyancy-induced convection and sedimentation may inhibit crystal growth, whereas in microgravity, convection and sedimentation are significantly reduced, allowing for the creation of structurally better and larger crystals.

The absence of sedimentation means that the protein crystals do not sink to the bottom of their growth container as they do on Earth. The crystals hover within the surrounding liquid (this argument exhibits a theoretical kinship with the corresponding subject of complex fluids already treated in Section 1.4.4). Consequently, they are not as likely to be affected by other crystals growing in the solution.

Moreover, because convective flows are also greatly reduced in microgravity, crystals grow in a much more quiescent environment, which may also be responsible for the improved structural order of space-grown crystals. It is known, in fact, that the morphological evolution of the crystals can exhibit high sensitivity to the motion of fluid in the surrounding environment.

As anticipated, the central goal of microgravity protein crystal growth experimentation is to obtain better crystals. However, it is not limited to this aspect and also tries to determine the basic science that controls how proteins interact and order themselves during the process of crystallization.

The fundamental knowledge provided by the space experiments with regard to the forces that control crystal growth (the transport of molecules to the face of a growing nucleus or crystal, and the frequency with which the molecules orient and attach themselves to the growing surface, i.e. mass transport in the liquid phase and attachment kinetics) will, in fact, also have implications for protein crystal growth processes on Earth.

In the light of the above arguments, "Protein Crystal Engineering in Space" can be defined as the field that studies the way to achieve the best possible conditions for the crystallization of the macromolecular substances needed for progress in biotechnology, pharmaceuticals, and drugs.

In addition to its medical and biotechnological significance, protein crystallization, however, has much in common and may provide an insight into crystallization phenomena that occur in a variety of systems. Furthermore, it exhibits a theoretical kinship with the complex of problems that come under the heading "order out of chaos."

In the complete absence of any convective disturbance and in some circumstances (crystals formed within special gels trying to simulate microgravity on the ground), in fact, protein crystallization leads to the formation of some intriguing ordered (characterized by a certain degree of periodicity in space) patterns of solid deposits. The ordered spatial distribution of these deposits and the related formation mechanisms exhibit interesting analogies with the category of nonlinear phenomena known as Liesegang patterns. These aspects also represent a significant part of the research currently carried out on protein growth mechanisms as well as a challenging task for numerical simulation. In Section 1.4.1, some attention has been devoted to nonlinear systems that exhibit strong "sensitivity to initial conditions." This is not an exclusive prerogative of fluid systems. Protein crystallization can also be affected by such a behavior as will be discussed in Chapter 6 (for additional and intriguing analogies between the nonlinear dynamics of fluids and crystallizing proteins the reader may consider some mathematical aspects concerning the behavior of "fractals" and disordered systems in the book of Bunde and Havlin, 1996).

1.6.2 *Tissue engineering*

The other focus of biotechnology in microgravity is cell- and tissue-culturing experiments. The goal of this research is to grow cells on a tissue in near-weightlessness that otherwise is unachievable on Earth. The medical benefit of microgravity tissue and culture engineering may lead to new research models in cellular and molecular biology. These studies are also developing new tissues for potential transplant operations, thus the concept of "Tissue Engineering in Space" is also nearing reality.

Tissue engineering is usuallly regarded as a new field that enables tissue equivalents to be created from isolated cells in combination with biomaterials and bioreactor culture vessels.

It provides a starting point for understanding the nature of tissue growth under real and idealized conditions. The selection of tissue replacement, the mechanisms affecting cell function and growth, the effects of environmental factors (such as gravity) on the body, cancer treatment and bioproduct production are some examples of questions that can eventually be addressed by culturing functional tissue in bioreactors. In future, moreover, as discussed earlier, the constructs obtained by these techniques will be used for treatment of damaged human tissues and/or healthy organs. Tissue engineering, in fact, has also been defined as "The use of naturally occurring and/or synthetic materials in conjunction with cells to create biological substitutes to serve as functional tissue replacements" (in fact, it often studies the process of building tissue like constructs ex vivo that will have eventual use for implantation in vivo; the ultimate target of a tissue-engineered product is to replace or augment existing tissue function).

Bioreactors are devices employed to culture tissues as diverse as human cancers and cartilage for both research and implantation purposes. The tissue to be cultured is placed within the reactor, which includes a means for supplying (liquid) culture medium to the growing tissue, providing it with nutrients and simultaneously removing waste products from its vicinity.

Multiple types of reactors exist, including perfusion vessels as well as spinner flasks, however, the best environment has been proven to be the NASA rotating vessel. The attention of the researchers in recent years has been focused in particular on this system (the so-called Rotary Culture Method) since this technique has proven to be the most practical solution to provide a suitable culture environment supporting 3-D tissue assemblies.

The rotating bioreactor is a horizontally rotating cylinder completely filled with fluid culture medium. Under these conditions, the fluid rotates at the same rate as its container (the cylinder), resulting in continuous suspension of the biological particles. The conditions in the bioreactor are very conducive to engineering tissues from individual cells and for simulating microgravity. It allows researchers to produce breakthrough results by creating three-dimensional cell cultures that are much more similar to tissues found in the human body (even if one of the major limitations to current tissue engineering is still the inability to vascularize the tissue in any manner).

The rotating bioreactor has applications on the ground as a simulator of microgravity, i.e. it provides some of the physical operating conditions commonly associated with microgravity such as randomization of the gravity vector and allowing cells to fall at terminal velocity (endless free-fall). Production of larger specimens, however, requires true microgravity. On Earth, samples may become so large (about 1 cm) that they are no longer suspended. In long space missions, large-sample growth in bioreactors can be studied in order to understand the extended growth.

1.7. Numerical simulation as a useful tool to reduce the expensive experiments in space

Some words recur in the foregoing discussions: "microgravity" (that seems to favor growth and many technological processes) and "fluid-dynamics" (that seems to somehow influence the final quality of the specimens regardless of their inorganic or organic nature). As illustrated in Sections 1.5 and 1.6, in fact, many techniques have been developed to obtain crystals and materials exploiting different processes, as melt solidification, precipitation from supersaturated solutions, etc., and despite differences, the presence of a fluid phase that feeds the growing specimen is a common feature of all the techniques. Adequate thermal and/or chemical stimuli have to be provided to drive the fluid phase out of equilibrium and to sustain the process. Therefore, the fluid phase is in nonuniform conditions: compositional and/or thermal gradients are present and correspondingly mass and/or heat transport can take place.

A common feature is also the presence of "interfaces" (solid/melt, liquid/air, protein crystal/mother liquor, living tissue/culture medium) that seem to somehow affect with their own characteristics (surface tension, surface chemical kinetics of growth, velocity and morphological evolution of the advancing fronts) the phenomena discussed in the foregoing pages.

It is thus on these issues (many of which are still unresolved or poorly understood) that the present book is mainly focused upon.

The volume covers many different aspects of material science and in particular, the behavior of inorganic (semiconductors and metal alloys) and organic (protein crystals) materials as well as the case of living biological tissues. In particular, attention is mainly devoted to the most promising and innovative technological processes especially conceived by the investigators to benefit from the conditions provided by space platforms and facilities and/or to take advantage on Earth of the new illuminating knowledge provided by microgravity experimentation.

Within this context, working engineering models are introduced and discussed that can be easily employed in technological applications, while providing a rigorous and rich learning environment for deeper understanding and effective numerical treatment of phenomena encountered in microgravity and/or unmasked by this environment (the text has been designed with the needs of both engineers and researchers in mind).

The potentialities of the numerical methods are shown in terms of prototype applications and validation is obtained through comparison with available and relevant experimental results. The volume contains numerous examples incorporating problem-solving techniques, demonstrations to illustrate topical material and appropriate references to supplementary material.

Of course, an important part of the text subject coverage is devoted to computational fluid dynamics (CFD). This branch of fluid dynamics complements experimental and theoretical fluid dynamics by providing an alternative cost-effective means of simulating real processes. It offers the means of testing theoretical advances for conditions often unavailable experimentally or having a prohibitive cost.

Along these lines, it should be stressed how in recent years, improvements in computer hardware performance have occurred hand in hand with a decreasing hardware cost. Consequently, for a given numerical algorithm and problem, the relative cost of a computational simulation has decreased significantly, historically. Moreover, paralleling the improvement in computer hardware has been the improvement in the efficiency of computational algorithms (and current improvements in hardware cost and computational algorithm efficiency seem to show no obvious sign of reaching a limit).

All these factors combine to make numerical computations increasingly cost effective in contrast to the cost of performing experiments in space that continues to increase (also in terms of human lives, see the recent Columbia disaster). The analysis by means of numerical simulation is a very inexpensive and feasible alternative and/or can provide a significant shrinkage of the cost of planned microgravity experiments. Preparation of such experiments requires great accuracy. It is generally desirable that a space research project be undertaken only after as much knowledge and experience has been gained through ground-based research as is practical to ensure that research conducted in space to be as effective as possible.

In the light of the foregoing arguments, particular attention is devoted to:

1. advanced processes for the production of materials that would benefit from the (real or simulated on Earth) microgravity environment (e.g., the Floating-Zone technique for the case of semiconductor materials; immiscible liquid couples for the case of metal alloys, the counterdiffusion technique for the case of macromolecular crystals, the rotary culture vessel technology for the biological tissues, etc.)
2. typical microgravity (oscillatory residual gravity, g-jitter, thermocapillary flow, etc.) and on the ground (buoyancy forces, sedimentation phenomena, etc.) disturbances that could adversely affect the final quality of the specimens and to possible strategies to contain their effects;
3. novel (computational fluid-dynamics) numerical techniques recently introduced to handle these problems. In turn, the rich variety of heretofore unseen results and theory obtained using these tools are discussed in an exhaustive and systematic manner.

Since, due to the intrinsic nature of the production processes, the problem often deals with the tracking of advancing or moving interfaces, "moving-boundary" methods are very relevant within this context.

The book is conceived to demonstrate both the potential and the challenges of fluid-dynamic mathematical modeling of materials science for a broad variety of situations and fields. It is organized in different chapters, each chapter dealing with a different aspect of the aforementioned topics, providing the necessary background information (i.e. fundamental concepts, literature, past history, and current status, information on the experimental techniques, etc.), focusing on recent advances and describing in detail some original numerical methods and the insights into the physics provided by the related simulations. Thus the physics is never lost and the mathematical procedures are strengthened through repetition of examples.

The order with which the topics are presented runs as follows:

Typical microgravity disturbances (*Chapter 2*) The microgravity environment is discussed in terms of typical phenomena, model equations, scaling and order of magnitude analyses. The basic features of buoyancy, Marangoni and vibrational convection are illustrated in detail. This chapter is intended to prepare the reader for the fully elaborated topics treated in the subsequent chapters.

Solidification of liquid metals and alloys (*Chapter 3*) This chapter deals with some phenomena still overlooked by engineers working in the metallurgy and foundry fields that can significantly affect the industrial process related to the solidification of alloys. Immiscible liquid couples and liquid–liquid systems with miscibility gap are considered and novel numerical methods specially designed to investigate these topics are introduced.

The text begins with a presentation of the noncoalescence and nonwetting phenomena and moves through focused and relevant examples clarifying many physical concepts (such as the existence of an immiscible fluid film responsible for nonwetting and the drop-pushing behavior due to combined solidification and nonwetting effects).

Then other mechanisms that can cause spatial separation of the minority and majority phases (migration of drops and bubbles driven by buoyancy and/or Marangoni effects) are introduced step by step with examples of growing complexity being treated as the chapter progresses.

After a brief discussion of the physics, general numerical techniques are presented that can take into account simultaneously all these effects. It is shown how the proposed methods can provide fundamental understanding of how drop migration due to residual gravity and/or Marangoni effects can interact cooperatively with coalescence to promote macroscopic phase separation.

Finally, some behaviors dealing with nonequilibrium conditions and dissolution processes are discussed (solutal effects and some instabilities of the related flow).

Growth of semiconductors in space (*Chapter 4*) This chapter deals with the floating zone (FZ) technique and similar techniques currently used to obtain semiconductor crystals with high purity.

The chapter begins with a somewhat educational introduction of models, equations and numerical techniques and then moves on through a deep analysis of the various parameters affecting the problem with emphasis on the instability boundaries and the structure of the three-dimensional flow under different conditions and situations. The first part, being focused on the philosophy of modeling, is intended for the engineer who wants to learn more about the FZ process and the related numerical treatment, the second part is devoted to the state-of-the-art for the specialist.

The numerical simulations of these aspects usually requires parallel (i.e. multi-processor) computations. For this reason in the first part, many pages are devoted to the introduction of a specially designed, robust and reliable parallel algorithm.

In the second part generalized correlations are developed, while providing a rigorous framework for deeper understanding and effective treatment of relevant phenomena encountered in the FZ process. The difference between the half-zone and the full zone and the related possibility to use one or the other configuration to capture the FZ fundamental hydrodynamics are elucidated.

The attention is focused on microgravity fluid-dynamic aspects since high-quality material processing is expected to be enabled in the space environment. Moreover, since most of the available experimental data come from on the ground experimentation, some effort is spent to elucidate the behavior of the floating zone under normal-gravity conditions.

Macromolecular (protein) crystal growth (Chapters 5 and 6) Nucleation phenomena, surface kinetics, morphological stability of organic crystals grown from solution and again related new numerical methods and results.

These chapters exhibit a strong interdisciplinary nature and deal with advances in protein (crystal) engineering models and computational methods as well as with novel results on the relative importance of "controlling forces" in macromolecular crystal growth. The attention is focused in particular on microgravity fluid-dynamic aspects.

The modeling of these processes leads to the introduction of a group of equations, strictly related, from a mathematical point of view, to the "kinetic conditions" used to model mass transfer at the crystal surface as well as to the level of detail required by the analysis ("local" or "global"); i.e. diversification of the model occurs according to the desired scale length. If the "local" evolution of the crystal surface is the subject of the investigation (distribution of the local growth rate along crystal faces, shape instabilities, onset of surface depressions due to diffusive and/or convective effects, etc., i.e. all those factors dealing with the local history of the shape) the method must provide microscopic and morphological details. For this case a "kinetic-coefficient-based" Volume of Fluid Method is presented taking into account the possibility of anisotropic (faceted) surface-orientation-dependent growth. On the contrary, if the size of the crystals is negligible with respect to the size of the reactor (i.e. if they are small and undergo only small dimensional changes with respect to the overall dimensions of the cell containing the feeding solution), the shape of the crystals is ignored and the proposed approach relies directly on an algebraic formulation of the nucleation events and on the application of an *integral* form of the mass balance kinetics for each protein crystal. The applicability and the suitability of the different models are discussed according to some worked examples dealing with microgravity conditions.

Also in these chapters a deductive approach is followed with systems of growing complexity (increasing number of crystals simultaneously present in the reactor) being treated as the book progresses.

Chapters 5 and 6 represent quite an exhaustive attempt to help organic crystal growers to discern the complex interrelations among the various parameters under one's control (that are not independent of one another) and to elaborate rational guidelines relating to physical factors that can influence the probability of success in crystallizing protein substances in space.

Organic tissues in bioreactors (*Chapter 7*) The attention is focused, in particular, on the Rotary Culture System since, as discussed in Section 1.6, this technique has proven to be the most practical solution to provide a suitable culture environment supporting 3-D tissue assemblies.

Some effort is devoted to demonstrate how numerical simulations can be used as a useful tool to discern the environmental factors affecting the surface kinetics of growing biological tissues (mechanisms of absorption of nutrients available in the culture-medium and subsequent generation of new cells and production of an extracellular matrix) as well as their relative importance in determining growth.

For the first time, in fact, mathematical models of the tissue surface kinetics and related sensitivity to the fluid-dynamic shear stress are introduced and used to elucidate very recent experiments.

The interplay between the increasing size of the tissue and the structure of the convective field is discussed showing that this interaction is essential in determining the time evolution of the tissue shape. The size of the growing specimen plays, in fact, a "critical role" for the intensity of convection and the related shear stresses. Convective effects, in turn, are found to impact growth rates, tissue size and morphology, as well as the mechanisms driving growth.

CHAPTER 2

Fundamental concepts, mathematical models and scaling analysis for the microgravity environment

2.1. Products and thermo-fluid-dynamic disturbances

As emphasized in Chapter 1, many materials science processes are deeply affected (in a detrimental way) by gravitational convection and sedimentation, and hence exploitation of microgravity environments in physical sciences is motivated in most cases by the establishment of quiescent or quasiquiescent regimes. Generally, convective phenomena promote in-homogeneity and defects, while diffusive transport in the fluid phase should yield to more uniform conditions along the growing interfaces and allow a better control of the material process. For this reason, microgravity conditions is usually regarded as an optimal environment. As previously explained, typical experiments that would benefit from these conditions are organic and inorganic crystal growth, solidification processes, production of alloys and many others. From a theoretical point of view, the goal of all the microgravity experimentations is the obtainment of a product (be it a material, a metal alloy, a semiconductor, a protein, etc.) and it is expected that this product (symbolically denoted by P) will somehow be "better" that those achievable on the ground at 1g (P_l).

Along these lines a Microgravity Relevance factor (MR) could be defined just in terms of the difference of these two products ($\Delta P_l = P_l - P$, Monti et al., 2001).

In practice, however, only if this difference is large enough (in an appropriate scale) can one claim that the experiment is worth doing in microgravity (instead of at 1g). In the early phases of the microgravity experimentation it was generally believed that it was sufficient just to bring samples in space and to process them under microgravity conditions to obtain materials with a great improvement of their physical properties. Unfortunately this was not the case: Convective disturbances potentially affecting technological processes of interest, in fact, are not completely suppressed in the microgravity environment, preventing the establishment of the desired diffusive transport in the fluid phases.

As discussed in Section 1.2, all facilities for microgravity experimentation suffer from some degree of residual acceleration and g-jitters; moreover, surface-tension forces, overshadowed or inhibited by gravity on Earth can become dominant in space (see the aspects treated in Section 1.4).

Usually the materials-production techniques are characterized by heat and mass transfer processes in fluid phases in the presence of density gradients, caused by thermal gradients (due to heat exchange or to latent-heat release during a phase change) or by concentration gradients arising, for example, from the rejection or the incorporation of the solute at the surface of macromolecular crystals. The combined action of these gradients and of the above-mentioned forces can result in the onset of different types of convective disturbances in the microgravity environment.

Therefore, if a "purely diffusive" (d) reference state is considered for comparison, one can similarly define another symbolic difference $\Delta P_d = P - P_d$ between the product that can be achieved in the real microgravity environment and the product (P_d) that could be (ideally) achieved in such an ideal condition.

The corresponding parameter for processes carried out on the ground is $\Delta P_g = P_1 - P_d$. This difference tends to be very large.

Within this theoretical context, the most significant microgravity processes are those characterized by large "values" of ΔP_1, and small values of ΔP_d. Similarly the most interesting processes on Earth are those with values of ΔP_g as small as possible. In the light of the above arguments it is obvious that $\Delta P_1 = O(\Delta P_g)$ and $\Delta P_d \ll \Delta P_g$ in many circumstances.

When convective disturbances exist on the platform, a distorted (with respect to a purely diffusive regime) thermo-fluid-dynamic (TFD) field is established; if these disturbances are "small" (often this is the distinguishing feature between a microgravity environment and the Earth's surface) the TFD can be thought to consist of the previous TFD_d plus a relatively small contribution. Having defined the disturbance ΔP_d and assumed the dependence P(TFD), one can always linearize and get $\Delta P_d = P - P_d = K(TFD - TFD_d) = K\varepsilon$ and relate the overall product difference to the thermo-fluid-dynamic field disturbances ($\varepsilon = TFD - TFD_d$). Such a linearization, in general, is not possible for the parameter ΔP_g.

It is noteworthy how most of the time it is the ratio between the convective term over the diffusive one in the scalar transport equations that gives a measure of the TFD disturbance ε(TFD) (in the limit of $\underline{V} \to \underline{0}$; $TFD \to TFD_d$ and $\varepsilon(TFD) \to 0$). The order of magnitude of this ratio is measured by the Peclet number ($Pe = V/V_d$) where V_d is the diffusive velocity ($V_d = \alpha/L$ in the case of thermal processes, where α is the thermal diffusivity; $V_d = D/L$ in the case of solutal processes, where D is the species diffusion coefficient and L is a reference length).

In conclusion, an estimation of the quality of the final product (via the related evaluation of the order of magnitude of the TFD disturbance) can be done by an appropriate evaluation of the order of magnitude of the convective velocity, that in turn depends on the intensity of the surface-tension forces and/or on the acceleration level: 9.81 m/s^2 on the ground, steady and/or oscillatory accelerations prevailing on the considered microgravity platform in space.

For these reasons the following sections are devoted to a critical and concise analysis of these types of convection. A significant (somewhat educational) description is provided to illustrate their genesis, the governing nondimensional parameters, the scaling properties as well as their structure, their stability behavior and their possible bifurcations to different patterns of symmetry in space as well as on the ground.

Such a preliminary analysis is intended to familiarize the reader with these aspects while providing a rigorous framework propaedeutical to a better comprehension of a number of intriguing phenomena and practical applications presented in the subsequent chapters.

2.2. Buoyancy convection and the Boussinesq model

In the presence of steady gravity (on the ground conditions $g = g_o$ or steady acceleration levels on microgravity platforms) the velocity, temperature, and concentration fields can be determined by the differential balance equations (for mass, momentum, energy, and mass species).

Following the usual Boussinesq approximation for incompressible fluids (see, e.g. Boussinesq, 1903; Chandrasekhar, 1981), the physical properties are assumed constant, except for the density ρ in the generation term in the momentum equation (see Eq. (2.2b)), which is assumed to be a linear function of temperature and concentration:

$$\rho = \rho_o[1 - \beta_T(T - T_o) - \beta_C(C - C_o)] \tag{2.1}$$

where β_T and β_C are the thermal and solutal expansion coefficients, respectively, T_o and C_o reference values for temperature and concentration, respectively.

In other words, the continuity equation may be reduced to the vanishing of the divergence of the velocity field, and variations of the density can be ignored in the momentum equation, except insofar as they give rise to a gravitational force.

The approximation is extremely accurate for many flows, and makes mathematics and physics simpler.

The equations in the dimensional form read:

$$\underline{\nabla} \cdot \underline{V} = 0 \tag{2.2a}$$

$$\frac{\partial \underline{V}}{\partial t} + \underline{\nabla} \cdot [\underline{V}\,\underline{V}] + \frac{1}{\rho_o}\underline{\nabla} p = \nu \nabla^2 \underline{V} + [\beta_T(T - T_o) + \beta_C(C - C_o)]\underline{g} \tag{2.2b}$$

$$\frac{\partial T}{\partial t} + \underline{\nabla} \cdot [\underline{V}T] = \alpha \nabla^2 T \tag{2.3a}$$

$$\frac{\partial C}{\partial t} + \underline{\nabla} \cdot [\underline{V}C] = D \nabla^2 C \tag{2.3b}$$

where V is the velocity field, T the temperature, C the concentration, and p the pressure (ν is the kinematic viscosity).

The identification of the relevant parameters in the momentum and energy (or species) equations requires that these equations are posed in a nondimensional form through the choice of relevant reference quantities.

2.2.1 *Thermogravitational convection*

In the case of thermal problems (solutal effects absent) usually the nondimensional form results from scaling the lengths by a reference distance (L) and the velocity by the energy diffusion velocity $V_d = V_\alpha = \alpha/L$; the scales for time and pressure are, respectively, L^2/α and $\rho\alpha^2/L^2$. The temperature, measured with respect to T_o, is scaled by ΔT. This approach leads to:

$$\underline{\nabla} \cdot \underline{V} = 0 \tag{2.4}$$

$$\frac{\partial \underline{V}}{\partial t} = -\underline{\nabla} p - \underline{\nabla} \cdot [\underline{V}\,\underline{V}] + Pr\, \nabla^2 \underline{V} + Pr \cdot Ra\, T \underline{i}_g \tag{2.5}$$

$$\frac{\partial T}{\partial t} = -\underline{\nabla} \cdot [\underline{V} T] + \nabla^2 T \tag{2.6}$$

and the relevant nondimensional numbers are the well-known Prandtl and Rayleigh numbers:

$$Pr = \frac{\nu}{\alpha} \tag{2.7}$$

$$Ra = \frac{g\beta_T \Delta T L^3}{\nu\alpha} \tag{2.8}$$

where the Rayleigh number also measures the magnitude of the buoyancy velocity V_g to the thermal diffusive velocity, where V_g reads

$$V_g = \frac{g\beta_T \Delta T (L)^2}{\nu} \tag{2.9}$$

Moreover,

$$Gr = Ra/Pr = \frac{g\beta_T \Delta T L^2}{\nu^2} \tag{2.10}$$

is the Grashof number that represents the ratio of buoyant to molecular thermal transport (only two of the Gr, Ra, and Pr numbers are independent).

2.2.2 *Solutogravitational convection*

In the case of compositionally inhomogeneous liquids (solutal problems) usually the nondimensional form results from scaling the lengths by a reference distance (L) and the velocity by the solute diffusion velocity $V_d = V_D = D/L$; the scales for time and pressure are, respectively, L^2/D and $\rho D^2/L^2$. The concentration, measured with respect to C_o, can be scaled by its initial value (i.e. $C = (C_{dim} - C_o)/C_o$). This approach leads to:

$$\underline{\nabla} \cdot \underline{V} = 0 \tag{2.11}$$

$$\frac{\partial \underline{V}}{\partial t} = -\underline{\nabla} p - \underline{\nabla} \cdot [\, \underline{V}\,\underline{V}\,] + Sc\, \nabla^2 \underline{V} + Sc \cdot Ra\, C \underline{i}_g \tag{2.12}$$

$$\frac{\partial C}{\partial t} = -\underline{\nabla} \cdot [CT] + \nabla^2 C \tag{2.13}$$

and the relevant nondimensional numbers are:

$$Sc = \frac{\nu}{D} \tag{2.14}$$

$$Ra = \frac{g \beta_C C_o L^3}{\nu D} \tag{2.15}$$

where the solutal Rayleigh number measures the magnitude of the buoyancy velocity V_g to the solutal diffusive velocity, where V_g reads

$$V_g = \frac{g \beta_C C_o (L)^2}{\nu} \tag{2.16}$$

Moreover,

$$Gr = Ra/Sc = \frac{g \beta_C C_o L^2}{D^2} \tag{2.17}$$

is the Grashof number that in this case represents the ratio of buoyant to molecular transport (i.e. diffusion) of solute.

Note that in practice to obtain a reliable estimation of the order of magnitude of the buoyancy velocity the effective concentration gradient ΔC driving the solutal flow should be used in place of C_o in Eq. (2.16).

2.2.3 *The distortions*

For both the thermal and solutal cases, it is known that when gravity is opposite to $\underline{\nabla}\rho$ the residual-g has no effect (does not induce convective flows) for any value of its magnitude (this effect is well-known on the ground: for instance, fluids uniformly heated from above do not exhibit convective motion); when gravity is concurrent to $\underline{\nabla}\rho$ (fluid heated from below), then TFD distortions arise only if the critical conditions for the onset of convection are exceeded (i.e. if the Rayleigh number is larger than the critical one $Ra_g = Ra_{cr}$, so that instability sets in). Finally, when gravity is orthogonal to $\underline{\nabla}\rho$ (fluid heated from the side) then TFD distortions arise for any value of the residual-g (and are somewhat proportional to the Rayleigh number, see, e.g., Fig. 2.1 for the case of a two-dimensional square test cell filled with a liquid having a Prandtl number $Pr = 15$ and heated from the side).

Figure 2.1b in particular shows that the maximum nondimensional velocity in the field (V_{max}/V_α) increases according to the square root of the Rayleigh number. This is a notable difference with respect to the case of surface-tension-driven flows where (V_{max}/V_α) varies almost linearly with the corresponding characteristic (Marangoni) number as will be discussed in Section 2.3.

2.2.4 *Historical developments and recent contributions*

The features of buoyancy flows are generally well-known (see, e.g., Lappa, 2004i). The onset of gravity-induced convection for different heating conditions (from the side or from below) and for various geometrical configurations, in fact, is a basic problem in many heat-transfer systems in technical applications and has been widely investigated in the literature. For instance, the behavior of fluids uniformly heated from below falls into the category of phenomena known under the heading of "Rayleigh–Bènard" problem that is one of the most intensively studied

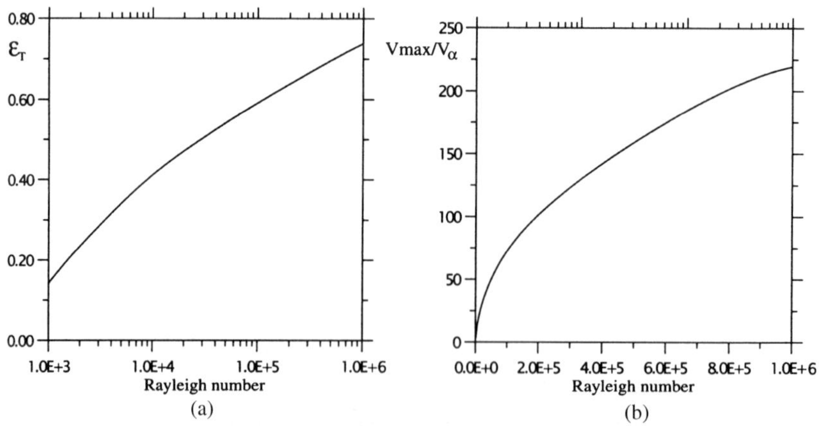

Fig. 2.1 TFD disturbances versus Rayleigh number, silicone oil 1 cs, $Pr = 15$: (a) $\varepsilon_T = \max(T - T_d)/\Delta T$, (b) nondimensional maximum velocity.

hydrodynamical systems. Also, the case of configurations "heated from the side" has instigated much research.

Rayleigh–Bènard problem The physics of the Rayleigh–Bènard instability is well understood. Most of these studies, however, considered infinite systems, i.e. the stability of the quiescent state of nonconfined fluid layers heated from below.

Three cases of the boundary conditions are possible, in principle, for this configuration and have been investigated in the literature as well: (1) stress-free isothermal horizontal boundaries, (2) no-slip isothermal horizontal boundaries, and (3) stress-free thermally insulated upper and no-slip isothermal lower boundary. Cases (1) and (2) are classical and the critical Rayleigh numbers for the onset of convection are $Ra_{cr}=657$ and $Ra_{cr}=1707$, respectively. In the third case, $Ra_{cr}=669$. It is also known that the onset of convection in all three problems is caused by two-dimensional perturbations of the diffusive state (see, e.g., the discussions in Gelfgat, 1999a).

If systems confined laterally by rigid sidewalls are considered, even in containers of large horizontal dimensions, however, the lateral walls can have a significant influence on the flow pattern that develops when the Rayleigh number exceeds its critical value. Thus, the results for the corresponding infinite layer cannot, in general, be used to make predictions about either the detailed structure or the stability of the roll pattern in practical situations.

It has been clearly illustrated by many investigators, in fact, that there are several modes of the most dangerous perturbation that replace each other when the aspect ratio A (ratio of the width and of the height of the finite layer) is varied (Gelfgat, 1999a). Within this context it should also be mentioned that the increase of the aspect ratio A generally results in the increase of the number of two-dimensional rolls within the finite-size rectangular container (see Fig. 2.2).

Several studies have also elucidated that for fixed values of A and Rayleigh numbers in the range $Ra=Ra_{\infty}+O(2/A)$ where Ra_{∞} is the critical Rayleigh number for the corresponding infinite layer, there exists a class of finite-amplitude

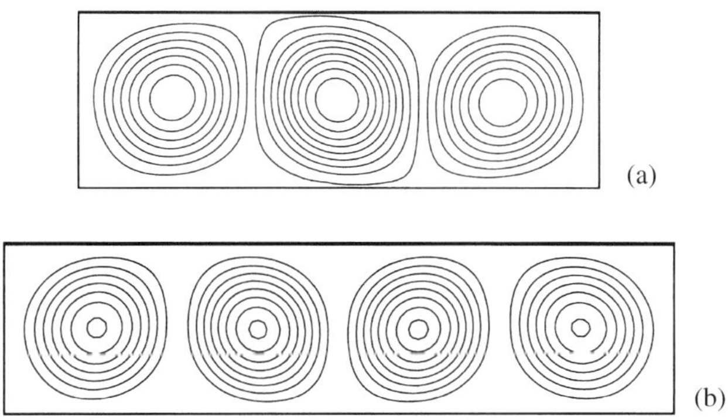

Fig. 2.2 Cavity heated from below and cooled from above ($Pr=0.01$, silicon, Rayleigh–Bènard problem), $Ra=3\times10^3$ (Ra based on the depth of the cavity): (a) $A=3$, (b) $A=4$.

steady-state two-dimensional "phase-winding" solutions that correspond physically to the possibility of an adjustment in the number of rolls in the container as the local value of Rayleigh number is varied (Daniels, 1984).

For the three-dimensional enclosure, similarly to the two-dimensional case, there are several modes of the most dangerous perturbation that replace each other when the aspect ratio is varied. However, the spectrum of possible perturbations is more complicated in the three-dimensional case. A larger variety of perturbations is obviously caused by the three-dimensional geometry. An additional complexity is caused by the variety of the symmetries of the problem. Along these lines, an interesting parametric study of the Rayleigh–Bènard instability in rectangular two-dimensional and three-dimensional boxes has been carried out recently by Gelfgat (1999a). As an example of the complexity of the 3D problem, the reader will find *seven* distinct branches of possible supercritical states for the Rayleigh–Bènard convection in a cube in the article of Pallares et al. (1999).

Because of the light shed on the problem by these and previous studies, most of the scientific community is currently aware of the importance that the control of the boundary conditions and the geometry of the container may have on the threshold values and observed flow patterns. Nonlinear effects, geometrical constraints, and the multiplicity of solutions are all essential ingredients of this type of flow.

Compared to the buoyant flow between parallel planes of infinite extent and the case of rectangular enclosures, the circular cylindrical problem has received less attention.

It is known (see Touihri et al., 1999a and references therein), however, that the structure of the emerging buoyancy flow, which is given by the first primary threshold, depends on the aspect ratio A (height/diameter). For the case of adiabatic sidewalls, in particular, the flow is axisymmetric for $A<0.55$ ($m=0$) and asymmetric for larger values of A ($m=1$, where m is the number of disturbance nodes in the azimuthal direction). The transition between axisymmetric and asymmetric modes occurs around $A=0.72$ if conductive lateral boundaries instead of adiabatic walls are considered. The thresholds for this primary (steady) instability are independent of the Prandtl number (the critical Rayleigh number behaves merely as a decreasing function of the aspect ratio).

If the Rayleigh number is further increased, the flow can undergo a subsequent transition to even more complicated flow patterns. This aspect is still a subject of investigation.

Wanschura et al. (1996) made a complete analysis of this secondary instability. But the study was limited to a range of aspect ratio for which the primary threshold corresponds to a steady axisymmetric flow. In this case the unstable mode for the secondary transition is generally a steady mode $m=2$, but it can also be a steady mode $m=1$, 3, or 4 according to the aspect ratio and the Prandtl number ($m=1$ and 3 for $Pr=0.02$ whereas $m=4$ appears for $Pr=1$). They showed that the secondary threshold increases with the Prandtl number and also depends strongly on the aspect ratio. They also elucidated how for large Prandtl numbers, the axisymmetric flow becomes unstable due to the classical thermal (Rayleigh–Bènard) instability mechanism, whereas for small Prandtl numbers (liquid metals) the secondary instability is inertial (hydrodynamic) in nature.

Lateral heating Also "the heated from the side" problem has received much attention over the years.

The motivation for such an interest is strictly related to the manufacture of bulk semiconductor crystals (see, e.g., Carruthers, 1977). In particular, Hurle (1966) showed that an oscillatory buoyancy convection was responsible for the presence of striations in crystals growing from melts. Accordingly, the problem of onset of oscillatory instability has been extensively studied.

The simplest geometry, among several crystal growth technologies, has the horizontal configuration of the Bridgman method (Dupret and Van der Bogaert, 1994; Monberg, 1994), where the convective flow is induced by the horizontal component of the temperature gradient (see also Section 1.5.2).

A synthesis of available stability analysis results concerning the simplified case of horizontal fluid layers of infinite extent limited from below and from above by parallel planes (with temperature changing linearly as a function of the horizontal coordinate) obtained by different research groups, is given in the article of Gershuni et al. (1992). It is shown therein that, if a velocity profile with an inflection point in the center of the layer section is assumed to be representative of the basic flow in this configuration (Fig. 2.3), then three perturbing mechanisms (two-dimensional hydrodynamic, three-dimensional oscillatory helical wave, and two-dimensional Rayleigh modes) are possible: the hydrodynamic mode corresponds to a shear instability and is related to the formation of vortices on the frontier (i.e. close to the midsection of the layer) of the two opposing flows characterizing the basic state (i.e. close to the aforementioned inflection point); the Rayleigh mode follows from the presence of zones of potentially unstable stratification near the upper and

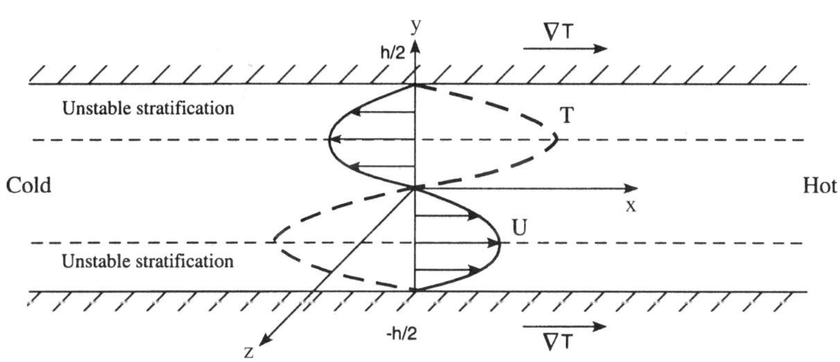

Fig. 2.3 Basic-state velocity and temperature profiles for an infinite horizontal liquid layer confined by two rigid walls.

lower horizontal boundaries (induced by the aforementioned basic flow) that makes possible the onset of instability of the Rayleigh–Bènard type therein (steady longitudinal rolls (SLR): in practice, this flow exhibits strong analogies with the convective case already discussed before in the case of uniform heating from below); finally, the helical wave mode is due to the onset of a pair of gravitational traveling waves perpendicular to the basic flow i.e., along the z axis in Fig. 2.3 (this instability appears in the form of oscillating longitudinal rolls (OLR) and is basically a three-dimensional phenomenon).

These three mechanisms occur in delimited intervals of Prandtl number: For small Prandtl numbers ($0 < Pr < 0.14$) the most dangerous is the plane (2D) hydrodynamic mode (this instability is initially stationary but onset of oscillatory convection is possible for large values of the temperature gradient); in a narrow range of Prandtl numbers ($0.14 < Pr < 0.44$) instability is caused by the OLR mode; for $Pr > 0.44$ the source of instability is transferred to the Rayleigh mode which remains the most dangerous up to extremely large Pr. Naturally this mode does not exist in the case of absence of unstable stratification zones, like the case of layer with horizontal insulated boundaries (see Kuo and Korpela, 1988) where only the hydrodynamic and the OLR modes occur in the range of small Pr.

For practical purposes, the simplest model of the convective flow in the horizontal Bridgman crystal growth method (HB) is the flow in a laterally heated rectangular enclosure of finite extent. This model has been intensively studied both experimentally and numerically (see the very rich reference list in Bontoux et al., 1986), and is used as a convenient benchmark problem for particular values of the aspect ratio and the Prandtl number (GAMM Workshop: Roux, 1990).

A quite exhaustive parametric study of multiple steady states, their stability, onset of oscillatory instability, and some supercritical unsteady regimes of convective flow in a laterally heated rectangular cavity of finite extent with adiabatic upper and lower boundaries has been carried out recently by Gelfgat et al. (1999b, c) for $Pr = 0.015$.

Many possible distinct branches of steady-state flows (pertaining to the aforementioned 2D-hydrodynamic mode) were found for this configuration. A complete study of stability of each branch was performed for the aspect ratio of the cavity varying continuously from 1 to 11. The results were represented as stability diagrams showing the critical parameters (critical Grashof number and the frequency at the onset of the oscillatory instability) corresponding to the transitions from steady to oscillatory states, appearance of multiroll states, merging of multiple states and backwards transitions from multiroll to single-roll states (Fig. 2.4, for instance, shows the multiple steady states of convection in a laterally heated cavity with $A = 4$).

Like the case of Rayleigh–Bènard convection, these studies have also been extended to the case of cylindrical geometry.

Two possible configurations have been considered in the literature: (1) Horizontal cylinder with differentially heated end walls, (2) Laterally heated vertical cylinders.

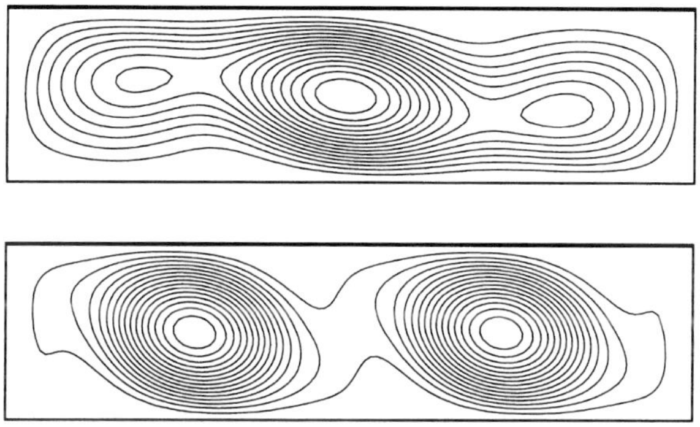

Fig. 2.4 Multiple steady states of convection in a laterally heated cavity with $A = 4$ (cold and hot sides on the left and on the right of each frame, respectively, upper and lower boundaries with adiabatic conditions). $Pr = 0.01$, silicon. Single- and two-roll steady-state flows for $Ra = 1 \times 10^3$ (a), and $Ra = 2 \times 10^3$ (b), respectively (*Ra* based on the depth of the cavity). For $A = 4$, two branches of possible steady state exist.

For the first case, in the past, flow predictions have often been made by using an asymptotic analytical approximation in the core (Bejan and Tien, 1978), or by assuming a two-dimensional solution for the plane of symmetry. Three-dimensional computations by Bontoux et al. (1986) and Viviani et al. (1996) showed, however, that in reality the buoyancy flow in this configuration is highly three-dimensional.

The case of a laterally heated vertical configuration is even more interesting for the aims of the present book, since, like the case of a rectangular container, it has an extensive background application in the field of bulk crystal growth.

As discussed in Section 1.5.2, many bulk crystal growth processes are carried out in axisymmetric geometric configurations.

Deviation from axisymmetric wall temperature conditions in vertical cylinders was investigated by Pulicani et al. (1992) as a possible source of nonaxisymmetric convection observed during vertical directional solidification experiments (VB). However, even under axisymmetric external conditions, the axisymmetric melt flows frequently become unstable and bifurcate to asymmetric steady or oscillatory states. The instabilities of melt flow, lead to the appearance of temperature oscillations and asymmetric flow patterns that, in their turn, lead to the aforementioned inhomogeneities in the structure of the growing crystals.

Some interesting results concerning the stability of buoyant axisymmetric convection in a vertical cylinder with a parabolic temperature profile at the sidewall have been recently presented by Gelfgat et al. (2000) for $0 \leq Pr \leq 0.05$ and $1 \leq A \leq 4$. This configuration was studied as a relevant model of the vertical Bridgman (Crochet et al., 1989; Favier, 1990) or liquid encapsulated melt zone technique (Lopez et al., 1999). The critical parameters corresponding to a transition from the steady axisymmetric (basic state) to the three-dimensional asymmetric (steady or oscillatory) flow pattern were computed and it was elucidated that the instability of the flow is three-dimensional for the whole range of governing parameters studied. In particular, the axisymmetric flow in relatively shallow cylinders tends to be oscillatorily unstable via a hydrodynamic Hopf bifurcation of the circulating flow, while in tall cylinders the instability sets in due to a steady bifurcation caused by the Rayleigh–Bènard mechanism. In the first case the three-dimensional perturbation has the form of a traveling wave; accordingly the oscillatory perturbation patterns (as well as the oscillatory component of the flow) rotate around the axis of the cylinder with the angular velocity f/m (where f is the frequency of the oscillations).

This analysis also predicted the possible existence of very complex three-dimensional behaviors for fixed values of the aspect ratio with the possibility of an adjustment in the number of azimuthal disturbance nodes as the value of the Rayleigh number is varied in a restricted range: For instance, for $A=1$ the critical Rayleigh numbers, that correspond to the azimuthal modes with $m=2$ and m in the range [5–10] (m is the number of disturbance nodes in azimuthal direction like the case of cylinders heated from below), are relatively close and grow rapidly with the increase of Pr. For this reason for $Pr<0.02$ a complex interaction of several azimuthal modes at relatively large supercriticalities is possible. In particular for $Pr=0.01$ the most dangerous mode with $m=4$ is excited at $Ra \cong 3 \times 10^3$, then the $m=2$ mode would be excited at $Ra \cong 3.7 \times 10^3$ and at $Ra \cong 4.5 \times 10^3$ many additional modes would be excited.

The detailed stability results obtained for these hydrodynamic models of crystal growth methods should lead, in principle, to the precise definition of operating parameters that may provide a stable melt flow pattern under the required technological conditions in the presence of normal or residual gravity. Some examples dealing with the case of solutal buoyancy convection and its relevance for the case of crystal growth from solution will be discussed in Chapters 5 and 6.

2.3. Some aspects of Marangoni flow

Microgravity gives the possibility to avoid some limitations related to the ground environment, e.g., the aforementioned buoyancy-driven convection that is the most important cause of imperfections and crystal defects. Moreover, in zero-g conditions, it is possible to form very large floating liquid volumes. As also anticipated in Chapter 1, such floating configurations are often used for processes that benefit from the absence of containers (the absence of contact with solid walls, in fact, prevents contamination and other undesired effects). During recent years, the availability of sounding rockets and orbiting laboratories such as the Spacelab, has made possible microgravity experiments and manufacturing processing with large melt/air surfaces, which could not be performed on Earth under normal-gravity conditions.

However, the presence of the aforementioned free-melt/gas interfaces at different temperatures (an intrinsic consequence of the solidification processes), that in many cases offers a number of advantages in terms of final purity of the materials, induces another type of undesired flow, the so-called Marangoni convection. This convective flow (as mentioned in Sections 1.4.1 and 1.4.3) is basically a surface-tension-driven and gravity-independent phenomenon. The next sections provide the necessary mathematical models.

2.3.1 *The genesis and relevant nondimensional numbers*

The surface tension σ for many cases of practical interest exhibits a linear dependence on temperature and concentration:

$$\sigma = \sigma_o - \sigma_T(T - T_o) - \sigma_C(C - C_o) \tag{2.18}$$

where σ_o is the surface tension for $T = T_o$ and $C = C_o$ (T_o and C_o are reference values), $\sigma_T = -d\sigma/dT > 0$ (σ is a decreasing function of T) and $\sigma_C = -d\sigma/dC$.

If a nonisothermal free surface is involved in the considered process (for instance, liquid metal solidification), then surface-tension forces $\underline{\nabla}_s\sigma$ ($\underline{\nabla}_s$ derivative tangential to the interface) arise that must be balanced by viscous stresses in the liquid $\underline{S}\cdot\underline{n}$ ($\underline{n}$ unit vector perpendicular to the free interface, $\underline{S}$ stress tensor). This balance yields:

$$\mu\frac{\partial \underline{V}}{\partial n} = -\sigma_T \underline{\nabla}_s T \tag{2.19}$$

and in the nondimensional form (scaling the lengths by a reference distance L, the velocities by the energy diffusion velocity and the temperature by a reference ΔT):

$$\frac{\partial \underline{V}}{\partial n} = -Ma\, \underline{\nabla}_s T \tag{2.20}$$

where $Ma = \sigma_T \Delta T L/\mu\alpha$ (μ being the dynamic viscosity) is the so-called Marangoni number. This condition enforces a flow by tangential variation of the surface tension. The motion immediately results whenever a temperature gradient exists, no matter how small (thermocapillary convection). The surface moves from the region with a low surface tension to that with a high surface tension. The viscosity transfers this motion to the underlying fluid, i.e. the flow penetrates into the bulk through viscous coupling to the motion at the interface (Scriven and Sternling, 1960).

Prior to the space program, this phenomenon had been ignored in investigations of materials processing on Earth. In microgravity, the reduced level of buoyancy-driven convection allowed Marangoni convection (named after the Italian physicist Carlo Giuseppe Matteo Marangoni (1840–1925)). Once it became recognized, it was found to be significant in some Earth-based processes as well.

The following nondimensional numbers are of particular importance (see, e.g., Kuhlmann, 1995):

$$Pr = \frac{\nu}{\alpha} \tag{2.21}$$

$$Re = \frac{\sigma_T \Delta T L}{\rho \nu^2} \tag{2.22}$$

$$Ca = \frac{\sigma_T \Delta T}{\sigma_o} \tag{2.23}$$

$$Ma = Re \cdot Pr = \frac{\sigma_T \Delta T L}{\mu \alpha} \tag{2.24}$$

The Reynolds number (Re) measures the magnitude of the tangential stress $\sigma_T \Delta T/L$ to the viscous stress $\rho\nu^2/L^2$; the Prandtl number (Pr) measures the rate of momentum diffusion ν to that of heat diffusion α, and the Marangoni number is $Ma = Re \cdot Pr$. Only two of these three numbers are independent.

The capillary number ($Ca = Re \cdot Oh$, $Oh = \rho\nu^2/\sigma_o L$ being the Ohnesorge number) gives the relative change of the surface tension due to temperature variations. It involves the mean surface tension σ_o and it is an important measure of the dynamic deformability of the free surface. The interface location is determined, in fact, by the hydrodynamic and the Laplace pressure. The relative importance of both is given by the ratio of the relative magnitudes ($\sigma_T \Delta T/L$) and σ_o/L. This ratio is the above-mentioned capillary number $Ca = \sigma_T \Delta T/\sigma_o$. If $Ca \to 0$ (i.e. $Ca \ll 1$) the dynamic surface deformation can be neglected.

In such a case, the liquid/gas interface of liquid layers and open cavities is usually assumed to be fixed, flat and plane in the xz plane (Fig. 2.5b); thus Eq. (2.20) reads (u and w being the nondimensional velocity components along x and z, respectively):

$$\frac{\partial u}{\partial y} = -Ma \frac{\partial T}{\partial x} \tag{2.25a}$$

$$\frac{\partial w}{\partial y} = -Ma \frac{\partial T}{\partial z} \tag{2.25b}$$

for a cylindrical free interface (liquid-bridge problem) the same condition reads:

$$\frac{\partial u}{\partial r} = -Ma \frac{\partial T}{\partial z} \tag{2.26a}$$

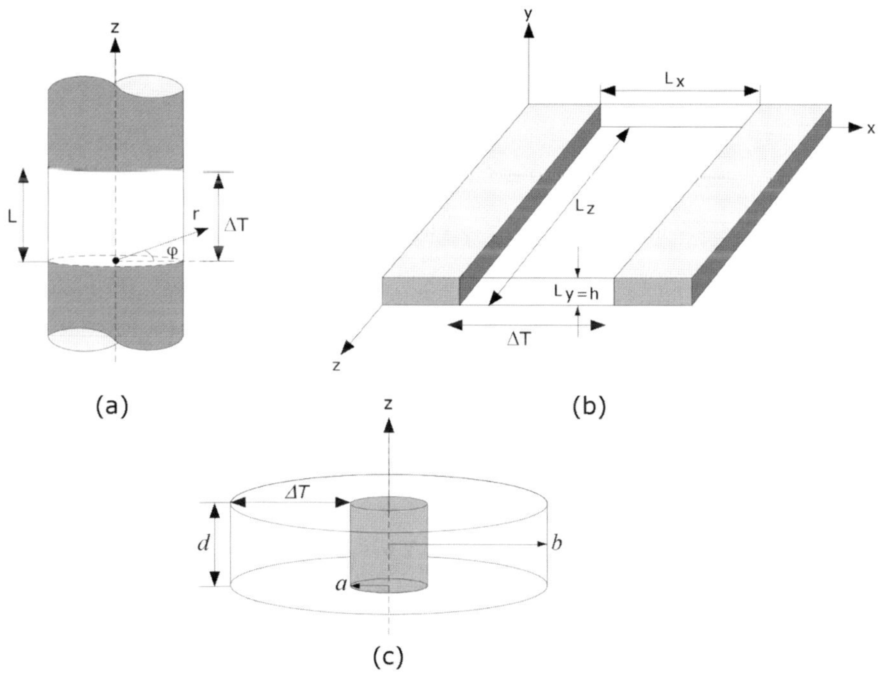

Fig. 2.5 Types of Marangoni convection: (a) liquid bridge, (b) rectangular layer or slot, (c) annular geometry.

$$r\frac{\partial V_\varphi}{\partial r} - V_\varphi = -Ma\frac{\partial T}{\partial \varphi} \tag{2.26b}$$

u and V_φ being the velocity components along z and along the azimuthal angle, respectively (z is the axis of the cylinder, r the radial coordinate, Fig. 2.5a).

Features of interest are the fundamental scales of velocity and temperature that determine the strength of the flow.

It has been noted by Rybicki and Floryan (1987) that the appropriate scaling of the dimensional velocity is the Marangoni velocity:

$$V_{Ma}^{\mathrm{T}} = \frac{\sigma_{\mathrm{T}}\Delta T}{\mu} \tag{2.27}$$

Then the Marangoni number can also be seen as the measure of the relative importance of the Marangoni and of the thermal diffusion velocities.

Moreover, for small Reynolds and Prandtl numbers the temperature field scales like $Re \cdot Pr \cdot \Delta T$.

In many situations, the capillary number is small and the flow-induced surface deformations can be expanded in a power series of Ca to yield the leading-order surface deflection δ_o, which turns out to be of $O(Ca \cdot L)$.

To summarize, in dimensional form the scaling for small *Re*, *Pr* and *Ca* reads:

$$V = \mathrm{O}\left(\frac{\sigma_{\mathrm{T}}\Delta T}{\mu}\right) \tag{2.28}$$

$$\varepsilon_{\mathrm{T}} = \max(T - T_{\mathrm{d}}) = \mathrm{O}(Re \cdot Pr \cdot \Delta T) = O(Ma \cdot \Delta T) \tag{2.29}$$

$$\delta_{\mathrm{o}} = \mathrm{O}(Ca \cdot L) \tag{2.30}$$

where T_{d} is the reference diffusive temperature field (i.e. the temperature distributions that would be established in the absence of convection) and ε_{T} is the corresponding temperature disturbance.

According to Eq. (2.18) Marangoni convection can also arise as a consequence of concentration gradients; in this case it is usually referred to as "solutocapillary" convection.

2.3.2 *Microzone facilities and microscale experimentation*

Despite the presence of buoyancy convection that in many cases obscures the Marangoni flow it must be pointed out how some insights into this type of convection can also be obtained by experimental investigation on Earth, if the typical scale of the liquid volume is reduced. Usually this kind of study and the related facility are referred to as "microscale" experimentation and "microzone apparatus," respectively.

In practice, simulated microgravity can be obtained on the ground by reducing the characteristic length of the system under investigation. This condition, of course, does not really cancel gravity. In practice, through this procedure it is possible to emphasize surface-tension effects with respect to buoyancy-induced ones.

In fact, the ratio of the Rayleigh (see Section 2.2 for the related definition) to the Marangoni number (that measures the driving actions of the gravity and of the surface-tension imbalance) grows quadratically with the linear dimension of the liquid zone making thermal buoyancy convection less important for small liquid volumes.

$$\frac{Ra}{Ma} = \frac{g\rho\beta_{\mathrm{T}}}{\sigma_{\mathrm{T}}} L^2 = \frac{1}{W} \tag{2.31}$$

If this ratio is kept small (*W* is kept large) by reducing *L* (few millimeters) the Marangoni effect is emphasized in comparison with the buoyancy effect.

The reduction of the scale also has another purpose. It can be used to avoid a large curvature of the liquid/gas interface. Such a deformation is usually somewhat proportional to the so-called Bond number, *Bo*:

$$Bo = \frac{\Delta\rho g L^2}{\sigma} \tag{2.32}$$

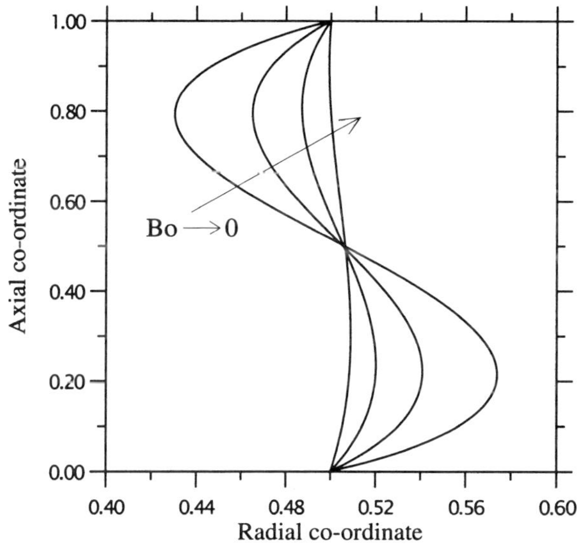

Fig. 2.6 Shape of the liquid-bridge surface as a function of the Bond number (silicone oil 1 cs, normal-gravity conditions): Bond numbers in order from the most arched profile to the flattest one are $Bo = 11.5$ ($L = 0.5$ cm), 7.36 ($L = 0.4$ cm), 4.14 ($L = 0.3$ cm), 1.84 ($L = 0.2$ cm), 0.46 ($L = 0.1$ cm). $Bo = 0$ would correspond to the cylindrical shape (zero-g conditions).

where $\Delta\rho$ is the density jump between the liquid and the surrounding gas ($\Delta\rho \cong \rho_{\text{liquid}}$).

The parameter *Bo* represents the ratio of internal hydrodynamic pressure to surface-tension force.

If this number is sufficiently small, the fluid/fluid interfaces behave (approximately) not much differently with respect to the case of zero-g. As an example, Fig. 2.6 shows the shape of a silicone oil liquid bridge with a fixed aspect ratio ($A = 1$) for different values of this number (corresponding to different heights L of the liquid column). It is shown that the shape tends to the cylindrical interface (microgravity conditions) when the Bond number tends to zero.

2.3.3 *Past history and current status*

Marangoni convection has been the subject of increasing interest in recent years with regard to many different geometrical configurations. It was learned that this surface-tension-driven convection could not only be vigorous, but could also become asymmetric, oscillatory and even turbulent. Like the case of thermogravitational convection described in Section 2.2, the direction of the imposed ΔT plays a critical role: if the externally imposed ΔT yields imposed temperature gradients that are primarily perpendicular to the interface, the basic state is static with a diffusive temperature distribution and motion ensues with the onset of instability when ΔT

exceeds some threshold; if the externally imposed ΔT yields imposed temperature gradients that are primarily parallel to the interface, as anticipated, in these cases motion occurs for any value of ΔT. When the temperature gradient acting along the free surface reaches a critical value, various types of instabilities can be initiated in the system.

The problem concerning the thin fluid layers uniformly heated from below falls into the category of phenomena known under the heading of "Marangoni–Bènard" convection. This phenomenon, is characterized by the well-known and suggestive presence of aesthetic hexagonal cells (see, e.g., Fig. 2.7); it is also known that at a certain distance (in terms of value of the Marangoni number) from the threshold of primary instability, square convection cells rather than the seemingly all-embracing hexagons become the persistent dominant features of this type of convection (Fig. 2.8).

The related mechanisms, however, are not relevant for the aims of this book and are no longer discussed within the present section (even if some theoretical relevance could be outlined with regard to the Czochralski technique). The reader will find the state-of-the-art in the recent article of Eckert et al. (1998), and in the book of Colinet et al. (2001) and references therein.

On the contrary, Marangoni convection driven by differentially heated free surfaces has very important effects during the semiconductor crystal growth carried out by a number of techniques; for instance, the Czochralski (see, e.g., Hurle and Cockayne, 1994; Nakamura et al., 1999), the floating zone and the "open-boat" (Hurle, 1994) methods. An undesired dopant distribution, in fact, has been reported in many circumstances. Also in these cases (see Section 2.2.4 for the corresponding problem of buoyancy flow) the dopant is found concentrated in regular banded structures called "striations," which are thought to be caused by an oscillatory convection. For these reasons much attention has been devoted to flows in open cavities, liquid bridges, and annular configurations used respectively, as models of the open-boat horizontal melt growth, of the floating-zone method and of the Czochralski technique (see also Section 1.5.2 for additional details about the basic features of these growth processes).

The state-of-the-art for the floating-zone problem and the corresponding historical developments will be discussed in Chapter 4 that is entirely devoted to these aspects. The first part of the present section is limited to a somewhat introductory historical perspective for the open-boat problem (HB), namely an upper surface-tension-driven flow, which implies Marangoni convection in a rectangular container with an insulated bottom wall and differentially heated side walls (the features of buoyancy convection for this problem have been treated in Section 2.2.4).

Along these lines, it is worth starting this discussion by mentioning the milestone linear stability analysis of Smith and Davis (1983). They considered a fluid layer with infinite extent, exhibiting a free surface supporting Marangoni stresses (with a

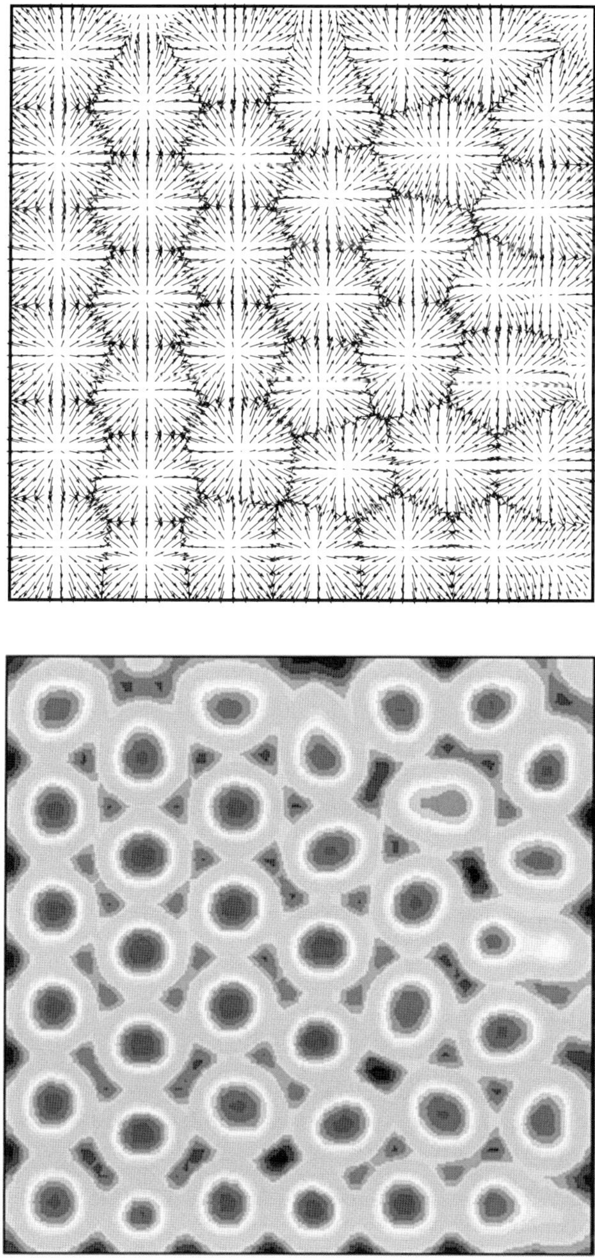

(a)

(b)

Fig. 2.7 Hexagonal cells due to Marangoni–Bènard convection, observable on the free surface of a layer of silicone oil 1 cs covered by air, having depth 1 mm and horizontal extension $L_x = L_z = 2$ cm ($Ma = 300$, Ma based on the depth L_y of the layer, zero-g conditions): (a) velocity field, (b) temperature distribution.

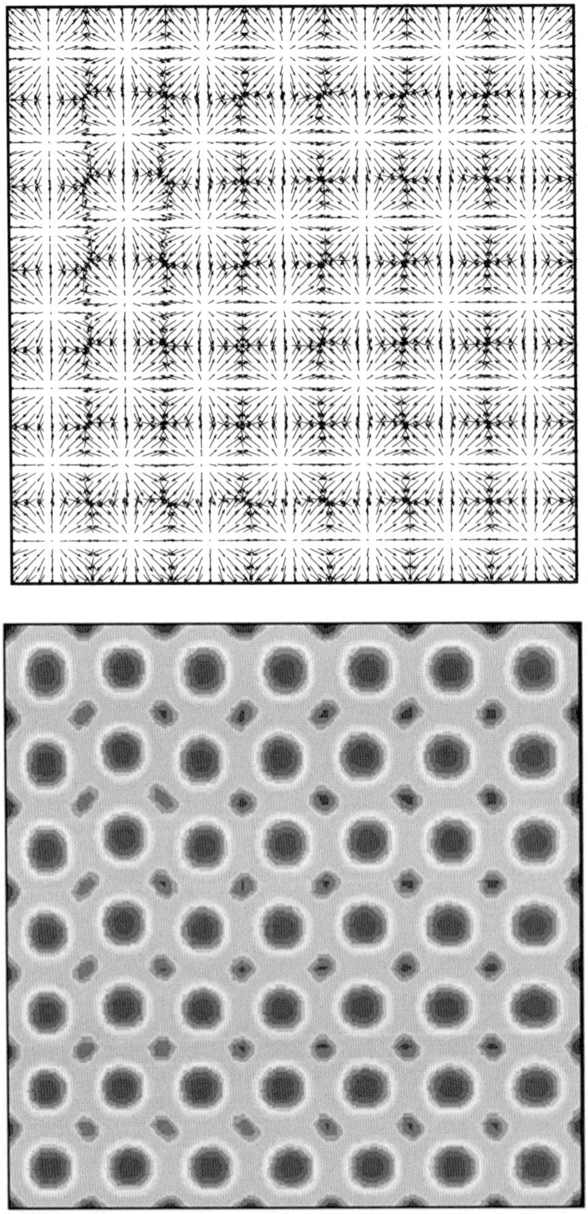

(a)

(b)

Fig. 2.8 Square cells due to Marangoni–Bènard convection, observable on the free surface of a layer of silicone oil 1 cs covered by air, having depth 1 mm and horizontal extension $L_x = L_z = 2\,\text{cm}$ ($Ma = 600$, Ma based on the depth L_y of the layer, zero-g conditions): (a) velocity field, (b) temperature distribution.

temperature gradient along x, see Fig. 2.5b), without gravity nor heat exchange to the ambient and found two types of thermal instabilities in the case of undeformable interface: stationary multicellular flow (SMC) and oblique hydrothermal waves (HTW). The first takes the form of stationary longitudinal rolls that become unstable in much the same way as the classical Marangoni layer heated from below and discussed earlier (however these rolls do interact with the underlying shear flow; the energy driving the instability comes from the vertical temperature gradient induced by a balance between the horizontal convection and vertical conduction). The second thermal instability takes the form of propagating waves (deriving their energy from the imposed horizontal temperature gradient when the Prandtl number of the liquid is small and from the vertical temperature gradient when the Prandtl number is large).

The occurrence of the SMC or HTW modes of convection was found to depend on the Prandtl number and on the basic flow considered (with or without return-flow profile).

Two parallel-flow solutions were assumed, in fact, as possible basic states, the first referred to as "linear-flow solution" (Fig. 2.9a) and second referred to as "return-flow solution" (Fig. 2.9b). For the return flow, the hydrothermal waves correspond to the preferred mode of instability.

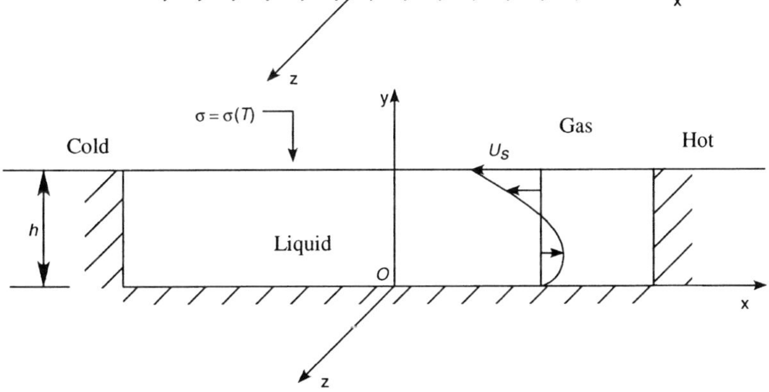

Fig. 2.9 Basic-state velocity profile for an infinite horizontal liquid layer bounded from below by a rigid wall: (a) linear flow solution, (b) return-flow solution.

This oscillatory mechanism was disclosed for the first time in this study. Smith and Davis (1983) together with the subsequent weakly nonlinear analysis of Smith (1986) elucidated that the oblique unsteady hydrothermal wave is basically a three-dimensional phenomenon exhibiting an angle of propagation relative to the basic state that is a function of *Pr*: the angle of propagation of this oblique hydrothermal wave is nearly perpendicular to the basic state for low-*Pr* materials (i.e. the disturbance propagates almost exactly in the spanwise direction, i.e. perpendicular to the surface flow and along z, see Fig. 2.5b) and nearly parallel to the basic state for high-*Pr* materials (i.e. the disturbance propagates almost exactly in the upstream direction).

Most of the subsequent analyses on the instabilities of Marangoni flow (also concerning quite different systems as liquid bridges, annular configurations introduced as models of the Czochralski method, bubbles surrounded by liquid, etc.) rely on these landmark studies. Even if limited to the case of absent end-effects, in fact, significant information was provided in terms of intrinsic mechanisms of the Marangoni flow instability, as explained in the following text.

For small Prandtl numbers ($Pr \ll 1$, semiconductor melts and liquid metals) when the Marangoni number is increased sufficiently, oscillatory behaviors should occur only if the extension (along z) perpendicular to the temperature gradient (along x) is sufficiently large, whereas in the opposite situation only stationary rolls should appear (replacing a single elongated initial vortex, see, e.g., Fig. 2.10); in this case, in fact, the hydrothermal wave predicted by Smith and Davis (1983), responsible for possible oscillatory flow, has a significant component perpendicular to the main flow and is therefore suppressed by 2D models and/or computations.

On the contrary, for large *Pr* the existence of an oscillatory flow should be possible even if the extension along z is very small since in this case the hydrothermal wave should be almost two-dimensional.

These theoretical arguments have received partial validation by a number of numerical and experimental studies carried out for finite-size open cavities and shallow liquid layers in the subsequent years.

For instance, the nearly two-dimensional nature of the hydrothermal wave for high Prandtl numbers is probably the reason why two-dimensional numerical computations (see, e.g., Peltier and Biringen, 1993) easily capture oscillatory flow in open containers and/or finite extended layers. Xu and Zebib (1998) found two-dimensional supercritical oscillatory bifurcations over a wide range of Prandtl numbers ($1 < Pr < 13.9$) and aspect ratios A (ratio of the width to the height of the finite layer). They also noted that the critical threshold tends to the corresponding value predicted by Smith and Davis (1983) for the hydrothermal wave case, in the limit as $A \to \infty$.

Among other things, numerical studies about laterally confined layers and cavities also elucidated the control that the boundary conditions and the geometry of the container may have on the threshold values and observed flow patterns. For instance, Xu and Zebib (1998) found that time-dependent motion is only possible

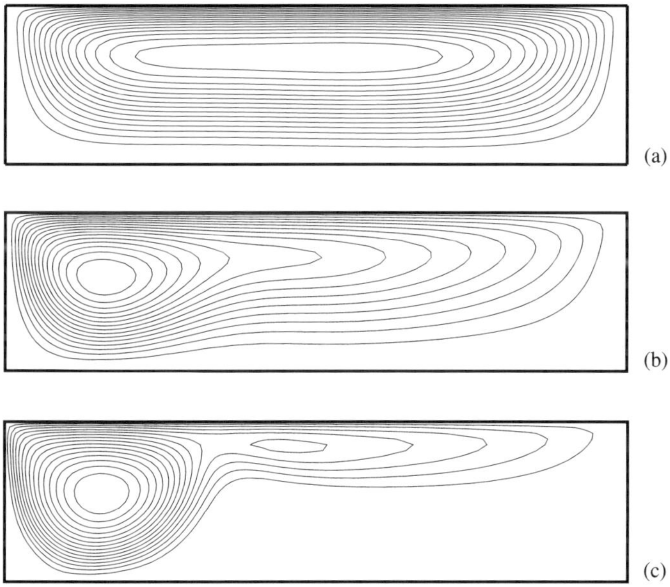

Fig. 2.10 Structure of pure Marangoni convection for $Pr = 0.01$ (silicon) and open cavity with aspect ratio $A = 4$ (cold and hot sides on the left and on the right of each frame, respectively, upper and lower boundaries with adiabatic conditions): (a) $Ma = 10$, (b) $Ma = 100$, (c) $Ma = 1000$ (Ma based on the width of the cavity).

if the aspect ratio exceeds a critical value that increases with decreasing Prandtl number.

They also considered the three-dimensional case. For this case, the sidewalls located along z have a damping effect on oscillations and hence increase the magnitude of the critical Ma. It was also illustrated how the existence of "spanwise standing waves" can reduce this critical Ma when the extension of the finite layer is increased along z.

For the opposite case of small Prandtl number, it is worth mentioning the two-dimensional numerical simulations of Ben Hadid and Roux (1990). In agreement with the predictions of Smith and Davis (1983) their computations did not capture oscillatory instabilities of the pure Marangoni flow in finite cavities and small Pr.

For $Pr = 0.015$ and $Pr = 0.02$, however, oscillations were found for mixed buoyant–Marangoni convection by means of the two-dimensional numerical computations by Mundrane and Zebib (1994). Apparently in contrast to the results of Ben Hadid and Roux (1990), in practice, these simulations showed that the presence of a buoyancy convection can alter the mechanisms of Smith and Davis (1983) and that this can lead to the onset of a two-dimensional oscillatory flow (superimposed on the initial stationary rolls) also for the low values of the Prandtl number.

Most of the experimental results for these configurations, are related to the case of transparent liquids (high Prandtl number) for which direct observation is possible in on the ground conditions. Since the problem of buoyant–thermocapillary convection in cavities is governed by a relatively large number of nondimensional parameters, and there is consequently a large number of different types of flow that can be found in this system, these results give disjoint glimpses of a wide variety of qualitatively and quantitatively different results in widely different parts of parameter space.

For instance, the existence of thermocapillary instabilities, and in particular of oscillations, has been demonstrated in numerous experiments in the case of real three-dimensional configurations (planar rectangular pools designed to most closely mimic the conditions of the Smith and Davis analysis), but only a few of them have provided any direct evidence of hydrothermal waves. In many experiments, the basic flow was observed to destabilize first against a stationary multicellular instability (transition from steady unicellular flow to steady multicellular flow with longitudinal rolls embedded in the main flow all rotating in the same direction) before exhibiting oscillatory behaviors.

The existence of these oscillatory patterns not corresponding to the propagation of hydrothermal waves was observed by Villers and Platten (1992), De Saedeleer et al. (1996) and many others for $Pr \gg 1$.

Riley and Neitzel (1998) provided the first experimental demonstration of oblique hydrothermal waves in a rectangular pool $L_x = 530$ mm, $L_z = 550$ mm for silicone oil for sufficiently thin layers ($h < 1.25$ mm). But for thicker layers, a steady multicellular flow appeared that becomes time dependent when the temperature gradient is increased.

The available results in the case of transparent liquids have been summarized in the recent analysis of Burguete et al. (2001) where it has been shown how, depending on the height of the liquid and on the extension along z, the two-dimensional basic flow destabilizes into oblique traveling waves or longitudinal stationary rolls: the first phenomenon seems to be favored for small thickness of the layer and a large extension of the pool along z; vice versa the latter tends to become dominant for thick layers and small extension along z. In this case, the oscillatory behavior that occurs when the thermal gradient is further increased, is characterized by the oscillation of the multicellular pattern.

These results are in partial agreement with the findings of Smith and Davis (1983), the discrepancies probably being due, like the case of small Prandtl number, to the effect of buoyancy forces.

For possible explanation of the experimental results obtained on the ground and deviation with respect to the theory for pure Marangoni flow, the reader may consider the overview of Gershuni et al. (1992), already cited in Section 2.2.4 with regard to the case of infinite layers bounded from above and from below by rigid walls as well as the analysis of Parmentier et al. (1993) (in practice, the former analysis considers conducting boundary conditions whereas the latter addresses the case of adiabatic horizontal surfaces).

According to Gershuni et al. (1992) when the upper boundary is free, if the Marangoni flow is weak ($W \cong 0.1$) the mechanisms leading to flow instability are essentially the same as in the pure buoyancy case (the aforementioned 2D steady or oscillatory hydrodynamic mode for $Pr < 0.045$, the 3D oscillatory longitudinal rolls for $0.045 \leq Pr \leq 0.85$ and the 2D Rayleigh mode for $Pr > 0.85$) whereas for the case of dominating the Marangoni convection ($W > 1$), the hydrodynamic mode is suppressed (the inflection point disappears, Fig. 2.11) and the most dangerous instability is either the 3D-oscillating mode (OLR induced by waves of gravitational type) for $Pr < 1$ or the stationary longitudinal rolls (SLR Rayleigh mode) for $Pr > 1$. With regard to the latter mode of convection, the recent stability analysis of Mercier and Normand (1996) in particular has clearly shown in the case of a liquid with $Pr = 7$, the existence of a transition between stationary rolls and traveling waves when W exceeds a value W_o which is a function of the surface heat exchange with the ambient. This analysis provides quite a definitive explanation for the transition between hydrothermal traveling waves and the stationary rolls that is observed on Earth for transparent media when the depth of the fluid pool is increased.

Within the context of the present section it is also worthwhile to stress how, besides the case of two- and three-dimensional rectangular layers, slots and elongated cavities, the mechanism of hydrothermal-wave propagation, disclosed by the analysis of Smith and Davis (1983) in the case of pure Marangoni flow, has some outstanding applications also for cylindrical configurations with free lateral surfaces. Even though, as anticipated above, these aspects will be treated in Chapter 4, for the sake of completeness it should be mentioned here that the concept of hydrothermal wave is at the root of the explanations provided by the investigators for the oscillatory (Hopf) bifurcation that characterizes liquid bridges in the case of transparent liquids (some fundamental information is also provided in the next section).

Finally, let us mention that traveling waves with characteristics similar to the hydrothermal waves have also been observed in other experimental configurations.

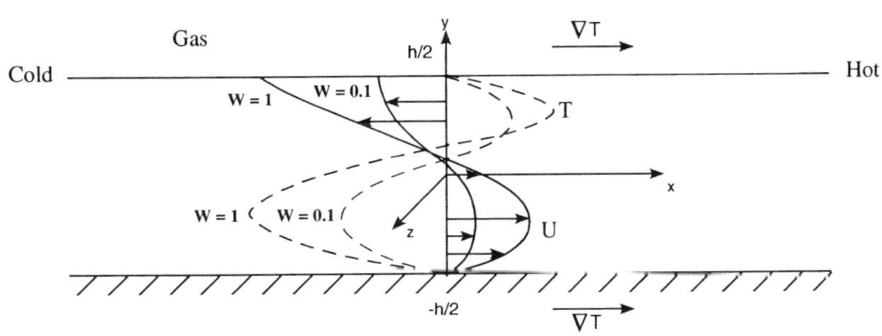

Fig. 2.11 Basic-state velocity and temperature profiles as a function of W, for an infinite horizontal liquid layer bounded from below by a rigid wall and from above by a free surface.

Kassemi and Rashidnia (1999), for instance, detected the azimuthal hydrothermal waves traveling in the liquid around air bubbles trapped by gravity below heated walls (additional details about this intriguing phenomenon will be provided in Chapter 3).

The azimuthal propagation of hydrothermal wave-like structures has also been reported in the case of annular shallow layers used as models of the Czochralski technique and subject to radial horizontal temperature gradients (outer portion of the annulus heated with respect to the inner portion). For instance, such phenomena have been clearly observed by Schwabe (2002) in the case of a fluid with Prandtl number $Pr = 7$ both under microgravity and under normal-gravity. Furthermore the presence of hydrothermal waves of the type predicted by Smith and Davis (1983) in the case of a small Prandtl number fluid has been proven for the first time in the recent numerical investigation of Li et al. (2004) (see Fig. 2.12a). These authors also showed that when the thickness of the low-*Pr* layer is increased, transition from traveling waves to multicellular oscillatory flow (see Fig. 2.12b) occurs.

2.4. Structure of buoyancy and Marangoni convection and of mixed flows

As previously explained, buoyancy convection arises, in the presence of a gravity acceleration, due to density gradients induced by temperature differences applied to the liquid. In the environment provided by microscale experimentation (if the liquid volume is small enough) Marangoni flows, induced by the temperature dependence of the surface tension along the free surface of the liquid, tend to prevail over buoyancy convection.

Typically, Marangoni convection is a surface phenomenon and is confined to regions near the free surface, whereas buoyancy drives fluid motion in the bulk of the specimen. Buoyancy forces are, in fact, volume driving actions (see the body force source term in Eq. (2.2b)), whereas Marangoni stresses are surface driving actions (see Eq. (2.19)). Marangoni flow can be inhibited in the absence of free fluid/fluid surfaces but is a gravity-independent phenomenon (it does not disappear even if zero-g conditions are considered); on the contrary buoyancy convection can arise regardless of the presence of any free liquid surface (it is induced by the presence of accelerations).

Under normal-gravity conditions and in the presence of free surfaces, if the size of the system is not limited to a few millimeters both Marangoni and buoyancy convection play an important role in determining the structure of the flow and can interact in a very complex way even if the flow is steady.

This section is devoted to a brief description of the structure of buoyancy and Marangoni flows when they are separated and, in particular, of related mixed flows for some typical and very common geometrical configurations used by the investigators as paradigm models for fundamental research or as simplified models

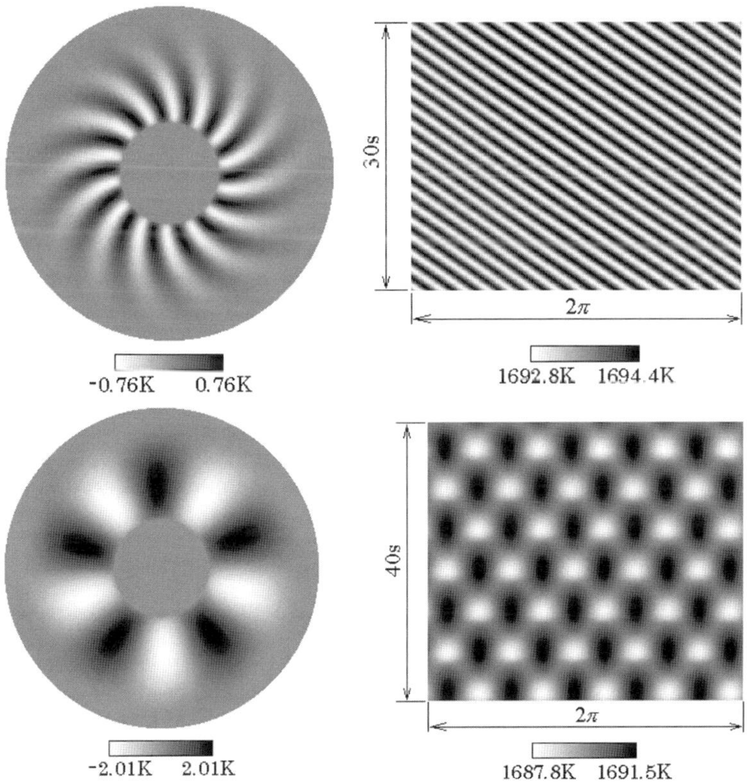

Fig. 2.12 Annual shallow layers – Snapshot of surface temperature fluctuation and space-time diagram of surface temperature along a circumference at $r = 20\,\text{mm}$ for a layer with $a = 15\,\text{mm}$, $b = 50\,\text{mm}$, $Pr = 0.01$, silicon, adiabatic conditions at the top and bottom surfaces (after Li et al., 2004): (a) $d = 1.5\,\text{mm}$, $Ma = 10.67$ (Ma based on the reference length $L = d^2/(b-a)$), the fields correspond to the propagation along the clockwise azimuthal direction of a hydrothermal wave, (b) $d = 8\,\text{mm}$, $Ma = 151.8$ (Ma based on the reference length $L = d^2/(b-a)$), the fields correspond to the presence of radially extended rolls periodically alternating their azimuthal sense of rotation.

of real technological material processes. Interacting surface-tension-driven and gravitational flows are very relevant for a number of material processes carried out on Earth.

The analysis progresses with the aid and support of both the experimental results and the numerical simulations for a better representation of the structure of convection and moves through very focused examples and situations. Transparent liquids (large Prandtl number) are considered for direct experimental visualization of the flow.

It is shown that convection can be very complex and three-dimensional even if the geometry of the problem is very simple.

2.4.1 *The open cavity*

The geometry of the problem used for the demonstrative experiments (see Fig. 2.13) is a rectangular liquid volume ($L_x = L_z = L = 5$ cm, height $L_y = 7.5$ cm and a square cross-section 5 cm × 5 cm) filled by liquid with a free flat surface.

The bottom wall and the windows in front and behind are made of a transparent material (Perspex) with a very low thermal conductivity, to reduce the heat exchange to the ambient. Each vertical wall is split into three parts: a copper wall (height $H = 5$ cm), adjacent to the volume liquid, is thermally insulated from an upper copper wall (height $h = 5$ mm) by a slab of Perspex 2 cm high. The temperatures T_i ($i = 1, 2, 3, 4$) of the four copper walls (T_1 and T_3 at the left side and the opposite T_2 and T_4 at the right side) are controlled independently by circulating water from four independent thermostats. The surface heaters at temperature T_1 and T_2 are thermally insulated from their corresponding volume heaters at temperature T_3 and T_4. The thickness of the surface heaters is 5 mm to generate a temperature gradient (between T_2 and T_1) parallel to the free surface in a thin layer of liquid only.

A similar configuration was used in the landmark experiments of Schwabe and Metzger (1989).

In this model configuration, the liquid can be subject to temperature gradients parallel and perpendicular to the free surface. The gradient parallel to the free surface generates a surface-tension gradient that forces the free surface to move; the gradients

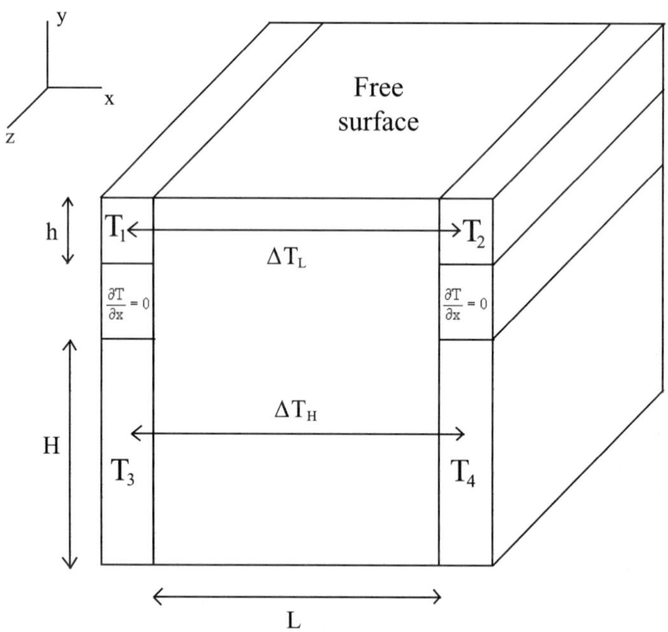

Fig. 2.13 Sketch of the paradigm model configuration for the on the ground investigation of interacting buoyancy and Marangoni convection (Savino et al., 2001a).

perpendicular to the free surface are responsible for the buoyancy forces that generate motion in the bulk. It is obvious, however, that the tension forces on the free surface will have some influence on the convective structure in the bulk and vice versa the buoyancy-driven roll in the bulk will influence the flow on the free surface.

It is worthwhile to stress how the present configuration is specifically designed to allow the buoyant forces to be varied independently from the Marangoni forces. This is achieved by the application of a temperature gradient along the free surface that could be different from that in the bulk; in this way one may vary almost independently (1) the magnitude and direction of the buoyant-convection effects, measured by the appropriate Rayleigh number $Ra=g\beta_T\Delta T_H H^3/\nu\alpha$ and (2) the Marangoni convective effects, measured by the appropriate Marangoni number $Ma=\sigma_T\Delta T_L L/\mu\alpha$ where $\Delta T_L(T_2-T_1)$ and $\Delta T_H(T_4-T_3)$ are the temperature differences between the surface and bulk heaters, respectively.

The gradients ΔT_L and ΔT_H can be applied both parallel in the bulk and along the free surface, or antiparallel, or one can be set equal to zero to investigate the separate effects of buoyancy and surface-tension forces (i.e. $\Delta T_L\neq 0$ and $\Delta T_H=0$ or vice versa).

The experiments are carried out on the ground. The liquid used in the experiments is silicone oil with a kinematic viscosity of 10 cs ($Pr=105$), that has well-known thermo-physical properties (particularly the surface-tension derivative with temperature).

Since the aspect ratios (L_x/L_y and L_z/L_y) are of O(1) the flow is steady and quite stable in the range of the parameters investigated (as already illustrated in Section 2.3, high Prandtl number flows in cavities are strongly stabilized by the presence of close sidewalls both along x and along z; see Xu and Zebib, 1998).

The flow field is visualized by tracers (density-matched ecospheres) in different vertical planes, illuminated with a 1-mm thick light sheet, and on the free surface. A PIV (particle image velocimetry technique applied to the tracers motion, recorded using a video camera and a video recorder) is used to obtain the velocity field.

An infrared thermocamera (Agema 900) is used to detect the temperature distributions on the free surface. The thermocamera sensor operates in the infrared 8–12 μm band, which allows one to perform accurate temperature measurements since the used liquid is opaque in this infrared band.

The numerical simulations are carried out by means of a primitive variable method that will be described in detail in the last section of this chapter.

Four different basic situations can be considered: (1) $\Delta T_H>0$, $\Delta T_L=0$ (dominant buoyancy flow); (2) $\Delta T_H>0$, $\Delta T_L>0$ (Marangoni and buoyant flows concurrent); (3) $\Delta T_H>0$, $\Delta T_L<0$ (Marangoni and buoyant flows counteracting); (4) $\Delta T_H=0$, $\Delta T_L>0$ (no imposed bulk temperature gradient and only surface-tension forces driving the convection).

Case (1) In this case, the presence of a destabilizing vertical temperature gradient ($T_4-T_2>0$) is responsible for a complex three-dimensional flow pattern. The hot fluid heated by the wall at temperature T_4 is carried towards the free surface (see Fig. 2.14) and gives rise to hot spots that can be observed directly near the hot side of the free interface.

An analysis carried out considering different values of the temperature gradient imposed in the bulk (ΔT_H) shows that for $\Delta T_L=0$ the number m of these surface spots is a function of the parameter ΔT_H. For $\Delta T_H=2.5$ K there are two spots on the free surface, three spots for $\Delta T_H=5$ K, four spots for $\Delta T_H=10$ K, five spots for $\Delta T_H=15$ K and six spots for $\Delta T_H=20$ K (see Fig. 2.15). This means that the field exhibits an increasing degree of complexity when the Rayleigh number is increased.

Case (2) Figure 2.16 shows the streamlines of the motion and the vector plots in the case $\Delta T_H=\Delta T_L=15$ K, when the surface and bulk temperature gradients are concurrent. The flow-field organization mainly consists of a combined convection roll around the whole cavity driven by the joint action of surface tension and buoyancy forces.

The convection roll is extended over the entire depth of the rectangular volume; a surface-tension-driven roll is present with its own recirculation in the upper third of the liquid zone and a buoyancy-driven circulation around the whole cavity. The hot fluid adjacent to the heated wall at temperature T_4 in the bulk (see Fig. 2.13) is carried toward the free surface at a rather high position by the buoyancy forces (due to the fact that $T_2<T_4$) and then accelerated by the Marangoni surface forces towards the cold wall at temperature $T_1<T_2$. When the fluid reaches the surface heater at temperature T_1, since $T_1<T_3$, it is cooled and accelerated downwards again in the bulk by the buoyancy forces. This explains the convective structure visible in the midplane of the rectangular cavity.

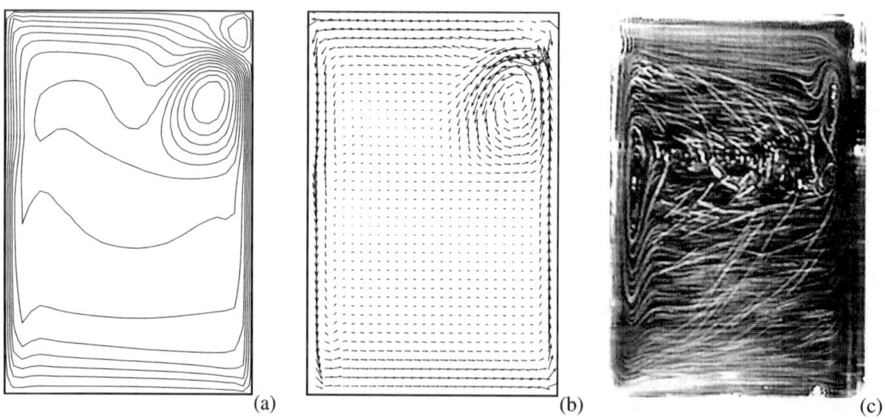

Fig. 2.14 Velocity distribution and streamlines in the vertical midplane ($z=2.5$ cm) for $\Delta T_H=15$ K and $\Delta T_L=0-V_{max}=0.20$ cm/s (a, b numerical; c experimental).

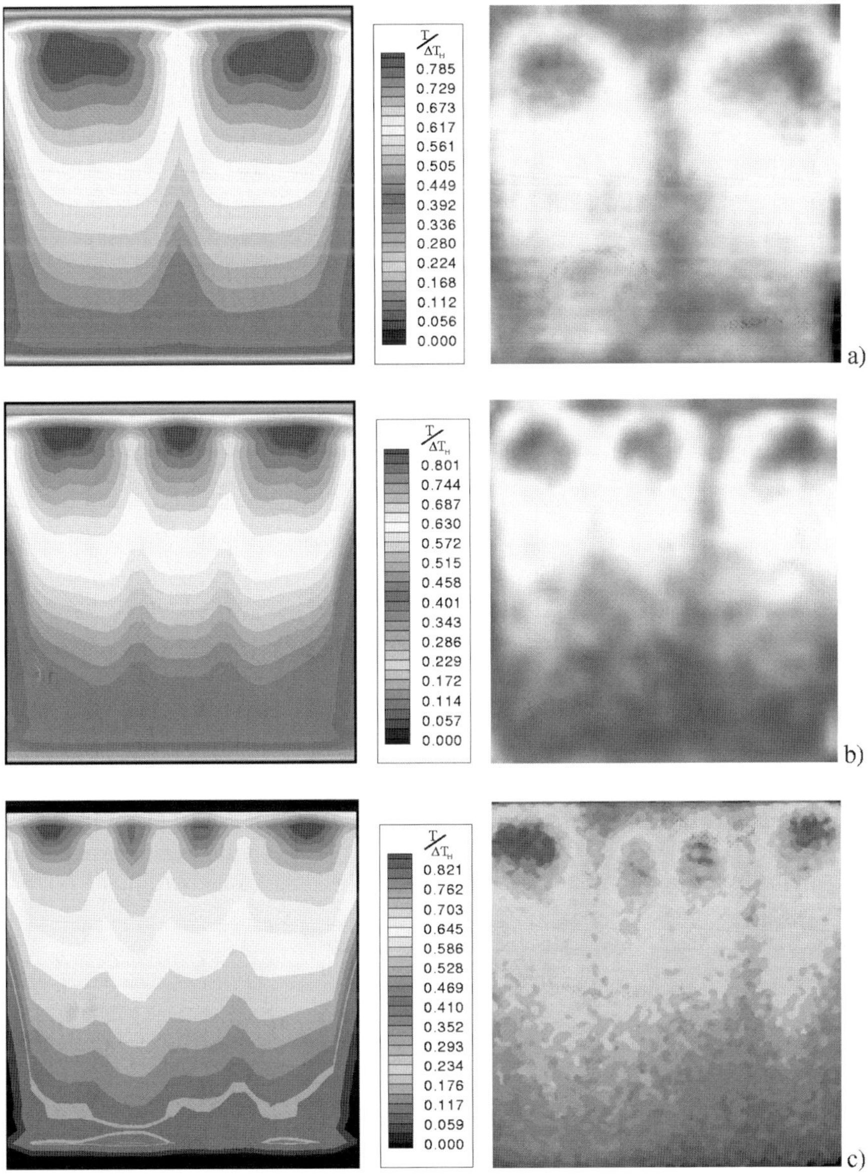

Fig. 2.15 Temperature distribution on the free surface for $\Delta T_L = 0$ and (a) $\Delta T_H = 2.5\,K$, (b) $\Delta T_H = 5\,K$, (c) $\Delta T_H = 10\,K$. On the left, numerical results, on the right, corresponding experimental temperature fields.

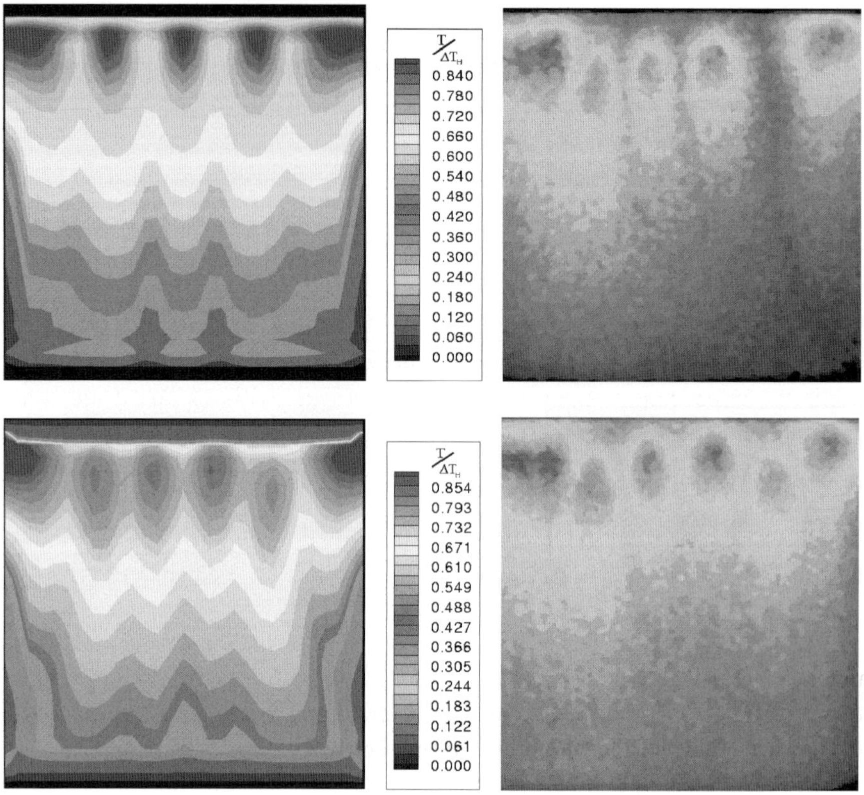

Fig. 2.15 (continued): Temperature distribution on the free surface for $\Delta T_L = 0$ and (d) $\Delta T_H = 15$ K, (e) $\Delta T_H = 20$ K, On the left, numerical results, on the right, corresponding experimental temperature fields.

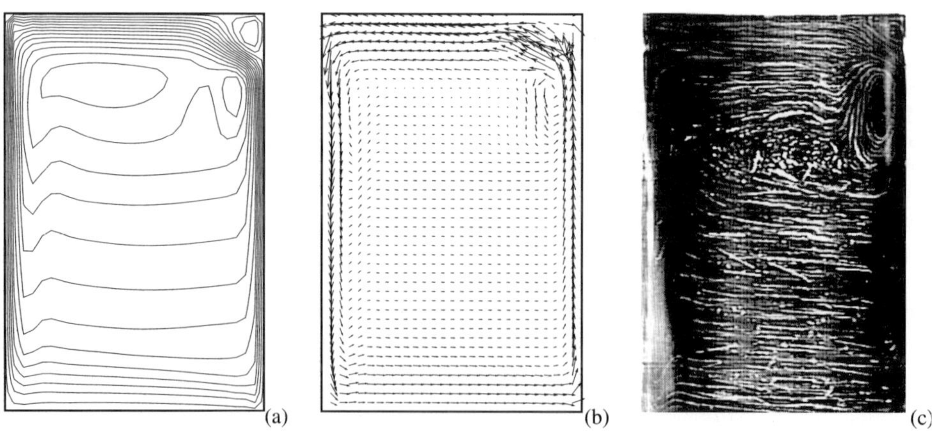

Fig. 2.16 Velocity distribution and streamlines in the vertical midplane ($z = 2.5$ cm), for $\Delta T_H = \Delta T_L = 15$ K, $V_{max} = 0.24$ cm/s (a, b, numerical; c, experimental).

Owing to the surface forces that accelerate the fluid from the side at temperature T_2 to the side at temperature T_1, the surface temperature spots are stretched, resulting in the presence of temperature fingers protruding from the hot side toward the cold side of the free interface (see Fig. 2.17, a: numerical, b: experimental, where six different temperature spots are visible on the free surface).

Since for $\Delta T_H = 15\,K$, the number of spots is $m=6$ in the case $\Delta T_L = 15\,K$ and $m=5$ in the case $\Delta T_L = 0\,K$, it can be argued that a general dependence law for m should take into account both ΔT_H and ΔT_L, i.e. $m = f(\Delta T_H, \Delta T_L)$.

Case (*3*) When the surface temperature gradient is reversed ($\Delta T_L < 0$), Marangoni and buoyant forces become opposing, a surface-tension-driven roll appears in the upper part of the cavity, counter-rotating with respect to the volume buoyancy-driven roll. Figure 2.18 show the velocity distribution and streamlines on the vertical midplane for $\Delta T_H = -\Delta T_L = 15\,K$. There is a fully separated surface-tension-driven roll in the upper third of the liquid zone rotating clockwise, whereas the buoyancy-driven roll in the lower two-thirds of the fluid volume rotates counterclockwise. In the contact region of the two rolls the flow has the same direction for both rolls.

In this case no surface temperature spots are evident. This behavior can be explained by the fact that the hot fluid adjacent to the bulk heater at temperature T_4 cannot reach directly the upper free surface (where it could give rise to hot spots) due to the existence of the surface counter-rotating convective cell. The fluid adjacent to the surface heater at temperature $T_2 < T_4$ is accelerated downwards by the buoyancy forces and it does not allow the hot bulk fluid (near the bulk heater at temperature T_4) to reach the surface directly. This hot fluid is carried out in the return flow between the two vortex cells.

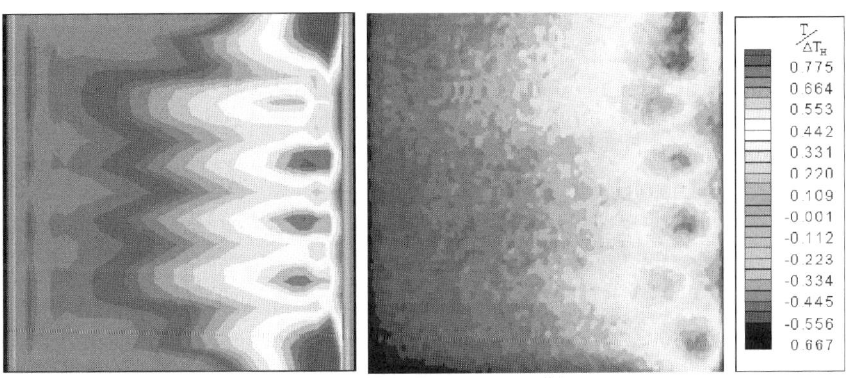

(a) (b)

Fig. 2.17 Temperature distribution on the free surface, for $\Delta T_H = \Delta T_L = 15\,K$ (a, numerical; b, experimental).

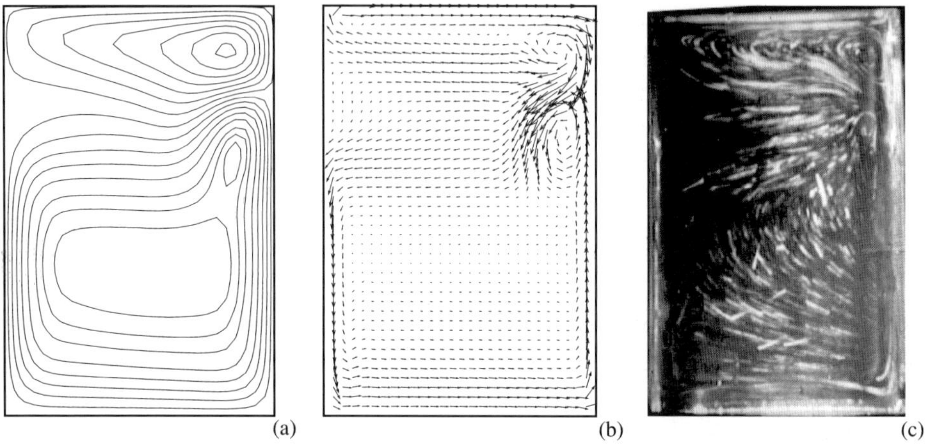

Fig. 2.18 Velocity distribution and streamlines in the midplane ($z = 2.5$ cm), for $\Delta T_H = -\Delta T_L = 15$ K, $V_{max} = 0.17$ cm/s (a, b numerical; c experimental).

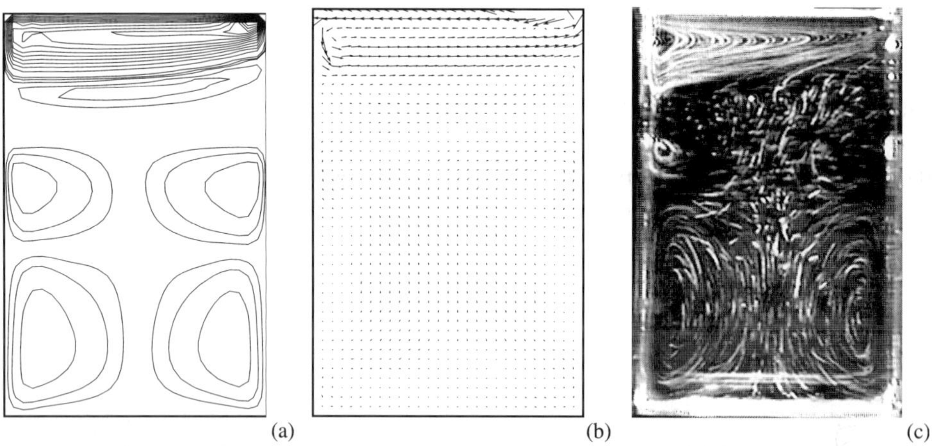

Fig. 2.19 Velocity distribution and streamlines in the midplane ($z = 2.5$ cm), for $\Delta T_H = 0$, $\Delta T_L = 15$ K ($T_1 = T_3$ and $T_2 > T_4$), $V_{max} = 0.12$ cm/s (a, b numerical; c experimental).

Case (*4*) Different situations may occur in the case $\Delta T_H = 0$ $\Delta T_L > 0$ according to whether $T_1 = T_3$ and $T_2 > T_4$ or $T_1 < T_3$ and $T_2 = T_4$.

In the case $T_1 = T_3$, $T_2 > T_4$ a very thin Marangoni cell is confined near the free surface (see Fig. 2.19) and the fluid in the bulk can be considered at rest. The influence of the buoyancy forces on the flow pattern is negligible and the flow is dominated by the Marangoni effect. The pattern is characterized by a full separation of the surface-tension-driven roll from the bulk and beneath it only induced convection rolls are present.

These weak counter-rotating vortex cells are induced by continuity in the interior of the cavity. Such a behavior was already pointed out in the case of two-dimensional configurations by Rybicki and Florian (1987) who found the emergence of several layers of vortices, with the strength of each layer decreasing approximately exponentially with the distance from the surface.

The average temperature of the free surface ($T_m = 1/2(T_1 + T_2)$) is significantly above $T_3 = T_4$. The mean temperature in the surface convection roll is higher than the mean temperature in the bulk fluid. Although the buoyancy is not actively propelling convection, it is the reason for the observed separation stabilizing hot liquid on top of colder (heavier) liquid. This separation should not occur in a microgravity experiment; there the pure surface-tension-driven roll should penetrate fully into the liquid.

In the case $T_1 < T_3$, $T_2 = T_4$ the flow pattern is more complex (see Fig. 2.20) and is characterized by the combined effect of buoyancy forces and surface-tension forces: the hot fluid adjacent to the surface heater at temperature T_2 is carried towards the side at temperature T_1 due to the surface-tension forces. When the fluid reaches the cold side, it is cooled and accelerated downwards by the buoyancy effect. Due to the fact that the average temperature of the bulk fluid is greater than T_1, the buoyancy forces give rise to a secondary inner vortex with the opposite sense of rotation that counteracts the cold fluid coming from the surface. The secondary buoyancy-driven roll occupies the lower third of the liquid zone and it is placed near the bulk heater at temperature T_3. In the contact region of the primary and of the secondary roll, the two rolls meet to generate the return flow. This last case is particular significant since it shows how, on Earth, buoyancy forces can influence, in a very unexpected way, a Marangoni flow.

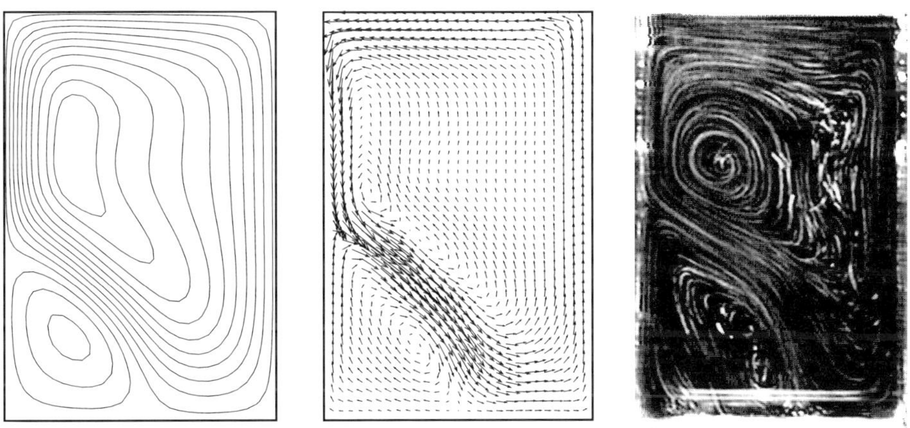

(a) (b) (c)

Fig. 2.20 Velocity distribution and streamlines in the midplane ($z = 2.5$ cm), for $\Delta T_H = 0$, $\Delta T_L = 15$ K ($T_1 < T_3$ and $T_2 = T_4$), $V_{max} = 0.21$ cm/s (a, b, numerical; c, experimental).

2.4.2 *The liquid bridge*

Like the open cavity, also, this axisymmetric configuration has been widely used in the literature as a relevant testbed for the investigation of pure Marangoni flow (in space) or of the interacting buoyancy and surface-tension-driven flows on Earth. Often organic liquids have been used for these analyses since these fluids are transparent to visible light, permitting optical access to the interior of the flows (see Fig. 2.21). The case of counteracting or concurrent buoyancy and surface-tension forces can be simply investigated by heating the bridge from above or from below, respectively.

It is known (Wanschura et al., 1996) that the onset of pure buoyancy convection in such a geometrical configuration is always three-dimensional. On the contrary, the pure Marangoni flow and the Marangoni flow weakly affected by buoyancy forces (microscale experimentation) are initially axisymmetric and undergo transition to oscillatory and/or three-dimensional flow only if the Marangoni number is increased beyond a certain critical threshold.

Since the symmetry of the flow is successively broken when the temperature difference between the supporting disks is increased, leading to a wealth of different spatiotemporal patterns (these aspects will be discussed in detail in Chapter 4 that is almost entirely devoted to the investigation of this subject) this configuration has been often used by many researchers interested in fundamental fluid-dynamic research as a pattern-forming dynamical system in dissipative media (see also Section 1.4.1).

Hereafter, some examples dealing with conditions corresponding to the microscale experimentation are considered (height of the liquid columns of a few millimeters).

For the sake of clarity, first, the case of a stable and steady Marangoni flow is examined, in order to study the effect of gravity on the axisymmetric Marangoni

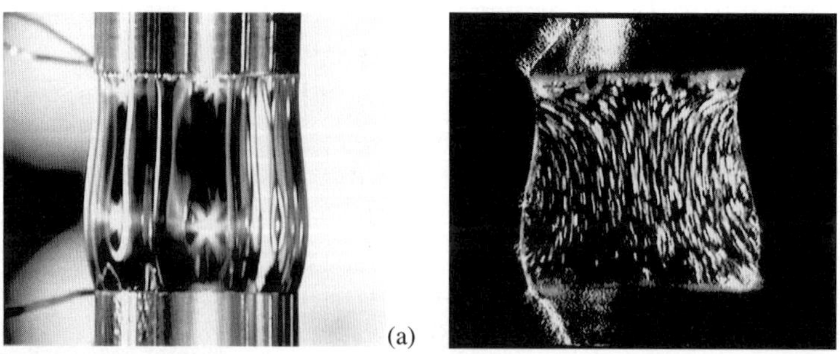

(a) (b)

Fig. 2.21 (a) A liquid bridge of opaque molten metal (tin) held between two stainless steel rods of 4 mm diameter (after Yang and Kou, 2001), (b) A liquid bridge of silicone oil sustained between two disks of 4 mm diameter (after Monti et al., 2000c, the liquid motion in the meridian plane is visualized using tracers scattering the light generated by a laser diode He–Ne with a wavelength of 635 nm, forming a light sheet; the laser beam is oriented orthogonal to the main optical path of a CCD camera).

convection, then the transition to an oscillatory three-dimensional state is briefly discussed.

In the light of the arguments pointed out in Section 2.2.3, in the absence of thermal radial gradients, and if thermocapillary forces are negligible, heating from above should yield a stable linear temperature distribution in the liquid bridge.

If thermocapillary effects are present, however buoyancy effects can modify the distribution of the temperature and velocity inside the bridge with respect to pure Marangoni flow.

The numerical results show, in fact, that at the center of the liquid bridge, the velocity in μg is larger than at 1g. The reason for this behavior is the fact that surface tension and buoyancy forces counteract in a liquid bridge heated from above.

When gravity changes from μg to 1g, the center of the vortex appears to move to the interface of the liquid bridge (the radial depth of the vortex cell is decreased, see Fig. 2.22). The reason is believed to be that the fluid near the cool disk is slowed down by the buoyancy forces; this effect reduces the radial extension of the convection roll.

The influence of the buoyancy effect can also be seen in the temperature field (Fig. 2.23).

In zero-g conditions a "cold finger" is present near the hot disk, which is created by the return flow that brings cool fluid away from the cold wall along the symmetry axis. The cold liquid is carried toward the hot surface at a rather high position.

In normal-gravity conditions, since the cold liquid, traveling in the return flow, is slowed down by the buoyancy forces, the temperature of this zone is weakened and consequently the radial temperature gradient near the hot disk is decreased.

When the bridge is heated from below, Marangoni and buoyancy flow act in the same direction. The velocity on the free surface increases with respect to heating from above (see the streamfunction Ψ in Fig. 2.22) and for very short liquid bridges other counter-rotating vortex cells induced by buoyancy appear in the interior of the bridge. These additional vortex cells are Rayleigh–Bènard convection rolls (they do not have any theoretical kinship with the vortex cells that appear in the interior of the liquid driven by Marangoni flow). The critical Rayleigh number for the occurrence of these phenomena is about 1700 as pointed out in Section 2.2 and the Rayleigh numbers corresponding to the conditions considered in the present example are larger than this value.

Due to an increase of the surface velocity (and of the return flow) and owing to the presence of other buoyancy vortex cells in the inner part of the liquid bridge, the temperature field is strongly deformed (see Fig. 2.23) with respect to the case of heating from above (this behavior is particularly evident for $A \leq 0.3$).

If the analysis for axisymmetric convection is extended to the case of the three-dimensional and oscillatory flow that arises when the Marangoni number is further increased, the effect of the combined effects of gravity and heating direction on the structure of the flow becomes even more important.

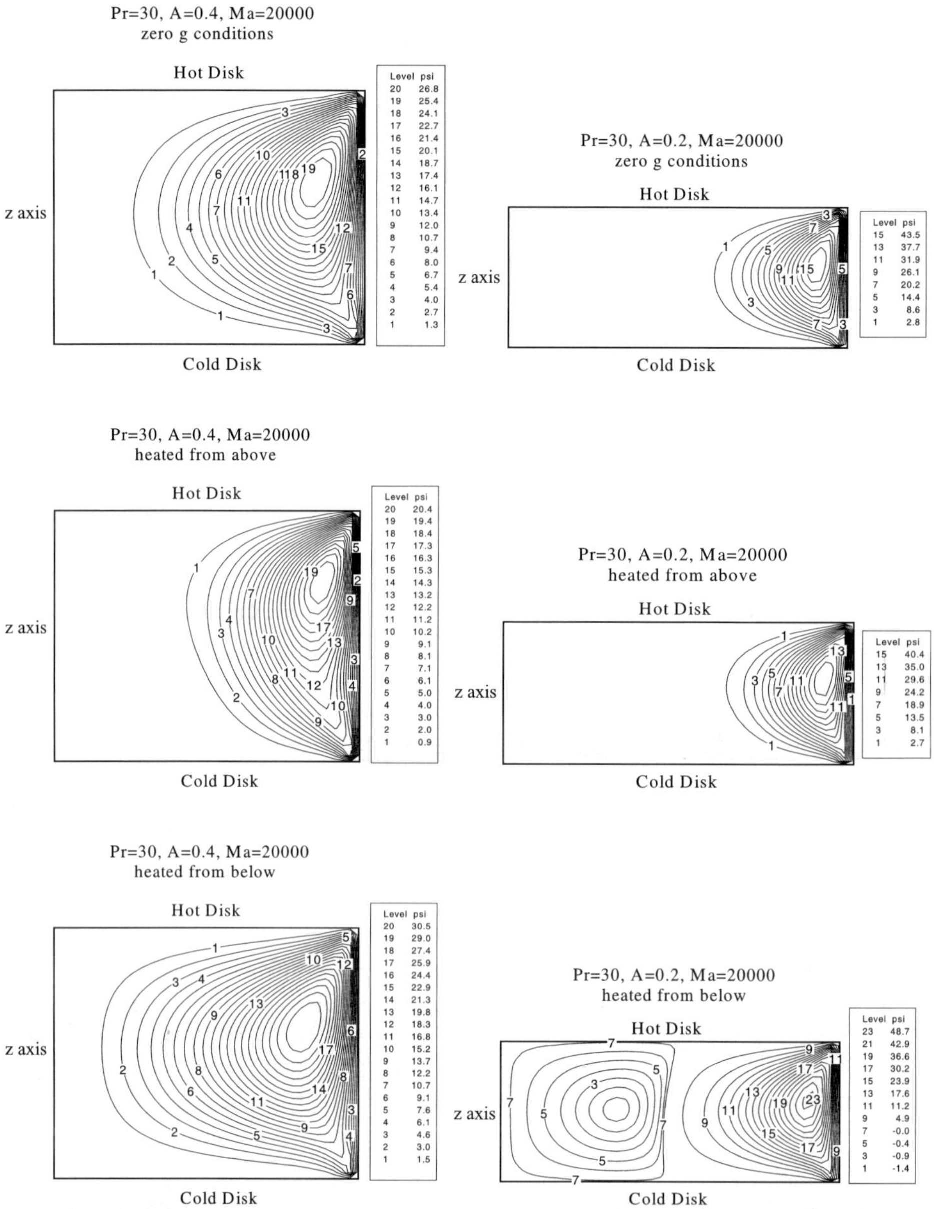

Fig. 2.22 Velocity field (A is the geometrical aspect ratio of the liquid bridge given by the ratio of its length and diameter, respectively) for different environments (0g and 1g) and heating conditions (bridge heated from above or from below on the ground). Bridges with height $L = 0.4\,\text{cm}$ and $L = 0.2\,\text{cm}$ are shown on the left and on the right of the page, respectively.

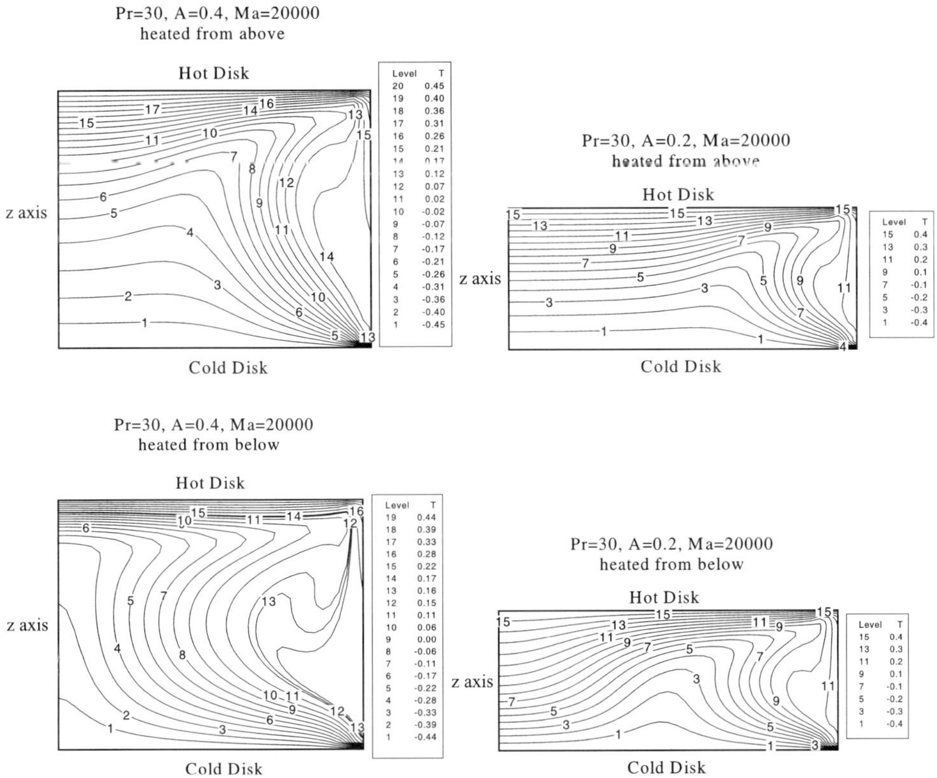

Fig. 2.23 Temperature field in normal gravity conditions (bridge heated from above or from below).

The structure of the 3D flow (characterized by the existence of m cold spots and m hot spots on the free surface with oscillatory behavior) completely changes according to whether the bridge is heated from above or from below. This behavior is evident by comparing the number of surface spots of Fig. 2.24 with Fig. 2.25, and of Fig. 2.26 with Fig. 2.27). In these figures the oscillation period τ is divided into four parts and the temperature field is shown in figures a, b, c, d corresponding to $t=0$, $\tau/4$, $\tau/2$, $3\tau/4$ or it is divided into two parts and the temperature field is shown in figures a, b corresponding to $t=0$, and $t=\tau/2$, respectively.

Also, these intriguing phenomena correspond to the propagation of hydrothermal waves as already elucidated for the case of liquid layers and open cavities. These aspects will be discussed in detail in Chapter 4. Within the context of the present section, however, it is worthwhile to point out that the main reason for the oscillatory instability for the liquid bridge of high Prandtl number fluid is the existence of the aforementioned "cold finger" near the hot disk, which is created by the return flow that brings cool fluid away from the cold wall along the symmetry axis. The cold liquid is carried toward the hot surface at a rather high position.

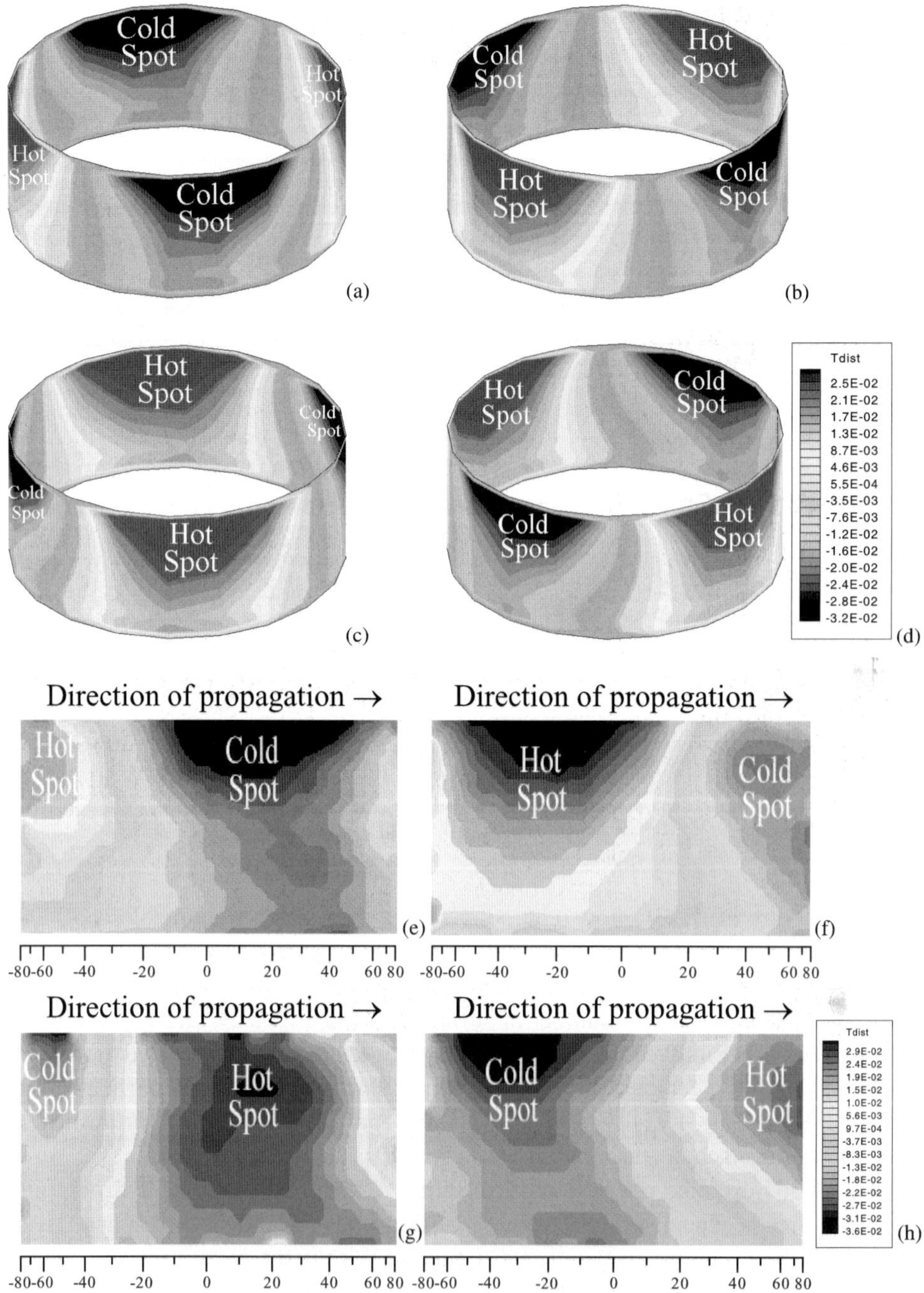

Fig. 2.24 Temperature disturbance on the bridge surface for $A_H = 0.4$ and traveling-wave regime ($Pr = 30$, $Ma = 3.6 \times 10^4$ bridge heated from above). The field is shown in frames a–d (3D computations) and frames e–h (experimental results, thermographic visualization), corresponding, respectively, to $t = 0$, $\tau/4, \tau/2, 3\tau/4$).

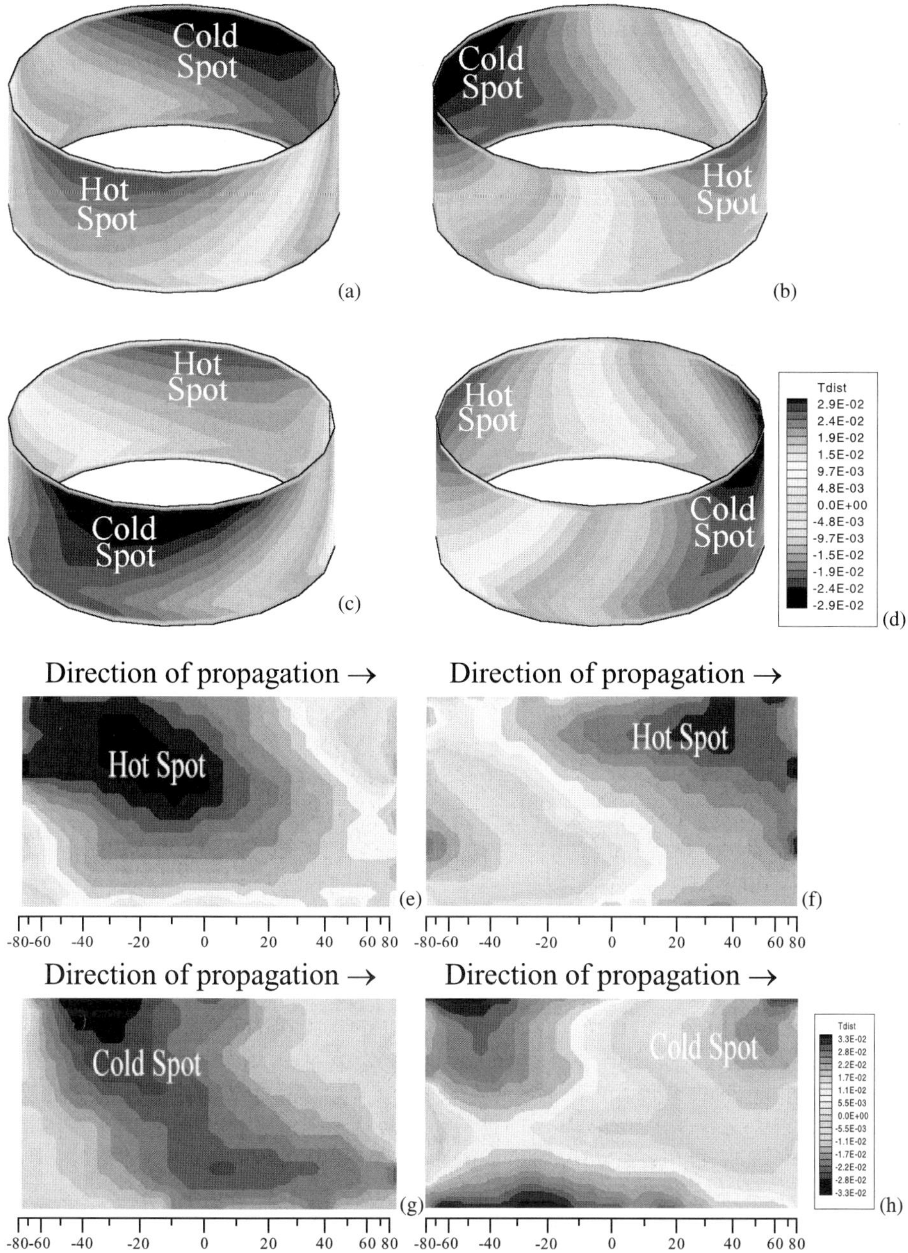

Fig. 2.25 Temperature disturbances on the bridge surface for $A_{\mathrm{H}} = L/D = 0.4$ and traveling-wave regime ($Pr = 30$, $Ma = 4.3 \times 10^4$ bridge heated from below). The field is shown in frames a–d (numerical results, 3D computations) and frames e–h (experimental results, thermographic visualization), corresponding, respectively, to $t = 0, \tau/4, \tau/2, 3\tau/4$).

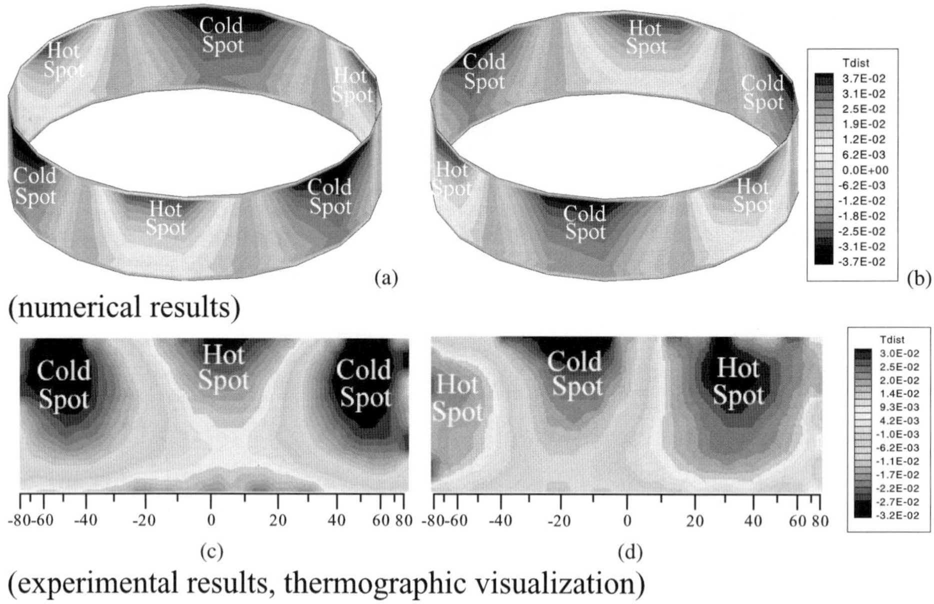

Fig. 2.26 Temperature disturbance on the bridge surface for $A_H = L/D = 0.25$ and standing-wave regime ($Pr = 30$, $Ma = 3.75 \times 10^4$ bridge heated from above, numerical and experimental results; the field is shown in a, b and c, d corresponding, respectively, to $t = 0$, and $t = \tau/2$).

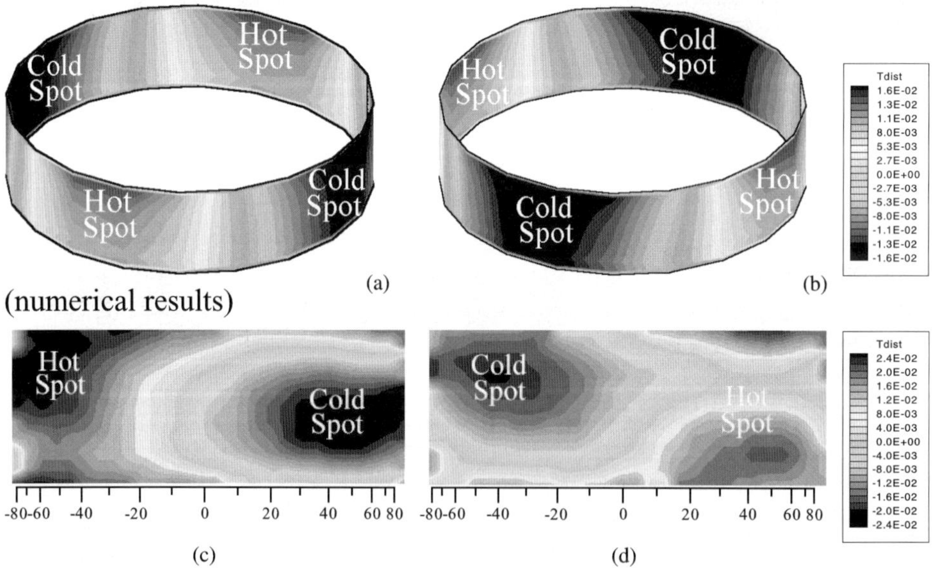

Fig. 2.27 Temperature disturbances on the liquid-bridge surface for $A_H = L/D = 0.25$ and standing-wave regime ($Pr = 30$, $Ma = 4.2 \times 10^4$ bridge heated from below, numerical and experimental results, the field is shown in a, b (numerical results) and c, d (experimental results) corresponding, respectively, to $t = 0$, and $t = \tau/2$).

This cold zone (close to the free surface) plays a crucial role in the mechanism leading to the amplification of surface-temperature disturbances and to the onset of surface hydrothermal waves. In previous results (Wanscura et al., 1995) it was pointed out that the instability of Marangoni flow in high Prandtl number liquids is related to the convective radial heat transport coupled with the Marangoni effect. The possible amplifications of surface temperature disturbances is based on the temporal interaction between the temperature distribution within the flow field (in particular the cold finger established inside the liquid bridge near the hot disk) and the mechanism with which the surface generates flow disturbances in response to temperature perturbations (this is why the surface temperature spots always appear near the hot disk as shown, e.g., in Figs. 2.24 and 2.25; therein the hot disk is located on the top of any frame).

In line with these arguments, the different behavior that the system exhibits in the oscillatory state according to the gravity direction is a simple consequence of the fact that gravity alters the aforementioned mechanisms by modifying the radial temperature gradients inside the bridge and the magnitude of the surface axial velocity as shown in Figs. 2.22 and 2.23 (Lappa et al., 2000).

The flow is stabilized by the effect of buoyancy forces with respect to zero-g conditions if the liquid is heated from above and further stabilization occurs if the imposed temperature gradient is reversed.

The reader may also consider the parametric experimental studies reported in Velten et al. (1991) for additional details about the azimuthal structure of the flow and the stability boundaries as a function of the heating direction and of the Prandtl number. They, in fact, studied the influence of gravity on the transition point, frequency and spatial structure of the flow field obtaining a large experimental database. The stabilizing effect of the heating from below condition was clearly observed over a large range of values of the Prandtl number ($Pr = 1, 7, 49$) and of the aspect ratio.

The weak stabilization of the bridge heated from above with respect to the situation of zero-g conditions may be explained by the weakened influence of the cold finger with respect to the case of zero-g conditions; in this case, in fact, as mentioned before, the temperature gradient along this cold radially elongated zone is weakened by the buoyancy effects, which slow down the fluid near the cool disk carried in the return flow.

In the latter case (heating from below) strong stabilization of the flow with respect to the situation of heating from above is due to the fact that small surface temperature disturbances induced near the hot disk from inside by the cold radially elongated zone have a weaker impact on the accelerations or deceleration of the main surface flow (the sensitivity of the free surface is decreased by the strong surface velocities).

The results of the linear stability analyses of Wanschura et al. (1997b) have also explained these behaviors in terms of a delicate balance between the stabilizing and destabilizing buoyancy effects. They suppose that since hydrothermal waves are characterized by strong axial vorticity while buoyant convection favors convection rolls with strong horizontal vorticity, therefore both the types of convection structures are incompatible in the sense that their respective transport mechanisms exclude each other yielding a stabilization of the basic state with respect to pure Marangoni flow.

The corresponding case concerning the effect of buoyancy forces in low-*Pr* liquid bridges is treated in Chapter 4.

2.5. Acceleration disturbances on the International Space Station

Purely diffusive ("d") environments do not exist in the real world; also the facilities for microgravity experimentation suffer from some degree of residual acceleration and g-jitters, arising mostly from orbiter maneuvers. The real microgravity environment for medium/long duration missions (i.e. orbital platforms) consists of a spectrum of accelerations at different frequencies ranging from zero (steady or quasisteady residual acceleration) to hundreds of Hertz (periodic g-jitters plus pulse like g-jitters).

For instance, the expected disturbances on the International Space Station are:

1. Steady (or quasisteady) residual-g. These include aerodynamic drag (1–$3 \times 10^{-7} g_o$), radiation pressure ($10^{-8} g_o$), micrometeorites impacts ($10^{-9} g_o$) and, for points distant from the center of mass, gravity gradient and rotation periodic with the orbit ($O(10^{-7})\ g_o$/m);
2. Periodic, high frequency, due to on-board machineries and natural frequencies excited by external forces ($10^{-6} < g/g_o < 10^{-2}$, $0.1\ \text{Hz} < f < 300\ \text{Hz}$).

These deviations from true microgravity are known to be the origin of convective contribution to mass transport.

The global effect of steady (or quasisteady) residual-g is of the order $O(10^{-6} g_o)$.

The predicted high-frequency g-jitter (e.g., recent NIRA 99 predictions for the US lab and for the ESA-COF) are usually reported in a plot of acceleration amplitudes vs. frequencies and compared with the System Allowable (so-called "ISS requirements curve," see Fig. 2.28).

Owing to the large extension of the ISS (see Section 1.2.6) and to the different vibrations induced by crew activities on the US Lab and on the ESA-COF module (see Fig. 2.30), the Space Station can be seen as an ensemble of microgravity platforms with substantial differences (in the convective disturbances) due to the different microgravity environments encountered by the experimental facility. In particular, the recent NASA models predict a value of the residual-g of the order of 0.5 μg for the US Lab and of about 1.6 μg for the ESA-COF. Figure 2.29 shows the velocity and temperature fields corresponding to these conditions for the two-dimensional square test cell (that is a typical reference configuration for the evaluation of TFD disturbances in the microgravity environment).

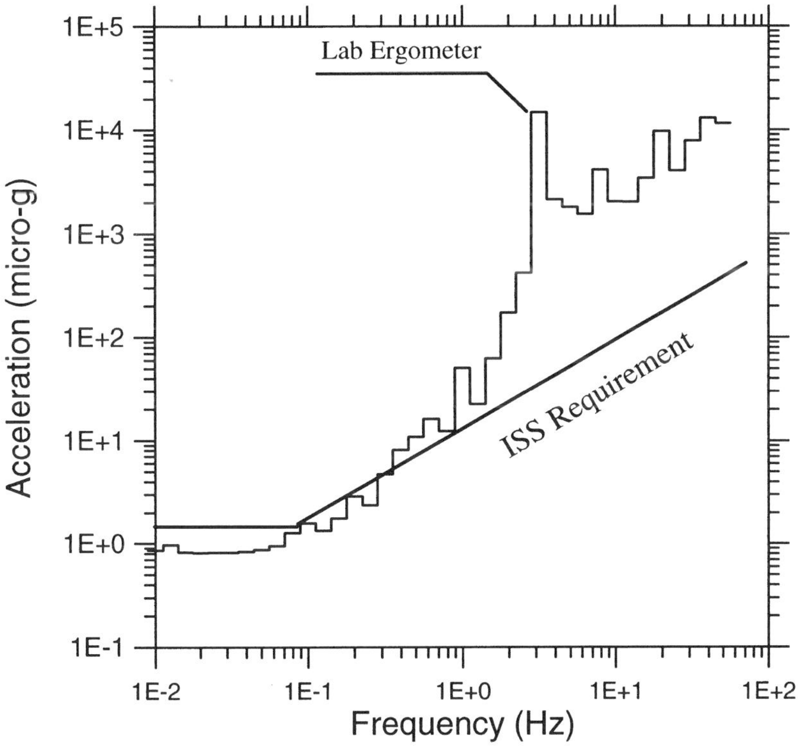

Fig. 2.28 Predicted high-frequency g-jitter reported in a plot of acceleration amplitudes vs. frequencies and compared with the System Allowable (so-called "ISS requirements curve").

On the other hand, the g-jitter disturbances are expected to be larger in the US Lab than in the European COF. It is essential to recognize that the main difference between the US Lab and the ESA COF is mainly due to the oscillation induced by the "Lab Ergometer" that is positioned inside the US Lab. In fact, the amplitude, at a frequency close to 3 Hz, induced by this equipment, is, respectively, about 150 μg for COF and 15 000 μg for the US Lab (see Fig. 2.28).

2.6. Oscillatory acceleration disturbances: g-jitters

2.6.1 *Equations and relevant parameters*

An oscillatory g-jitter includes all the periodic time-dependent accelerations that can be approximated by sinusoidal functions. The most common sources of these accelerations on the ISS are structural vibrations (e.g., at the fundamental natural frequencies), equipment operations, and crew activity (e.g., repetitive exercises that induce cyclic displacement of the position of the test cell).

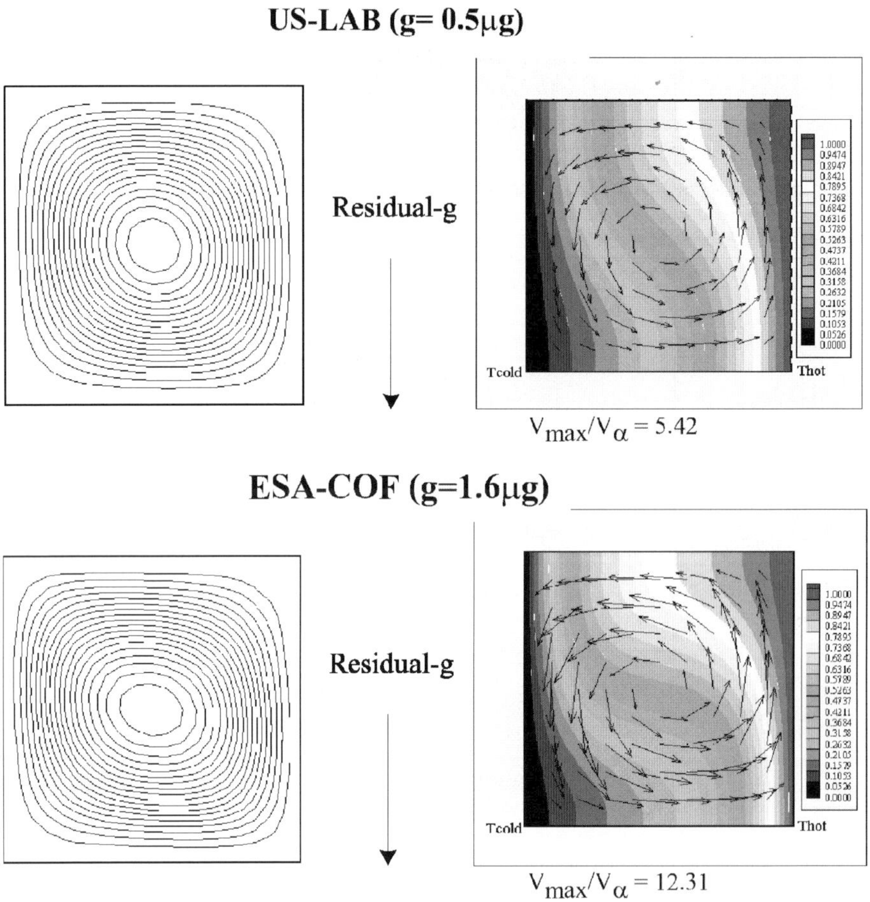

Fig. 2.29 Computed streamlines, vector plots and temperature distributions for the US-Lab and the ESA-COF, when only residual-g is present, $\Delta T = 50\,\text{K}$, silicone oil, $Pr = 15$ (steady state).

Disturbances induced in a fluid cell by a sinusoidal displacement

$$\underline{s}(t) = b \sin(\omega t)\underline{n} \tag{2.33}$$

where b is the amplitude and $\omega = 2\pi f$ (f is the frequency) induce an acceleration:

$$\underline{g}(t) = \underline{g}_{\omega} \sin(\omega t) \tag{2.34}$$

where $g_{\omega} = b\omega^2 n$.

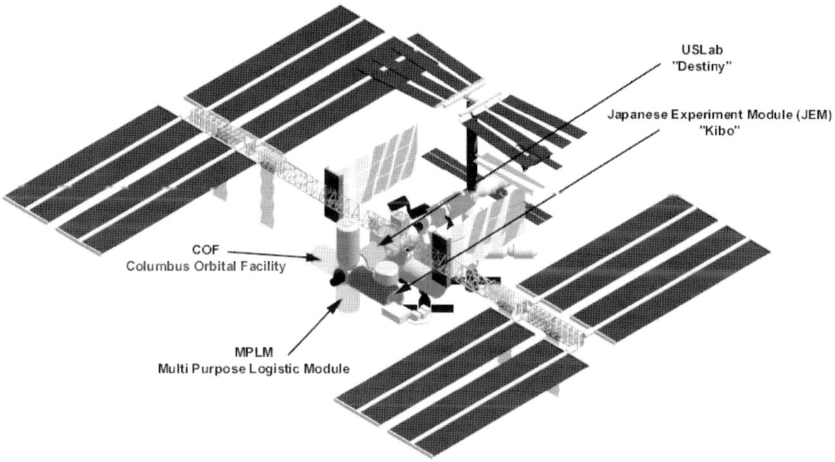

Fig. 2.30 Sketch of the ISS with the different facilities and modules (COF, US lab, JEM, MPLM).

Using the Boussinesq approximation, in nondimensional form the momentum equation reads:

$$\frac{\partial \underline{V}}{\partial t} = -\underline{\nabla} p - \underline{\nabla} \cdot [\,\underline{V}\,\underline{V}] + Pr\,\nabla^2 \underline{V} + Pr \cdot \frac{b\omega^2 \beta_{\mathrm{T}} \Delta T L^3}{\nu\alpha} T \sin\left(\frac{L^2\omega}{\alpha} t\right) \underline{n} \tag{2.35}$$

for thermal problems ($V_{\mathrm{d}} = \alpha/L$), and

$$\frac{\partial \underline{V}}{\partial t} = -\underline{\nabla} p - \underline{\nabla} \cdot [\,\underline{V}\,\underline{V}] + Sc\,\nabla^2 \underline{V} + Sc \cdot \frac{b\omega^2 \beta_{\mathrm{C}} C_{\mathrm{o}} L^3}{\nu D} C \sin\left(\frac{L^2\omega}{D} t\right) \underline{n} \tag{2.36}$$

for solutal problems ($V_{\mathrm{d}} = D/L$).

The above equations have been the subject of intensive research during the last decade (many scaling and order-of-magnitude analyses). Many theoretical and numerical studies have been dedicated to this topic (see, e.g., Monti et al., 1987; Schneider and Straub, 1989; Alexander, 1990; Monti and Savino, 1994a, b, 1995, 1996a, b; Savino, 1997; Savino and Monti, 1999a, b; Naumann, 2000).

It is well-known that relevant nondimensional parameters for these cases are the nondimensional frequency (Ω) and displacement (Λ):

$$\Omega = \frac{\omega L^2}{\alpha} \tag{2.37a}$$

$$\Lambda = b\,\frac{\beta_{\mathrm{T}} \Delta T}{L} \tag{2.37b}$$

for thermal problems,

$$\Omega = \frac{\omega L^2}{D} \tag{2.38a}$$

$$\Lambda = b\frac{\beta_C C_o}{L} \tag{2.38b}$$

for solutal problems.

The nondimensional equations, in fact, read:

$$\frac{\partial \underline{V}}{\partial t} = -\underline{\nabla} p - \underline{\nabla} \cdot [\,\underline{V}\,\underline{V}] + Pr\,\nabla^2 \underline{V} + \Lambda\Omega^2 T \sin(\Omega t)\underline{n} \tag{2.39}$$

$$\frac{\partial \underline{V}}{\partial t} = -\underline{\nabla} p - \underline{\nabla} \cdot [\,\underline{V}\,\underline{V}] + Sc\,\nabla^2 \underline{V} + \Lambda\Omega^2 C \sin(\Omega t)\underline{n} \tag{2.40}$$

for thermal and solutal problems, respectively.

2.6.2 *Fields decomposition*

Numerical solution of Eq. (2.39) or (2.40) with the additional ones for energy and solute transport (i.e. Eq. (2.6) or (2.13)) has shown that when soliciting the fluid cell by periodic accelerations, the diffusive temperature distribution $T_d(\underline{r}, t)$ and/or the concentration distribution $C_d(\underline{r}, t)$ are distorted and the difference (suitably defined) between the temperature (and/or concentration) distribution and the diffusive (ideal) one, can define the g-jitter disturbance (as also discussed in the first section of this chapter).

It is also known (see, e.g., Monti et al., 1998b, 2001; Savino and Lappa, 2003a) that the velocity field $\underline{V}$, induced by periodic $\underline{g}$ is made up by an average value $\overline{\underline{V}}$ plus a periodic oscillation of amplitude $\underline{V}'$ ($\underline{V}=\overline{\underline{V}}+\underline{V}'$) at the g-jitter frequency f or at frequencies that are multiple of f.

As a result of such a convective field, the scalar quantities (temperature and/or species concentration) are also distorted.

These distortions, in turn, are also made up by a steady plus an oscillatory contribution ($T=\overline{T}+T'$, $C=\overline{C}+C'$).

This behavior finally introduces the need for different definitions of the g-jitter disturbance parameters (one for the steady and the other for the oscillatory contribution, respectively, i.e. $\bar{\varepsilon}$ and ε').

With regard to these aspects, it is worthwhile to emphasize how different situations may occur, depending on the oscillation frequency.

For instance, it is known that on increasing the frequency there is a first regime characterized by a relatively large oscillatory velocity and oscillatory temperature

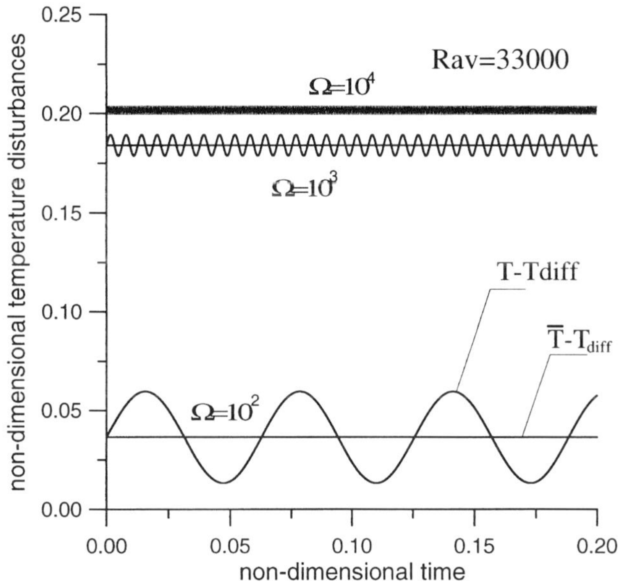

Fig. 2.31 Temperature signals as a function of time in a square test cell filled with a $Pr = 15$ liquid versus the nondimensional frequency Ω (at low frequency, there is a regime characterized by relatively large oscillatory temperature disturbances and relatively small time-average steady disturbances. On the contrary at high frequencies the oscillatory thermal disturbances are very small with respect to the steady ones induced by the time-average part of the velocity field).

(and/or species concentration) disturbances and relatively small time-average steady disturbances. On the contrary at high frequencies the oscillatory thermal (and/or solutal) disturbances are very small with respect to the steady ones induced by the time-average part of the velocity field (these behaviors are very evident in Fig. 2.31).

2.6.3 *The TFD disturbances*

In view of the arguments pointed out in Section 2.1, in the following the subscript "d" denotes the parameters corresponding to the purely diffusive conditions (e.g., the temperature $T_d(x, y, z, t)$). The thermal disturbances induced by any acceleration will therefore be defined at each point (x, y, z) and at any time (t) as:

$$\delta T = T(x, y, z, t) - T_d(x, y, z, t) \tag{2.41a}$$

In the case of numerical integration at the grid point "j" of the mesh:

$$\delta T_j(t) = T_j(t) - T_{j_d}(t) \tag{2.41b}$$

According to the foregoing discussion (Section 2.6.2) the local temperature can be split into a time-averaged steady contribution plus an oscillating part ($T_j = \overline{T}_j + T'_j$) so that the local temperature disturbance reads:

$$\delta T_j = T_j - T_{jd} = (\overline{T}_j - T_{jd}) + T'_j \tag{2.41c}$$

where $\overline{T}_j$ is the nondimensional local temperature (scaled by the characteristic temperature difference ΔT imposed across the cell) averaged over the period $2\pi/\omega$ and T'_j is the oscillatory part expressed in terms of the local amplitude of the temperature oscillation (ΔT_j):

$$T'_j = \Delta T_j f_j(\omega) \tag{2.42}$$

where $f_j(\omega)$ is a periodic, zero time-average function (e.g., $f_j(\omega) = \sin(\omega t + \varphi_j)$).

At this stage, in the light of the discussion in Section 2.6.2, two different overall thermal disturbance parameters (over all the grid points N) must be defined to characterize the disturbed state of the temperature field:

$$\bar{\varepsilon}_{\mathrm{T}} = \frac{1}{N}\sum_{j=1}^{N} |(\overline{T}_j - T_{jd})| \tag{2.43a}$$

$$\varepsilon'_{\mathrm{T}} = \frac{1}{N}\sum_{j=1}^{N} |(\Delta T_j)| \tag{2.43b}$$

If both $\bar{\varepsilon}_{\mathrm{T}}$ and $\varepsilon'_{\mathrm{T}}$ are $\ll 1$, one can assume that the purely diffusive regime is not disturbed by the presence of the onboard acceleration $g(t)$. Similar arguments can be pointed out for the solutal problem replacing T with C.

In many circumstances no comparison can be made between the two types of disturbances; the order of magnitude of $\bar{\varepsilon}_{\mathrm{T}}$ and $\varepsilon'_{\mathrm{T}}$ for the negligibility of the thermal disturbances usually is different.

In fact, the results that the disturbances depend on acceleration and frequency as g_ω^2/ω^2 (Monti and Savino, 1996b) or as g_ω/ω^2 (Monti et al., 1987) are both correct but refer to the two types of disturbances: the first refers to $\bar{\varepsilon}_{\mathrm{T}}$ and the second to $\varepsilon'_{\mathrm{T}}$:

$$\bar{\varepsilon}_{\mathrm{T}} = \mathrm{O}(g_\omega^2/\omega^2)$$
$$\varepsilon'_{\mathrm{T}} = \mathrm{O}(g_\omega/\omega^2)$$

at low frequencies, the main contribution is given by ε' (that for $\Omega \ll 1$ refers to quasisteady regimes); at relatively high frequencies ($\Omega \gg 1$) $\bar{\varepsilon} \gg \varepsilon'$.

2.6.4 *High frequencies and the thermovibrational theory*

As mentioned earlier, for the given experimental conditions there is a strong dependence of both types of disturbances (average and periodic) on the g-jitter frequency. The amplitudes of the periodic temperature and/or concentration disturbances tend quickly to decrease with frequency; conversely the average disturbances are less dependent on the frequency so that one expects the steady disturbances to prevail over the unsteady ones at high frequencies (and vice versa).

This encouraged the investigators to introduce strong simplifications in the analysis of the disturbances computation for the case of high-frequency g-jitters (the so-called Gershuni formulation or thermovibrational theory). According to this formulation (Gershuni et al., 1982; Gershuni and Zhukhovitskii, 1986), the time-averaged distortions can be simply computed (i.e. with much less computation time) by a simplified set of equations in terms of quantities averaged over the oscillation period.

Accordingly, till date, two methods for g-jitter analysis have been considered in the literature: (a) numerical solution of the full nonlinear and time-dependent Navier–Stokes equations with a time-dependent body force that gives rise to a time-dependent flow (i.e. Eqs. (2.4), (2.6), (2.35) for thermal problems or Eqs. (2.11), (2.13), (2.36) for solutal problems); (b) numerical solutions of the time-averaged field equations (Gershuni formulation) for the thermovibrational convection problem. The first case requires very lengthy computations since the time step for the integration of the equations must be (of course) smaller than the period τ of the oscillations; the second case is very cheap from a computational point of view but is restricted to the case of sufficiently small amplitudes ($\Lambda \ll 1$) and sufficiently large frequencies ($\Omega \gg 1$) of the g-jitter (ISS case). Under these conditions Gershuni and Lyubimov (1998) showed that, for a given Prandtl number, the steady (streaming) convection depends only on one relevant dimensionless parameter, the vibrational Rayleigh number:

$$Ra_{\mathrm{V}} = \frac{(b\omega\beta_{\mathrm{T}}\Delta TL)^2}{2\nu\alpha} = \frac{(\beta_{\mathrm{T}}\Delta TL)^2}{2\nu\alpha}\left(\frac{g_\omega}{\omega}\right)^2 = \frac{\Omega^2\Lambda^2}{2Pr} \tag{2.44a}$$

for thermal problems (thermovibrational convection) and

$$Ra_{\mathrm{V}} = \frac{(b\omega\beta_{\mathrm{C}}C_{\mathrm{o}}L)^2}{2\nu D} = \frac{(\beta_{\mathrm{C}}C_{\mathrm{o}}L)^2}{2\nu D}\left(\frac{g_\omega}{\omega}\right)^2 = \frac{\Omega^2\Lambda^2}{2Sc} \tag{2.44b}$$

for solutal problems (solutovibrational convection).

Under the assumptions of small amplitudes ($\Lambda \ll 1$) and large frequencies of the oscillatory accelerations ($\Omega \gg 1$), this formulation leads to a closed set of equations for the time-averaged quantities. The time-averaged continuity, energy and species equations remain unchanged; the time-averaged momentum equation must be rewritten as:

$$\frac{\partial \overline{\underline{V}}}{\partial t} = -\underline{\nabla}\,\overline{p} - \underline{\nabla}\cdot[\,\overline{\underline{V}}\,\overline{\underline{V}}\,] + Pr\,\nabla^2\overline{\underline{V}} + Pr\cdot Ra_{\mathrm{V}}\left[(\underline{w}\cdot\underline{\nabla}\overline{T})\underline{n} - \underline{w}\cdot\underline{\nabla}\,\underline{w}\right] \tag{2.45a}$$

where $\underline{w}$ is an auxiliary potential function satisfying the equations:

$$\underline{\nabla}\cdot\underline{w} = 0 \tag{2.45b}$$

$$\underline{\nabla}\wedge\underline{w} = \underline{\nabla}\overline{T}\wedge\underline{n} \rightarrow \nabla^2\underline{w} = -\underline{\nabla}\wedge\left(\underline{\nabla}\overline{T}\wedge\underline{n}\right) \tag{2.45c}$$

for thermal problems and

$$\frac{\partial \overline{\underline{V}}}{\partial t} = -\underline{\nabla}\,\overline{p} - \underline{\nabla}\cdot[\,\overline{\underline{V}}\,\overline{\underline{V}}\,] + Sc\,\nabla^2\overline{\underline{V}} + Sc\cdot Ra_{\mathrm{V}}[(\underline{w}\cdot\underline{\nabla}\overline{C})\underline{n} - \underline{w}\cdot\nabla\underline{w}] \tag{2.46a}$$

where

$$\underline{\nabla}\cdot\underline{w} = 0 \tag{2.46b}$$

$$\underline{\nabla}\wedge\underline{w} = \underline{\nabla}\overline{C}\wedge\underline{n} \rightarrow \nabla^2\underline{w} = -\underline{\nabla}\wedge\left(\underline{\nabla}\overline{C}\wedge\underline{n}\right) \tag{2.46c}$$

for solutal problems.

Hereafter, where not expressly stated, thermal problems will be considered (the treatment of the corresponding solutal problem being analogous).

2.6.5 *Primary and secondary patterns of symmetry*

The relative orientation of the imposed temperature or concentration gradients and of the direction of the periodic acceleration is a very sensitive parameter for the onset of convection and its strength.

In fact, the case of g-jitter disturbances perpendicular to the density gradient is usually referred to as "the worst case" whereas in the case of g-jitter disturbances parallel to the density gradient, g-jitter tends to maintain the initial diffusive conditions. It is worthwhile to point out how it is possible to provide a simple and immediate explanation for this effect on the basis of the above equations ((2.45) and (2.46)) obtained in the framework of the Gershuni formulation.

If the case of g-jitter disturbances parallel to the density (temperature and/or concentration) gradient is considered, the driving force appearing in the momentum equation vanishes: in fact, $\underline{\nabla}\cdot\underline{w} = 0$, $\nabla^2\underline{w} = 0$ and since $\underline{w} = 0$ on the boundary of the domain, the vector Poisson equation for $\underline{w}$ gives $\underline{w} = 0 \rightarrow \underline{V} = 0$ (diffusive conditions).

With regard to the structure of the convective field, for the sake of simplicity, it is convenient to provide such an analysis in the case of a two-dimensional square (Fig. 2.32) or rectangular cavities.

For the case of g-jitter disturbances perpendicular to the density gradient (worst case in terms of TFD disturbances) it is known (Gershuni et al., 1982) that if the Ra_V number is not too high, the flow is characterized by a four-vortices structure (Fig. 2.33a). It is symmetrical by reflection with respect to both $x=1/2$ and $y=1/2$ planes. This structure is replaced by a different pattern of symmetry if the orientation of the g-jitter is changed (Fig. 2.34 shows that for $\alpha \neq 90$ a single vortex pervasive throughout the test cell appears). Figure 2.35 shows the intensity of the flow (in terms of maximum of the streamfunction) versus the angle α. The maximum intensity occurs for $\alpha=45°$. For $\alpha=0°$ as previously elucidated, no flow arises and consequently the maximum value of the streamfunction is zero (Fig. 2.35).

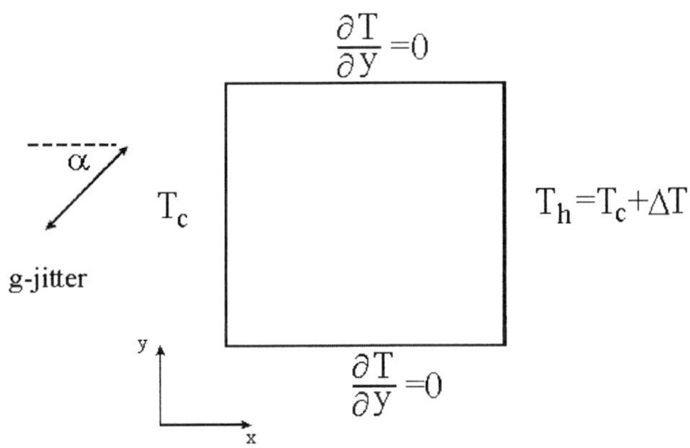

Fig. 2.32 Sketch of the test cell, its boundary conditions and orientation of g-jitter.

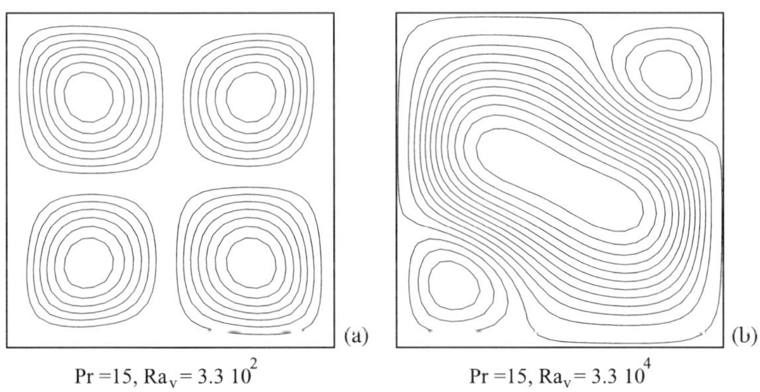

Fig. 2.33 Primary (a) and secondary (b) patterns of symmetry induced by g-jitter perpendicular to the imposed temperature gradient (thermovibrational convection).

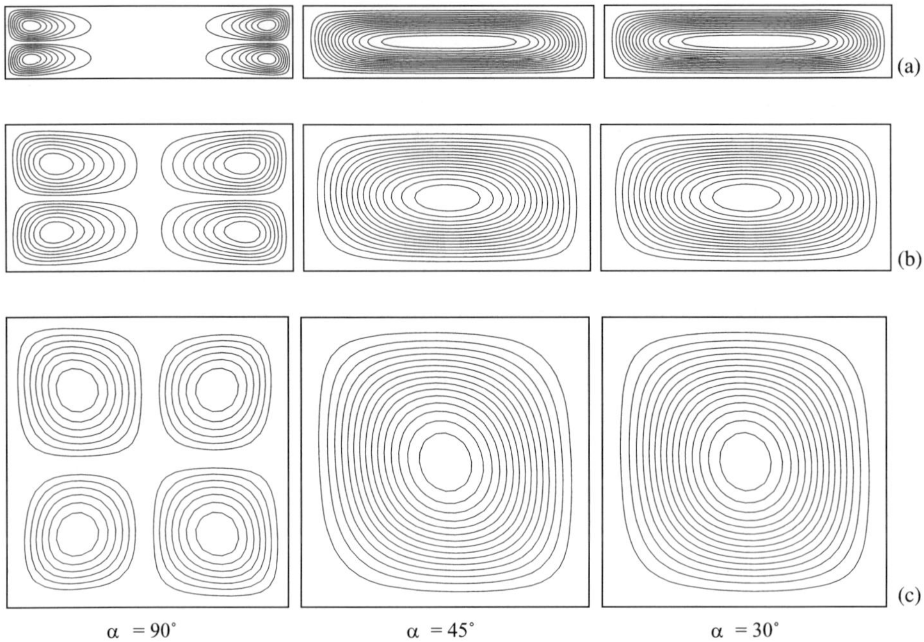

Fig. 2.34 Velocity field versus the geometrical aspect ratio and α; (a) $A=0.25$, (b) $A=0.50$, (c) $A=1.0$, silicone oil, $Pr=15$, $Ra_V=333$ (thermovibrational convection).

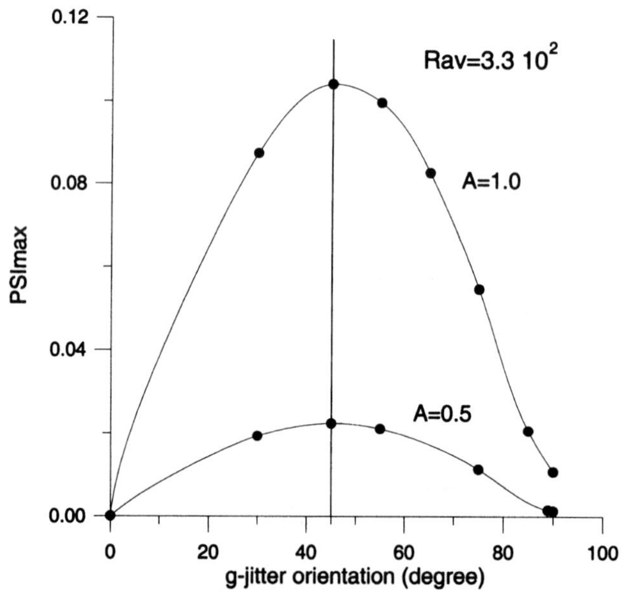

Fig. 2.35 Intensity of the flow (in terms of maximum of the stream function) versus the angle α, $Pr=15$ (thermovibrational convection).

Like the case of buoyancy and Marangoni flows (see Sections 2.2.4 and 2.3.3), thermovibrational convection can also undergo transition to patterns of symmetry with increasing degree of complexity when the characteristic number is increased (Lizée and Alexander, 1997). Figure 2.33b shows that for $\alpha = 90°$ fixed, if Ra_V is increased sufficiently a secondary stable regime appears possessing another symmetry and bifurcating from the main four-vortices one. It forms through fusion of two diagonal vortices (Fig. 2.33b). The secondary flow may be regarded as a one-vortex regime, because it consists of one main vortex formed as a result of fusion of a diagonal pair of vortices with the opposite direction of circulation.

2.6.6 *Extension to ranges of medium and low frequencies*

The identification of sets of reduced equations and associated ranges of validity also for the case of low and intermediate values of the frequency where the Gershuni (thermovibrational) theory cannot be applied is of potentially great importance for the scientific community. Even if the governing equations for Newtonian fluid dynamics, the unsteady Navier–Stokes equations, have been known for 150 years or more, the development of reduced forms of these equations is still an attractive area of research.

This is particularly true for the case of problems related to g-jitter, for which, in fact, current research is mainly focused on the construction and solution of the governing equations for the different categories of problems and the study of various approximations to those equations. The motivation for such an interest lies in the fact that the aforementioned approximations can be used to obtain analytical and/or mixed numerical/analytical solutions of the problem.

Regarding this aspect, note that, in general, the methods attempted by the different investigators to yield these simplified solutions fall into two categories: (a) specifying series expansions, substituting them in the partial differential equations and writing a simplified algebraic set of equations for the coefficients of the analytical expansions; (b) finding some set of alternative equations by mathematical transformation and manipulation (a relevant example is Gershuni's theory). For both cases, the probability of success is greatly improved by the simplification (by dropping some terms) of the initial set of nonlinear equations. By these techniques, the solution algorithm can be accelerated up to 100 times with respect to the case of the solution of the complete Navier–Stokes equations.

Reduced sets of equations are provided herein for $Pr = 15$ and two values of Ra_V (3.3×10^2 and 3.3×10^4 that correspond to the maximum values to be expected on the International Space Station, see Fig. 2.36). These different sets of equations, given by the omission of some terms, have been obtained through an order-of-magnitude analysis based on the direct numerical solution of the original Navier–Stokes equations (2.4), (2.6) and (2.35) by Savino and Lappa (2003a).

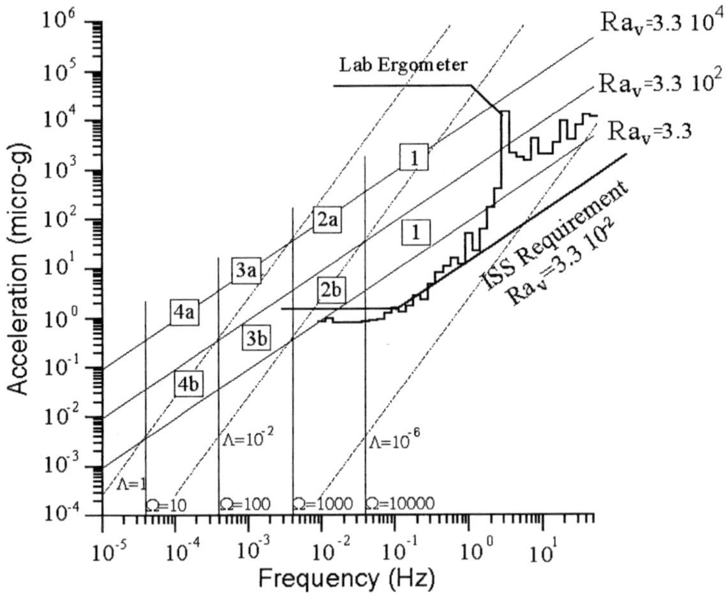

Fig. 2.36 Different regimes in the frequency–acceleration plane.

Rewriting the nondimensional equations in terms of the average values ($\underline{\overline{V}}$, $\overline{T}$) plus periodic oscillations ($\underline{V}'$, T'): $\underline{V} = \underline{\overline{V}} + \underline{V}'$, $T = \overline{T} + T'$, they read:

$$\underline{\nabla} \cdot \underline{\overline{V}} + \underline{\nabla} \cdot \underline{V}' = 0 \tag{2.47a}$$

$$\frac{\partial \underline{\overline{V}}}{\partial t} + \frac{\partial \underline{V}'}{\partial t} + (\underline{\nabla}\overline{p} + \underline{\nabla}p') + \underline{\nabla} \cdot [\underline{\overline{V}}\,\underline{\overline{V}}] + \underline{\nabla} \cdot [\underline{\overline{V}}\,\underline{V}'] + \underline{\nabla} \cdot [\underline{V}'\,\underline{\overline{V}}] + \underline{\nabla} \cdot [\underline{V}'\,\underline{V}']$$
$$= Pr\left(\nabla^2 \underline{\overline{V}} + \nabla^2 \underline{V}'\right) - \Lambda\Omega^2 \sin(\Omega t)\overline{T}\underline{n} - \Lambda\Omega^2 T' \sin(\Omega t)\underline{n} \tag{2.47b}$$

$$\frac{\partial \overline{T}}{\partial t} + \frac{\partial T'}{\partial t} + \underline{\nabla} \cdot [\underline{\overline{V}}\,\overline{T}] + \underline{\nabla} \cdot [\underline{\overline{V}}T'] + \underline{\nabla} \cdot [\underline{V}'\overline{T}] + \underline{\nabla} \cdot [\underline{V}'T']$$
$$= \nabla^2 \overline{T} + \nabla^2 T' \tag{2.47c}$$

If the values of $\underline{V}(t)$ and $T(t)$ are known at a given field point, the average values for V and T can be computed by:

$$\underline{\overline{V}} = \frac{\Omega}{2\pi} \int_0^{2\pi/\Omega} \underline{V}\,dt, \qquad \overline{T} = \frac{\Omega}{2\pi} \int_0^{2\pi/\Omega} T\,dt \tag{2.48}$$

The time-dependent parts $\underline{V}'$ and T' at each instant of time can therefore be defined as:

$$\underline{V}' = \underline{V} - \overline{\underline{V}}, \qquad T' = T - \overline{T} \tag{2.49}$$

If steady conditions are reached for the averaged fields, i.e.:

$$\frac{\partial \overline{\underline{V}}}{\partial t} = 0 \quad \text{and} \quad \frac{\partial \overline{T}}{\partial t} = 0 \tag{2.50}$$

at these conditions in terms of the following quantities:

$$\begin{aligned} &F_{V1} = \frac{\partial \underline{V}'}{\partial t}, \quad F_{V2} = \underline{\nabla}\overline{p}, \quad F_{V3} = \underline{\nabla} p', \quad F_{V4} = \underline{\nabla} \cdot [\overline{\underline{V}}\,\overline{\underline{V}}], \\ &F_{V5} = \underline{\nabla} \cdot [\overline{\underline{V}}\,\underline{V}'], \quad F_{V6} = \underline{\nabla} \cdot [\underline{V}'\overline{\underline{V}}], \quad F_{V7} = \underline{\nabla} \cdot [\underline{V}'\underline{V}'], \\ &F_{V8} = -Pr\,\nabla^2 \overline{\underline{V}}, \quad F_{V9} = -Pr\,\underline{\nabla}^2 \underline{V}', \quad F_{V10} = \Lambda\Omega^2 \sin(\Omega t)\overline{T}\underline{n}, \\ &F_{V11} = \Lambda\Omega^2 T' \sin(\Omega t)\underline{n} \end{aligned} \tag{2.51a}$$

$$\begin{aligned} &F_{T1} = \frac{\partial T'}{\partial t}, \quad F_{T2} = \underline{\nabla} \cdot [\overline{\underline{V}}\,\overline{T}], \quad F_{T3} = \underline{\nabla} \cdot [\overline{\underline{V}}T'], \quad F_{T4} = \underline{\nabla} \cdot [\underline{V}'\overline{T}], \\ &F_{T5} = \underline{\nabla} \cdot [\underline{V}'T'], \quad F_{T6} = -\nabla^2 \overline{T}, \quad F_{T7} = -\nabla^2 T' \end{aligned} \tag{2.51b}$$

Eqs. 2.47(b, c) read:

$$\sum_{i=1}^{11} F_{Vi} = 0 \tag{2.52a}$$

$$\sum_{i=1}^{7} F_{Ti} = 0 \tag{2.52b}$$

In the above equations, the generic term F_i present in Eq. (2.47), can be considered the sum of an average value $\overline{F}_i(x, y)$ plus a periodic oscillation $F_i'(x, y)$:

$$\overline{F}_i(x, y) = \frac{\Omega}{2\pi} \int_0^{2\pi/\Omega} F_i(x, y)\,\mathrm{d}t; \qquad F_i' = F_i - \overline{F}_i \tag{2.53}$$

By definition $\overline{F_i'} = 0$ so that:

$$\begin{aligned} &\overline{F}_{Vi} = 0 \qquad \text{for } i = 1, 3, 5, 6, 9, 10 \\ &\overline{F}_{Ti} = 0 \qquad \text{for } i = 1, 3, 4, 7 \end{aligned} \tag{2.54a}$$

and that,

$$\begin{aligned} F'_{Vi} &= 0 \qquad \text{for } i = 2, 4, 8 \\ F'_{Ti} &= 0 \qquad \text{for } i = 2, 6 \end{aligned} \tag{2.54b}$$

Only the nonlinear periodic oscillating terms, appearing in Eq. (2.52), that contain the product of oscillating quantities (F_{V7}, F_{V11}, F_{T5}) have nonzero values of the time average:

$$\begin{aligned} F_{V7} &= \overline{F}_{V7} + F'_{V7} \\ F_{V11} &= \overline{F}_{V11} + F'_{V11} \\ F_{T5} &= \overline{F}_{T5} + F'_{T5} \end{aligned} \tag{2.54c}$$

When averaging Eqs. (2.52a) and (2.52b) one gets:

$$\sum_{i=1}^{11} \overline{F}_{Vi} = 0 \tag{2.55a}$$

$$\sum_{i=1}^{7} \overline{F}_{Ti} = 0 \tag{2.55b}$$

that taking into account Eqs. (2.54a) and (2.54b) read:

$$\underline{\nabla} \cdot \underline{\overline{V}} = 0 \tag{2.56a}$$

$$(\underline{\nabla}\overline{p}) + \underline{\nabla} \cdot [\underline{\overline{V}}\,\underline{\overline{V}}] + \underline{\nabla} \cdot [\overline{\underline{V}'\,\underline{V}'}] = Pr\big(\nabla^2 \underline{\overline{V}}\big) - \Lambda\Omega^2\, \overline{T' \sin(\Omega t)\underline{n}} \tag{2.56b}$$

$$\underline{\nabla} \cdot [\underline{\overline{V}}\,\overline{T}] + \underline{\nabla} \cdot [\overline{\underline{V}'\,T'}] = \nabla^2 \overline{T} \tag{2.56c}$$

The order-of-magnitude analysis based on extensive numerical simulations carried out by Savino and Lappa (2003a) shows that different terms can be dropped in the averaged equations as well as in the complete equations according to the regime considered (in terms of Ra_{V} and frequency range).

First, the reduced sets of equations (in dimensional form) are shown for the case $Ra_{\mathrm{V}} = 10^5/3$ (regimes (a) in Fig. 2.36). Then the case $Ra_{\mathrm{V}}=10^3/3$ is considered (regimes (b)). T_{o} relates to reference conditions.

For $\Omega > 10\,000$ (regime 1a):

Complete equations

$$\frac{\partial \underline{V}'}{\partial t} + \frac{1}{\rho_{\mathrm{o}}} \underline{\nabla} p' = b\omega^2 \beta_{\mathrm{T}} (\overline{T} - T_{\mathrm{o}}) \sin(\omega t)\underline{n} \tag{2.57a}$$

$$\frac{\partial T'}{\partial t} + \underline{\nabla} \cdot [\underline{V}'\overline{T}] = 0 \tag{2.57b}$$

Averaged equations

$$\frac{1}{\rho_o}\underline{\nabla}\bar{p} + \underline{\nabla} \cdot [\overline{\underline{V}'\underline{V}'}] = \nu\nabla^2\overline{\underline{V}} + b\omega^2\beta_T \overline{T' \sin(\omega t)}\underline{n} \tag{2.58a}$$

$$\underline{\nabla} \cdot [\overline{\underline{V}}\overline{T}] = \alpha\nabla^2\overline{T} \tag{2.58b}$$

It is worthwhile to point out how the simplified set of equation for this case corresponds to that used by Gershuni to develop the thermovibrational theory discussed in Section 2.6.4 (see the discussion below).

Using the Hodge theorem the vector field $(\overline{T} - T_o)\,\underline{n}$ can be decomposed into a solenoidal part plus an irrotational part:

$$(\overline{T} - T_o)\,\underline{n} = \underline{w} + \underline{\nabla}\zeta \tag{2.59}$$

with $\underline{\nabla} \cdot \underline{w} = 0$ and $\underline{\nabla} \wedge \underline{w} = \underline{\nabla}\overline{T} \wedge \underline{n}$.

According to Eq. (2.59), Eq. (2.57a) can be separated into two equations (for the potential and the solenoidal part):

$$\frac{\partial \underline{V}'}{\partial t} = b\omega^2\beta_T \sin(\omega t)\underline{w} \tag{2.60a}$$

$$\frac{1}{\rho_o}\underline{\nabla}p' = b\omega^2\beta_T \sin(\omega t)\underline{\nabla}\zeta \tag{2.60b}$$

Equation (2.60a) can be easily integrated with respect to time, obtaining

$$\underline{V}' = \beta_T\underline{w}\int_0^t b\omega^2 \sin(\omega t)\,\mathrm{d}t = -\beta_T\underline{w}b\omega\cos(\omega t) \tag{2.61a}$$

substitution of Eq. (2.61a) into Eq. (2.58a) and integrating with respect to time yields:

$$T' = \beta_T\underline{w} \cdot \underline{\nabla}\overline{T}\int_o^t b\omega\cos(\omega t)\,\mathrm{d}t = \beta_T\underline{w} \cdot \underline{\nabla}\overline{T}b\sin(\omega t) \tag{2.61b}$$

taking into account Eq. (2.61) and that

$$\overline{-b\omega^2 \sin^2(\omega t)} = -\overline{b\omega^2 \cos^2(\omega t)} \tag{2.62}$$

Eq. (2.55) reduces to Eq. (2.45); this is an impressive example of the aforementioned possibility to use simplified equations (by dropping some terms) in order to introduce alternative sets of equations that can be solved with enhanced computational efficiency with respect to the original set of Navier–Stokes equations.

For $1000 < \Omega < 10\,000$ (regime 2a):

Complete equations

$$\frac{\partial \underline{V}'}{\partial t} + \frac{1}{\rho_o} \underline{\nabla} p' = b\omega^2 \beta_T (\overline{T} - T_o) \sin(\omega t) \underline{n} + \nu \nabla^2 \underline{V}' \tag{2.63a}$$

$$\frac{\partial T'}{\partial t} + \underline{\nabla} \cdot [\underline{V}' \overline{T}] + \underline{\nabla} \cdot [\underline{V}' T'] = \alpha \nabla^2 T' \tag{2.63b}$$

Averaged equations

$$\frac{1}{\rho_o} \underline{\nabla} \overline{p} + \overline{\underline{\nabla} \cdot [\underline{V}' \underline{V}']} = \nu \nabla^2 \underline{\overline{V}} + b\omega^2 \beta_T \overline{T' \sin(\omega t) \underline{n}} \tag{2.64a}$$

$$\underline{\nabla} \cdot [\underline{\overline{V}}\, \overline{T}] + \overline{\underline{\nabla} \cdot [\underline{V}' T']} = \alpha \nabla^2 \overline{T} \tag{2.64b}$$

For $100 < \Omega < 1000$ (regime 3a):

Complete equations

$$\frac{1}{\rho_o} \underline{\nabla} p' = b\omega^2 \beta_T (\overline{T} - T_o) \sin(\omega t) \underline{n} + b\omega^2 \beta_T T' \sin(\omega t) \underline{n} + \nu \nabla^2 \underline{V}' \tag{2.65a}$$

$$\frac{\partial T'}{\partial t} + \underline{\nabla} \cdot [\underline{V}' \overline{T}] + \underline{\nabla} \cdot [\underline{V}' T'] = \alpha \nabla^2 T' \tag{2.65b}$$

Averaged equations

$$\frac{1}{\rho_o} \underline{\nabla} \overline{p} + \overline{\underline{\nabla} \cdot [\underline{V}' \underline{V}']} = \nu \nabla^2 \underline{\overline{V}} + b\omega^2 \beta_T \overline{T' \sin(\omega t) \underline{n}} \tag{2.66a}$$

$$\overline{\underline{\nabla} \cdot [\underline{V}' T']} = \alpha \nabla^2 \overline{T} \tag{2.66b}$$

For $10 < \Omega < 100$ (regime 4a):

Complete equations

$$\frac{1}{\rho_o} \underline{\nabla} p' = b\omega^2 \beta_T (\overline{T} - T_o) \sin(\omega t) \underline{n} + b\omega^2 \beta_T T' \sin(\omega t) \underline{n} + \nu \nabla^2 \underline{V}' \tag{2.67a}$$

$$\underline{\nabla} \cdot [\underline{V}' \overline{T}] = \alpha \nabla^2 T' \tag{2.67b}$$

Averaged equations

$$\frac{1}{\rho_o}\underline{\nabla}\overline{p} = \nu\nabla^2\underline{\overline{V}} + b\omega^2\beta_T\,\overline{T'\sin(\omega t)}\underline{n} \tag{2.68a}$$

$$\overline{\underline{\nabla}\cdot[\underline{V}'T']} = \alpha\nabla^2\overline{T} \tag{2.68b}$$

Hereafter, the regimes corresponding to $Ra_V = 10^3/3$ are denoted as the (b) regimes: for $\Omega > 10\,000$ (regime 1b) the simplified set of equation corresponds to the Gershuni ones as for $Ra_V = 10^5/3$.

For $1000 < \Omega < 10000$ (regime 2b):

Complete equations

$$\frac{\partial \underline{V}'}{\partial t} + \frac{1}{\rho_o}\underline{\nabla}p' = b\omega^2\beta_T(\overline{T} - T_o)\sin(\omega t)\underline{n} + \nu\nabla^2\underline{V}' \tag{2.69a}$$

$$\frac{\partial T'}{\partial t} + \underline{\nabla}\cdot[\underline{V}'\overline{T}] = \alpha\nabla^2 T' \tag{2.69b}$$

Averaged equations

$$\frac{1}{\rho_o}\underline{\nabla}\overline{p} + \overline{\underline{\nabla}\cdot[\underline{V}'\underline{V}']} = \nu\nabla^2\underline{\overline{V}} + b\omega^2\beta_T\,\overline{T'\sin(\omega t)}\underline{n} \tag{2.70a}$$

$$\underline{\nabla}\cdot[\underline{\overline{V}}\,\overline{T}] + \overline{\underline{\nabla}\cdot[\underline{V}'T']} = \alpha\nabla^2\overline{T} \tag{2.70b}$$

For $100 < \Omega < 1000$ (regime 3b):

Complete equations

$$\frac{1}{\rho_o}\underline{\nabla}p' = b\omega^2\beta_T(\overline{T} - T_o)\sin(\omega t)\underline{n} + \nu\nabla^2\underline{V}' \tag{2.71a}$$

$$\frac{\partial T'}{\partial t} + \underline{\nabla}\cdot[\underline{V}'\overline{T}] + \underline{\nabla}\cdot[\underline{V}'T'] = \alpha\nabla^2 T' \tag{2.71b}$$

Averaged equations

$$\frac{1}{\rho_o}\underline{\nabla}\overline{p} + \overline{\underline{\nabla}\cdot[\underline{V}'\underline{V}']} = \nu\nabla^2\underline{\overline{V}} + b\omega^2\beta_T\,\overline{T'\sin(\omega t)}\underline{n} \tag{2.72a}$$

$$\overline{\underline{\nabla}\cdot[\underline{V}'T']} = \alpha\nabla^2\overline{T} \tag{2.72b}$$

For $10 < \Omega < 100$ (regime 4b):
Complete equations

$$\frac{1}{\rho_o}\underline{\nabla} p' = b\omega^2\beta_T(\overline{T} - T_o)\sin(\omega t)\underline{n} + \nu\nabla^2\underline{V}' \tag{2.73a}$$

$$\underline{\nabla}\cdot[\underline{V}'\overline{T}] = \alpha\nabla^2 T' \tag{2.73b}$$

Averaged equations

$$\frac{1}{\rho_o}\underline{\nabla}\overline{p} = \nu\nabla^2\overline{\underline{V}} + b\omega^2\beta_T\,\overline{T'\sin(\omega t)\underline{n}} \tag{2.74a}$$

$$\overline{\underline{\nabla}\cdot[\underline{V}'T']} = \alpha\nabla^2\overline{T} \tag{2.74b}$$

2.7. Mixed buoyant/thermovibrational flows

The case of combined buoyancy (induced by steady residual-g) and thermovibrational convection is of great interest since it corresponds to the real situation occurring on the ISS (see the discussion in Section 2.5).

This topic is made even more significant by the fact that it is possible to take advantage of the interplay between these two types of convection to induce a mitigation of the resulting convective disturbances within the fluid cell.

As discussed in Section 2.2.3, in the absence of g-jitter, when the residual-g is opposite to $\underline{\nabla}\rho$ the residual-g has no effect (does not induce convective flows) for any value of its magnitude; when the residual-g is concurrent to $\underline{\nabla}\rho$ then TFD distortions arise only if the critical conditions for the onset of convection are exceeded.

The effect of high-frequency vibrations, like that of steady acceleration, strongly depends on the direction of the vibration relative to the density gradient. In particular, as illustrated in Section 2.6.5, vibrations parallel to the density gradient tend to maintain initial diffusive conditions. In fact, as pointed out by Birikh et al. (1993), the mean vibration force is a bulk driving action induced by temperature gradients normal to the vibration axis (i.e. thermovibrational convection arises when the isotherms are not perpendicular to the vibration axis). A local feature of this force is that, if temperature distortions, with respect to the purely diffusive case, are induced by another type of convection, average vibrational flows arise in such a way as to permit the isotherms to turn and again become perpendicular to the vibration direction.

The above considerations suggest that any undesired convective disturbances can be weakened by appropriate orientation of the experiment cell with the density gradient along the residual-g or along the g-jitter according to the prevailing effect.

The influence of the facility orientation is discussed herein for the same study case considered in the previous sections, consisting of a test cell bounded by rigid walls, completely filled with liquid, in the presence of a prescribed temperature difference. Although the quantitative results are restricted to this particular study case, conclusions could be extended to different classes of microgravity experiments (e.g., Bridgman crystal growth, directional solidification, see Naumann, 1996 and Fedoseyev and Alexander, 2000) in which the temperature or the concentration gradient direction is well defined, and for which the facility orientation can be properly changed to reduce undesirable convection disturbances).

Figures 2.37, 2.38, and 2.39 summarize the numerical results for the reference study case with $Pr = 15$. Figure 2.37 compares the computed streamlines, vector plots and temperature distributions for the US-Lab and the ESA-COF environments, when only g-jitter is present with direction perpendicular to the temperature gradient (i.e. in the unrealistical situation that the residual-g is zero and the cell is subjected only to the vibrations predicted for the US Lab or for the ESA-COF).

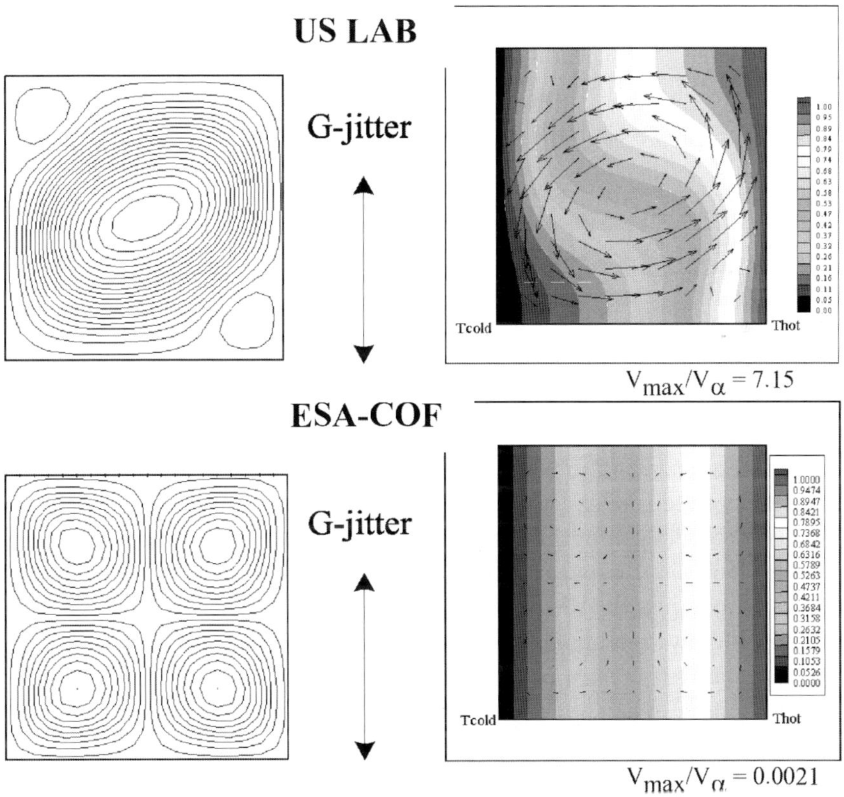

Fig. 2.37 Computed streamlines, vector plots and temperature distributions for the US Lab and the ESA-COF, when only g-jitter is present, $\Delta T = 50\,\text{K}$, silicone oil, $Pr = 15$ (time-averaged code, steady state).

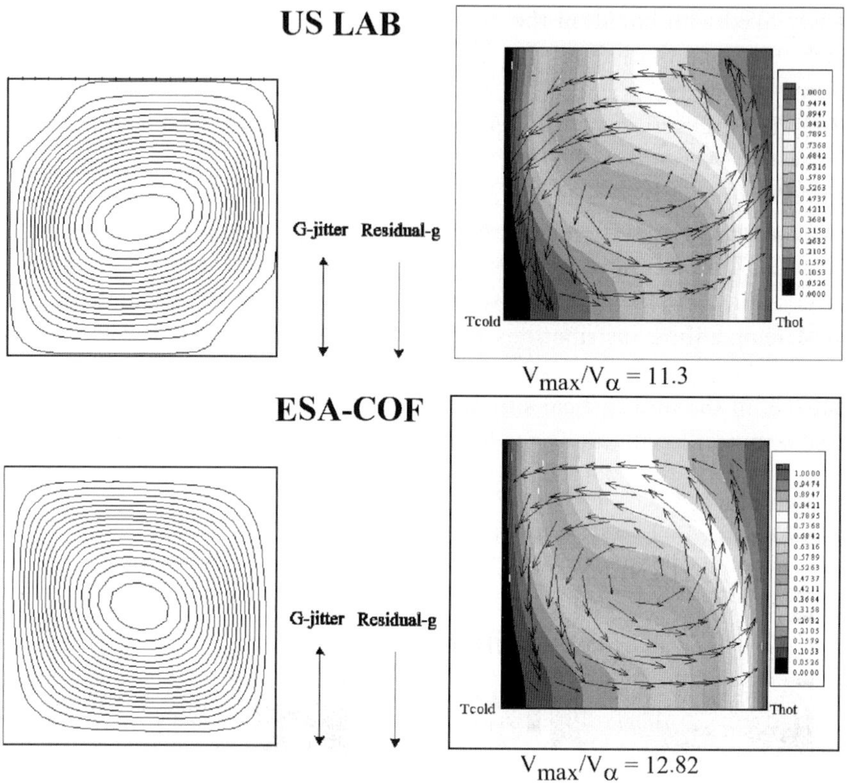

Fig. 2.38 Computed streamlines, vector plots and temperature distributions for the US Lab and the ESA-COF, when residual-g and g-jitter are both present (both perpendicular to the imposed temperature gradient).

Since the predicted vibrations are larger for the US Lab module (see Section 2.5 and Fig. 2.28) the convective disturbances and the temperature distortions are larger for the US Lab. Accordingly, for the latter case the flow exhibits the secondary pattern of symmetry already discussed in Section 2.6.5 (on the contrary for the US-Lab module the flow is characterized by the primary structure with four vortex cells).

Figure 2.29 shows the numerical results obtained in the other extreme situation that only residual-g is present (0.5 μg for the US Lab and 1.6 μg for the ESA-COF) without any g-jitter. The situation is reversed, i.e. the convective disturbances are large in the ESA-COF, due to the larger value of the residual gravity.

The results obtained assuming that both residual-g and g-jitter are present with direction perpendicular to the temperature gradient are shown in Fig. 2.38. In this case the overall convective disturbances are comparable in value, but in the US Lab the major cause of disturbances is the g-jitter, on the contrary in the ESA-COF it is the residual-g.

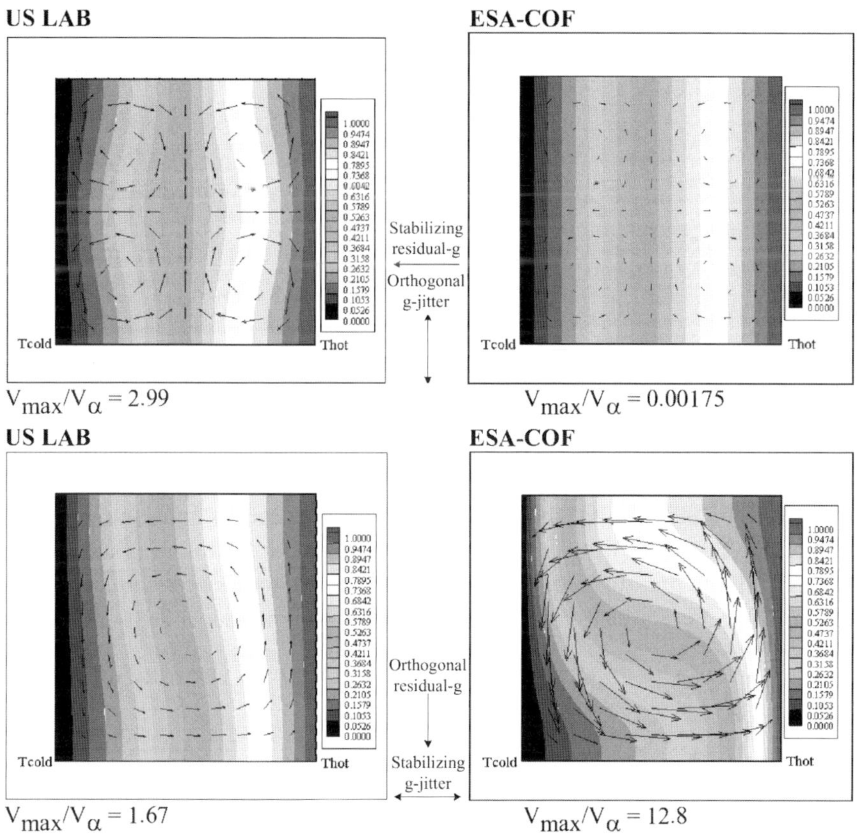

Fig. 2.39 Vector plots and temperature distributions for the US Lab and the ESA-COF, when residual-g and g-jitter have different orientations.

If the residual-g and g-jitter are not parallel, as anticipated, orienting the experimental cell with the density gradient along the residual-g or along the g-jitter can help to mitigate the convective disturbances.

Figure 2.39 show the different thermo-fluid-dynamic fields obtained by changing the orientation of the residual-g and g-jitter vectors. The TFD distortions in the US Lab are minimized when the cell is oriented with the density gradient parallel to the g-jitter direction (main source of disturbances); in this case the stabilizing effect of the g-jitter can also be beneficial in minimizing the destabilizing effect of a residual-g parallel to the temperature gradient (i.e. like a cell heated from below on Earth).

In the European COF the convective disturbances are minimized only when the residual-g vector is antiparallel to the temperature gradient (i.e. stabilizing), even in the presence of an orthogonal g-jitter.

These simple demonstrative simulations illustrate how the combined quasisteady residual-g and the oscillatory g-jitter may be used to be beneficial to approach pure diffusion conditions.

In particular, for the reference case considered in the present study (the microgravity environment provided by the ISS), the numerical results indicate that, for the different possible configurations, corresponding to different relative orientations between residual-g and g-jitter, one should be able to properly orient the experiment to minimize the convection disturbances according to the known disturbances predicted for the different modules of the ISS.

2.8. Solution methods for the incompressible Navier–Stokes equations

The last section of this chapter is devoted to the introduction of a reliable and robust numerical technique for the solution of the sets of equations introduced in the previous sections (continuity, momentum, species and energy equations).

With regard to this aspect it is worthwhile to point out how a number of methods are available in the literature for the solution of the Navier–Stokes equations in their incompressible and complete form (multidimensional, time-dependent and nonlinear equations).

2.8.1 *Vorticity methods*

In some cases the equivalent forms of the momentum equation (i.e. equivalent in the continuum, but generally not equivalent when spatially discretized) can be used to advantage. These alternative equations are all derived from the "primitive" equation from differentiation (in part) and often after additional insights regarding incompressible flow. They are also often used as the starting point for generating alternative numerical solution methods (Gresho, 1991).

For instance, for two-dimensional cases, streamfunction–vorticity (Ψ–ζ) methods have enjoyed widespread use during recent years owing to computational efficiency and simplicity. The major advantage of these methods relies on the possibility to solve the equations leaving aside problems related to the pressure gradient that is usually regarded as a 'strange' object by numerical investigators. For the same reason velocity–vorticity methods have also been used for three-dimensional problems. In such methods the presence of the pressure gradient is removed through application of the curl operator to the momentum equation (the curl of the gradient of a scalar quantity, in fact, is zero). Applying the curl operator and using the fact that both $\underline{V}$ and $\underline{\omega} = \underline{\nabla} \wedge \underline{V}$ are div-free:

$$\frac{\partial \omega}{\partial t} + \underline{V} \cdot \underline{\nabla}\, \underline{\omega} = \underline{\omega} \cdot \underline{\nabla}\, \underline{V} + Pr\, \nabla^2 \underline{\omega} + \underline{\nabla} \wedge \underline{F}_g \qquad (2.75a)$$

The curl of $\underline{\nabla} \wedge \underline{V} = \underline{\omega}$ leads to another higher-order equation, a Poisson equation for the velocity:

$$\nabla^2 \underline{V} = -\underline{\nabla} \wedge \underline{\omega} \tag{2.75b}$$

which is still a kinematic equation. These two equations are at the basis of velocity–vorticity formulations (see, e.g., Farouk and Fusegi, 1985).

Some numerical investigators (see, e.g., Aziz and Hellums, 1967) have also tried an approach based on the introduction of the 3D streamfunction defined as

$$\underline{\nabla} \wedge \underline{\Psi} = \underline{V} \tag{2.76a}$$

known as the so-called 3D vorticity/vector-potential formulation. Accordingly, the dimensionless equations governing the evolution of ω and Ψ are Eqs. (2.75a) and (2.76b), respectively:

$$\underline{\nabla} \wedge \underline{\nabla} \wedge \underline{\Psi} = \underline{\omega} \tag{2.76b}$$

These methods on the one hand have enjoyed a reasonable success for the aforementioned reasons, but on the other hand have been characterized by some conceptual difficulties mainly coming from a noncomplete knowledge of the boundary conditions making the problem well posed from a computational/theoretical point of view (Gresho, 1991).

2.8.2 *Primitive variables methods*

A different class of methods not needing mathematical manipulation and transformation (i.e. the introduction of the vorticity through application of the curl operator), however, has been simultaneously developed. In such a class the pressure gradient is not eliminated. Rather it is used for introducing (with the aid of the continuity equations) an additional equation that leads to a closed set of equations. The major advantage offered by this class of methods (see, e.g., the MAC method introduced by Harlow and Welch, 1965) relies on the possibility to solve the problem directly in terms of the primitive variables (i.e. velocity and pressure) of the fluid. Superimposed on this, the problem related to the boundary conditions that make the problem well posed has been investigated to a certain extent leading to widely recognized theoretical results (Gresho and Sani, 1987).

For instance, forward differences in time can be used to discretize the partial differential equations in their primitive variable form (of course, this is only an example used herein for the convenience of the reader, usually high-order temporal schemes should be used to obtain reliable accuracy), obtaining:

$$\begin{aligned} \underline{V}^{n+1} &= \underline{V}^n + \Delta t\left[-\underline{\nabla}\cdot(\underline{V}\,\underline{V}) + Pr\,\nabla^2\underline{V}\right]^n - \Delta t\,\underline{\nabla} p^n + \Delta t\,\underline{F}_g^n \\ \rightarrow \underline{V}^{n+1} &= \underline{R}(\underline{V}^n) - \Delta t\,\underline{\nabla} p^n \end{aligned} \tag{2.77}$$

where the operator $\underline{R}$ reads $\underline{R}(\underline{V}) = \underline{V} + \Delta t\left[-\underline{\nabla}\cdot(\underline{V}\,\underline{V}) + Pr\,\nabla^2\underline{V}\right] + \Delta t\underline{F}_g$, where $\underline{F}_g$ takes into account gravity (steady or oscillatory) forces (n superscript denotes time step).

The pressure field can be computed by solving a Poisson equation resulting from the divergence of the momentum equation with the help of the continuity equation $\underline{\nabla}\cdot\underline{V}^{n+1} = 0$:

$$\nabla^2 p^n = \frac{1}{\Delta t}\,\underline{\nabla}\cdot\underline{R}^n \tag{2.78}$$

This formulation requires (for retaining coherence with respect to Eq. (2.78), i.e. for obtaining a well-posed problem) that at the physical boundaries of the computational domain the pressure satisfies conditions obtained projecting the operator $\underline{R}$ along the direction perpendicular to the boundary itself, i.e.:

$$\frac{\partial p}{\partial n} = \frac{1}{\Delta t}\underline{R}\cdot\underline{\hat{n}} \tag{2.79}$$

$\hat{n}$ being the unit vector pointing into the liquid domain.

Gresho and Sani (1987) in particular were the first to demonstrate that the proper boundary condition (BC) is simply the Neumann BC obtained by applying the normal component of the momentum equation on the frontier of the domain (i.e. Eq. (2.79)) because this is the only BC that can always assure that $\underline{\nabla}\cdot\underline{V} = 0$ in the domain.

Subtracting the Poisson equation from the divergence of the momentum equation this reads:

$$\frac{\partial(\underline{\nabla}\cdot\underline{V})}{\partial t} = 0. \tag{2.80}$$

This equation is important since it shows that $\underline{\nabla}\cdot\underline{V} = 0$ remains zero in the computational domain if it is zero at the initial instant, i.e. $\underline{\nabla}\cdot\underline{V}_o = 0$.

Equations (2.77) and (2.78) can be used for a time-marching procedure where p^n is computed as a function of $\underline{V}^n$ (Eq. (2.78)) and $\underline{V}^{n+1}$ is computed as a function of $\underline{V}^n$ and p^n (Eq. (2.77)). This approach was referred to as the Consistent Pressure Poisson Equation CPPE in Gresho and Sani (1987). If this CPPE formulation is employed and the initial velocity field is div-free, according to Eq. (2.80) the solution will be div-free for all time.

On the contrary, the initial condition for vorticity formulations is poorly understood: there are no constraints and therefore the problem is often ill posed (Gresho, 1991). There are no BCs on the vorticity, but the value (which is often poorly behaved) of the vorticity on the boundary does play a major role in most vorticity-based methods. Whereas vortex sheets are little more than a minor inconvenience in primitive variable formulations, they are often the principal focus of attention in vorticity-based methods, partly because they are computationally cumbersome.

Vorticity is a challenging variable to compute. Usually more grid points are needed in the boundary layer with a vorticity method with respect to a primitive variables method to get the same accuracy and the prescription of an effective solid-surface boundary condition is often a weak point.

Some difficulties arise, however, also with the primitive variable formulation due to the complexity of the operator $\underline{R}$ that requires demanding manipulations for spatial discretization. For this reason a simplified variation has been proposed, the so-called SMAC method. Among other things this method also allows one to impose homogeneous boundary conditions for the pressure rather than the complex nonhomogeneous ones (2.79).

The SMAC method for the solution of the incompressible Navier–Stokes equations proceeds as a type of fractional step method by first writing a modified momentum equation and then updating the velocity field using the computed pressure to account for the continuity equation. More precisely, at each time step, an intermediate velocity field $\underline{V}^*$ is found without the knowledge of the correct pressure field, and therefore no incompressibility condition is enforced. The intermediate velocity field is then corrected by a second step in which a pressure equation is solved and then the computed pressure is used to produce a divergence-free velocity field.

In practice, according to the SMAC method, the computation of the velocity field at each time step is split into two substeps. In the first, an approximate velocity field $\underline{V}^*$ corresponding to the correct vorticity of the field, but with $\underline{\nabla} \cdot \underline{V}^* \neq 0$, is computed at time $(n{+}1)$ neglecting the pressure gradient in the momentum equation, i.e.

$$\underline{V}^* = \underline{V}^n + \Delta t[-\underline{\nabla} \cdot (\underline{V}\,\underline{V}) + Pr\,\nabla^2 \underline{V}]^n + \Delta t \underline{F}_{\mathrm{g}}^n \tag{2.81}$$

In the second substep, the pressure field is computed by solving a Poisson equation obtained, imposing that $\underline{\nabla} \cdot \underline{V}^{n+1} = 0$:

$$\nabla^2 p^n = \frac{1}{\Delta t} \underline{\nabla} \cdot \underline{V}^* \tag{2.82}$$

Finally, the velocity field is updated using the computed pressure field to account for continuity:

$$\underline{V}^{n+1} = \underline{V}^{*} - \Delta t \, \underline{\nabla} p^{n} \tag{2.83}$$

If the effective velocity boundary conditions are used for the solution of Eq. (2.81) then $\underline{V}^{*}$ need not be corrected on the boundary (i.e. $\underline{V}^{n+1} = \underline{V}^{*}$); this leads to homogeneous boundary conditions for the pressure equation, i.e.

$$\frac{\partial p}{\partial n} = 0 \tag{2.84}$$

By comparison of Eqs. (2.78) and (2.82) and (2.79) and (2.84) it is evident how the SMAC methods introduces significant simplifications with respect to the MAC formulation without sacrificing accuracy or well-posed conditions.

In practice, this method makes use of the Hodge decomposition theorem, which states that any vector function $\underline{V}$ can be decomposed into a divergence-free part $\underline{V}^{*}$ plus the gradient of a scalar potential $\Re$, i.e. $\underline{V} = \underline{V}^{*} + \underline{\nabla}\Re$ where for the SMAC method $\underline{\nabla}\Re = -\underline{\nabla} p/\Delta t$. Accordingly the quantity p in Eq. (2.82) is often called the pseudopressure to distinguish it with respect to the physical pressure that can be obtained by solving Eq. (2.78). If the physical pressure is needed Eq. (2.78) still must be solved with the BC (2.79) when the SMAC formulations is used.

The Poisson equation (2.82) is an elliptic equation and can be solved with a SOR (Successive Over Relaxation) iterative method. Equation (2.81) is instead a parabolic equation and can be solved using explicit or implicit schemes (the use of explicit schemes is recommended in the present book since they facilitate a parallel implementation of the code based on a grid partition strategy, see, e.g., Section 4.4 for further details).

Also, the species and energy equations are parabolic in nature (T or C at time $n+1$ can be simply computed as a function of the fields at time n):

$$T^{n+1} = T^{n} + \Delta t[-\underline{\nabla} \cdot (\underline{V} T) + \nabla^{2} T]^{n} \tag{2.85}$$

If g-jitters are present and their properties (in terms of frequency and amplitude) fall within the range of applicability of Gershuni's formulations, the source term in the momentum equation reads:

$$\underline{F}_{g}^{n} = Pr \cdot Ra_{V}[(\underline{w} \cdot \underline{\nabla} T)\underline{n} - \underline{w} \cdot \underline{\nabla}\, \underline{w}]^{n} \tag{2.86}$$

where the auxiliary potential function $\underline{w}$ must be obtained by solving the equation:

$$\nabla^{2} \underline{w}^{n} = -\underline{\nabla} \wedge (\underline{\nabla} T^{n} \wedge \underline{n}) \tag{2.87}$$

that is elliptic in nature and can be solved with the same numerical procedures used for Eq. (2.82).

Note that both MAC and SMAC methods require staggered grids to guarantee reliable coupling of velocity and pressure (see Fletcher, 1991 for further details of this aspect). Staggering means that the velocity components are computed on the faces of the control volumes/computational cells used for pressure, temperature and concentration (these quantities occupy the center of such volumes), i.e. that the control volumes/computational cells used for the velocity components are shifted with respect to the one used for pressure, temperature and/or concentration. Figure 2.40 show some examples.

The numerical schemes for the spatial discretization of the equations (central differences, upwind schemes, etc.) are no longer discussed in the present chapter (it is out of the scope of the present analysis; the reader will find all the details in the excellent books of Fletcher, 1991; Hirsch, 1994; Deville et al., 2002 that are entirely devoted to such aspects).

Rather, in the spirit of the book, the available space is devoted herein to the discussion of some special computational problems that arise when flows with free fluid/fluid interfaces (Marangoni, buoyancy and/or mixed conditions) and/or solid/liquid solidification fronts are considered.

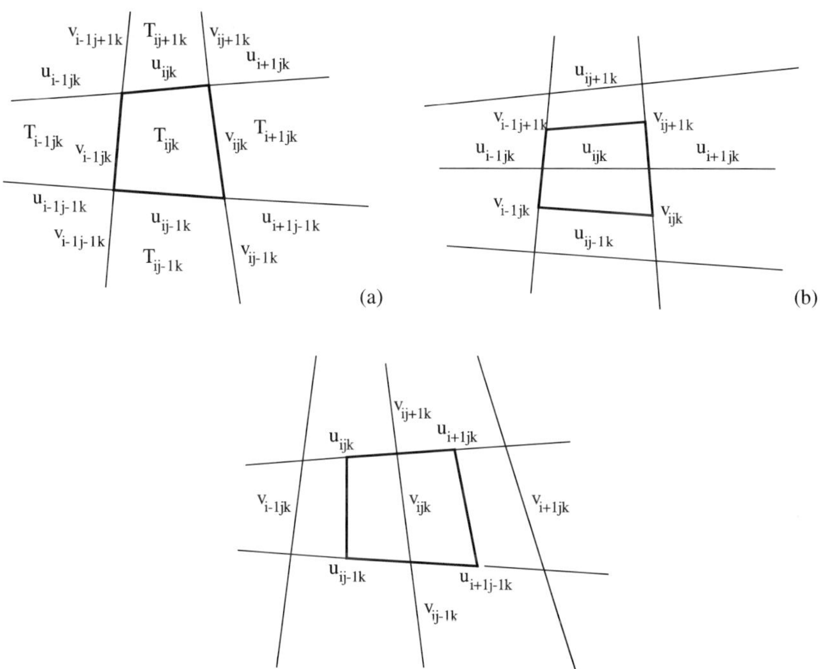

Fig. 2.40 Staggering for the different variables.

The existence and numerical treatment of momentum, thermal or solutal boundary layers and of vorticity singularities must be regarded as critical issues for the appropriate simulation of these flows. In particular, on the one hand, the boundary layers require special care for mesh generation and grid convergence (i.e. for obtaining a grid-independent solution); on the other hand, the presence of numerical singularities introduces additional need for focusing on conditions that make the problem well posed from a computational point of view (Lappa, 2002a).

2.8.3 *Numerical treatment of boundary layers*

Grid quality is a critical factor for the study of problems concerning the Marangoni and buoyancy flows (pure or mixed) and of their steady states, stability properties and possible bifurcations. The complexity of the flow field (in many cases three-dimensional and time-dependent, as illustrated for instance in Section 2.4) and the large number of parameters involved require very accurate computations to predict the behavior of the system.

This is particularly true when the flow is characterized by the presence of some regions (hereafter referred to as boundary layers) where extremely steep gradients of velocity and/or temperature and/or concentration are present.

In these cases a larger number of computational points must be used in these portions of the domain to locally enhance the resolution. These points are generally distributed according to stretching functions that allow a gradual, somewhat smooth increase of the density of points along a given direction. This smooth increase has a two-fold purpose: it is required for linking regions where normal resolution is necessary (fixed spatial step) with the aforementioned boundary layers as well as to avoid numerical instabilities; it is known, in fact, that step changes in the points distribution could cause serious problems in terms of stability of the computational algorithm.

In the following, for the sake of simplicity and for educational purposes, one-dimensional stretching functions are considered. Among other things, it is worthwhile to point out how these functions have enjoyed a widespread use in the literature owing to simplicity and efficiency in the treatment of problems where strong gradients occur. For these reasons they are briefly discussed in the present book as well.

In these functions the dependent and independent variables are usually expressed in normalized form. For a one-dimensional stretching function applied along a segment AB, the appropriate normalized independent variable varying within the range [0, 1] along the aforementioned segment is denoted herein by η ($\eta = 0 \rightarrow$ A, $\eta = 1 \rightarrow$ B).

A possible one-dimensional stretching function is (see Fletcher, 1991)

$$s = \wp\eta + (1 - \wp)\left\{1 - \frac{\tanh[Q(1-\eta)]}{\tanh Q}\right\} \tag{2.88}$$

where $\wp$ and Q are the parameters controlling grid-points distribution ($\wp$ effectively provides the slope of the distribution close to $\eta=0$; Q controls the departure from the linear s versus η; small values of Q cause small departures from linearity; however, also if $\wp$ is close to unity the departure from linearity is small).

Then the grid cells can be distributed along the direction ξ according to simple laws such as:

$$\xi=\xi_{\mathrm{A}}+s(\xi_{\mathrm{B}}-\xi_{\mathrm{A}}) \tag{2.89}$$

where ξ is the direction perpendicular to the boundary layer and ξ_{A} and ξ_{B} (i.e. $\xi_{\mathrm{B}}-\xi_{\mathrm{B}}$) characterize its width.

The choice of the proper values of $\wp$ and Q (or of other similar parameters if other categories of stretching functions are used, see, e.g., the one-dimensional Poisson grid-generators in Knupp and Steinberg, 1994) ensuring rapid solution convergence when the number of points is increased, is usually not trivial and requires extensive numerical experimentation.

Also, the directions (ξ) along which the stretching process should be applied are not a priori data of the problem and require numerical experimentation and iterative procedures (first computations with coarse and uniform grids are carried out, then critical information for mesh generation and points clustering is gained from these preliminary computations).

A first step towards such a critical knowledge, however, is also represented by the Order of Magnitude Analysis (OMA) whose general formulation was derived by Napolitano (see, e.g., Russo and Napolitano, 1984) and applied to Marangoni, buoyancy and combined free convection.

The aforementioned analysis, in fact, allows the individuation of the ranges of the nondimensional characteristic numbers (defined in Sections 2.2 and 2.3) and of the conditions for which boundary layers are present. The knowledge of the nature (momentum, thermal, solutal) and of the scale factors governing the evolution of the aforementioned layers is propaedeutical to the appropriate choice of the regions and directions where mesh stretching is necessary as well as to the definition of the related stretching functions.

The results of the Napolitano's OMA are summarized in Tables 2.1 and 2.2 (the case of compositionally homogeneous media is evidently obtained by formally setting $Sc=0$).

Table 2.1 Governing equations and Order of Magnitude Analysis.

Pr	Sc	χ	Governing equation
$\leq$O(1)	$\leq$O(1)	1	Momentum
> O(1)	$\leq$O(1)	Pr	Energy
$\leq$O(1)	> O(1)	Sc	Species
> O(1)	> O(1)	Max{Pr, Sc}	Energy or species

Table 2.2 Length scale factor and velocity scaling.

Test on	ℓ	Velocity scaling
$\chi Re \leq O(1)$	1	V_{Ma}
$\chi Re > O(1)$	$(\chi Re)^{-1/3}$	$V_{Ma}(\chi Re)^{-1/3}$
$\chi Gr \leq O(1)$	1	V_g
$\chi Gr > O(1)$	$(\chi Gr)^{-1/4}$	$V_g(\chi Gr)^{-1/2}$

Therein ℓ is a length scale factor at most of order one: if $\ell = 1$ no boundary layers are present for the considered phenomenon, if $\ell \neq 1$ boundary layers exist and their thickness scales according to ℓ.

Consider, for instance, the case of a pure thermal Marangoni flow ($Sc = 0$) and a very small value of the Prandtl number ($Pr \ll 1$, e.g., a liquid metal or a semiconductor melt). For this case Table 2.1 shows that the existence of boundary layers is possible with respect to momentum and gives $\chi = 1$. Correspondingly, in Table 2.2, two cases are possible $Re \leq O(1)$ with $\ell = 1$ and $Re > O(1)$ with $\ell = (Re)^{-1/3}$. In the second case (it is the usual situation arising in material processing) the existence of a boundary layer is predicted with a thickness changing according to $(Re)^{-1/3}$ (as Re increases it becomes thinner).

This means that for the case of liquid metals and semiconductor melts nonuniform grids stretched close to the free surface should be used.

This zone plays a critical role in the computations since therein a Marangoni boundary layer is present with extremely steep radial gradients of velocity induced by the driving force acting on the free surface. Direct numerical experimentation shows that this is particularly true for very large values of the Re number ($O(Ma) > 10^2$) as shown in Fig. 2.10 (see Section 2.3.3).

The results provided by the OMA in this case become even more important if one considers that, on the basis of Table 2.1, no thermal boundary layers should be present (or if present should be weak): "Momentum" in the last column means, in fact, that the prominent features of the field are related to this quantity and that other (thermal and/or solutal) effects should be negligible in terms of steepness of the related gradients (hereafter the word "prominent" is used to denote only the features of the TFD field that exhibit steep gradients). From a physical point of view the absence of thermal boundary layers follows from the large thermal diffusivity of liquid metals that spreads thermal gradients over wide regions.

These are the reasons for which, for relatively small values of the Marangoni number ($Ma = O(10)$), grid-refinement studies simply show that for the case of liquid metals ($Pr \ll 1$), the numerical simulation of the problem is not adversely affected by the use of uniform grids. For this case, in fact, in the light of the aforementioned arguments, both momentum and thermal boundary layers are absent or weak and grid convergence is achieved quite easily.

As anticipated, when *Ma* becomes large, stretching functions clustering the computational points close to the free surface should be chosen. Since the isotherms tend to be compressed by the flow towards the cold wall, some points should also be accumulated close to this boundary.

Strong thermal gradients, however, are expected for the case of high Prandtl number (transparent) liquids; extensive numerical experimentation shows that for $Pr \gg 1$, in fact, very thin thermal boundary layers appear close to the heated (or cooled) sides. For this case special care (nonuniform grids and/or regularization conditions) has to be provided in order to capture the detailed structure of the thermal and velocity fields (in particular the velocity distribution close to the corners) and to obtain grid-convergence (see the next section for this subject).

This situation is also predicted in principle by the OMA results in Tables 2.1 and 2.2. In fact, for $Pr > O(1)$ the prominent features of the field are related to its thermal structure ("Energy" appears in the last column of Table 2.1); in this case $\chi = Pr$ and the thickness of the thermal boundary layers scales according to $\ell = (Pr \cdot Re)^{-1/3} = (Ma)^{-1/3}$. Such a behavior is qualitatively confirmed by the numerical results shown in Fig. 2.41. The temperature fields clearly exhibit bands close to the heated (or cooled) walls where the isotherms tend to be compressed. The width of these layers is reduced as *Ma* increases, leading to an increasing demand for points in these regions (see, e.g., Fig. 2.42).

2.8.4 *Viscous singularities and regularization functions*

Confined Marangoni flows (for instance, in open cavities or in liquid bridges) are characterized by the presence of a singularity where free and solid surfaces meet. The singularity is particularly evident if one considers the vorticity distribution (see Fig. 2.43 and the genesis of the vorticity singularity).

Since the vorticity is given by $\zeta = \partial v/\partial x - \partial u/\partial y$, it reduces to $\zeta = -\partial u/\partial y > 0$ along the free surface (where $v = 0$) that behaves as a source of positive vorticity. The no-slip conditions on the solid boundaries generate instead a negative vorticity $\zeta = \partial v/\partial x < 0$ ($u = 0$). Accordingly, there is a jump discontinuity for the vorticity in the surface corners (in practice, it is a jump discontinuity for the shear stress from which the more proper denomination of "viscous singularity" arises).

This singularity is very important as it affects the flow at any contact free surface/solid boundary. It can seriously jeopardize the physical consistency of the numerical solution and also retard convergence, and sometimes lead to oscillations that may result in grid-divergence. In the light of these arguments, a physically relevant modeling of the contact free surface/solid boundary appears as a prerequisite of any numerical simulation.

In practice, any computational treatment of these flows must filter this singularity: Depending on the circumstances this task can be accomplished implicitly by using

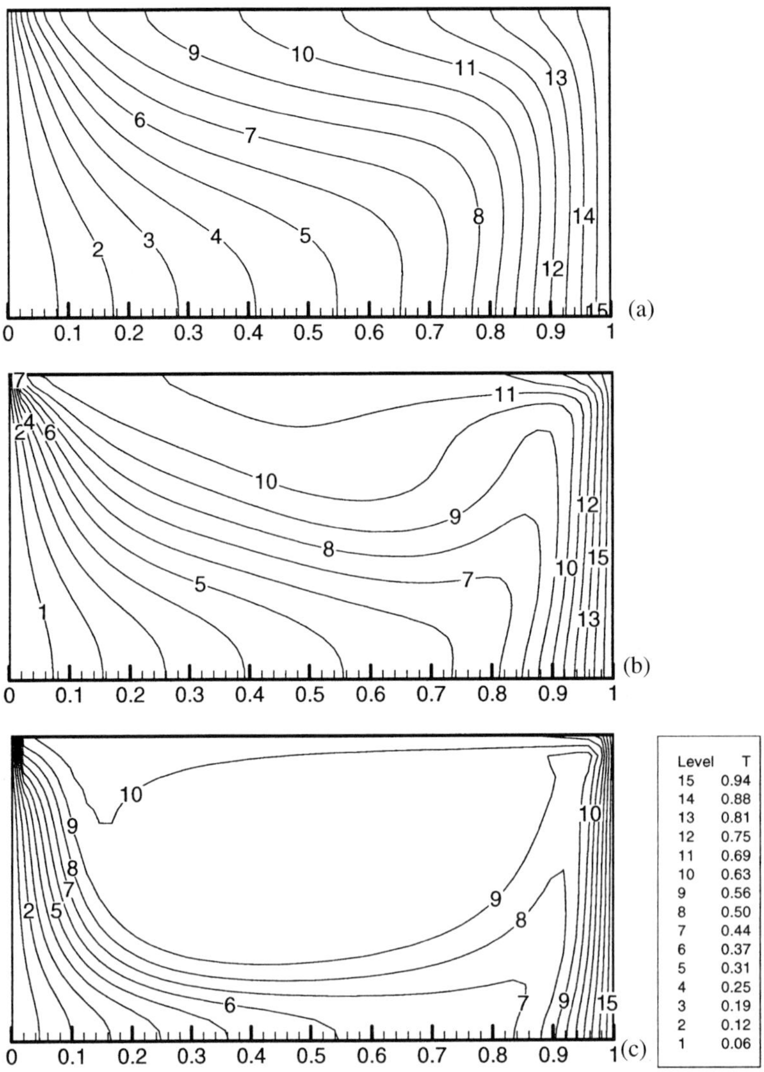

Fig. 2.41 Open cavity problem, $Pr=15$, $A=2.0$, temperature distribution, cold wall on the left side, hot wall on the right side: (a) $Ma=10^3$, (b) $Ma=10^4$, (c) $Ma=10^5$.

finite precision methods or explicitly by modifying the boundary conditions of the problem (the so-called regularization, Kasperski and Labrosse, 2000).

Usually such a regularization is not necessary for low Prandtl number fluids ($Pr \ll 1$) since in this case the singularity is implicitly bypassed by the finite precision spatial discretization that is based on local approximation of the derivatives. If the problem, for instance, is solved with the very commonly employed, second order in space, finite-difference scheme, the resulting numerical solution will not be affected

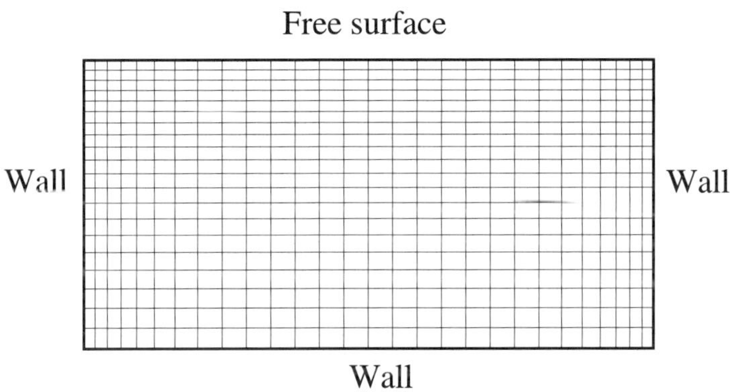

Fig. 2.42 Example of grid for the open-cavity problem obtained combining one-dimensional stretching function along x and y ($\wp = 2$ and $Q = 1$).

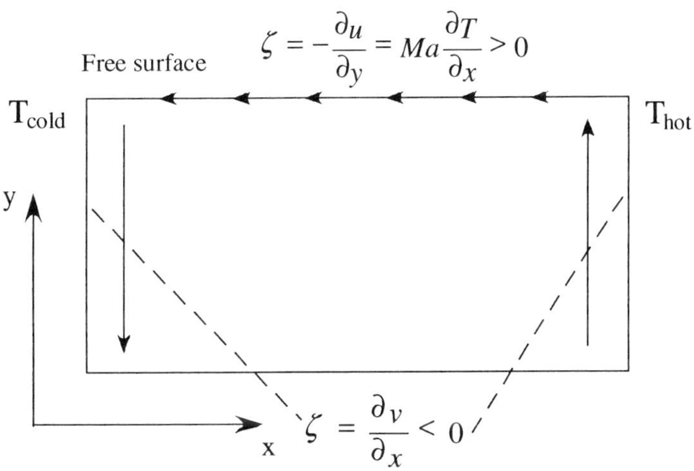

Fig. 2.43 Sketch of the conditions responsible for the viscous singularities (u and v are the velocity components along x and y, respectively).

by any singular contribution thanks to the filtering properties of the local approximation of the differential operators.

In practice, the finite-difference schemes automatically act to regularize, through some filtering mesh-dependent function, the initially singular formulation. The numerical solution converges, asymptotically with the mesh refinement, toward the solution of a continuous problem (Kasperski and Labrosse, 2000).

On the contrary, the explicit regularization is a critical requirement, for high Prandtl number liquids ($Pr > 1$) and in particular for large values of the Marangoni number. For these cases grid convergence cannot be obtained or occurs only if very dense (often prohibitive for three-dimensional simulations) computational meshes are used.

As previously explained, for these liquids strong thermal boundary layers appear close to the heated (or cooled) boundaries with thickness scaling according to $\ell = (Pr \times Re)^{-1/3} = (Ma)^{-1/3}$.

The boundary layer close to the cold wall in particular is responsible for the presence of a large number of isotherms confined in a very thin region close to the free surface. Such a compression in turn is responsible (via the surface Marangoni boundary condition) for the presence of a velocity peak therein. Often the magnitude of the velocity in this point diverges as the mesh is refined.

These are sufficient reasons for considering the explicit treatment of the singularity as essential regarding the physical relevance of the results.

The regularization approach, rather than relying on the intrinsic filtering properties of the numerical scheme (which in this case are not sufficient), tries to get an explicit regular continuous formulation of the problem by an appropriate modeling of the underlying physical regularity.

The physical consistency requires, in particular, that experiments performed with a frozen geometry should exhibit small scales, on the free surface, in the vicinity of the solid boundaries within which the physical regularity is satisfied at the scale of the continuous medium. From a mathematical point of view, this condition is achieved if the Marangoni stresses vanish at the solid walls (supposed to be located at $x = \pm 1$) while keeping a continuous behavior. Following Kasperski and Labrosse (2000) all these elements together can be taken into account, in a very simplified way, by introducing the regularizing function

$$f_n(x) = \left(1 - x^{2n}\right)^2 \tag{2.90}$$

in the boundary conditions (2.20), i.e.:

$$\frac{\partial \underline{V}}{\partial n} = -Ma\, f_n(x)\, \underline{\nabla}_s T \tag{2.91}$$

where n is the so-called filtering parameter.

A measure of the filtering length, the distance over which the regularization acts, is defined by $\delta(n)$ such that $f_n(x = 1 - \delta(n)) = 0.95$, which leads to

$$\delta(n) = 1 - 0.0253^{1/2n} \tag{2.92}$$

i.e. the higher is n, the smaller is the filtering scale.

This length scale behaves as a new free parameter with respect to which the convergence of the results must be assessed.

It must be decreased (increasing n) until an asymptotic behavior appears that is a good indication of the physical relevance of the model and of the obtained flow. In other words for a given problem, preliminary convergence tests have to be carried out for the appropriate choice of n. These tests generally show that the appropriate n

behaves as an increasing function of the Marangoni number (see, e.g., Kasperski and Labrosse, 2000).

In the case of solidification problems (nonfrozen geometries), the use of the regularization condition (2.90) near the melt/solid fronts follows directly from the physical behavior of the liquid therein; the viscosity, in fact, must tend to an infinite value on the freezing front (i.e. $Ma \rightarrow 0$ locally).

CHAPTER 3

Dispersed droplets and metal alloys

3.1. Introduction

In recent years binary immiscible liquid–liquid systems, and partially miscible liquid pairs, have been finding increasing interest in fundamental studies of interphase mass transfer due to their relevance for technological applications. This relevance is motivated in particular by problems related to the case of metal alloys. These materials include structural materials, many types of composites, electrical conductors, and magnetic materials.

The properties of these products are linked to their crystalline and chemical structure; for example, the mechanical strength and the corrosion resistance of many alloys are determined by their internal structure, which develops as the metal or alloy solidifies from its molten state.

Research in this area is primarily concerned with advancing the understanding of the related processing so that the structure and, ultimately, the properties of the alloy, can be controlled as the materials are originally formed. In particular by removing the influence of gravity, scientists are more closely observing the influential processes in structure formation that occur during solidification. Understanding how the method is used to produce, or process, a material affects its structure and its properties will help us to make improved materials for many purposes.

Within this context, of particular interest is the case of immiscible alloys (for additional information and theoretical background, see the learning environment provided in Section 1.5.1) where a number of different phenomena dealing with the dynamics of dispersed particles occur (drop coalescence and wetting, sedimentation and flotation, Marangoni migration, etc.).

For the immiscible alloys, in fact, the two-phase liquid consists of dispersed drops in a matrix liquid and the quality of the dispersion alloys is defined by the degree of homogeneity of the minority-phase distribution. It is well known that the strength of these alloys can be improved if a uniform dispersion of fine particles in the matrix is achieved (Ratke, 1993).

The particle distribution in the matrix is one of the major microstructural features which determines the properties and performances of these materials. Optimization of these systems therefore can be seen as a matter of searching, as systematically as possible, for the conditions that lead to the aforementioned uniform dispersion of

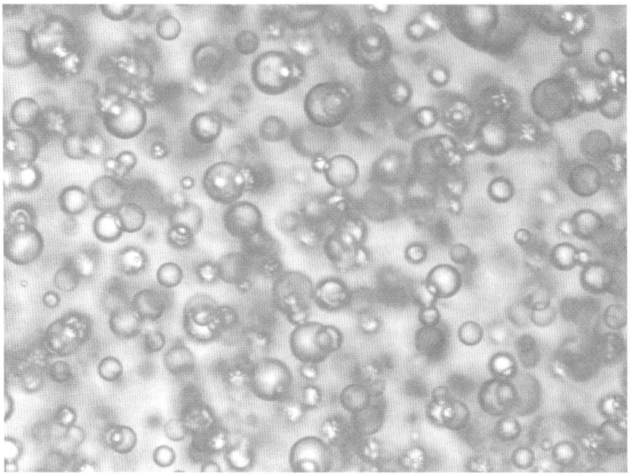

Fig. 3.1 Dispersion of fine particles in the matrix (visualization obtained using transparent model liquids, e.g., silicone oil in poli-isobutilen).

the drops (Fig. 3.1). In practice, this corresponds to a study of the effects that can move the system away from the desired condition. A detailed knowledge of such effects may help researchers in preventing them.

For instance, in the presence of a gravity field (i.e. on the ground), as the density of the phases is different, the minority phase may experience buoyancy (Prinz and Romero, 1993), i.e. rapid spatial separation of the alloy components may occur through sedimentation or flotation.

Superimposed on this is the fact that during alloy solidification, the melt is subject to temperature gradients. These gradients can be responsible for several additional complex and intriguing phenomena. In a gravity field, they lead in most cases to buoyancy-driven convective flows and regardless of the presence of gravity (i.e. also under microgravity conditions) they produce a Marangoni stress at the interface between the two liquids.

These stresses, in turn, are responsible for important phenomena known as wetting and coalescence prevention. The noncoalescence of drops due to the Marangoni convection was observed for the first time in space. Following this observation a number of theoretical, numerical and experimental studies was initiated on the ground that have shown that both noncoalescence and nonwetting are caused by Marangoni flows that entrain surrounding fluid in the gap between two drops (or one drop and a solid surface) forming an immiscible fluid film that prevents intimate contacts between molecules that belong to the adjacent surfaces (see, e.g., Dell'Aversana et al., 1996; Dell'Aversana and Neitzel, 1998, 2004; Monti et al., 1998d, 2000a; Neitzel and Dell'Aversana, 2002; Savino et al., 2003). These mechanisms are expected to play a very important role in the aforementioned dynamics of solidifying alloys with a minority phase dispersed in an immiscible external liquid matrix.

Marangoni stresses are also responsible for another important effect, the so-called Marangoni migration: due to the surface Marangoni stress distribution induced by thermal effects, free liquid drops surrounded by an immiscible liquid migrate towards the hot region. For this reason often the anticipated structure of the uniformly distributed intrusions of the minority phase in the external matrix after solidification does not occur even if the experiments are carried out under microgravity conditions. Many authors have studied this aspect in detail in the case of immiscible liquids (see, e.g., Young et al., 1959; Subramanian and Balasubramaniam, 2001; Subramanian et al., 2001).

Despite the aforementioned different worthy contributions to the understanding of the behavior and the dynamics of the liquid–liquid systems as a model of industrial solidification processes, there are still many aspects that need further investigation. The basic information gleaned from these future studies will help scientists to develop better models to predict the properties of metals and metal alloys. Those, in turn, will lead toward ways to improve metal processing increasing its efficiency, reducing its cost, or perhaps creating new alloys with interesting properties.

Along these lines, the present chapter deals with a number of curious and intriguing phenomena that occur when liquid droplets are dispersed in an external immiscible or partially miscible matrix. A deductive approach is followed with systems of growing complexity being treated as the book progresses.

Many of these phenomena have been unmasked by recent (last decade) experimentation carried out in the microgravity environment and are still poorly known or ignored by professionals in the metallurgy and the foundry field. In the spirit of the book, relevant numerical methods are introduced for the accurate simulation (and better understanding) of these effects and numerical results are used to provide insights into the physics of problems, where direct experimental visualization is not possible.

3.2. Coalescence and wetting prevention by Marangoni effect

Research studies on wetting and coalescence prevention by the Marangoni flows originated by observations made during the Spacelab mission D2 (1992). The astronaut (payload specialist) working on the Fluid Physics Module (FPM) and the PI (Professor R. Monti, University of Naples, Italy) on the ground observed that it was difficult to create a liquid bridge by putting in contact two drops sitting on two circular disks before isothermal conditions were reached. Following this observation a number of theoretical, numerical and experimental studies were initiated (Fukai et al., 1993, 1995; Anderson and Davis, 1995; Braun et al., 1995).

Till date the analyses have been mainly concerned with the unusual behavior of noncoalescing liquid drops (see, e.g., Monti and Savino, 1996c, 1997; Savino and

Monti, 1997; Dell'Aversana and Neitzel, 1998), as well as with the intriguing phenomenon of a drop of liquid that does or does not wet a solid surface, according to the temperature difference between the solid and the drop (see, e.g., Monti et al., 1998c).

In order to investigate these topics, some model paradigm configurations have been introduced. Such configurations (with hemispherical drops sustained by cylindrical disks, see Fig. 3.2) offer significant advantages to the investigators: the disks can be translated along the vertical symmetry axis changing the relative distance between the hanging and the sitting drops or between a drop and the underlying flat surface (Fig. 3.2).

In the first case, when two drops are brought into contact, coalescence can be inhibited if a sufficiently large temperature difference is imposed between the supporting disks (Fig. 3.3). In the second situation, if the pendant drop and the solid surface are at the same temperature, or if the wall is warmer than the drop, the liquid spreads over the surface (Fig. 3.4a). On the contrary, if the drop is heated and/or the wall is cooled, then the drop does not wet the solid, even if pressed against the surface, but is deformed similarly to an elastic spherical balloon (Fig. 3.4b).

The behavior in Fig. 3.4b has an extensive application background in the field of immiscible alloys as will be elucidated in Section 3.4. Before starting to deal with this topic, however, in this section another problem is discussed, i.e. the case of a pendant drop approaching and touching a free surface of a miscible liquid at a different temperature.

The preliminary analysis of this case is necessary for the understanding of a number of phenomena that will be treated in the following on the basis of a somewhat deductive approach. Moreover, it is relevant for the aims and the scope of

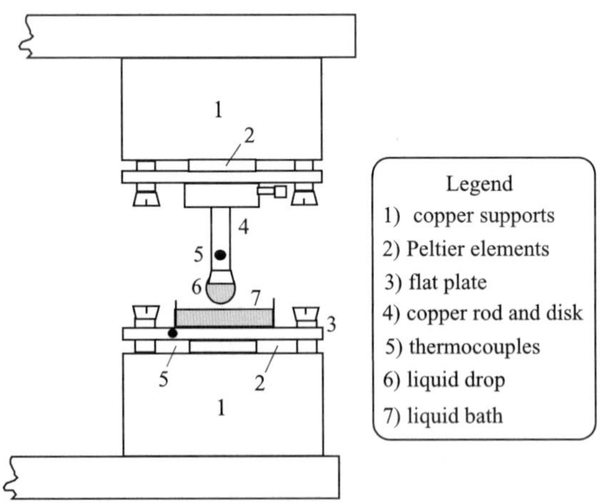

Fig. 3.2 Typical configuration used by researchers for experimental investigation of nonwetting phenomena.

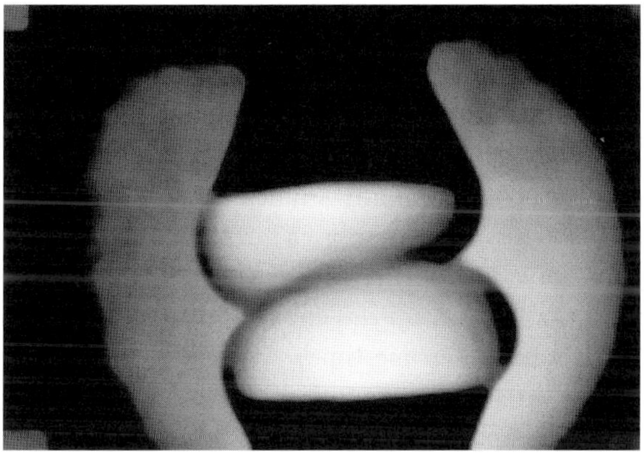

Fig. 3.3 Noncoalescence of two drops of silicone oil in presence of a temperature difference.

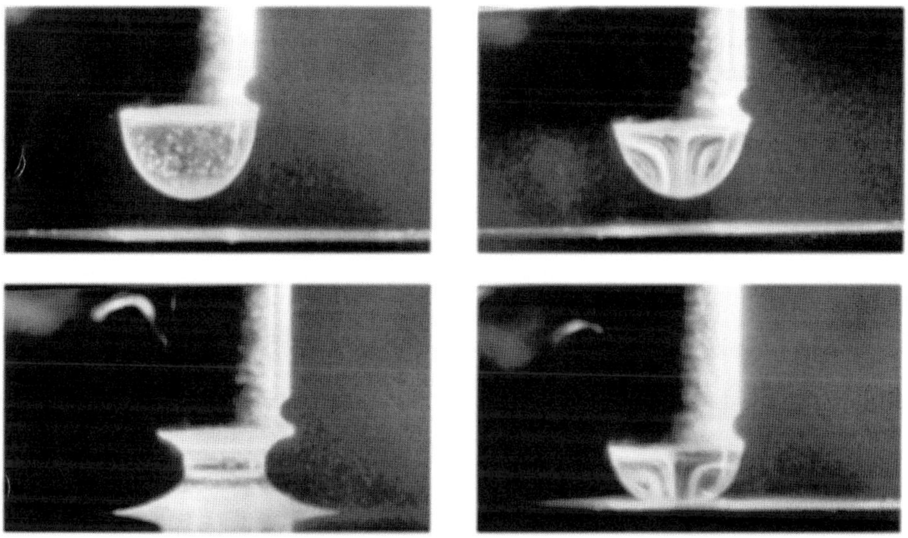

a) b)

Fig. 3.4 Wetting prevention by Marangoni effect – the drop is of silicone oil 5 cs, the radius of the copper disk sustaining the drop is $R = 1.5$ mm: (a) $\Delta T = 0$, (b) $\Delta T = 25$ K.

the present book since it often arises in the soldering processes, when a drop of a melted metal alloy is put in contact with the same liquid (Yost et al., 1993). The topic is also very important to understand some still poorly known phenomena dealing with free liquid-metal jets falling in a gas environment (and approaching an underlying pool of the same melt), which are used in casting and fabrication processes related to metal alloys. Often, due to fluid-dynamic instabilities (see Section 3.6), the free-falling jets are broken into large (macroscopic) drops and/or small drops of liquid metal may be ejected from the jet free surface. Due to the noncoalescence phenomena, the aforementioned droplets can float (see, e.g., Fig. 3.5) over the underlying liquid surface created by pouring the melt into a mould.

These curious behaviors were initially studied for some metal alloys (e.g., Sn–Pb or Sn–In at the eutectic compositions). The experiments pointed out that, if isothermal conditions prevail, the drop spreads over the liquid surface in times of the order of milliseconds. On the contrary, if the drop is hot or cold with respect to the liquid, wetting is prevented (for some time). Currently, however, these phenomena are often investigated with transparent liquids (e.g., silicone oils with different viscosities, see Table 3.1) since for these liquids, small temperature differences are sufficient and a direct observation is possible.

As pointed out earlier, if a pendant drop is moved towards a solid surface under isothermal conditions, the liquid spreads immediately over the solid, forming a liquid bridge. If this experimental sequence is repeated after establishing a temperature difference between the disk and the solid wall, no wetting occurs and the drop can be squeezed against the surface.

Similarly, if a pendant drop is formed on the circular disk and is moved towards the surface of a miscible liquid, by translating the disk downwards, when the disk is at the same temperature of the liquid pool no motion arises in the liquid phases; at these conditions, if the drop is put in contact with the bath surface, by moving the

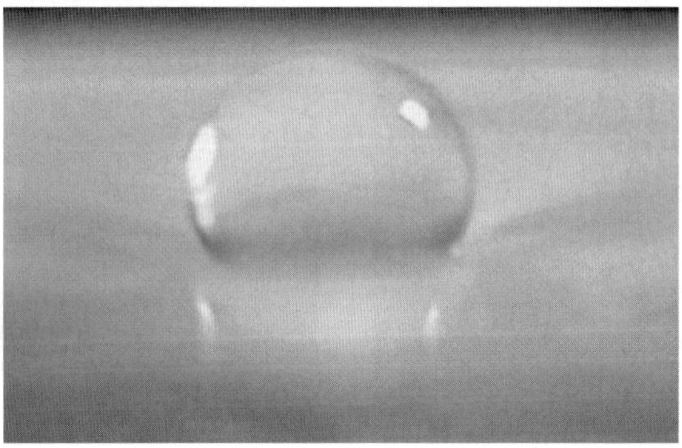

Fig. 3.5 Noncoalescing droplet floating on a pool of the same liquid under normal gravity.

Table 3.1 Physical properties of silicone oils.

Silicone oil (cs)	ρ (kg m^{-3})	ν (m^2 s^{-1})	α (m^2 s^{-1})	σ (dyne m^{-1})	σ_T (dyne m^{-1} K^{-1})
1	0.816×10^3	1.0×10^{-6}	5.98×10^{-8}	17.4×10^2	6
3	0.880×10^3	3.0×10^{-6}	6.40×10^{-8}	18.9×10^2	6
5	0.910×10^3	5.0×10^{-6}	6.83×10^{-8}	19.7×10^2	6
10	0.930×10^3	1.0×10^{-5}	7.43×10^{-8}	20.1×10^2	6

disk down, the drop is engulfed by the liquid surface in times of the order of milliseconds.

If this experimental sequence is repeated after establishing a temperature difference between the disk and the liquid pool (by heating the pendant drop and/or by cooling the pool with respect to the ambient temperature), the engulfment is prevented even if the drop is pressed on the liquid surface and enters the liquid pool. Figure 3.6 shows a sequence of photographs illustrating the behavior of a drop of silicone oil 3 cs over a pool surface of the same liquid in the presence of a temperature difference $\Delta T = 5\,\mathrm{K}$. The disk supporting the drop is, in this case, warmer than the liquid bath.

In this case, Marangoni flows are present in both the liquid drop and the liquid bath, due to the surface-tension imbalance induced by the surface temperature differences, as shown in Fig. 3.7, where a light-cut technique has been adopted to show the motion of tracers in a symmetry section (see Section 2.3 for additional information about the genesis of these flows and related mathematical models). In particular, since the surface tension is a decreasing function of the temperature, the velocity along the drop surface is downward (opposite to the imposed temperature gradient), i.e. it is directed from the upper disk towards the symmetry axis. By continuity, the flow inside the drop, close to the symmetry axis, is directed from the contact region towards the upper disk.

In the liquid pool, since the region of contact is heated by the drop, a temperature gradient is established in the radial direction on the free horizontal surface of the bath and hence the liquid is convected from the contact region towards the peripheral regions. However, since the contact region is heated by a few degrees centigrade, and the radial temperature gradient established is small compared to the ΔT acting along the drop surface, the velocities in the liquid bath tend to be small.

Precise laboratory experiments can be performed to evaluate the minimum value of the temperature difference (referred to as the critical temperature difference, ΔT_c) that prevents the engulfment.

The experimental sequence, in principle, is the following: a relatively large temperature difference is imposed, under this condition the drop can be pressed against the surface until the disk on which the drop is pending is at the same level of the horizontal pool surface; the drop is maintained under this position and the temperature difference is progressively reduced (with a sufficiently small temperature ramp, in order to minimize unsteady effects); when the temperature difference reaches a critical value (ΔT_c), the drop is engulfed.

3.3. A fluid dynamic model of coalescence prevention

The aforementioned intriguing noncoalescence behaviors can be explained according to a numerical model based on the assumption that a thin immiscible fluid film is formed between the contacting surfaces (liquid–solid in the case of drop/solid

a) d) b) e) c) f)

Fig. 3.6 Photographs illustrating the behavior of a liquid drop approaching a liquid pool at different temperature. (a): $t=0$, (b): $t=1$ s, (c): $t=3$ s, (d): $t=4$ s, submerging drop; (e): $t=6$ s, (f): $t=7$ s, withdrawal of the drop. The liquid in the drop and in the pool is silicone oil 3 cs, the radius of the copper disk sustaining the drop is $R=1.5$ mm, the drop is warmer than the pool and $\Delta T=5$ K (Monti et al., 1998d).

configuration, liquid–liquid in the case of drop/liquid pool). This immiscible sheet is due to the fact that, under certain conditions, a Marangoni flow is generated convecting ambient air between the surfaces. This effect forms an air gap that prevents the molecules belonging to the opposite surfaces from contacting each other.

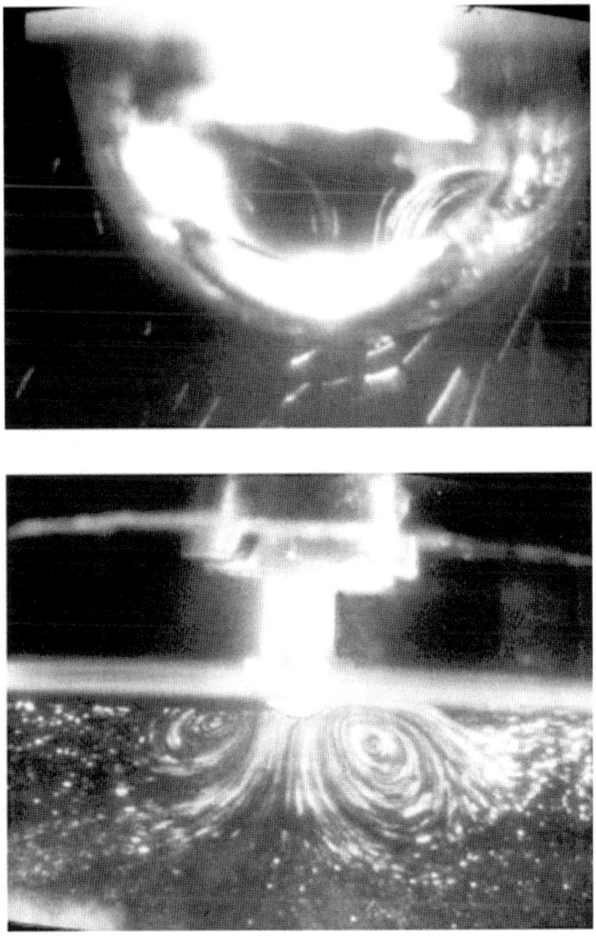

a)

b)

Fig. 3.7 Photographs illustrating the motion of tracers in the liquid drop and in the liquid pool with a light-cut technique. The liquid in the drop and in the pool is silicone oil 3 cs, the radius of the copper disk sustaining the drop is $R = 1.5\,\mathrm{mm}$, the drop is warmer than the pool and is completely immersed in it, $\Delta T = 5\,\mathrm{K}$.

Along these lines, the present section introduces numerical methods able to elucidate the experimental observations both qualitatively and quantitatively.

The model of the separating air film, for the drop-bath system that prevents wetting, can be summarized as follows.

If the hanging drop is hot and the bath is cold, then the liquid along the drop surface is directed from the upper disk towards the symmetry axis, dragging the nearby external air with it by viscous forces. Thus the ambient air is entrained between the two surfaces and this entrainment effect is assumed to be responsible for the presence of the air film. For wetting to occur, this thin air lubrication channel must be removed or must become sufficiently thin so that the liquid at the bath sides can come into molecular contact.

In practice, in the bath the Marangoni flow along the contact region is instead directed from the symmetry axis towards the horizontal free surface so that it opposes the aforementioned entrainment effect (the air is convected outside the film); however, because of the smaller velocity along the liquid bath surface compared to the one along the drop surface (pointed out in Section 3.2), the entrainment effect driven by the droplet surface-velocity distribution prevails.

Thus the pressure in the air film changes from the ambient value at the horizontal liquid surface to a maximum value in correspondence of the axis of symmetry.

According to the present model, this pressure increase is necessary to balance two different contributions (to balance the droplet weight and to deform the pool surface):

$$\Delta p_{\text{air film}} = \frac{\rho g}{R} + \frac{2\sigma}{R} \tag{3.1}$$

where $\rho g/R$ and $2\sigma/R$ represent the hydrostatic and the deformation contributions, respectively (ρ and σ are the liquid density and surface-tension, respectively).

In light of the above arguments, the numerical simulation of these aspects must include the solution of the flow and temperature fields in the liquid drop and in the liquid bath and the evaluation of the velocity and pressure distributions in the air film between the drop and the bath.

The shape of the drop can be assumed to be hemispherical (this has been proved to be a good approximation in the case of sizes of a few millimeters and it is supported by the experimental visualization of the surface shapes). The surface of the bath can be assumed to be planar and horizontal except in the contact region where a spherical shape having almost the same radius of the drop is considered (Fig. 3.8). The air film is assumed to have constant thickness (h) throughout its angular extension.

The equations governing the temperature and the velocity fields in the liquid drop, in the liquid bath and in the air film are the continuity, the Navier–Stokes and energy equations:

$$\underline{\nabla} \cdot \underline{V} = 0 \tag{3.2}$$

$$\frac{\partial \underline{V}}{\partial t} + \underline{V} \cdot \nabla \underline{V} + \frac{1}{\rho} \nabla p = \nu \nabla^2 \underline{V} + \beta_T (T - T_o) \underline{g} \tag{3.3}$$

$$\frac{\partial T}{\partial t} + \underline{V} \cdot \underline{\nabla T} = \alpha \nabla^2 T \tag{3.4}$$

where ν is the kinematic viscosity and α the thermal diffusivity, β_T is the thermal expansion coefficient, T_o is the reference (ambient) temperature and g is the gravity acceleration vector.

The boundary conditions for the drop are shown in Fig. 3.8 and include: the no-slip conditions for the velocity and prescribed constant temperatures on the upper support; a surface-momentum balance and energy conservation along the contact surface in the region of constant thickness h.

The boundary conditions for the liquid bath include: the Marangoni condition for the velocity and the adiabatic condition along the plane surface contacting the external ambient; the surface-momentum balance and energy conservation along the contact surface in the region of thickness h.

In Fig. 3.8, σ_T is the derivative of the surface tension with the temperature, n and s denote the directions orthogonal and tangential to the drop surface, u and v are the velocity components along r and θ; the subscripts (1), (2) and (a) refer to the drop, bath and air phases, respectively.

The Eqs. (3.2–3.4) rewritten in polar coordinates for the two-dimensional axi-symmetric case and in a dimensionless form for each phase (using the radius R, α/R,

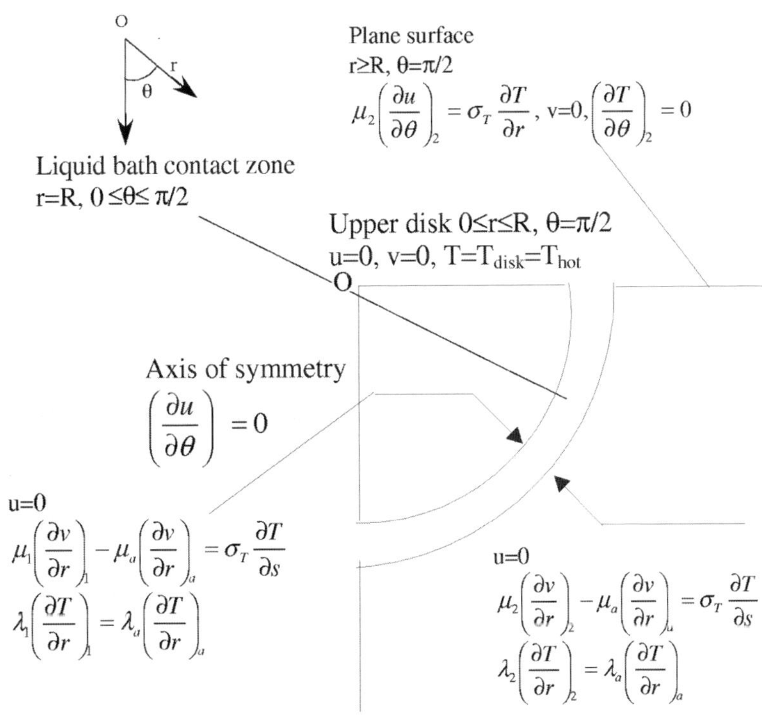

Fig. 3.8 Drop/pool system – configuration and boundary conditions.

R^2/α, $\rho\alpha^2/R^2$, to scale the lengths, the velocity, the time and the pressure, respectively) read (Monti et al., 1998d):

$$\frac{\partial(r^2u)}{\partial r}+\frac{r}{\sin\theta}\frac{\partial(v\sin\theta)}{\partial\theta}=0 \tag{3.5a}$$

$$\begin{aligned}\frac{\partial u}{\partial t}+u\frac{\partial u}{\partial r}+v\left(\frac{1}{r}\frac{\partial u}{\partial\theta}-\frac{v}{r}\right)=\\ -\frac{\partial p}{\partial r}+Pr\left(\nabla^2u-2\frac{u}{r^2}-\frac{2}{r^2\sin\theta}\frac{\partial(v\sin\theta)}{\partial\theta}\right)+Pr\cdot Ra\,T\cos\theta\end{aligned} \tag{3.5b}$$

$$\frac{\partial v}{\partial t}+u\frac{\partial v}{\partial r}+v\left(\frac{1}{r}\frac{\partial v}{\partial\theta}+\frac{u}{r}\right)=-\frac{1}{r}\frac{\partial p}{\partial\theta}+Pr\left(\nabla^2v+\frac{2}{r^2}\frac{\partial u}{\partial\theta}-\frac{v}{r^2\sin^2\theta}\right)+Pr\cdot Ra\,T\sin\theta \tag{3.5c}$$

$$\frac{\partial T}{\partial t}+u\frac{\partial T}{\partial r}+v\frac{1}{r}\frac{\partial T}{\partial\theta}=\nabla^2T \tag{3.5d}$$

where

$$\nabla^2=\frac{1}{r^2}\frac{\partial}{\partial r}\left(r^2\frac{\partial}{\partial r}\right)+\frac{1}{r^2\sin\theta}\frac{\partial}{\partial\theta}\left(\sin\theta\frac{\partial}{\partial\theta}\right)$$

and the temperature, measured with respect to the ambient temperature, is scaled by the temperature difference imposed between the drop and the bath (ΔT). The Prandtl number is $Pr=\nu/\alpha$ and the Rayleigh number is defined in terms of the drop radius by $Ra=g\beta_T\Delta TR^3/\nu\alpha$. Another characteristic nondimensional parameter is introduced by the dimensionless form of the boundary conditions along the liquid–gas interface: the Marangoni number, defined as $Ma=\sigma_T(\Delta T)R/\mu\alpha$ (see also Section 2.3).

The flow in the air film between the drop and the liquid surface can be modeled as a duct flow in an axisymmetric channel. The velocity profiles on the contact region interfaces obtained by the numerical solution of the flow inside the drop and inside the bath can be assigned as the boundary conditions for this channel (i.e. the flow in the air intermediate region is computed a posteriori after solving the flow fields in the drop and in the liquid bath).

The thickness h of the air film separating the liquid drop from the other surface must be considered as an input data for the numerical model: its order of magnitude is usually provided by experimental investigation: (a) a direct visualization of the air gap by means of a background illumination; (b) an interferometric technique that provides interference fringes related to the change of the film thickness in the contact region (the detailed discussion of these techniques is out of the scope of the present book, see Dell'Aversana et al. (1997) for additional details).

For instance, the experimentally detected thickness of the air film for the conditions shown in Figs. 3.6 and 3.7 is $h = 3\,\mu\text{m}$.

Equation (3.5) for the different phases can be solved according to the SMAC method described in Section 2.8 based on the introduction of a nonphysical pseudopressure. The distribution of the physical pressure in the air film requires additional solution of Eq. (2.78).

As an example, the next sections describe some typical applications.

3.3.1 *Droplet warmer than the liquid pool*

Figure 3.9 show the computed isotherms (Fig. 3.9a) and the streamlines (Fig. 3.9b) in the case of a drop of silicone oil 3 cs, a liquid bath of the same viscosity silicone oil and a temperature difference $\Delta T = 5\,\text{K}$ (disk warmer than the ambient).

The streamlines in the drop and in the liquid bath show the presence of a clockwise vortex cell. In particular, as previously shown by experimental results (Fig. 3.7a), since the surface-tension gradient is opposite to the imposed temperature gradient, the flow along the drop surface is directed from the hot support towards the symmetry axis.

In the liquid bath, the temperature gradient acts in the radial direction and the flow on the free plane surface is directed from the contact zone towards the peripheral regions so that the fluid on the liquid bath contact surface moves in the opposite direction with respect to the flow on the drop surface.

The core of the vortex cell in the bath is shifted in the radial direction with respect to the cell in the drop and there is a very good agreement between the experimental flow configuration shown in Fig. 3.7b and the numerical one in Fig. 3.9b.

The corresponding pressure contours in the air film are shown in Fig. 3.10. For better illustration of the pressure field, in this figure different scales have been used along the major extension of the channel (the θ-direction) and its thickness (the scale along the radial direction is greatly enlarged). The tangential viscous stresses in the air film are responsible for the change of the pressure, which gradually increases from the ambient value at the edge of the contact region to the maximum value on the symmetry axis (stagnation point).

The hydrostatic pressure in the liquid pool and the pressure increase in the air film versus the azimuthal coordinate (θ) are reported in Fig. 3.11. It is shown that, for a film thickness $h = 3\,\mu\text{m}$ (corresponding to experimental observations), the maximum pressure difference on the axis ($\Delta p \simeq 400\,\text{dyne cm}^{-2}$) balances, to a good approximation, the pressure necessary to keep the drop submerged in the liquid pool, i.e. the pressure jump necessary to deform the bath surface ($2\sigma/R = 266\,\text{dyne cm}^{-2}$) plus the hydrostatic pressure ($\Delta p = 120\,\text{dyne cm}^{-2}$). This procedure provides mathematical validation for the theory proposed herein, explaining the

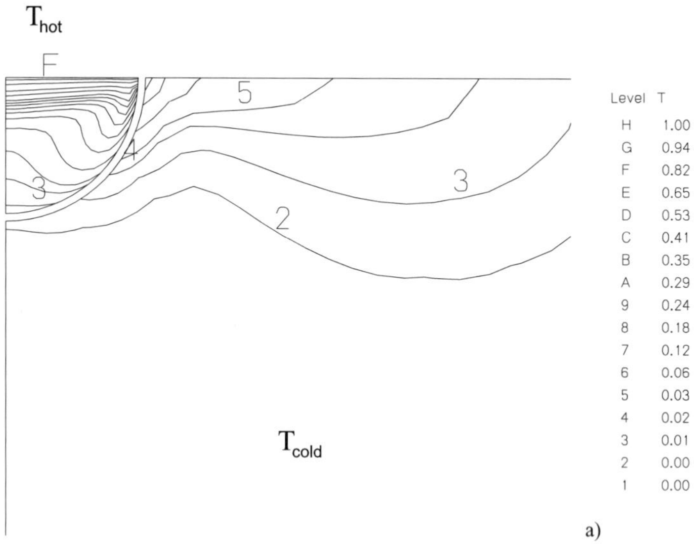

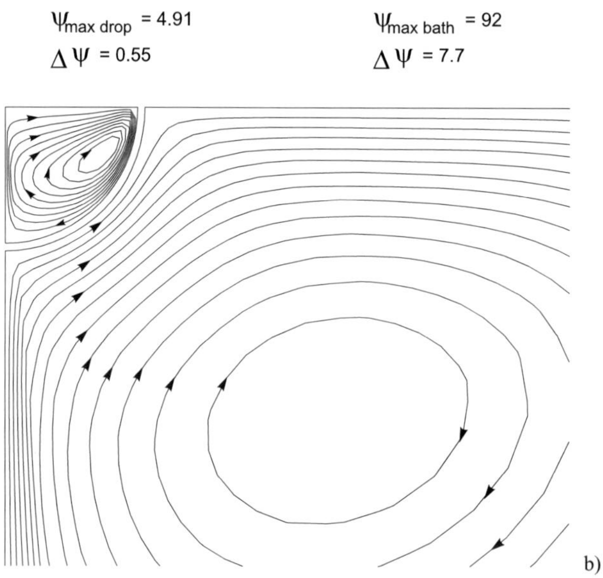

Fig. 3.9 Configuration 3 cs – 3 cs – $\Delta T = 5$ K: (a) computed isotherms, (b) streamlines.

nonwetting phenomena. Simulations carried out for conditions corresponding to the experimentally determined critical temperature difference and the thickness of the air film, in fact, prove that pressure equilibrium is attained for such experimental conditions.

Similar mathematical models based on the existence of an air film separating the liquid–liquid or liquid–solid surfaces and on the corresponding entrainment effect

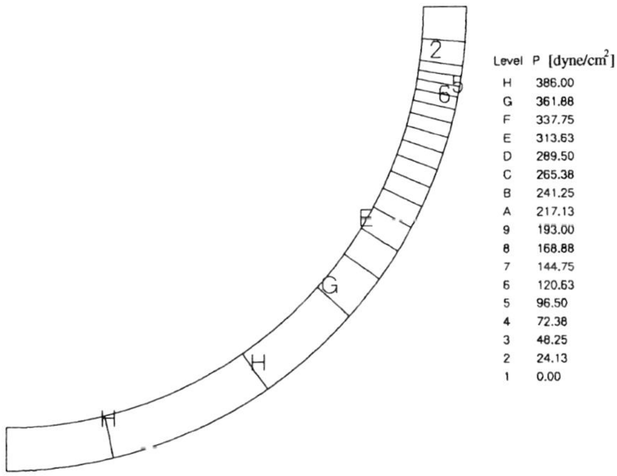

Fig. 3.10 Pressure contours in the air film for the same conditions of Figs. 3.6 and 3.7 ($h = 3\ \mu$m).

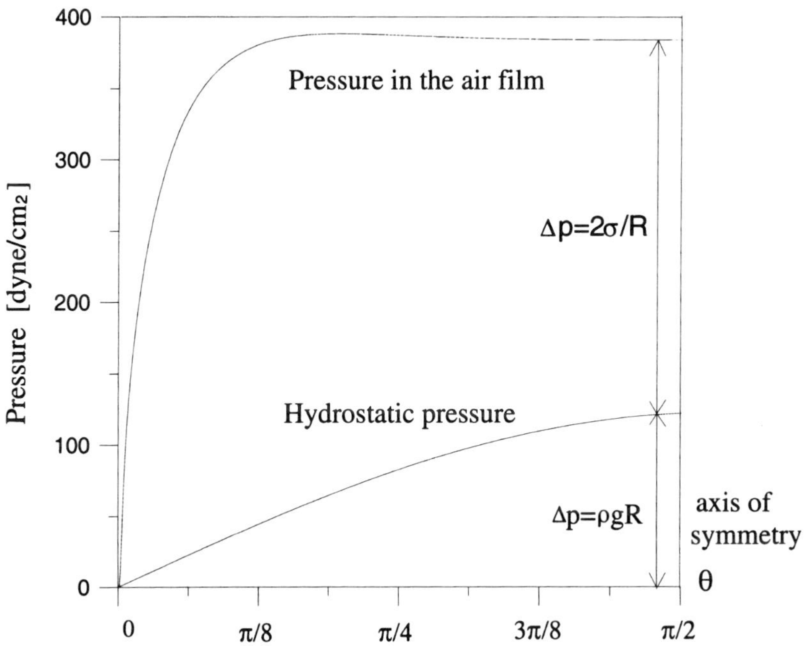

Fig. 3.11 Pressure distribution in the air film for the same conditions of Figs. 3.6 and 3.7 ($h = 3\ \mu$m).

driven by surface-tension forces, can also be used to elucidate the behaviors of drop–drop or drop–solid configurations.

In the case of a system drop/solid-surface for instance, the explanation is even simpler (Fig. 3.12). In this case, in fact, the nonslip conditions prevail at the solid

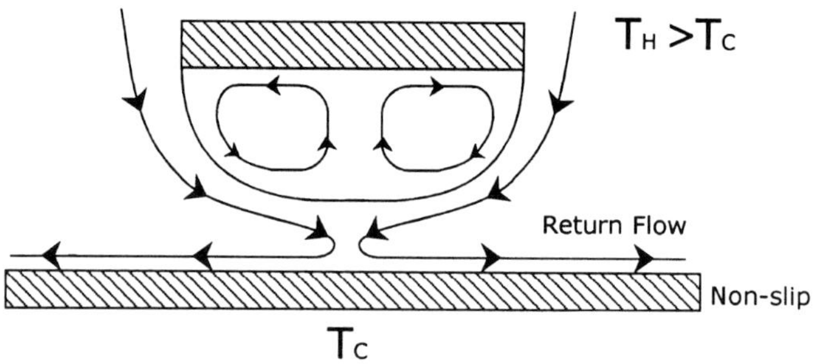

Fig. 3.12 Sketch of the mechanism leading to pressure increase for the system drop/wall.

surface. The entrainment effect is driven by the velocity distribution on the drop surface, whereas air is transported outside only by continuity (return flow). This leads to an increase of pressure in the region between the drop and the wall and the efficiency of the pressure-increase mechanism is even better with respect to the case of the drop/pool system (the removal action of air driven by the lower surface being absent, the withdrawal action is weakened with respect to the drop/pool system).

The foregoing discussion is a good example of the capabilities of the proposed fluid-dynamic model and the associated numerical method. Among other things, the next section demonstrates how they can be used to estimate the critical temperature difference when it is unknown.

3.3.2 *Reverse conditions*

Figure 3.13 shows the computed isotherms (Fig. 3.13a) and streamlines (Fig. 3.13b) for the same liquids and the same temperature difference as Fig. 3.6 ($\Delta T = 5$ K), when the temperature of the disk supporting the drop is smaller than the temperature of the liquid pool (reverse conditions).

The experimental evidence in this case shows a behavior similar to that observed when the temperature of the disk is higher than the ambient temperature (i.e. no wetting and possible full immersion of the liquid drop in the pool).

According to the streamlines, the liquid bath is dominated by the presence of a counterclockwise vortex cell (the computed stream function in the liquid pool is negative, Fig. 3.13b). The temperature distribution shows that the free surface of the bath is characterized by a strong temperature gradient, while the temperature gradient on the drop surface is almost negligible. This implies that the motion is mainly governed by the Marangoni effect on the bath surface that convects cold fluid from the peripheral regions towards the contact zone. Therefore, for this case, the flow on the contact zone from the liquid pool side is responsible for the entrainment action in the air film, while the flow on the drop surface is negligible.

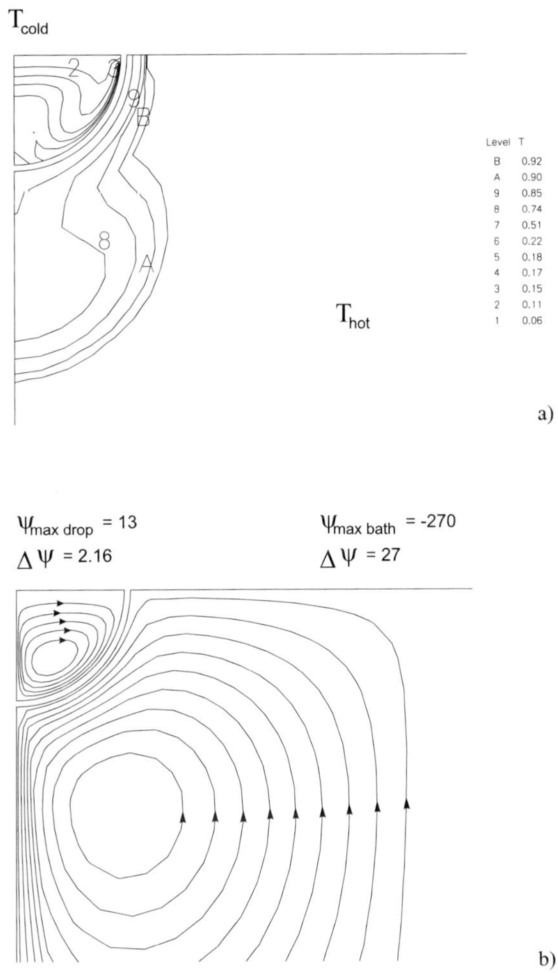

Fig. 3.13 Computed isotherms (a) and streamlines (b) for the configuration 3 cs – 3 cs, $\Delta T = 5\,\text{K}$ in the case of drop colder than the liquid pool (reverse conditions).

Owing to the strong temperature gradients on the free surface of the bath, the entrainment action is stronger compared to the one that sets in the case of Figs. 3.6 and 3.7 (when the drop is warmer than the ambient air and the entrainment of air is induced by the flow on the drop surface). This is confirmed by the computed pressure distribution in the air film (Fig. 3.14). Assuming the same thickness and the same temperature difference ΔT considered in Fig. 3.11, the pressure difference in the case of a drop cooled is, in fact, $\Delta p = 640\,\text{dyne}\,\text{cm}^{-2}$ and this value is larger than the value computed when the drop is warmer than the liquid pool.

According to these results, the critical temperature difference, in this case, is expected to be smaller than in the case in which the drop is warmer than the ambient. This numerical finding is in agreement with the experimental results. In fact, the

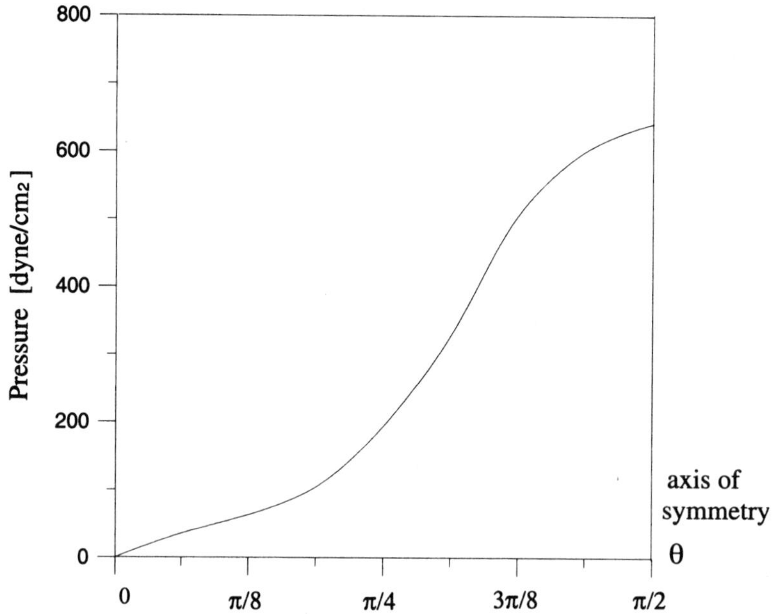

Fig. 3.14 Pressure distribution in the air film for the same conditions of Fig. 3.13.

measured critical temperature difference, for a configuration with a drop of silicone oil 3 cs and a pool of the same liquid is $\Delta T = 5$ K for a drop warmer than the ambient and $\Delta T = 3.1$ K for a drop colder than the ambient.

3.4. Free droplets in liquid matrices: typical phenomena

This subject is of great relevance in the field of materials science in microgravity where, as illustrated in the introduction of this chapter, a number of different phenomena are not well understood.

Within this context of particular interest, for the reasons already illustrated in Section 1.5.1, is the case of the so-called "immiscible alloys." These alloys exhibit two immiscible liquid phases within a certain temperature range.

In particular, for a monotectic system a homogeneous solution exists only at temperatures above the so-called critical temperature T_c. When $T < T_c$ the melt starts to decompose into two liquid phases. Above the so-called critical temperature T_c both components are completely miscible in the liquid state. Below this temperature two liquids of different composition and density coexist in equilibrium.

As anticipated, these special alloys (e.g., In–Al, Al–Pb, etc.) can be fully exploited only if a uniform finely dispersed minority phase can be preserved in the solid state (e.g., materials for self-lubricating bearings with a soft dispersed phase in a high mechanical strength matrix; lead or indium dispersions in an aluminum matrix are the most promising).

In the Earth's gravitational field, the usual differences in density cause rapid spatial separation of the alloy components through sedimentation or flotation; in practice the different densities of these liquid phases leads to the formation of two layers (see, e.g., Fig. 3.15). This prevents the homogeneous distribution of particles in the matrix using simple and inexpensive casting processes. Therefore, the industrial exploitation of such alloys has been limited so far.

To avoid these limitations, some investigators carried out experiments in the full weightless conditions provided by space platforms. In the early phase of microgravity-based research, it was suggested, in fact, that it should be possible to easily solidify materials containing uniform dispersions of the second-phase gas bubbles, liquid drops, or solid particles, because of the absence of sedimentation. Included were the monotectic and the peritectic systems where the second phase tends to separate by sedimentation on Earth. Although microgravity sometimes yielded improvements in such dispersions, this was usually not true. Materials science experiments dealing with metallic alloys processing in microgravity resulted in a number of disappointing results (see Fig. 3.16 and Ratke et al., 1996). A number of attempts to obtain, under such conditions, a homogeneous distribution in various alloys and model materials also unexpectedly yielded separated phases and sometimes greater phase separation occurred in microgravity than on Earth.

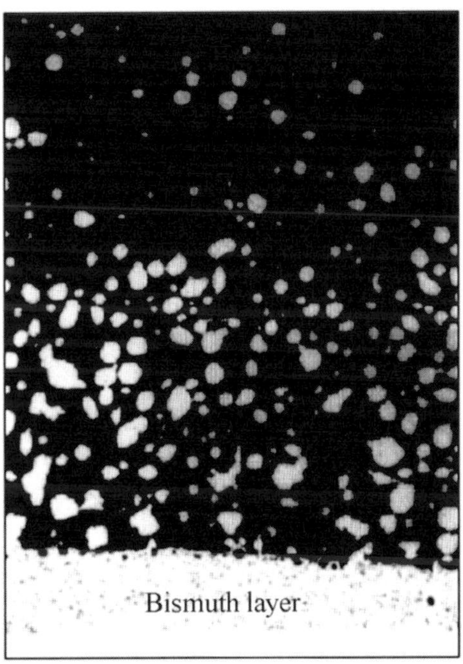

Fig. 3.15 Metallographic section of an AlSiBi-alloy rapidly cooled from about 1000 °C to room temperature under normal gravity.

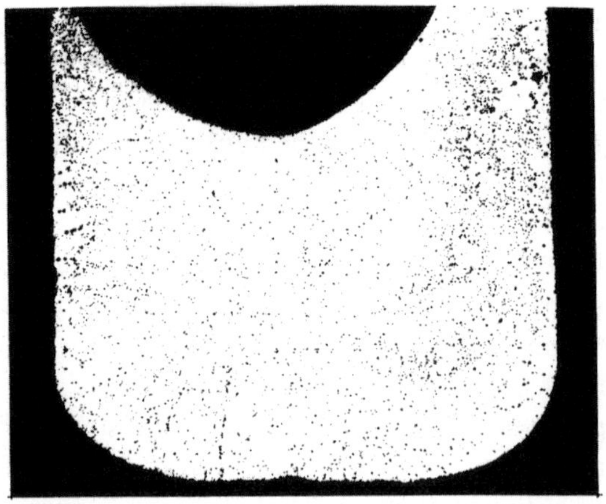

Fig. 1a : Sample 7, Al-Bi 7 vol.%, directionally solidified. 20 x.

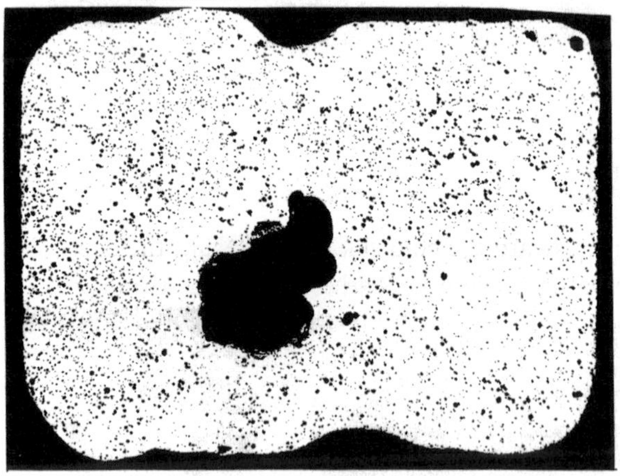

Fig. 1b: Sample 2, Al-Bi 7 vol.%, solidified in the isothermal region. 20 x.

Fig. 3.16 Examples of separation occurred during directional and radial solidification processes in microgravity (from H.U. Walter; ESA SP-219 1984, p. 47).

The anticipated structure of uniformly distributed intrusions of the minority phase in the external matrix after solidification was not found.

In many cases, the minority phase accumulated in the hot region during the solidification process. This means that the minority phase (initially uniformly dispersed in the majority phase) moved to the region that solidified later (one end of the cartridge, for directional solidification or the internal region when cooling takes place at the external walls, Fig. 3.16).

Failure to obtain uniform dispersions in space as well as on the ground is usually referred to as "the phase-separation problem." Sometimes it is also referred to as "coarsening of the dispersion" (where "phase separation" stands for "spatial" separation of the phases; the thermodynamic counterpart of the process, i.e. the origin of decomposition, in fact, is a well-known and essentially understood phenomenon since at least mid-1930s). In the light of the current knowledge of the physics of the microgravity environment, this may be attributed to several reasons; among them: (1) pushing of the second phase by the freezing interface, (2) sedimentation due to the presence of residual accelerations, and (3) surface-tension-driven Marangoni motion of the drops.

The so-called drop-pushing effect is basically a nonwetting phenomenon and will be discussed in the next section.

The genesis of the Marangoni migration is discussed in the following: the differences in surface tension caused by temperature variation (thermocapillarity) cause drop migration in microgravity. In the normal-gravity experiments, this contribution is nearly always overshadowed by buoyancy, except for very small drops subjected to relatively high temperature gradients, and except for drops as dense as the continuous phase. The surface tension at the interface between the drop and the surrounding liquid is typically a decreasing function of temperature, as illustrated in Section 2.3. As a result, the temperature gradient applied across the drop creates a surface tension-gradient at the interface. It is this latter gradient that causes the liquid adjacent to the interfaces to flow in the direction of increasing surface tension (i.e. toward the cold side). As a result, the drop propels itself toward the higher-temperature regions. These gradients relate directly to the local temperature gradients in the matrix during cooling and solidification and are therefore independent of gravity.

This subject was a curiosity when first studied by Young et al. (1959), but has received impetus from the advent of space-flight programs. This is because the related mechanisms (owing to the intrinsic nature of Marangoni stresses) are independent of gravity and tend to be as dominant in the reduced-gravity environment as gravity is on Earth.

Beginning with a few articles in the 1970s, some recent books have been entirely devoted to this topic (Subramanian and Balasubramanian, 2001; Subramanian et al., 2001).

Marangoni motion of drops and bubbles in a temperature gradient can lead to undesirable spatial separation of the phases both on Earth and in low-gravity environments.

If properly understood, however, thermocapillary effects could be used in a positive way to control multiphase fluid transport and phase separation. The ideal would be to minimize the sedimentation caused by (normal or residual) gravity by balancing it with the Marangoni migration effect. If the droplets reach a state of dynamic equilibrium, then stable dispersion in a solidifying matrix becomes attainable. This concept has been partially tested and new materials, for example an aluminum alloy matrix with evenly distributed inclusions of lead or bismuth, are being tested as candidates for advanced sliding bearings in car engines.

Further improvements along these lines, however, require the capability to predict the magnitude and the mutual interaction of both these effects (gravitational and Marangoni migration) over a wide range of conditions and situations, i.e. the introduction of a numerical method that can capture these behaviors is needed.

Before embarking on the description and design of such a method, the next sections describe some basic concepts and analytical solutions available in the literature. The fundamental dependencies and behaviors that can help the reader in understanding the phenomena under investigation are discussed. Moreover, a significant effort is being provided to illustrate the state-of-the-art of these topics.

3.4.1 *Solidification and drop pushing by Marangoni effect*

Experiments carried out on the ground with microscale facilities and model liquids have shown that if the ambient gas (air) surrounding the droplets is replaced by an immiscible liquid, the Marangoni effects at the liquid–liquid interface (e.g., silicone oil and fluorinert) can play the same role as in the liquid–gas systems, leading to suppression of coalescence or wetting.

All that is found for drops in air holds for drops submerged in an immiscible liquid matrix. The situation is more complex but the explanation is basically the same.

In view of these arguments, the modeling (and numerical treatment) of these phenomena falls into the same category of numerical techniques described in Section 3.3.

It should be emphasized, however, that while nonwetting phenomena in liquid–gas systems are of particular interest for the design of foundry reactors or in the soldering processes, the same phenomena in the case of immiscible liquid couples have extensive application background in problems directly related to the separation of immiscible alloys (see also Wilcox and Regel, 2001 and references therein). The findings on wetting and coalescence prevention elucidated in the earlier sections suggest, in fact, an additional explanation for the phase separation that occurs during an alloy solidification (see also Savino et al., 2001b).

Consider, for instance, two immiscible liquids (phases (1) and (2)) with one of the two phases (for instance, phase (2)) solidifying due to temperature changes.

Figure 3.17 (it is assumed that the solidification temperature of the minority phase is higher than the solidification temperature of the primary phase) shows how, owing to an entrainment effect similar to that elucidated in Section 3.3.1 (for the case of a liquid droplet approaching a solid surface), wetting between the drop and the advancing solid front could be prevented, leading to a pushing (due to pressure increase in the intermediate region) of the drops towards the hot zone.

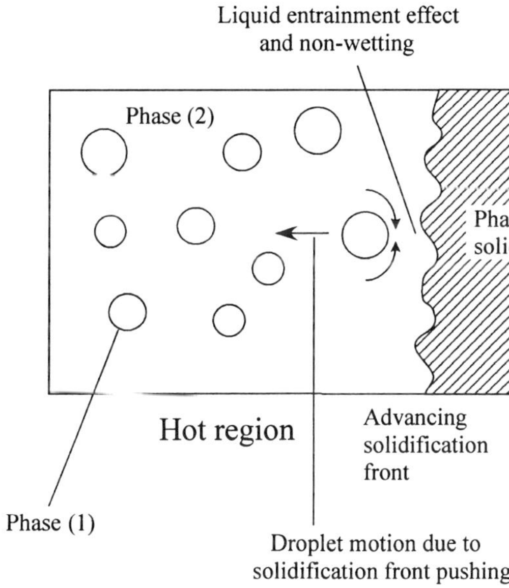

Fig. 3.17 Sketch of drop pushing due to solidification and nonwetting behavior.

In other words, the pressure distribution on the drop results in a force that pushes it far from the cold wall and might be the reason for the separation of the minority phase during the solidification of immiscible alloys, which tends to be clustered in the vicinity of the hot side.

This phenomenon (still poorly known or ignored by most experts in the metallurgy field) exhibits some interesting analogies with the mechanism (already known) with which *solid* particles (or very viscous drops) can be repelled by advancing solid fronts. As recognized by Shangguan et al. (1992), Feuillebois (1993), the repulsion of a small particle by a freezing front can occur under particular conditions (see also Azouni et al., 1990 and Stefanescu et al., 1990, for experimental data).

The related mechanism involves the increase of pressure in the liquid between the particle and the front. Such an increase is driven by the lubrication forces when the distance between the particle and the solid becomes sufficiently small.

This mechanism is absent or very weak in the case of a liquid drop (unless it has a very large viscosity) since the "mobility" of its surface allows the fluid between its surface and the solidification front to be squeezed outward with much less resistance than for the rigid-particle case. As previously elucidated, however, if the immiscible liquids exhibit a Marangoni effect at the interface, the aforementioned entrainment effect driving fluid in the intermediate region determines a pressure increase therein. This pressure increase can also be stronger than that related to solid particles or very viscous drops approaching the solidification front. The drop will be repelled even if it is characterized by a very low viscosity. Since the magnitude of the surface Marangoni flow is a decreasing function of the drop viscosity, the lower the viscosity, the larger will be the repulsive force. The available numerical computations (Monti et al., 2000a) have shown, in particular that this force is an increasing function of the drop diameter and an inverse function of the distance between the drop and the solid surface as well as an inverse function of its viscosity, as mentioned above.

3.4.2 *Droplet sedimentation*

Since the study of the motion of liquid droplets has many applications in industry, numerous studies on spherical liquid droplets are found in the literature. One well-known text in this area is a book published by Clift et al. (1978) where an astonishing amount of information dealing with the fluid mechanics of these aspects as well as with heat- and mass-transfer processes can be found (mathematical models, analytical and pseudoempirical relationships about slow viscous flow past sedimenting spheres, fluid spheres at high Reynolds number, ellipsoidal fluid particles, deformed fluid particles of large size, wall effects, formation and breakup of fluid particles, particle dissolution, etc.). In particular, the analytical relationships for the terminal velocity of a sedimenting fluid sphere are reported for both cases of low and high values of the Reynolds number.

The terminal velocity of a sphere with radius R (phase (1)) in the creeping flow (convective transport is neglected in the momentum equation and the Reynolds number is sufficiently low that there is no wake at the rear of the fluid spherical particle) is given by:

$$U_T = \frac{2}{3}\frac{gR^2(\rho_1 - \rho_2)}{\mu_2}\left(\frac{1 + (\mu_1/\mu_2)}{2 + 3(\mu_1/\mu_2)}\right) \tag{3.6}$$

where μ_1 and μ_2 are the viscosity of the drop and of the external fluid, respectively (it was obtained by Clift et al. (1978), using a simplified version of the Navier–Stokes equations and equating the total drag to the net gravity force $4\pi R^3 \Delta\rho g/3$). Interesting information concerning the development of shape instabilities for the drop undergoing gravity-induced motion is reported in the landmark study of Koh and Leal (1989) in the limit as $Re \rightarrow 0$.

For large values of the Reynolds number ($4 < Re < 100$) a quite reliable estimate of the terminal velocity can be obtained by solving the following pseudoempirical equations

$$C_D = 3.05\frac{\left[783(\mu_1/\mu_2)^2 + 2142(\mu_1/\mu_2) + 1080\right]}{(60 + 29(\mu_1/\mu_2))(4 + 3(\mu_1/\mu_2))}Re^{-0.74} \tag{3.7a}$$

$$C_D\frac{1}{2}\rho Re^2\frac{\pi}{4} = g\frac{4}{3}\pi R^3\Delta\rho \tag{3.7b}$$

where $Re = 2U_T R/\nu_2$.

An important limitation of the book of Clift et al. (1978) however, is that only those phenomena in which particle–particle interactions are of negligible impact are treated. Hence a direct application of the book is limited to single-particle systems or very dilute suspensions.

3.4.3 *Marangoni migration*

There is a vast body of literature dealing with the subject of a gravity-driven motion of drops and related heat- and mass-transfer phenomena. However, most of the work on the motion of drops due to interfacial tension gradients is relatively recent.

It is known (Young et al., 1959) that in the limit of a zero Marangoni number (i.e. with negligible convective-transport effects) the droplet velocity U within an unbounded fluid medium in the laboratory coordinate frame is related to the uniform temperature gradient $\partial T/\partial x$ by

$$U_S = -\left[\frac{2R\sigma_T|_{\mathrm{drop}}}{\mu_2(\lambda_1/\lambda_2 + 2)(3\mu_1/\mu_2 + 2)}\right]\frac{\partial T}{\partial x} \tag{3.8}$$

In this expression, $\sigma_{T,\mathrm{drop}}$ represents the variation of interfacial tension with temperature at some reference temperature (e.g., that at the center of the drop), μ_2 and λ_2 are the viscosity and the thermal conductivity of the external fluid respectively. Equation (3.8) is valid as long as the drop remains sufficiently small that convective transport of heat and momentum can be neglected compared to molecular transport of these quantities, i.e. for small values of the Reynolds number, $2\rho_2 U_S R/\mu_2$, and of the Marangoni number, $Ma = Re \cdot Pr = 2U_S R/\alpha_2$, where ρ_2 and α_2 denote the density and thermal diffusivity of the external fluid, respectively.

Similar analyses have been published in the limit when the Reynolds (*Re*) and Marangoni (*Ma*) numbers are both zero. Most of these studies were based on the so-called quasistationary approximation, where the velocity and the temperature fields are considered to be fully established at every moment of time, while the inertia and the convective effects are neglected. Under this approximation, the temperature field can be found independently of the flow field and this greatly facilitates the solution of problems.

An analysis dealing with the opposite case of very high values of *Re* and *Ma* has been presented recently by Balasubramanian and Subramanian (2000). They analyzed the problem of the steady migration of a drop in a continuous phase when subjected to a temperature gradient under conditions such that the inertial terms in the momentum equation and the convective-transport terms in the energy equation dominate over the corresponding molecular-transport terms ($Ma \to \infty$ and $Re \to \infty$ for which they are able to obtain analytical solutions). The migration velocity of the drop was obtained by these authors in the framework of a potential-flow theory by equating the rate at which work is done by the thermocapillary stress to the rate of the viscous dissipation of energy; the method of matched asymptotic expansions was employed to solve the conjugate heat-transfer problem in the two phases.

In physical terms, they find that in the limit as $Ma \to \infty$ the velocity of a drop is proportional to the square of the temperature gradient and the cube of the radius of the drop:

$$U_S = \frac{4(\sigma_T)^2 R^3 h(\delta)}{\mu_2^2 \alpha_1 (2 + 3(\mu_1/\mu_2))^2 (1+\delta)^2} \left(\frac{\partial T}{\partial x}\right)^2 \tag{3.9}$$

where,

$$\delta = \sqrt{\frac{\alpha_1}{\alpha_2}\frac{\lambda_2}{\lambda_1}}$$

and ($h(\delta) = 5.68 \times 10^{-3}$ for $\delta = 0$; 6.42×10^{-3} for $\delta = 0.5$; 7.75×10^{-3} for $\delta = 5$; 8.26×10^{-3} for $\delta \to \infty$).

These results are important since they provide the functional dependencies as well as a quite reliable estimation of the velocity of migration.

However, it should be pointed out that all these studies considered nondeformable bubbles or drops with a fixed spherical shape and/or were obtained for limiting cases ($Re, Ma \to 0$, $Re, \mathrm{Ma} \to \infty$) of the Reynolds and the Marangoni numbers.

3.4.4 *Droplet interaction and coalescence*

In real applications fluid drops are not isolated. For this reason, the problem is made even more complicated and interesting by the possibility that the drops also collide and coalesce with each other as they move (these aspects often are referred to as collision and coagulation events). Moreover, their motion may be influenced by the presence of neighboring particles.

Owing to the experimental difficulties in investigating the fluid-dynamics of nontransparent liquids (metal alloys are opaque) a number of mathematical models and methods have appeared over the last few years for a numerical/theoretical analysis of these aspects when they are disjoint or partially combined.

For instance, conditions favorable for the occurrence of coalescence were initially investigated within the framework of the lubrication theory (e.g., Davis et al., 1989). The hydrodynamic force resisting the relative motion of two unequal drops was determined for the Stokes flow conditions and in the absence of surface Marangoni effect. The drops were assumed to be in near-contact and to have a sufficiently high interfacial tension that they remain spherical. The squeeze flow in the narrow gap between the drops was analyzed using the aforementioned lubrication theory. Depending on the ratio of drop viscosity to that of the continuous phase, and also on the ratio of the distance between the drops to their radii, different possible flow situations were pointed out, corresponding to nearly rigid drops, drops with partially mobile interfaces, and drops with fully mobile interfaces.

The shape of the droplets was assumed undeformable in these analyses. In reality, the related interaction can cause drop deformation even in the case of negligible convective transport and can also substantially alter the migration and sedimentation velocities (owing to deformation and "wake" effects).

Axisymmetric thermal-wake interaction of two bubbles in a uniform temperature gradient at large Reynolds and Marangoni numbers was studied by Balasubramanian and Subramanian (1999). Their analysis considered the bubbles moving in the direction of the temperature gradient and assumed to interact axisymmetrically via the influence of the thermal wake of the leading bubble on the trailing bubble. This analysis is very important since it was proven that the thermal wake of the leading bubble can induce a nonmonotonic temperature field on the surface of the trailing bubble. The effective temperature gradient on the trailing bubble is weakened and hence its migration speed is reduced compared to the case when it is isolated.

Similar behaviors are also expected in the case of the interacting drops. The temperature gradient in the wake will be weaker than the applied gradient. The thermal wake field of the leading drop will wrap around the trailing drop having a significant impact on its motion.

In the aforementioned landmark analysis, shape deformation was neglected and results were obtained in the limit as $Ma \rightarrow \infty$ (flow potential theory).

Most of the available theoretical studies of the interaction of bubbles and droplets in the course of their Marangoni migration were performed under the assumption of nondeformable drops (zero capillary number).

The literature on the thermocapillary motion of deformable drops, in fact, is limited, in contrast to a somewhat related problem concerning the interaction of drops undergoing a buoyancy-driven motion. For the latter case, numerous recent studies revealed a rich variety of interaction patterns of deformable drops depending on the Bond number and the initial configuration of the system (see, e.g., Manga and Stone, 1993; Rother et al., 1997; Lavrenteva and Nir, 2003).

The deformation and the motion of interacting droplets was investigated by these authors in the framework of the boundary-integral techniques (with mesh adaptation and stabilization) applicable to the case of slow viscous motion (for the state-of-the-art see, e.g., Zinchenko et al., 1997, 1999; Bazhlekov et al., 2004).

As a recent example of such computations, the reader may consider the axisymmetric buoyancy-driven interaction of a leading drop and a smaller trailing drop treated by Davis (1999). He demonstrated that the trailing drop elongates considerably due to the hydrodynamic influence of the leading one and that later, depending on the governing parameters, the drops may either separate and return to a spherical shape or the trailing drop may be captured by the leading one (coalescence), or one of the drops may break up.

For the Marangoni motion, the effect of deformability was studied mostly by a perturbation technique assuming small deformations; some recent analyses within the framework of integral methods are due to Zhou and Davis (1996) and Rother et al. (2002).

From the foregoing it is evident that current understanding is primarily limited to the motion of single drops and bubbles, or at most two interacting drops and bubbles, which are often assumed to remain spherical (a limit which requires that interfacial tension forces are large compared to viscous and pressure forces). Moreover, all the studies dealing with the boundary-integral method and deformable drops are still limited to the case of very viscous flow ($Re \rightarrow 0$).

Initial progress on the simulation of many drops has been made by combining the theoretical results available for neighboring drops (discussed in the text above) with economical multipole techniques previously used in multiparticle conductivity (see e.g., Zinchenko, 1994). On such a philosophy are based the so-called "population methods" (see, e.g., Davis et al., 1993; Diefenbach et al., 1993 and the more recent contributions Wu et al., 2003; Zinchenko and Davis, 2003).

With these methods, drops (a large number, i.e. ensemble of droplets) are assumed to be in relative motion due to either gravitational sedimentation or thermal Marangoni migration. Possible collisions are predicted using a "trajectory analysis" to follow the relative motion of pairs of drops. The trajectory analysis, in turn, is based on governing equations formulated for conditions of small Reynolds number (negligible inertia) and on estimation of the sedimentation or (Marangoni) migration velocity provided by Eq. (3.6), Eq. (3.8) or similar analytical relationships.

When the drops become sufficiently close, they are assumed to interact with each other due to hydrodynamic disturbances. This hydrodynamic disturbance, as mentioned above, is modeled in the light of information obtained by means of a separate microscopic approach to the problem.

For instance, drops with low viscosity are assumed to become aligned and coalesce due to their shape deformations as observed and described by Manga and Stone (1993), whereas drops with modest viscosity become stretched and may break as a result of hydrodynamic interactions and drops with large viscosity tend to be swept around larger ones.

Therefore, these methods rely on the fundamental insights provided by earlier analyses (lubrication theory for spherical drops and boundary-integral methods for the deformable case) in the case of only two interacting drops. Such information is used by population methods to model from a macroscopic point of view, interaction of droplets at microscopic scale length (local interaction). For this reason, droplets pertaining to the initial distribution are categorized according to their size (drop size categories), and as a result of drop collisions and coalescence, the drop size spectrum in a dispersion is allowed to change over time with respect to the initial distribution. In practice, this change in the drop size distribution represents a macrophysical problem that is solved using "population dynamics equations." From a macroscopic point of view, drops are treated as isolated, microscopic quantities compared to field variables like temperature, but the aforementioned equations include the kernels which contain the information regarding the interaction at microscopic scale between two drop size categories (this information, as discussed above, being provided a priori by solving separately a microphysical problem involving only a limited number of droplets). Finally the coalescence events are handled at a macroscopic scale length as instantaneous unions of drops at the center of mass, replacing them with a new drop with their combined volume.

Such a philosophy allows a simple and an efficient treatment of the problem from a computational point of view and has been applied successfully to situations in which the physical phenomena of interest have a large length scale with respect to the average droplet size. However, vital information is lost about the effective microscopic evolution of the phenomena (drop deformation, shape instabilities, i.e. all those factors dealing with the local history of the shape or other "local" effects). Moreover the applicability of these methods is limited by the assumptions at the basis of the methods and information used to build the aforementioned kernels ($Re \rightarrow 0$).

With regard to all these numerical techniques and the possible approaches to the problem discussed in this section, it should be pointed out that quasisteady (very viscous) or nonviscous models are often inadequate, and that a fully transient and non-simplified analysis would be necessary in order to properly describe and interpret the effective behavior of drops in real experiments and processes. An outstanding need is for new scientific approaches not plagued by any limitation or simplification of the governing model equations, applicable to the case of low, high, as well as intermediate values of the Reynolds number, and allowing the treatment of multiple drops and related interplay, deformation and coalescence from a "local" point of view without resorting to dual-scale models. The present book seeks to meet the outstanding need by introducing moving-boundary methods able to predict drop interactions and deformations as well as collision and coagulation events in temperature gradients, with or without gravity being present over a wide range of conditions.

It can capture in a single numerical treatment and without limiting assumptions both macroscopic aspects (i.e. the macrophysical problem, heretofore treated in terms of population dynamics) and microscopic details (i.e. the microphysical problem, heretofore treated within the framework of boundary integral methods and/or under the assumption of nondeformable drops and/or under the assumptions of infinite Reynolds number and fluid-dynamic potential theories).

These methods have only recently been made sufficiently powerful to meet these objectives, including fully three-dimensional simulations of multiple drops.

3.5. VOF – Volume of Fluid Method and moving drops

Currently the effects described in Section 3.4 can be simulated in terms of direct numerical solution of the Navier–Stokes and energy equations in their complete nonlinear, viscous and coupled form. This has been made possible by the recent developments in moving-boundary methods.

Historically, pioneering work on the VOF (Volume of Fluid or Volume of Fraction, also known as the Volume Tracking) methods goes back to Noh and Woodward (1976), but with Hirt and Nichols (1981) and their SOLA-VOF code, the method became widely used.

As discussed by Rider and Kothe (1998), however, these techniques still possess solution algorithms that are too often perceived to be heuristic and without adequate mathematical formalism. A part of this misconception lies in the difficulty in applying standard hyperbolic PDE (partial differential equations) numerical analysis tools, which assume algebraic formulations, to methods that are largely geometric in nature (hence, the more appropriate term *volume tracking*). To some extent the lack of formalism in volume-tracking methods, manifested as an obscure underlying methodology, has impeded progress in evolutionary algorithmic improvements and application in particular to the case of immiscible metal alloys.

Despite some results being available for the case of rising (1 g conditions) and interacting bubbles (Esmaeeli and Tryggvason, 1998, 1999), in fact, the numerical simulation of these aspects in the case of drops sedimenting due to gravity or migrating owing to Marangoni forces can still be regarded as an open task. Despite the widespread use of these methods, the VOF-based simulations dealing with these particular aspects are still very rare in the literature (rare and excellent efforts have been provided by Gueyffier et al., 1999; Han and Tryggvason, 1999; Mehdi-Nejad et al., 2003 for the case of droplet sedimentation; and Haj-Hariri et al., 1997 and Tryggvason et al., 2001 for the case of Marangoni migration).

The present section is devoted to the introduction of a general-purpose VOF method able to deal with the drop motion due to a density difference under the normal (or residual) gravity as well as with the Marangoni-migration phenomena (or both cases). In the light of the above discussion, particular care is devoted to the formalism.

3.5.1 *The variable material properties approach*

The classical VOF formulation (see the excellent overviews of Rudman, 1997; Rider and Kothe, 1998; Scardovelli and Zaleski, 1999) relies on the fact that two or more fluids (phases) are not interpenetrating. For each additional phase added to the model a variable is introduced: the volume fraction of the phase in the computational cell. In each control volume (computational cell), the volume fractions of all phases must sum to an unity. The fields for all variables and properties are shared by the phases and represent volume-averaged values, as long as the volume fraction of each of the phases is known at each location. Thus the variables and the properties in any given cell are either purely representative of one of the phases, or representative of a mixture of phases depending upon the volume fraction values. Hereafter, for the sake of simplicity the case of interest for the purposes of the present book is considered (i.e. only two immiscible phases). If the volume fraction of phase (1) in the cell is denoted as ϕ, the following three conditions are possible:

$$\begin{aligned} \phi &= 0 \text{ (the cell is empty of fluid (1))} \\ \phi &= 1 \text{ (the cell is full of fluid (1))} \\ 0 < \phi &< 1 \text{ (the cell contains the interface between the fluids)} \end{aligned}$$

Based on the local values of ϕ, the appropriate properties and variables are assigned to each control volume within the computational domain. If χ denotes the generic fluid property (e.g., density ρ, dynamic viscosity μ, thermal conductivity λ, specific heat coefficient C_p, etc.) the corresponding value in each cell is given by:

$$\chi = \chi_1\phi + \chi_2(1-\phi) \tag{3.10}$$

which means that the concept of mixed properties is used to interpret the cells containing multiple fluids.

Accordingly, a single momentum equation is solved throughout the domain and the resulting velocity field is shared among the phases, i.e. the governing equations are written for the whole computational domain and the different phases are treated as a single fluid with variable material properties:

$$\underline{\nabla} \cdot \underline{V} = 0 \tag{3.11}$$

$$\frac{\partial(\rho\underline{V})}{\partial t} = -\underline{\nabla}p - \underline{\nabla}\cdot[\rho\underline{V}\,\underline{V}] + \underline{\nabla}\cdot\left[\mu(\underline{\nabla}\,\underline{V} + \underline{\nabla}\,\underline{V}^{\mathrm{T}})\right] + \underline{F}_{\mathrm{g}} + \underline{F}_{\sigma} \tag{3.12}$$

$$\frac{\partial\rho C_{\mathrm{P}}T}{\partial t} = [-\underline{\nabla}\cdot(\rho C_{\mathrm{P}}\underline{V}T) + \underline{\nabla}\cdot(\lambda\underline{\nabla}T)] \tag{3.13}$$

where the source terms in Eq. (3.12) take into account gravity ($\underline{F}_{\mathrm{g}}$) and surface-tension ($\underline{F}_{\sigma}$) effects, respectively.

The phase variable is advected according to a simple transport equation:

$$\frac{\mathrm{D}\phi}{\mathrm{D}t} = \frac{\partial\phi}{\partial t} + V\cdot\underline{\nabla}\phi = 0 \tag{3.14}$$

The interfaces are tracked in the volume-tracking methods by evolving fluid volumes forward in time with solutions of the above advection equation.

Since at any time in the solution, exact interface locations are not known (i.e. a given distribution of volume data does not guarantee a unique interface), interface geometry must be inferred, based on local volume data and the assumptions of a particular algorithm, before the interfaces can be reconstructed. The reconstructed interface is then used to compute the volume fluxes necessary to integrate the volume-evolution equation (3.14) above, i.e. for better computation of the convective contribution $\underline{V} \cdot \underline{\nabla}\phi$ (these aspects are discussed in Section 3.5.3).

This approach is also known as the single-region formulation since there is no need (from a computational point of view) to distinguish the different phases, i.e. to use different computational domains (this is the great advantage offered by such formulations, see also Section 5.4 for further details on this classification).

Along these lines the reader should be informed of the fact that from a general point of view, two broad strategies exist to deal with interface calculations. One is to use deformable meshes based on a finite volume or finite-element representation or body-fitted coordinates. The other strategy is to keep the mesh fixed, and to use a separate procedure to describe the position of the interface. From this perspective, the main advantage offered by fixed uniform grids is the great simplicity they afford in the treatment of the bulk fluid regions, away from the interfaces. A further advantage of fixed-grid methods is to avoid the remeshing that may be necessary whenever interface motion deforms the grid exceedingly.

3.5.2 *The continuum surface force and stress models*

The surface-tension model recommended in the present book is the Continuum Surface Force (CSF) proposed for the first time by Brackbill et al. (1992). With this model, the addition of the surface-tension effects to the VOF calculation results in a source term in the momentum equation, i.e. $\underline{F}_{\sigma}$ in Eq. (3.12).

Surface tension (see also Section 1.4) arises as a result of attractive forces between the molecules in a fluid. Consider an air bubble in water, for example. Within the bubble, the net force on a molecule due to its neighbors is zero. At the surface, however, the net force is radially inward, and the combined effect of the radial components of force across the entire spherical surface is to make the surface contract, thereby increasing the pressure on the concave side of the surface. The surface tension is a force, acting only at the surface, that is required to maintain equilibrium in such instances. It acts to balance the radially inward intermolecular attractive force with the radially outward pressure gradient force across the surface. In regions where the two fluids are separated, but one of them is not in the form of spherical bubbles, the surface tension acts to decrease the area of the interface. The explanation is basically the same in the case of a liquid–liquid system formed by immiscible fluids.

In practice, the well-known Young–Laplace equation can help the reader to understand the origin of the source term in Eq. (3.12) modeling surface-tension forces. Consider, for instance, the case where the surface tension is constant along the surface and where only the forces normal to the interface are present (according to the theoretical arguments pointed out in Section 2.3 this special case occurs if the isothermal conditions are considered and/or the liquid is compositionally homogeneous). The pressure jump across the surface depends on the surface-tension coefficient σ and the surface curvature as measured by the two radii in orthogonal directions R_1 and R_2:

$$p_2 - p_1 = \sigma\left(\frac{1}{R_1} + \frac{1}{R_2}\right) \tag{3.15a}$$

$$\text{i.e.,} \quad \Delta p = \sigma K \tag{3.15b}$$

where K is the curvature.

In the light of the foregoing arguments, the surface-tension can be written simply in terms of a pressure jump across the surface. This force in turn can be expressed as a volume force in Eq. (3.12) using the divergence theorem or a similar artifice to replace the surface force with a volume force. The end of the story is that the interfacial surface forces can be incorporated as body forces per unit volume in the momentum equations rather than as boundary conditions. Instead of a surface-tensile-force boundary condition applied at a discontinuous interface of the two fluids, a volume force can be used that acts on fluid elements lying within a transition region of finite thickness. This also means that the CSF formulation makes use of the approach that discontinuities can be approximated, without increasing the overall error of approximation, as continuous transitions within which the fluid properties vary smoothly from one fluid to the other over a distance of O(h), where h is a length comparable to the resolution of the computational mesh. Surface tension, therefore, is felt everywhere within the transition region through the volume force included in the momentum equations. A similar mathematical treatment is possible for the contribution related to surface-tension gradients along the interface (Marangoni stress).

It is known, in fact, that the expression for the stress jump $\underline{f}$, across the interface is given by (see the elegant formalism of Haj-Hariri et al., 1997).

$$\underline{f} = \left[\sigma K \underline{\hat{n}} - \frac{\partial \sigma}{\partial T}\left(\underline{\underline{I}} - \underline{\hat{n}}\,\underline{\hat{n}}\right) \cdot \underline{\nabla} T\right] \tag{3.16}$$

where $\underline{\hat{n}}$ is the unit vector perpendicular to the fluid/fluid interface, K is the curvature and $\underline{\underline{I}}$ is the identity matrix:

$$\underline{\hat{n}} = \frac{\underline{\nabla}\phi}{|\underline{\nabla}\phi|} = (\alpha, \beta) \tag{3.17}$$

$$K = -\underline{\nabla} \cdot \underline{\hat{n}} = \frac{1}{|\underline{n}|}\left[\frac{\underline{n}}{|\underline{n}|} \cdot \underline{\nabla}|\underline{n}| - \underline{\nabla} \cdot \underline{n}\right] \tag{3.18a}$$

K is formally a second derivative of ϕ, in cylindrical coordinates (x, y, being the coordinates along the symmetry axis and the radial direction, respectively) it reads:

$$-K = \frac{\partial}{\partial x}\left[\frac{\partial\phi/\partial x}{\sqrt{(\partial\phi/\partial x)^2 + (\partial\phi/\partial y)^2}}\right] + \frac{\partial}{\partial y}\left[\frac{\partial\phi/\partial y}{\sqrt{(\partial\phi/\partial x)^2 + (\partial\phi/\partial y)^2}}\right]$$
$$+ \frac{1}{y}\frac{\partial\phi/\partial y}{\sqrt{(\partial\phi/\partial x)^2 + (\partial\phi/\partial y)^2}} \tag{3.18b}$$

but it is usually computed according to Eq. (3.18a) since direct discretization of second and higher derivatives can lead to major inaccuracies for low-order methods, and becomes rather complicated at high order.

The surface force per unit interfacial area, should be added in the momentum equation as $\underline{f}\delta$ where δ is the Dirac-pulse function used to localize the force explicitly at the interface. Using some theoretical artifices (as previously discussed) it can formally be replaced by its volume-distributed counterpart, $\underline{F}$, which satisfies.

$$\int_{-\infty}^{\infty} \underline{F}(s)\,\mathrm{ds} = \int_{-\infty}^{\infty} \underline{f}(s)\delta(s)\,\mathrm{ds} \tag{3.19}$$

where (s) is a level-set function denoting the normal distance from the interface (the interface corresponds to $s=0$) and δ(s) is the Dirac delta function with its singularity on the interface. Rigorously speaking, f is not defined for nonzero s and must be extended appropriately (e.g., it can be assigned a value of zero for nonzero s because of the presence of the delta function). A comparison of the two integrals in Eq. (3.19) suggests $\underline{F}=\underline{f}\delta$(s). Although this expression is of little value in its present form, much can be gained from it if a mollified delta function is used, consistent with the smearing of the interface. Recognizing that

$$\int_{-\infty}^{\infty} \delta(s)\,\mathrm{ds} = \int_{-\infty}^{\infty} \underline{\hat{n}} \cdot \underline{\nabla}\phi\,\mathrm{d}s = 1 \tag{3.20}$$

the mollified delta function can be defined as $|\underline{\nabla}\phi|$ which, in turn, leads to $\underline{F}=\underline{f}|\underline{\nabla}\phi|$.

Thus the second source terms in Eq. (3.12) reads:

$$\underline{F}_\sigma = \left[\sigma K\underline{\hat{n}} - \frac{\partial\sigma}{\partial T}\left(\underline{\underline{I}} - \underline{\hat{n}}\,\underline{\hat{n}}\right) \cdot \underline{\nabla}T\right]|\underline{\nabla}\phi| \tag{3.21}$$

where, as previously discussed in the light of the Young–Laplace equation, the first term in the square parentheses is the normal stress contribution responsible for the shape equilibrium in the perpendicular direction, whereas the second is the Marangoni shear stress along the surface (whose genesis has already been discussed in Section 2.3) that can be responsible for the Marangoni migration phenomena.

It is important to highlight how reliable computation of both K and $\hat{\underline{n}}$ is not easy and special care must be devoted to such a computation. Usually the unit normal-vector results from the gradient of a smoothed phase field ϕ (the so-called mollification).

The smoothed VOF function can be computed for instance by convolving ϕ with a B-spline of degree l (Brackbill et al., 1992). The smoothed VOF function is given by:

$$\hat{\phi}_{i,j} = \frac{9}{16}\phi_{ij} + \frac{3}{32}\left(\phi_{ij+1} + \phi_{ij-1} + \phi_{i+1j} + \phi_{i-1j}\right) + \frac{1}{64}\left(\phi_{i+1j+1} + \phi_{i+1j-1} + \phi_{i-1j+1} + \phi_{i-1j-1}\right) \tag{3.22}$$

This formula may be applied iteratively by multiple passes through the mesh for increased degrees of smoothing.

For further details on these aspects see, e.g., the excellent works of Rider and Kothe (1998) and Gueyffier et al. (1999).

To summarize:

$$\underline{F}_\sigma = \left[\sigma K\hat{\underline{n}} - \frac{\partial\sigma}{\partial T}\left(\underline{\underline{I}} - \hat{\underline{n}}\,\hat{\underline{n}}\right)\cdot\underline{\nabla}T\right]\left|\underline{\nabla}\hat{\phi}\right| = \sigma K\begin{bmatrix}\dfrac{\partial\hat{\phi}}{\partial x}\\ \dfrac{\partial\hat{\phi}}{\partial y}\end{bmatrix} - \frac{\partial\sigma}{\partial T}\begin{bmatrix}1-\alpha\alpha & \alpha\beta\\ \beta\alpha & 1-\beta\beta\end{bmatrix}\cdot\begin{bmatrix}\dfrac{\partial T}{\partial x}\\ \dfrac{\partial T}{\partial y}\end{bmatrix}\cdot\sqrt{\left(\frac{\partial\hat{\phi}}{\partial x}\right)^2 + \left(\frac{\partial\hat{\phi}}{\partial y}\right)^2} \tag{3.23}$$

where

$$\alpha = \frac{\partial\hat{\phi}}{\partial x}\Big/\sqrt{\left(\frac{\partial\hat{\phi}}{\partial x}\right)^2 + \left(\frac{\partial\hat{\phi}}{\partial y}\right)^2}, \qquad \beta = \frac{\partial\hat{\phi}}{\partial y}\Big/\sqrt{\left(\frac{\partial\hat{\phi}}{\partial x}\right)^2 + \left(\frac{\partial\hat{\phi}}{\partial y}\right)^2}$$

and $\hat{\phi}$ is the mollified phase-field variable.

The treatment of the source term $\underline{F}_g$ taking into account gravity is relatively simple, if compared with the corresponding effort provided for the modeling of $\underline{F}_\sigma$.

In fact, the first source term in Eq. (3.12) simply reads:

$$\underline{F}_g = -\rho\underline{g} = -\left(\rho_1\phi + \rho_2(1-\phi)\right) \rightarrow \tag{3.24a}$$

and since, according to the Boussinesq hypothesis

$$\begin{aligned} &\rho_i = \rho_{io}[1 - \beta_{Ti}(T - T_o)] \rightarrow \\ &\underline{F}_g = -\rho \underline{g} = -\{[\rho_{1o}\phi + \rho_{2o}(1-\phi)] - [\rho_{1o}\beta_{T1}\phi + \rho_{2o}\beta_{T2}(1-\phi)](T - T_o)\} \end{aligned} \tag{3.24b}$$

where the first and the second terms in the square parentheses can be responsible for sedimentation owing to the different density of the phases (e.g., $\rho_{1o} > \rho_{2o}$) and for possible onset of buoyant convection (in nonisothermal conditions), respectively.

Gravity can manifest itself in the process of drop sedimentation, in fact, in two ways. It has a direct effect on the drop velocity, through buoyancy, and in a more subtle (but not necessarily insignificant) way, through natural convection in the surrounding liquid. This convection can directly affect drop motion and also alter the temperature gradients at the interface.

As anticipated, the main advantage of the above formulation is the topological simplification resulting from the incorporation of the interfacial jump conditions into the bulk equations. This eliminates the rather involved task of generating grids for the interior and exterior domains that must be tracked and reconstructed after each iteration step (the so-called multiple-region formulations, see Section 5.4). With the VOF formulation discussed in the present book, a single grid can be generated without regard for the actual shape of the moving drop. However, as anticipated in Section 3.5.1 a reconstruction technique is required. Different procedures can be used in principle to reconstruct the interface (see the next section).

3.5.3 *Interface-reconstruction techniques*

The ϕ-field is the only phase information stored in the VOF methods. Approximate interface locations are found from a so-called interface reconstruction. This is needed, as anticipated, for advecting ϕ, for determining the local properties (density, viscosity, etc.) and for better graphical representation. In the earlier versions, usually called SLIC (Simple Line Interface Method), after Noh and Woodward (1976), interfaces were approximated with either a horizontal or vertical line in each cell. Certainly, such a reconstruction appears rather crude. Nevertheless, it has been used in many versions up to the present day, and many reasonable simulation results were obtained.

The SLIC method approximates interfaces as a piecewise constant, where the interfaces within each cell are assumed to be lines (or planes in three dimensions) aligned with one of the logical mesh coordinates. The Hirt–Nichols (H–N) scheme, as used in the SOLA-VOF code (Hirt and Nichols, 1981), is a variation of the piecewise constant method (piecewise constant/"stair-stepped" approximation). In the piecewise constant/stair-stepped methods, the interfaces are also forced to align with the mesh coordinates, but are additionally allowed to "stair-step" (align with more than one mesh coordinate) within each cell, depending upon the local distribution of discrete volume data.

A notable feature of the SLIC method is that its volume fluxes can be formulated algebraically, i.e. without needing an exact reconstructed interface position. The volume fluxes can be expressed as a weighted sum of the upwind and the downwind contributions, depending upon the orientation of the interface relative to the local flow direction. For this case interface reconstruction is used only for visualization purposes.

More accurate reconstructions are possible with PLIC (Piecewise Linear Interface Construction). The interface is approximated by a straight line of arbitrary orientation in each cell. Its orientation is found from the liquid distribution in the neighbor cells; given the volume fraction of one of the two fluids in each computational cell and an estimate of the normal vector to the interface, a planar surface is constructed within the cell having the same normal and dividing the cell into two parts each of which contains the proper volume of one of the two fluids.

This has several advantages: the fluxes of ϕ, with which the phase field ϕ is updated, can be determined more accurately, and essentially free of numerical diffusion. The fluid properties can be allocated accurately. Finally, the straight lines also give a graphical representation of good quality.

See the excellent overview of Rider and Kothe (1998) for additional details on these geometrical reconstruction techniques.

Figure 3.18 shows the reconstructed interfaces (shaded regions) for a drop of arbitrary radius (continuous line) using the SLIC and PLIC methods.

3.5.4 *Discretization*

The basic numerical algorithm for the solution of Eqs. (3.11–3.13) is the splitting method, already discussed in Section 2.8. For the case under investigation it must be modified to take into account the fact that two phases with different densities are present in the computational domain; it consists of the following two steps (for educational purposes, a first-order temporal integration scheme is used):

$$\underline{V}^{*} = \underline{V}^{n} + \Delta t \frac{1}{\rho^{n}} \left[-\underline{\nabla} \cdot [\rho \underline{V}\,\underline{V}] + \underline{\nabla} \cdot \left[\mu (\underline{\nabla}\,\underline{V} + \underline{\nabla}\,\underline{V}^{\mathrm{T}}) \right] + \underline{F}_{\mathrm{g}} + \underline{F}_{\sigma}^{t} \right]^{n} \tag{3.25}$$

where $\underline{F}_{\sigma}^{t} = [-(\partial\sigma/\partial T)(\underline{\underline{I}} - \underline{\hat{n}}\,\underline{\hat{n}}) \cdot \underline{\nabla} T] |\underline{\nabla}\phi|$

$$\underline{V}^{n+1} = \underline{V}^{*} + \Delta t \frac{1}{\rho^{n}} \left[-\underline{\nabla} p + \underline{F}_{\sigma}^{p} \right]^{n} \tag{3.26}$$

where $\underline{F}_{\sigma}^{p} = [\sigma K \underline{\hat{n}}] |\underline{\nabla}\phi|$ and the elliptic equation reads:

$$\underline{\nabla} \cdot \left[\frac{1}{\rho^{n}} \underline{\nabla} p \right] = \frac{1}{\Delta t} \underline{\nabla} \cdot \underline{V}^{*} + \underline{\nabla} \cdot \left[\frac{1}{\rho^{n}} \underline{F}_{\sigma}^{P} \right]^{n} \tag{3.27}$$

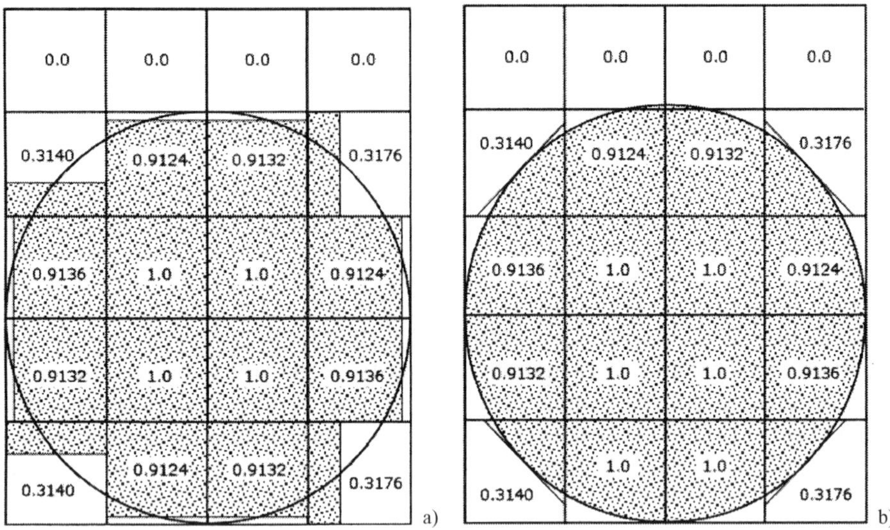

Fig. 3.18 Reconstructed interfaces (shaded regions) for a drop using the SLIC and PLIC methods. The piecewise constant approximation in SLIC forces the reconstruction to align with selected mesh logical coordinates, whereas the piecewise linear approximation in PLIC allows the reconstruction to align naturally with the interface. Numbers in the cells denote volume fractions. (a) SLIC reconstruction. (b) PLIC reconstruction.

During each time step, the intermediate velocity field V^* is calculated from Eq. (3.25), and used to obtain the pressure through an iterative solution of Eq. (3.27). Subsequently, the new pressure distribution is used in Eq. (3.26) to advance the velocity field to the next time step.

Once the new velocity field, V^{n+1}, is determined, the phase variable is advected according to Eq. (3.14) and the interface-reconstruction technique used; this step provides the new distribution ϕ^{n+1}.

Then, the temperature field is updated by using an explicit Euler time marching in Eq. (3.28):

$$T^{n+1} = \left\{[\rho C_P T]^n + \Delta t[-\underline{\nabla}\cdot(\rho C_P \underline{V} T) + \underline{\nabla}\cdot(\lambda\underline{\nabla}T)]^n\right\}/[\rho C_P]^{n+1} \tag{3.28}$$

and the migration velocity of the drop can be determined by calculating the average velocity within the drop phase.

In the next two sections, the role played by coalescence in changing the Marangoni migration velocity is discussed together with the possible influence of the thermal-wake effects for small and large values of the Prandtl number.

3.5.5 *Thermal-wake effects*

In the first part of this section, some numerical examples obtained with the VOF method are briefly illustrated in the case of a single drop migrating (owing to gravity

Table 3.2a Physical properties of the ethyl-alcohol (1) silicone-oil 1 cs (2) system.

ρ_1 (kg m^{-3})	0.86×10^3
ρ_2 (kg m^{-3})	0.816×10^3
ν_1 (m^2 s^{-1})	1.16×10^{-6}
ν_2 (m^2 s^{-1})	1.0×10^{-6}
C_{p1} (J kg^{-1} K^{-1})	2.11×10^3
C_{p2} (J kg^{-1} K^{-1})	2.05×10^3
λ_1 (J m^{-1} s^{-1} K^{-1})	1.8×10^{-1}
λ_2 (J m^{-1} s^{-1} K^{-1})	1.0×10^{-1}
σ (dyne m^{-1})	$\cong 5 \times 10^2$
σ_T (dyne m^{-1} K^{-1})	$\cong 1.5$

or Marangoni forces) in an immiscible external liquid. The couple ethyl-alcohol/silicone oil (Table 3.2a) is considered for the simulation since they are transparent and immiscible liquids. In particular, attention is focused on the disturbance induced in the temperature field by the migration processes.

Figure 3.19 show an ethyl-alcohol droplet with a diameter of 1 cm, and an initial velocity of 0 m s^{-1} in the x-direction that is released under the effect of a residual (steady) $g = 10^{-4}\, g_o$ (100 μg) in a tank (height 6 cm and radius 1.5 cm) filled with silicone oil 1 cs at rest having initially a nonuniform temperature distribution along x (temperature increasing towards the top). Marangoni effects and buoyancy convection are neglected for this simulation in order to discern the effect of the sedimentation process per se.

A few seconds after the release of the droplet the start of a weak wake formation behind the droplet is observed. This temperature distribution can be explained by the mechanism of heat transfer during the droplet-fall. As the droplet moves inside the oil tank, in fact, the temperature of oil particles that are in contact with the front of the droplet increases. These particles continue their path toward the back of the droplet during which they continuously gain energy. The related distortion induced in the temperature field can often be responsible for the onset of an undesirable buoyancy convection otherwise prevented in a liquid uniformly heated from above.

Figure 3.20 show the opposite situation in which in the absence of gravity (0g) the drop migrates towards the hot region owing to thermocapillary effects.

Initially, the temperature profile is linearly stratified ($\partial T/\partial x = 0.1$ K cm^{-1}), the drop ($R = 0.5$ cm) is of ethyl-alcohol and the surrounding matrix is silicone oil 1 cs as for the case of sedimentation.

It is worthwhile to point out how the temperature profile near the front pole of the drop is highly distorted with respect to the stratified temperature profile so that the drop does not migrate as in an unbounded medium of infinite extent subjected to a uniform temperature gradient that is assumed by many simplified analyses available in the literature.

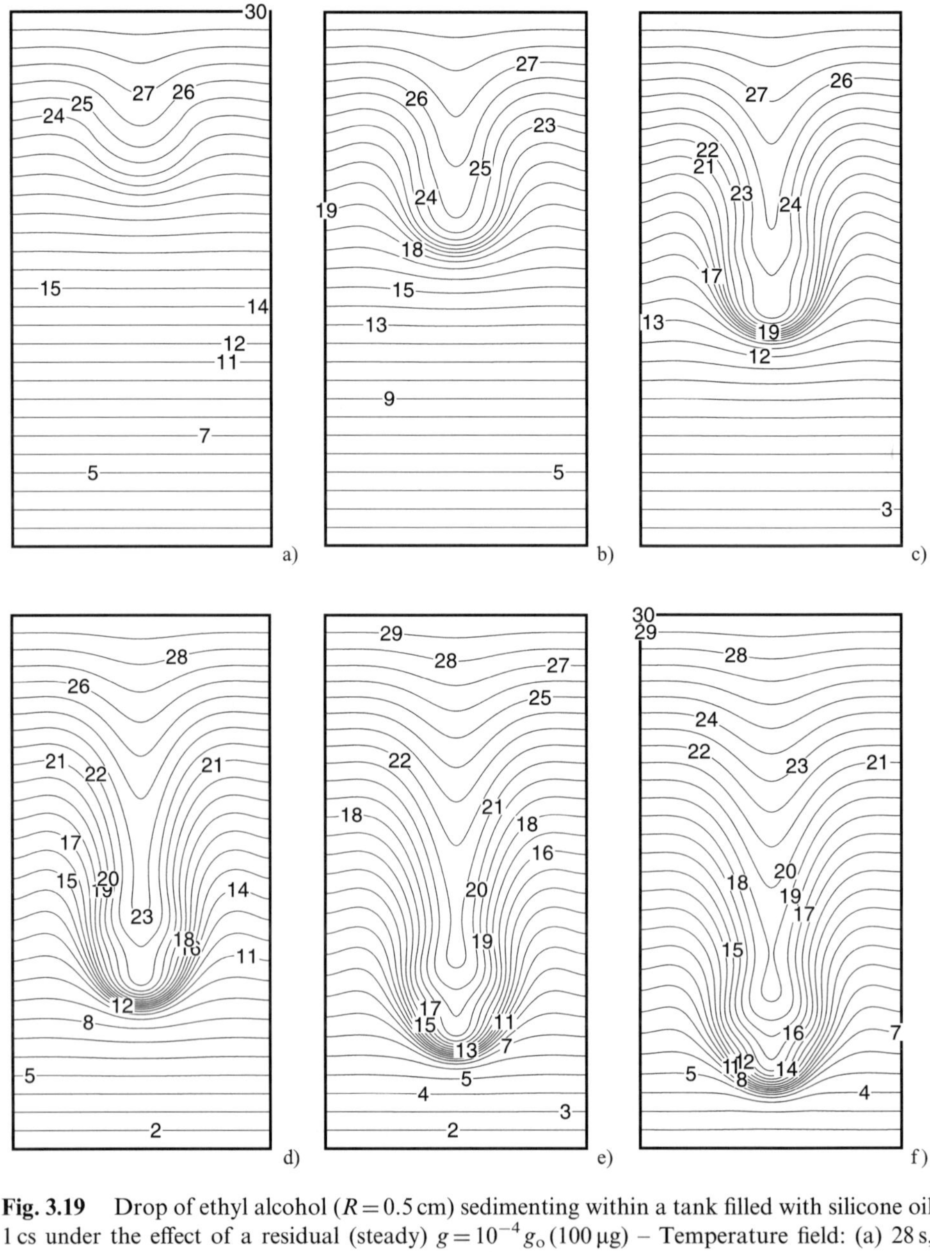

Fig. 3.19 Drop of ethyl alcohol ($R = 0.5$ cm) sedimenting within a tank filled with silicone oil 1 cs under the effect of a residual (steady) $g = 10^{-4} g_o$ (100 μg) – Temperature field: (a) 28 s, (b) 56 s, (c) 84 s, (d) 112 s, (e) 140, (f) 168 s (level 1 $\rightarrow T_{min}$, level 30 $\rightarrow T_{max}$).

Since the energy exchange by heat between the drop and the surrounding phase is very small (both the liquids have small values of the thermal conductivity), the drop tends to stay cold. Accordingly, the external isotherms encountered by the drop during the migration process do not penetrate within the drop and fold over it.

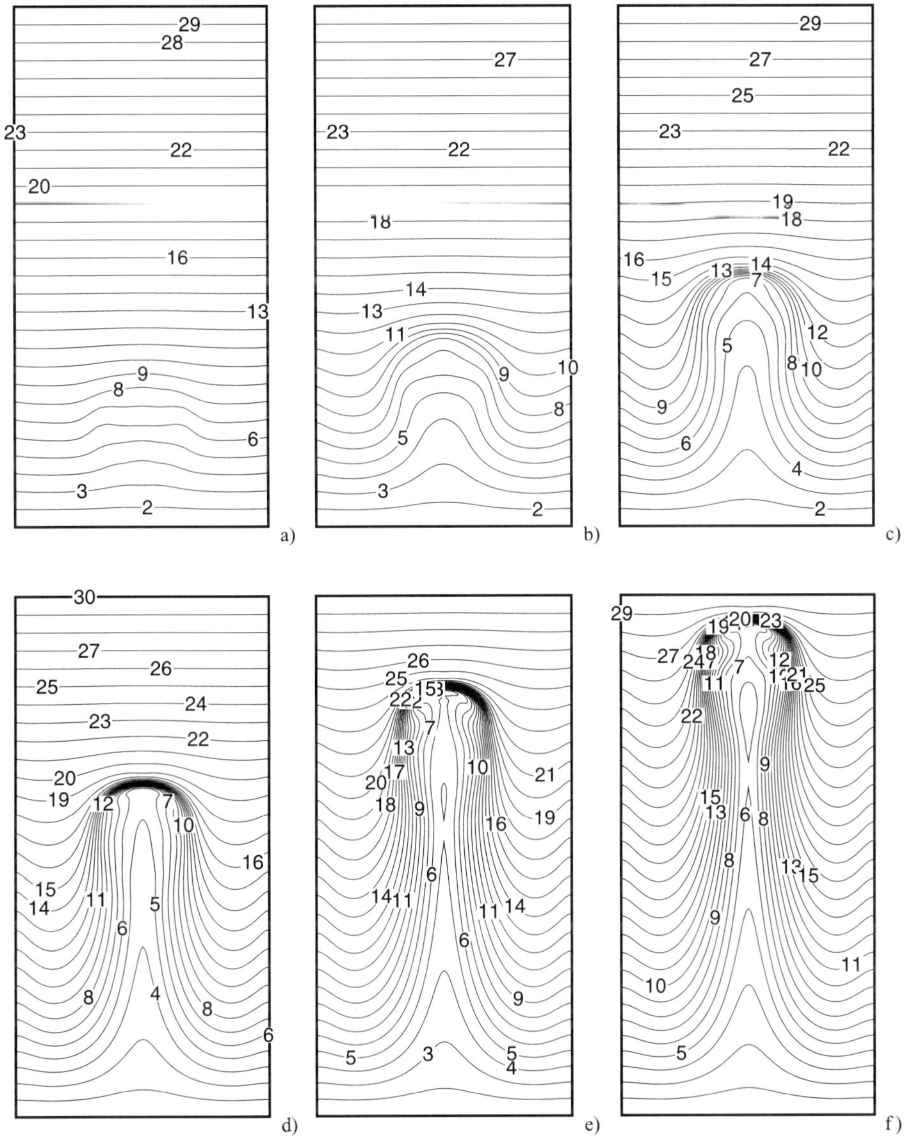

Fig. 3.20 Drop of ethyl alcohol ($R = 0.5$ cm) migrating in silicone oil 1 cs under the effect of temperature gradient $\partial T/\partial x = 0.1\ \mathrm{K\,cm^{-1}}$ – Temperature field: (a) 4.2 s, (b) 10.1 s, (c) 16 s, (d) 21.9 s, (e) 27.8, (f) 33.7 s (level 1 → $T = 0$°C, level 30 → $T = 0.6$ °C).

The drop pushes the isotherms and constrains them in a narrow thermal boundary layer surrounding the drop and having a steep normal temperature gradient.

Figure 3.20 also shows how the external fluid creates a cold wake that extends behind the drop and entrains the hot region (this wake can be responsible for the aforementioned effects of thermal-gradient weakening on a trailing drop).

These simple numerical examples show the basic potentialities of the method introduced in the previous sections.

3.5.6 *Collisional coagulation and phase separation*

At this stage, however, it is worthwhile to highlight some other relevant aspects of the problem related to phase separation and in particular dealing with the aforementioned role played by the coalescence phenomena (discussed in Section 3.4.4).

It should be kept in mind that as noncoalescence can play an important role in the mechanisms leading to phase separation (owing to the drop-pushing behavior elucidated in Section 3.4.1), coalescence phenomena can also deeply affect (in a different way) the same mechanisms; for instance, on the ground, the droplet coalescence strengthens the mechanisms leading to a spatial separation and a coarsening of the dispersion since heavier drops are formed with an increased velocity of sedimentation.

As described by Stokes' law, the sedimentation velocity of the heavy droplets increases as the square of the particle diameter (Eq. (3.6)), so large droplets settle much more rapidly than small ones. When droplets of different diameters and thus different sedimentation velocity collide, they may coagulate to form bigger droplets that settle even more rapidly.

In the absence of gravity, the effect is similar; drop coalescence in space leads to larger drops and this increases the velocity of Marangoni migration owing to the increase of the interface where the surface-tension stresses act propelling the drop (the readers may also consider eqs. (3.8) and (3.9) as representative of the effect of the radius on the Marangoni migration for the cases $Ma \rightarrow 0$ and $Ma \rightarrow \infty$: the migration velocity is proportional to R or to R^3 for small or large values of Ma, respectively).

Accordingly, coalescence phenomena of droplets increase the efficiency of the mechanisms of phase separation on the ground as well as in space. Also, these behaviors can be simulated in the framework of VOF methods.

As an example, Fig. 3.21 shows the subsequent stages of evolution of the coalescence process for two droplets of ethyl-alcohol having initially different radius ($R_A/R_B = 1.5$) in a liquid matrix of silicone oil 1 cs in isothermal conditions and the absence of gravity (0g) when they are sufficiently close to allow intimate contacts between molecules that belong to adjacent surfaces.

When the drops come into physical contact, they coalesce into a single, larger drop due to the tendency of interfacial tension to minimize the surface area of the contacted drops.

The boundary of the new larger drop obtained by the coalescence process, oscillates due to counteracting inertial and surface-tension forces. These oscillations, however, tend to be dumped by the viscous effects and finally a single spherical drop is obtained.

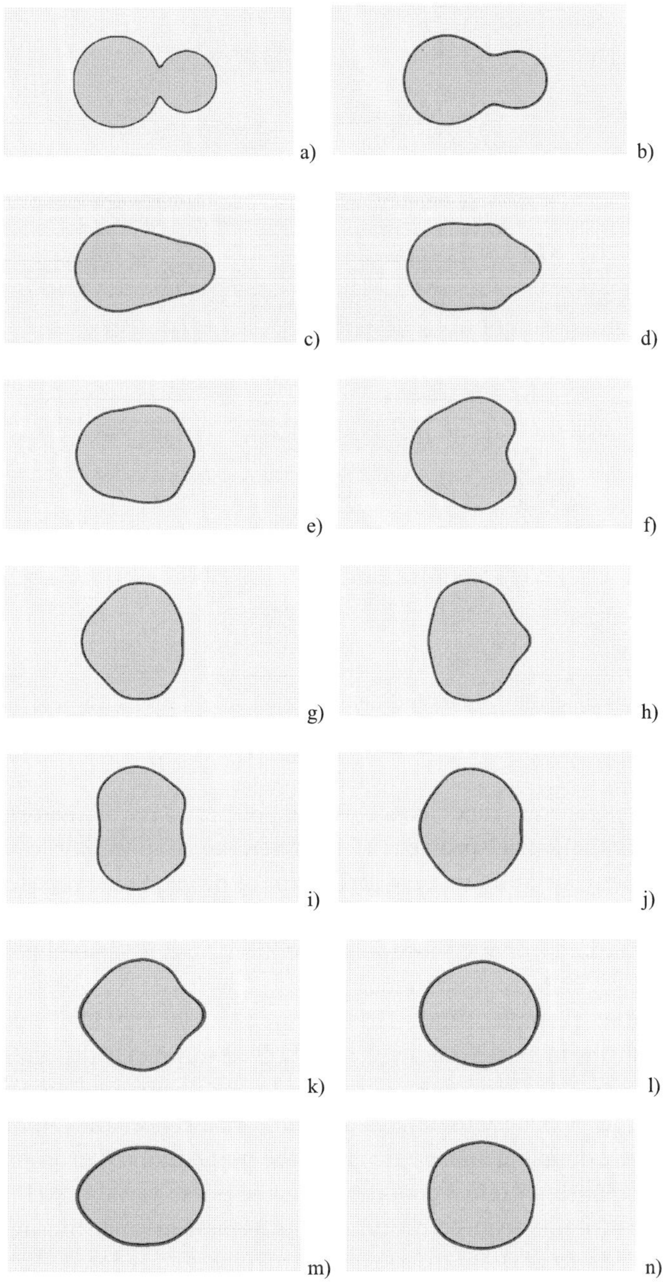

Fig. 3.21 Subsequent stages of evolution of the coalescence process for two droplets of ethyl-alcohol having initially different radius ($R_A = 0.6$ mm, $R_B = 0.4$ mm) in a liquid matrix of silicone oil 1 cs in isothermal conditions and absence of gravity (0g).

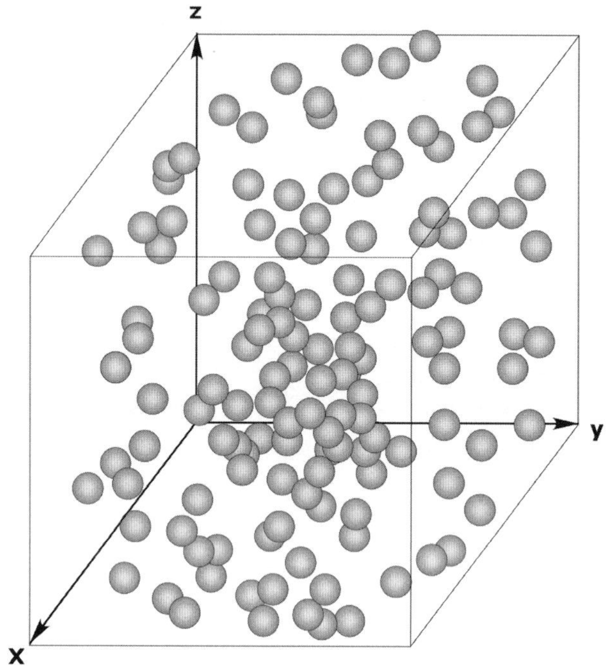

Fig. 3.22 Example of spatially inhomogeneous population of drops whose dynamics must be solved to predict phase-separation rates in problems related to alloys.

For nonhomogeneous and concentrated dispersions of deformable drops (Fig. 3.22 shows the typical problem of interest), the drop motion causes a spatial as well as temporal variation in the distribution of drops. Superimposed on this are the coalescence phenomena. The calculation of the separation rate is possible (for either the gravitational or thermocapillary motion), using three-dimensional VOF methods and very dense computational grids.

As a relevant example, Eqs. (3.11–3.14) for the spatially inhomogeneous population of drops in Fig. 3.23a have been solved to predict the phase-separation rates in the presence of a temperature gradient for a dispersion of drops at moderate concentrations, at which both migration and coalescence are important (zero-g conditions). The example demonstrates how the population of drops can be simply reduced to a proper set of initial conditions for the phase-variable ϕ (Fig. 3.23a) and the subsequent dynamics can be captured in a natural manner by simulation (Fig. 3.23b–f). The case of Zn(1)–Bi(2) system in particular is considered (Table 3.2b). The container is a square cell with side $L = 1$ cm filled with the melt, initially at $T = 750$ K. The walls of the container are supposed to be cooled with a ramping rate of $1\ \mathrm{K\,s^{-1}}$.

Figure 3.23 shows that all the bismuth drops migrate towards the center of the container. Simultaneously, during their run, some coalescence phenomena occur, i.e. they tend to agglomerate into a smaller number of larger particles.

This simulation provides a possible theoretical justification for the behavior shown in Fig. 3.16b.

Hence the application of the VOF methods to these problems can provide a fundamental understanding of how drop migration due to residual gravity and/or the Marangoni effects interact cooperatively with coalescence to promote a macroscopic phase separation of an immiscible dispersion (see also Lappa, 2004h).

These simulations could also help to resolve some critical issues, such as the roles of concentration, viscosity, interfacial tension, and other parameters on the coalescence process.

Among other things they also provide some interesting information about the role played by the thermal-wake effects.

As mentioned in Section 3.4.4, these aspects also represent an additional and a relevant part of the problem in the limit as $Ma \to \infty$.

The numerical simulations carried out herein for cases of technological interest, show that, in practice, different situations arise according to the Prandtl number.

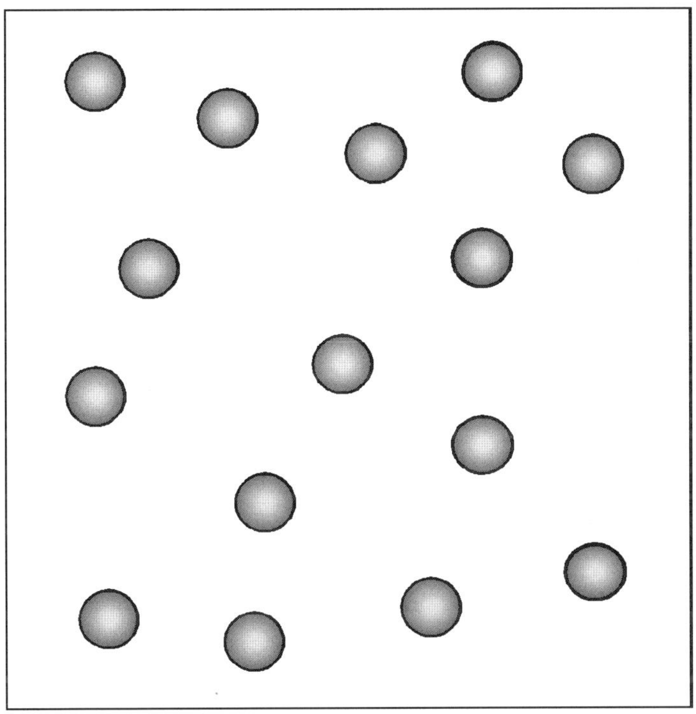

Fig. 3.23 Bismuth drops in a square container with side $L = 1$ cm filled with zinc melt, initially at $T = 750$ K. The walls of the container are supposed to be cooled with a ramping rate of $1\ \mathrm{K\,s^{-1}}$ (zero-gravity conditions).

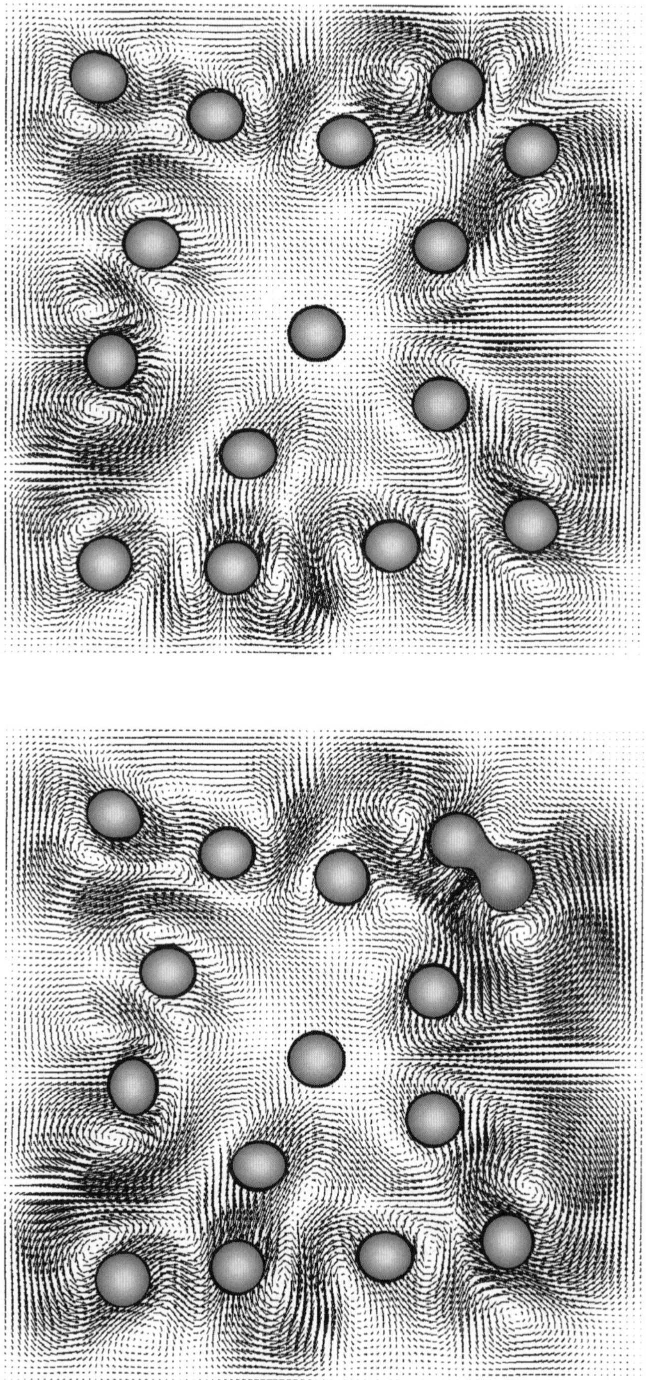

Fig. 3.23 Continued.

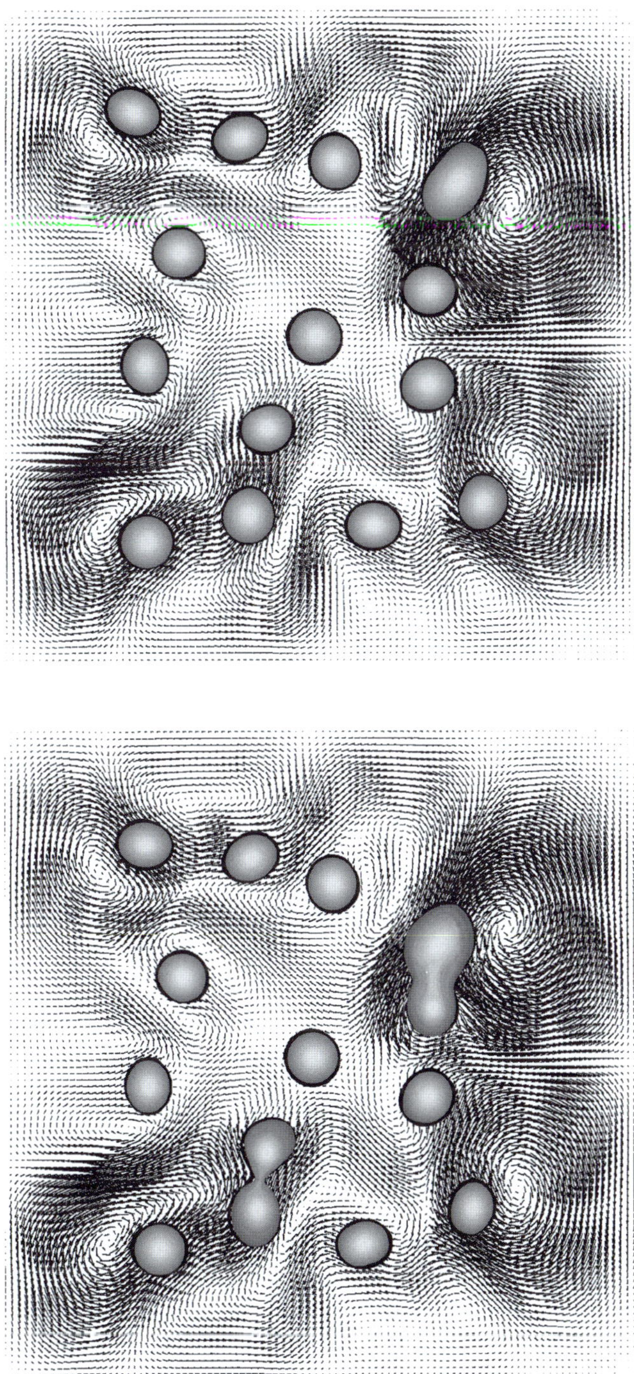

Fig. 3.23 Continued.

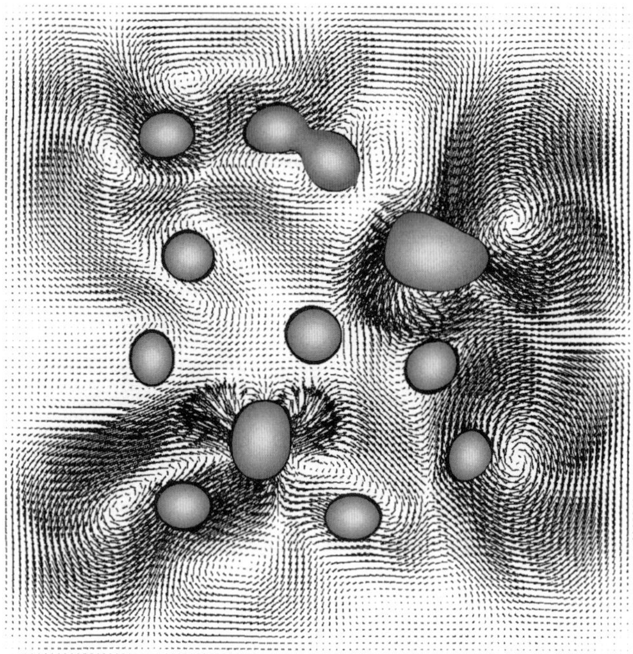

Fig. 3.23 Continued.

Table 3.2b Physical properties of the Zn(1)–Bi(2) system.

T_{m1} (K)	692.7
T_{m2} (K)	544.6
ρ_1 (kg m^{-3})	6.61×10^3
ρ_2 (kg m^{-3})	9.87×10^3
ν_1 (m^2 s^{-1})	0.49×10^{-6}
ν_2 (m^2 s^{-1})	0.136×10^{-6}
C_{p1} (J kg^{-1} K^{-1})	0.495×10^3
C_{p2} (J kg^{-1} K^{-1})	0.145×10^3
λ_1 (J m^{-1}s^{-1}K^{-1})	15.8
λ_2 (J m^{-1} s^{-1} K^{-1})	58.1
σ (dyne m^{-1})	$\cong 6 \times 10^4$
σ_T (dyne m^{-1}K^{-1})	$\cong 10$

If Pr is low ($Pr \ll 1$ in the case of liquid metals and semiconductors), thermal-wake effects are generally negligible and the only mechanisms playing an effective role in affecting the phase-separation phenomena are Marangoni motion of the drops and the coagulation events. This aspect is confirmed by the simulation for the case Zn–Bi illustrated earlier.

For such a case, in fact, the temperature field is almost diffusive. This follows from the fact that, owing to the large thermal diffusivity of the materials, temperature distortions induced by the motion of the drops are rapidly destroyed (they are spread over large regions).

If large Prandtl number fluids are considered (transparent and/or organic liquids), on the contrary, one can expect that thermal-wake effects become important. Owing to the small thermal diffusivity of these liquids, the distortions induced in the temperature field, in fact, are somehow "frozen" with respect to the times with which the drops migrate within the container. This aspect is confirmed by the simulations previously shown in Fig. 3.20.

Recently, these moving-boundary approaches have been extended to other types of problems that are very different with respect to those treated in the present chapter, e.g., the case of macromolecular organic growth from supersaturated solutions (due to the addition/incorporation of solute molecules (building blocks or growth units) to the crystal lattice, see Chapter 5) and the growth of biological tissues in bioreactors (see Chapter 7).

3.6. Absolute and convective instabilities in falling liquid jets

The literature on the convective instability of liquid jets is very rich. For instance, many analyses have appeared dealing with the liquid–gas or the liquid–vacuum systems (i.e. the case of liquid jets falling in an ambient gas or in vacuum). For these cases, the commonly observed manifestation of the onset of instability in a liquid jet with a sufficiently large velocity is the amplification of disturbances that are convected downstream to break up the jet into drops. This instability belongs to a general class of instability called "convective instability" that allows the disturbance to be convected only in the downstream direction. The literature on these phenomena is reviewed in many articles. It is interesting to point out how another class of instability also exists, called the "absolute instability" that permits the disturbance to propagate in both the downstream and upstream directions. Leib and Goldstein (1986a, b) were the first to demonstrate that, also in the absence of gravity and ambient gas, a liquid jet with a relatively small velocity may become absolutely unstable and that this behavior is due to surface-tension forces. Even in the presence of gas, with (Lin and Lian, 1989) or without gas viscosity (Lin and Lian, 1993), the absolute instability can occur.

Transition from absolute to convective instabilities has been considered by De Luca and Costa, (1997).

These behaviors have been the focus of intense investigation to develop recommendations on how to avoid unwanted hydrodynamic phenomena for the free surface of free-falling liquid metal jets, which are used in casting processes (e.g., the so-called "stationary" and "continuous" casting techniques, where the first is the pouring of molten metal into molds and allowing the metal to cool, and the latter is the production of sheet, rod, or other long shapes by solidifying the metal while it is being poured through an open-ended mold, respectively).

The undesirable hydrodynamic behaviors consist of excess rippling (formation of surface tension waves), deformation of the falling jet and droplet ejection at its surface (drops of liquid metal may be ejected from the jet free surface due to turbulent disturbances and shear instabilities); these droplets in turn, being at different (higher) temperature with respect to the underlying liquid can float over its surface according to the mechanisms elucidated in Section 3.2.

Recommendations about the instabilities of these jets have been developed by qualitative understanding of the physical mechanisms that cause the related phenomena (see, e.g., Wu and Faeth, 1995; Soderberg and Alfredson, 1998). This subject is not treated in depth within the present book, since very recently a volume entirely devoted to such aspects (Lin, 2003) has been published.

These arguments are cited within the context of the present chapter owing to the theoretical kinship with the aforementioned problem related to the design of foundry reactors for alloys and related manufacturing processes. But there is also another reason for citing such a topic here. Recently it has been demonstrated (see Konkachbaev and Morley, 2000) that the delicate mechanisms of the instabilities of free-falling liquid-metal jets and the related phenomenon of jet break-up and droplets ejection at the liquid surface can also be directly simulated in the framework of the VOF methods and Continuum Surface Force Model elucidated in Section 3.5 (see also Tryggvason et al., 2001).

3.7. Dissolution and solutal convection in liquid systems with miscibility gap

Within the context of the present chapter, a final aspect needs analysis and clarification: The dissolution and related convective phenomena that may occur under certain conditions when the liquid–liquid systems under consideration exhibit a miscibility gap.

As pointed out in Section 3.4 if an immiscible alloy is cooled below the critical temperature T_c, it tends to separate into two components at different compositions where each component exhibits a different concentration of the two pure substances initially used for the alloy (e.g., Al and In). In this condition no dissolution phenomena occur and the two phases behave as immiscible liquids (like oil and water, for example).

However, if the liquid representing the minority phase is directly injected into the majority phase and the two components are initially pure (e.g., pure Al and pure In), the dispersed particles tend to dissolve within the external matrix (Savino and Lappa, 2003b; Lappa et al., 2004b).

A dissolution process also occurs if droplets, formed in some regions by thermodynamic decomposition induced by cooling of the immiscible alloy, move (owing to Marangoni migration or gravitational sedimentation) to regions with different temperature. The phases, in fact, on the basis of the phase diagram (see, e.g., Fig. 3.24), have different equilibrium concentrations at different temperatures. A droplet

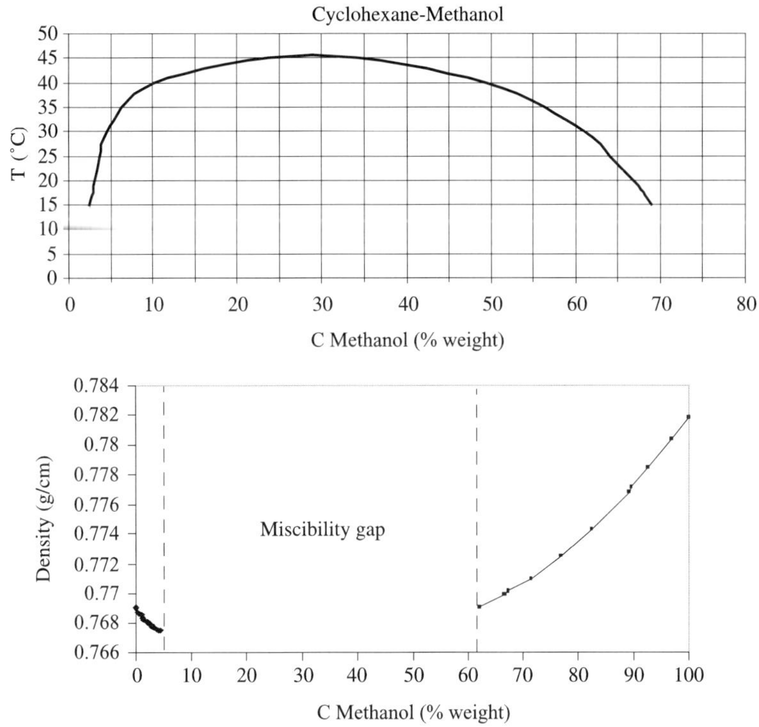

Fig. 3.24 Phase diagram (a) and density versus methanol concentration (b) for the cyclohexane–methanol system.

that is in equilibrium with the surrounding liquid in a certain region at a certain temperature will no longer be in such a stable condition if it moves rapidly to another zone with different temperatures. Therein, in fact, it will tend to return to equilibrium by exchanging mass with the surrounding phase decreasing or increasing its size. This process can affect coarsening kinetics (the spatial separation of the phases) for the same reasons discussed in Section 3.5.6. It, in fact, combines with coalescence to produce particle-size changes (see, e.g., the fundamental information provided by Ratke, 1987 and the numerical simulations of Beckermann et al., 1999). This mass exchange, among other things can lead to the onset of solutal convection.

These convective effects are still poorly known, owing to the experimental difficulties encountered when trying to discern the separate contributions provided by many concurrent and interacting mechanisms (the aforementioned thermodynamic separation with droplet nucleation and ensuing growth, sedimentation, Marangoni migration, coalescence, and related interplay of all these aspects). During material processing, the different effects cannot be separated; they act all together and the related mutual interference can be very complex.

This limitation, however, can be removed if a single droplet is considered. This fundamental situation is considered in the next sections.

The conditions corresponding to the aforementioned lack of equilibrium, in principle, can be reproduced by directly injecting the pure liquid representing the minority phase into the majority phase at a temperature below the critical one. If this is accomplished by means of a needle, accordingly, a single dissolving droplet anchored to the top of the needle is obtained: the coagulation events cannot occur since a single drop is present and the Marangoni migration as well as the sedimentation phenomena are prevented by the fact that the droplet is locked to its initial position.

In order to better reproduce conditions corresponding to real industrial processes (where the materials are formed by means of a thermal quench of the initially homogeneous mixture exhibiting a miscibility gap from temperatures above the gap to lower temperatures inside the gap) the dissolution process must also be investigated under the presence of an imposed temperature gradient.

Unfortunately immiscible metal alloys are not transparent to visible light and a direct visualization of flows and solutal phenomena is not possible. A miscibility gap in the liquid phase, however, is also found in many other different systems: binary mixtures of organic liquids, sulfides and silicates systems, glasses and liquid crystals. Further to the case of metal alloys, many industrial applications and processes are based on miscibility-gap and related phenomena.

According to the major difficulty in investigating liquid metals for their opaqueness, organic liquids are often used as model substances also to study these aspects. For instance, methanol and cyclohexane have been selected and investigated both in Earth and in Space laboratories since they are transparent liquids exhibiting a miscibility gap in the phase diagram; in addition, the small difference in density (0.769 $g\,cm^{-3}$ the cyclohexane, 0.782 $g\,cm^{-3}$ the methanol) allows one, in principle, to minimize buoyancy and sedimentation effects.

The thermosolutal-capillary migration of a dissolving liquid drop, initially composed of pure methanol, injected in a closed cavity filled of (initially) pure cyclohexane with differentially heated end walls under microgravity conditions was studied by Bassano (2003) using a finite-volume formulation and a level-set technique. The main goal of the analysis, with respect to previous analyses dealing with immiscible liquids, was to clarify at what extent the thermally induced drop migration is affected by the dissolution process.

It is worthwhile to stress how, however, the behavior of dissolving drops is even more interesting in the case of normal-gravity conditions, the industrial processes being usually carried out on the ground. On Earth, the dissolution of dispersed drops leads to solutal pluming phenomena that can exhibit a very intriguing behavior. In this case the interaction between the liquid drops and the surrounding miscible fluid is very complex. It is manifested by a fluid motion brought about by different coexisting mechanisms: Marangoni convection, generated by the interfacial stresses due to temperature or concentration gradients along the drop surface (thermocapillary or solutocapillary convection, respectively) and the gravitational convection driven by the volumetric buoyancy forces caused by thermally and/or solutally generated density variations in the bulk of the fluid. The next section introduces a VOF method specially designed for the analysis of these topics (Lappa et al., 2004a).

3.8. DDVOF – Volume of Fluid Method for dissolving drops

The DDVOF is a single-region formulation and allows a fixed-grid solution to be undertaken. It is therefore able to utilize standard solution procedures for the fluid flow and species equations directly, without resorting to mathematical manipulations and transformations (e.g., body-fitted curvilinear coordinates).

The model is based on the mass-balance equations. The diffusion of the species is governed by the equations (drop = phase 1 $\rightarrow \phi = 1$, matrix = phase 2 $\rightarrow \phi = 0$):

$$\frac{\partial C_1}{\partial t} = \phi\left[-\underline{\nabla}\cdot(\underline{V}C_1) + D\nabla^2 C_1\right] \tag{3.29}$$

$$\frac{\partial C_2}{\partial t} = (1-\phi)\left[-\underline{\nabla}\cdot(\underline{V}C_2) + D\nabla^2 C_2\right] \tag{3.30}$$

where D is the interdiffusion coefficient and C_1 and C_2 are the concentrations (mass fraction) of one of the two components (for instance, methanol) in the minority (drop) and majority (matrix) phases, respectively.

The species equations (3.29) and (3.30) can be solved throughout the computational domain including the drop and the matrix. The presence of the terms ϕ and $(1-\phi)$ ensures, in fact, that each equation characterizes a different phase. The initial constant concentrations in the drop and in the matrix are denoted by $C_{1(\mathrm{o})}$ and $C_{2(\mathrm{o})}$, respectively (for the case under investigation: $C_{1(\mathrm{o})} = 1$, i.e. drop of pure methanol and $C_{2(\mathrm{o})} = 0$, i.e. matrix of pure cyclohexane). As time passes, owing to molecular diffusion through the interface, the drop tends to be enriched with cyclohexane (i.e. C_1 becomes < 1), correspondingly, the majority external phase tends to be enriched with methanol (i.e. C_2 becomes > 0).

The behavior of the two phases is coupled through the equilibrium concentration values imposed on the two sides of the interface (this means that when a drop of pure methanol is injected into the cyclohexane liquid matrix, it is assumed that equilibrium at the liquid–liquid interface is attained instantaneously, see also Perez de Ortiz and Sawistowski, 1973). The aforementioned values come from the miscibility law (see Fig. 3.24) according to the local value of the temperature of the interface (for isothermal conditions, of course, the equilibrium values are constant along the interface).

At the interface $(0 < \phi < 1)$, the concentrations must satisfy the equilibrium conditions:

$$C_1|_i = C_{1(\mathrm{e})}(T) \tag{3.31a}$$

$$C_2|_i = C_{2(\mathrm{e})}(T) \tag{3.31b}$$

that is the equilibrium methanol concentrations at the consolute points of the miscibility diagram (Fig. 3.24a).

Equations (3.31) behave as "moving-boundary conditions." Note that in place of the "sharp" transition from one phase to the other that would characterize the multiple-region formulations (in that case the interface separating the bulk phases is a mathematical boundary of zero thickness), here (following the VOF philosophy) the phase field varies smoothly but rapidly through an interfacial region whose thickness is not zero. This region, defined by the mathematical conditions $|\underline{\nabla}\phi| \neq 0$, $0 < \phi < 1$, moves through the computational domain according to the behavior of the different phases (i.e. according to the behavior of ϕ).

The flow is governed by the continuity, and Navier–Stokes equations:

$$\underline{\nabla} \cdot \underline{V} = 0 \tag{3.32}$$

$$\frac{\partial(\rho \underline{V})}{\partial t} = -\underline{\nabla} p - \underline{\nabla} \cdot [\rho \underline{V}\,\underline{V}] + \underline{\nabla} \cdot \left[\mu(\underline{\nabla}\,\underline{V} + \underline{\nabla}\,\underline{V}^{\mathrm{T}})\right] + \underline{F}_{\mathrm{g}} \tag{3.33}$$

where

$$\rho = \rho_{1(\mathrm{o})}\phi + \rho_{2(\mathrm{o})}(1-\phi) \tag{3.34a}$$

$$\mu = \mu_1\phi + \mu_2(1-\phi) \tag{3.34b}$$

and in the last term of Eq. (3.33):

$$\begin{aligned} F_g = \Big\{ &\rho g[\beta_{\mathrm{C}1}(C_1 - C_{1(\mathrm{o})})]\phi + \rho g[\beta_{\mathrm{C}2}(C_2 - C_{2(\mathrm{o})})](1-\phi) \\ &+ \rho g[\beta_{\mathrm{T}1}(T - T_{\mathrm{ref}})]\phi + \rho g[\beta_{\mathrm{T}2}(T - T_{\mathrm{ref}})](1-\phi) \Big\} \end{aligned} \tag{3.34c}$$

the Boussinesq approximation being used to model the buoyancy forces ($\beta_{\mathrm{C}1}$ and $\beta_{\mathrm{C}2}$ are the solutal expansion coefficients related to the mutual interpenetration of the phases (1) and (2), $\beta_{\mathrm{T}1}$ and $\beta_{\mathrm{T}2}$, the thermal expansion coefficients).

The temperature distribution is governed by the energy equation:

$$\frac{\partial \rho C_{\mathrm{P}} T}{\partial t} = [-\underline{\nabla} \cdot (\rho C_{\mathrm{P}} \underline{V} T) + \underline{\nabla} \cdot (\lambda \underline{\nabla} T)] \tag{3.35}$$

where the thermal conductivity λ and the specific heat C_{P} are computed as

$$\lambda = \lambda_1\phi + \lambda_2(1-\phi) \tag{3.36a}$$

$$C_{\mathrm{P}} = C_{\mathrm{P}1}\phi + C_{\mathrm{P}2}(1-\phi) \tag{3.36b}$$

The core of the DDVOF method is its technique for adjourning ϕ. The tracking of the interface between the phases is accomplished by the solution of a special continuity integral equation for the volume of the liquid drop taking into account the release or absorption of a solute through the interface. Within this context the liquid drop is assumed to be a sphere of radius R increasing due to growth or decreasing due to dissolution (on the ground this assumption is reliable if the Bond number is sufficiently small, i.e. for small droplets).

If ζ denotes the velocity with which the interface moves through the reference frame, the overall jump mass balance reads $\delta[\rho(V_n - \zeta)] = 0$ where V_n is the fluid velocity component orthogonal to the free surface. Assuming $\rho_1 \cong \rho_2$ (see Table 3.3) the above equation reduces to $(V_{n1} - \zeta) = (V_{n2} - \zeta) \rightarrow V_{n1} = V_{n2}$. Moreover, $V_n = 0$ on both sides of the interface since there is no local macroscopic convective transport of fluid across the drop surface (the inward or outward motion of the interface being only based on molecular diffusion). Thus the specie balance jump equation $\delta[\rho(V_n - \zeta)C - \rho D(\partial C/\partial n)] = 0$ reads (see also Bassano, 2003):

$$-D\left.\frac{\partial C_2}{\partial n}\right|_i + C_2|_i\dot{m} = -D\left.\frac{\partial C_1}{\partial n}\right|_i + C_1|_i\dot{m} \tag{3.37a}$$

where $\dot{m}$ is the mass flow rate through the interface (n points towards the matrix; if the drop dissolves, $\dot{m}$ is positive and ζ is negative, i.e. $\dot{m} = -\zeta$ with $\zeta \cong (\mathrm{d}R/\mathrm{d}t)$); therefore:

$$D\left(\left.\frac{\partial C_2}{\partial n}\right|_i - \left.\frac{\partial C_1}{\partial n}\right|_i\right) = \zeta\,\Delta C \quad \text{with} \quad \Delta C = C_1|_i - C_2|_i \tag{3.37b}$$

that is, the difference between the diffusive solutal fluxes computed at the two different sides of the dissolving liquid–liquid surface must be balanced by the convective contribution due to the interface motion. In conclusion, integrating over the surface of the drop, one obtains for the time evolution of the radius

$$\frac{\mathrm{d}R}{\mathrm{d}t} = D\frac{1}{S}\oint \frac{1}{C_1|_i - C_2|_i}\left(\left.\frac{\partial C_2}{\partial n}\right|_i - \left.\frac{\partial C_1}{\partial n}\right|_i\right)\mathrm{d}l \tag{3.38}$$

Table 3.3 Physical properties of the cyclohexane–methanol system.

$D\,(\mathrm{m^2\,s^{-1}})$	2×10^{-9}
$\rho_1\,(\mathrm{kg\,m^{-3}})$	0.782×10^3
$\rho_2\,(\mathrm{kg\,m^{-3}})$	0.769×10^3
$\nu_1\,(\mathrm{m^2\,s^{-1}})$	1.03×10^{-6}
$\nu_2\,(\mathrm{m^2\,s^{-1}})$	1.275×10^{-6}
$\beta_{c1}\,(-)$	-4.62×10^{-2}
$\beta_{c2}\,(-)$	-4.95×10^{-2}
$\beta_{T1}\,(\mathrm{K^{-1}})$	1.14×10^{-3}
$\beta_{T2}\,(\mathrm{K^{-1}})$	1.14×10^{-3}
$C_{p1}\,(\mathrm{J\,kg^{-1}\,K^{-1}})$	2.51×10^3
$C_{p2}\,(\mathrm{J\,kg^{-1}\,K^{-1}})$	1.86×10^3
$\lambda_1\,(\mathrm{J\,m^{-1}\,s^{-1}\,K^{-1}})$	2.02×10^{-1}
$\lambda_2\,(\mathrm{J\,m^{-1}\,s^{-1}\,K^{-1}})$	1.25×10^{-1}
$\sigma_T\,(\mathrm{dyne\,m^{-1}\,K^{-1}})$	17.42

where S is the surface of the drop, and

$$\frac{\partial C}{\partial n} = \underline{\nabla} C \cdot \underline{\hat{n}} = \alpha \frac{\partial C}{\partial x} + \beta \frac{\partial C}{\partial y} \tag{3.39}$$

$$\underline{\hat{n}} = -\frac{\underline{\nabla}\phi}{|\underline{\nabla}\phi|} = (\alpha, \beta) \tag{3.40}$$

$$\alpha = -\frac{\partial \hat{\phi}}{\partial x} \Bigg/ \sqrt{\left(\frac{\partial \hat{\phi}}{\partial x}\right)^2 + \left(\frac{\partial \hat{\phi}}{\partial y}\right)^2}, \qquad \beta = -\frac{\partial \hat{\phi}}{\partial y} \Bigg/ \sqrt{\left(\frac{\partial \hat{\phi}}{\partial x}\right)^2 + \left(\frac{\partial \hat{\phi}}{\partial y}\right)^2} \tag{3.41}$$

The unit vector $\underline{\hat{n}}$ perpendicular to the interface results from the gradient of a smoothed phase field $\hat{\phi}$, where the transition from one phase to the other takes place continuously over several cells (typically 4 or 5 for a grid having 200×200 points) as also discussed in Section 3.5.2 (see Eq. (3.22), the smoothed phase field $\hat{\phi}$ is obtained by convolution of the unsmoothed field ϕ with an interpolation function). The interface orientation depends on the direction of the volume fraction gradient of the phase $\hat{\phi}$ within the cell, and that of the neighbor cell (or cells) sharing the face in question. Depending on the interface's orientation and on the side (phase (1) or phase (2)) on which computations are performed, concentration gradients must be discretized by forward or backward schemes. After the computation of the radius at the new instant $(n+1)$, then the distribution of the phase variable ϕ is reinitialized to take into account the volume change, i.e. $\phi = 1$ for $r < R$ and $\phi = 0$ for $r > R$.

Equations (3.29), (3.30), (3.32), (3.33), (3.35), and (3.38) represent a system of five partial differential equations and one ordinary differential equation whose solution governs the nonlinear behavior of the physical system under investigation.

Note that the present mathematical model and related numerical technique can be regarded as a very hybrid volume-tracking method.

In the "classical" VOF methods, the phase field variable ϕ is "advected" solving an appropriate partial differential transport equation (Eq. (3.14)); this formulation, as illustrated in Section 3.5, can be used for the solution of typical problems dealing with the migration of bubbles or drops posed in liquids. It relies on the fact that the fluids are not interpenetrating. For the present method, the equation governing the evolution of ϕ comes from mass-balance conditions rather than from transport. Moreover, interpenetration of the different fluids is allowed according to the coupled behavior of Eqs. (3.29), (3.30), and (3.38).

Furthermore, Eq. (3.38) provides the necessary coupling among the species and the momentum equations. The density and the dynamic viscosity of the liquid in Eq. (3.33), in fact, are computed according to the instantaneous distribution of ϕ.

An additional coupling between the species and the momentum equations is due to the volume force term (Boussinesq approximation) in Eq. (3.33).

The Marangoni condition at the drop interface reads (hereafter V_s and V_n denote the velocity components in the plane (x, y) parallel and orthogonal to the free surface, respectively):

$$\mu_1 \frac{\partial V_s}{\partial n}\bigg|_1 - \mu_2 \frac{\partial V_s}{\partial n}\bigg|_2 = \sigma_T \frac{\partial T}{\partial s} \tag{3.42}$$

where $V_s = \beta u - \alpha v$ (u and v are the velocity components along x and y respectively) and

$$V_n = \alpha u + \beta v = 0 \tag{3.43}$$

$$\frac{\partial V_s}{\partial n} = \alpha \frac{\partial V_s}{\partial x} + \beta \frac{\partial V_s}{\partial y} \tag{3.44}$$

$$\frac{\partial T}{\partial s} = \beta \frac{\partial T}{\partial x} - \alpha \frac{\partial T}{\partial y} \tag{3.45}$$

It is intersting to point out how, following Warren and Webb (1969), the sole (experimentally determined) parameter σ_T is needed to account for the dependence of the surface tension on both T and C (the surface equlibrium concentrations, in fact, are not free parameters but, in their turn, functions of T as shown by Fig. 3.24 and Eqs. (3.31)). The physical properties of the cyclohexane-menthanol system are shown in Table 3.3.

Depending on the interface's orientations and on the side (phase (1) or phase (2)) on which computations are performed, Eqs. (3.42–3.45) should be discretized by forward or backward schemes and allow, in prinicple, the computation of the u and v components satisfying the Marangoni shear stress balance along the drop surface. Within this context, however, it should be emphasized how the Marangoni effect could also be taken into account within the framework of the Continuum Surface Stress model described in Section 3.5.2, rather than as the cumbersome boundary conditions (3.42–3.45); for further details see, e.g., Lappa and Piccolo (2004) and Lappa et al. (2004b).

Equations (3.29), (3.30), (3.32), (3.33), (3.35) can be solved with the SMAC method (see Chapter 2).

Hereafter, some experimental results (supported by numerical simulations in the framework of the DDVOF method) are discussed as relevant examples of the curious dissolution phenomena that may be encountered on the ground.

First, the attention is focused on the case of an isothermal configuration, then the presence of a temperature gradient is considered and compared with the isothermal condition. It is worthwhile to stress how, in the case of constant temperature throughout the system, the onset of Marangoni stresses at the drop surface is prevented. In this case, in fact, neither the temperature nor the solutal gradients can exist at the interface separating the fluids (the equilibrium concentration values outside and inside the droplet, respectively, are constant along the surface). The comparison between the isothermal and the nonisothermal conditions, allows the role played by the Marangoni effect in the dissolution process and on the underlying convective mechanisms to be discerned.

3.9. Dissolution in isothermal conditions

According to the experimental results, if a droplet of methanol is formed on the tip of a capillary immersed into the cyclohexane (Fig. 3.25), in spite of the small differences in density between the drop and the surrounding matrix, a convective plume is clearly observed; the plume originates at the tip of the drop and is directed upward (Fig. 3.26).

In previous experiments with different liquid–liquid systems (aniline/water, isobutanol/water, ethilacetate/water, see Agble and Mendes Tatsis, 2000) similar convective plumes were also observed. In these studies, the systems basically consisted of sitting drops or of pendant drops. It was shown that the convective plume originated from the drop is directed downward or upward according to whether the liquid is heavier or lighter than that used for the matrix.

For the system under investigation, the droplet (methanol) is heavier than the surrounding matrix. Nevertheless, the experiments performed in this case (isothermal conditions) show an upward methanol flow, starting from the upper

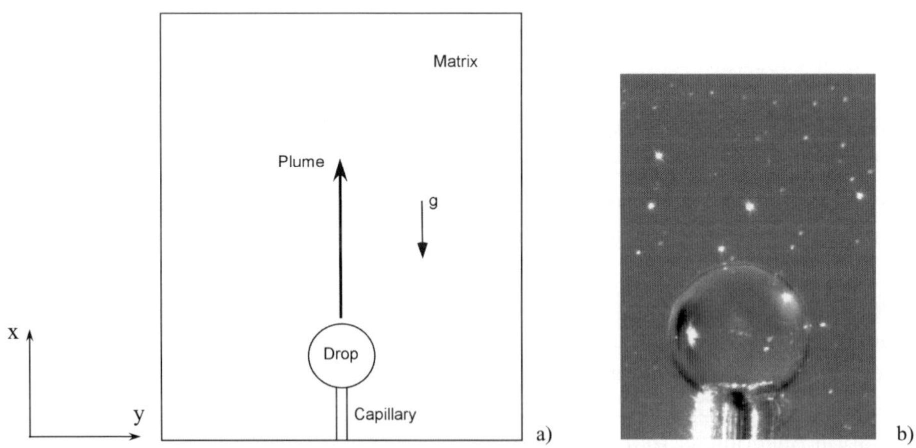

Fig. 3.25 (a) Sketch of the geometrical configuration; (b) liquid drop at the tip of the capillary tube.

point of the drop. This flow corresponds to the classical solutal Rayleigh–Taylor "plume" typically observed in buoyant phenomena. It is worth spending some words about the fact that such a plume is created in the host liquid (cyclohexane) despite this liquid exhibiting a smaller density with respect to the drop (methanol). Such an apparent counterintuitive behavior, in reality, is a consequence of the fact that the density of the cyclohexane–methanol system does not necessarily behave as an increasing function of the methanol concentration and can be a decreasing function for small values of this concentration (Fig. 3.24b). This means that at early stages of the dissolution process, the density of the majority phase (rich in cyclohexane) tends to become smaller even though methanol is heavier than cyclohexane.

When a drop of pure methanol is injected into the cyclohexane liquid matrix, equilibrium at the liquid–liquid interface is attained instantaneously (Perez de Ortiz and Sawistowski, 1973). Assuming negligible thermal effects (for instance, an average temperature of 25°C), according to the phase rule the concentrations at the interface will be C_1 and C_2 (inside and outside the droplet, respectively, see Fig. 3.24). Correspondingly, the mixture surrounding the droplet tends

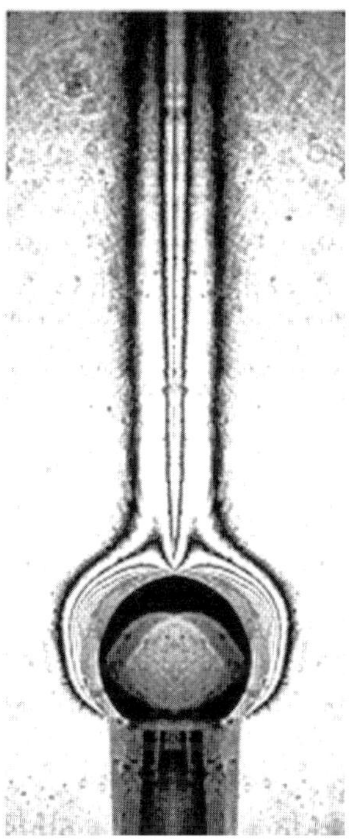

Fig. 3.26 Rising plume from a droplet of methanol dissolving in cyclohexane (uniform temperature distribution $T = 25\,^\circ\mathrm{C}$, initial drop volume 3 μl): Experimental results are shown here.

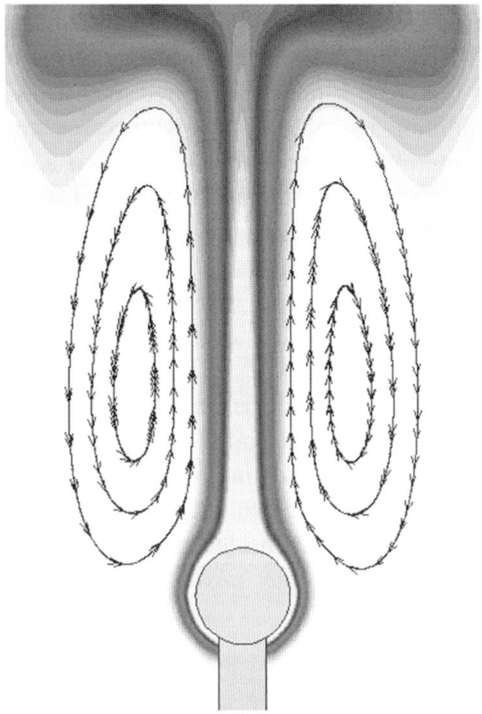

Fig. 3.27 Rising plume from a droplet of methanol dissolving in cyclohexane (uniform temperature distribution $T = 25\,^{\circ}\mathrm{C}$, initial drop volume 3 μl): Numerical results are shown here.

to be lighter than the external matrix and therefore a rising plume is originated (Fig. 3.27).

After an initial transient time (Fig. 3.28), a quite stable buoyant rising jet continuously carrying methanol towards the top of the test cell is created (Fig. 3.27).

This case is particularly significant and relevant within the context of this book since it demonstrates (in principle) that, owing to the particular density behavior of systems with a miscibility gap, drops sedimented to the bottom of the enclosure containing the immiscible alloy can give rise to rising solutal jets even if they are heavier than the surrounding liquid.

3.10. Dissolution in nonisothermal conditions

If a temperature gradient ($\partial T/\partial x$) is applied to the liquid matrix, by heating the top of the test cell and cooling the bottom, the regular release of methanol in the rising jet discussed above, is taken over by a new spatiotemporal behavior (Lappa et al., 2004a).

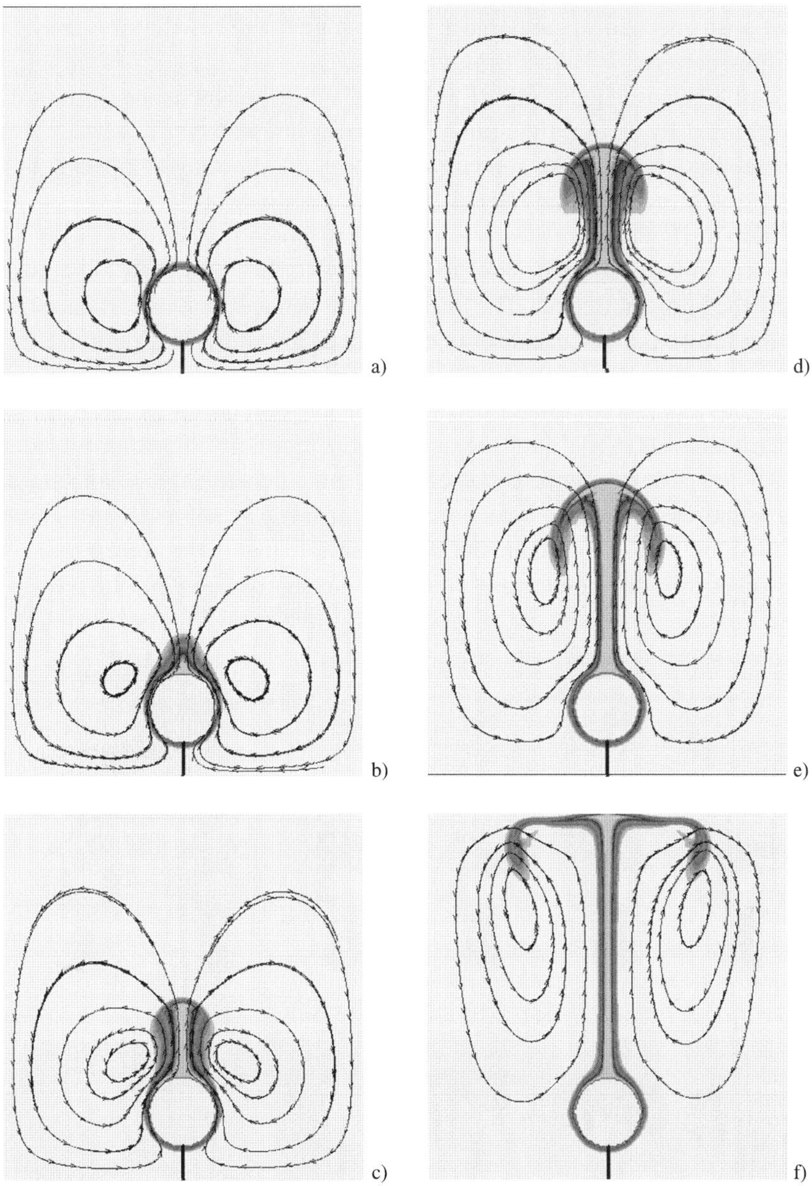

Fig. 3.28 Computed evolution of the concentration and flow-field patterns at the beginning of the dissolution process (droplet of methanol dissolving into a matrix of cyclohexane, initial volume of the droplet 3 μl). The time intervals are 8 s.

For sufficiently small values of the temperature gradient ($\partial T/\partial x$), the transport of methanol in the rising jet is similar to that obtained in the case of isothermal conditions, but if $\partial T/\partial x$ exceeds a certain critical value (for instance, for a droplet having an initial volume of 3 μl, the critical $\partial T/\partial x$ is about $1\,^{\circ}\mathrm{C\,cm^{-1}}$), the aforementioned mechanism undergoes a transition to an oscillatory axisymmetric complex flow pattern.

Instead of a steady regular plume, release of solute occurs in the form of discrete events, separated by certain time (and thus space) intervals.

The instability, in fact, leads somehow to a periodic shooting of a packet of lighter fluid in the surrounding liquid (Fig. 3.29).

This behavior appears in the generic meridian plane in the form of a couple of curls that are initially formed close to the drop surface and then are transported upward by the buoyancy effect. The phenomenon behaves as a disturbance traveling towards the top (i.e. a perturbation of the solutal and flow field periodically rising along the symmetry axis).

These behaviors are confirmed by the numerical simulations (Fig. 3.30).

The mathematical model and the associated numerical algorithm introduced in Section 3.8 are able to capture the complex time-dependent phenomena and to elucidate some experimental behaviors. For instance, the numerical simulations (Fig. 3.30) show that the curls in the interference-fringe pattern correspond to the presence of a toroidal convection roll periodically released at the peak of the dissolving drop.

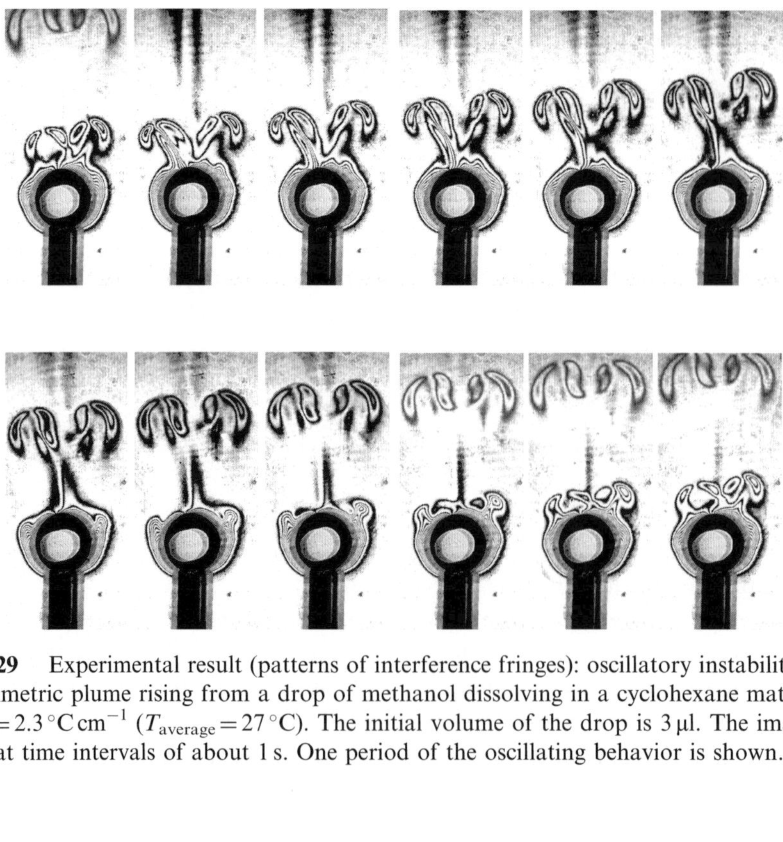

Fig. 3.29 Experimental result (patterns of interference fringes): oscillatory instability of the axisymmetric plume rising from a drop of methanol dissolving in a cyclohexane matrix with $\partial T/\partial x = 2.3\,^{\circ}\mathrm{C\,cm^{-1}}$ ($T_{\mathrm{average}} = 27\,^{\circ}\mathrm{C}$). The initial volume of the drop is 3 μl. The images are taken at time intervals of about 1 s. One period of the oscillating behavior is shown.

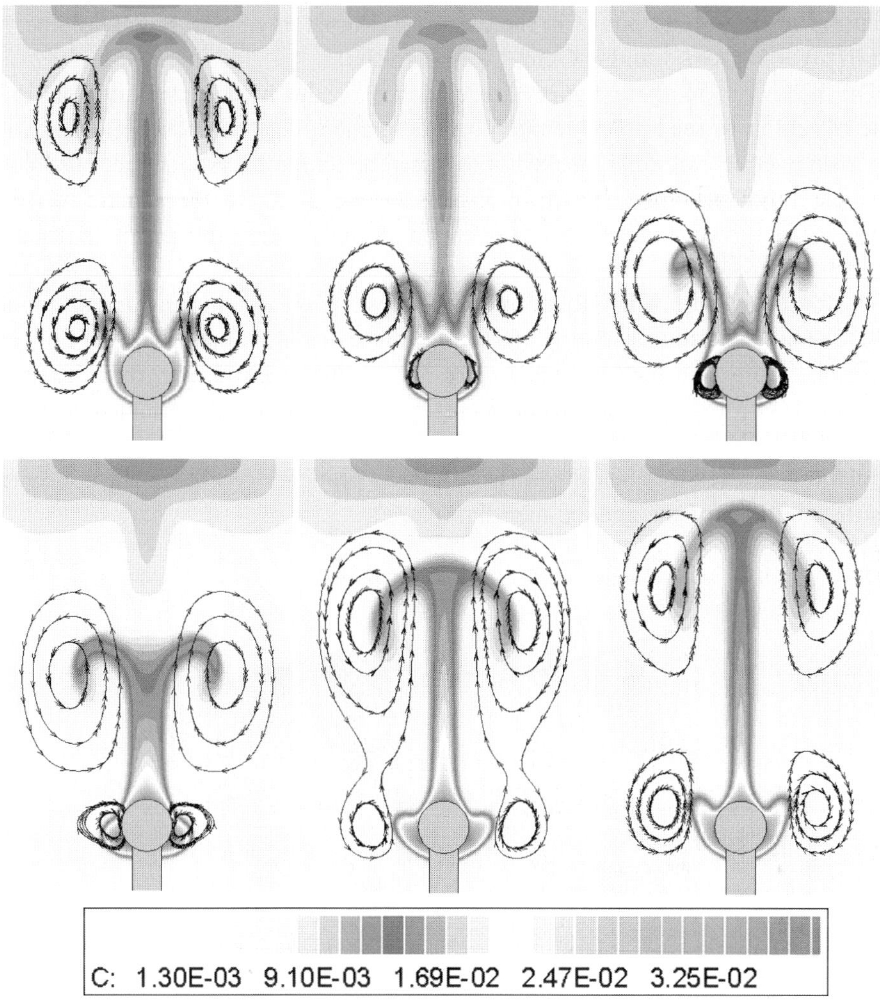

Fig. 3.30 Computed evolution of the unstable solutal plume: concentration and flow-field patterns for a 3 μl droplet of methanol dissolving into a matrix of cyclohexane with $\partial T/\partial x = 2.3\,^{\circ}\mathrm{C\,cm^{-1}}$ ($T_{\mathrm{average}} = 27\,^{\circ}\mathrm{C}$). One period of the oscillating behavior is shown (the period τ has been divided into 6 parts and the concentration and velocity fields are shown in figures a–f corresponding to $t = 0$, $\tau/6$, $\tau/3$, $\tau/2$, $2\tau/3$, $5\tau/6$, respectively, with $\tau \cong 19\,\mathrm{s}$).

If the analysis is extended to larger values of the imposed temperature gradient, as expected, the phenomena exhibit an increasing degree of complexity (for some theoretical background see the discussion on the instability of flow and related route to disorder in Section 1.4.1 and Lappa and Piccolo, 2004).

The periodic release of the aforementioned convection roll tends, in fact, to become very intricate. Figures 3.33 and 3.34 show intriguing computed stages of evolution of the solutal plume in the case of a droplet having an initial volume 5 μl and $\partial T/\partial x = 5\,^\circ\mathrm{C\,cm^{-1}}$ and $8^\circ\mathrm{C\,cm^{-1}}$, respectively.

The behaviors in these figures are evidence of the high sensitivity that these systems exhibit to the applied temperature difference. Very complex scenarios arise. The rising jet is not only broken into discrete temporal events, but also loses its single-finger structure (shown in Figs. 3.29 and 3.30). In the generic meridian plane, in fact, two somewhat wave-shaped branches appear (see, in particular, Fig. 3.31).

The frequency with which the toroidal vortices are released increases as a function of $\partial T/\partial x$ and, accordingly, many interacting rolls can simultaneously affect flow and solute distribution in the test cell. Figure 3.31 show, in particular, that the multicellular structure along the symmetry axis is featured by a periodic alternance of toroidal rolls with small and large diameter. This effect is responsible for the ondulations in the aforementioned two opposed branches (visible in the concentration distribution for the generic meridian plane).

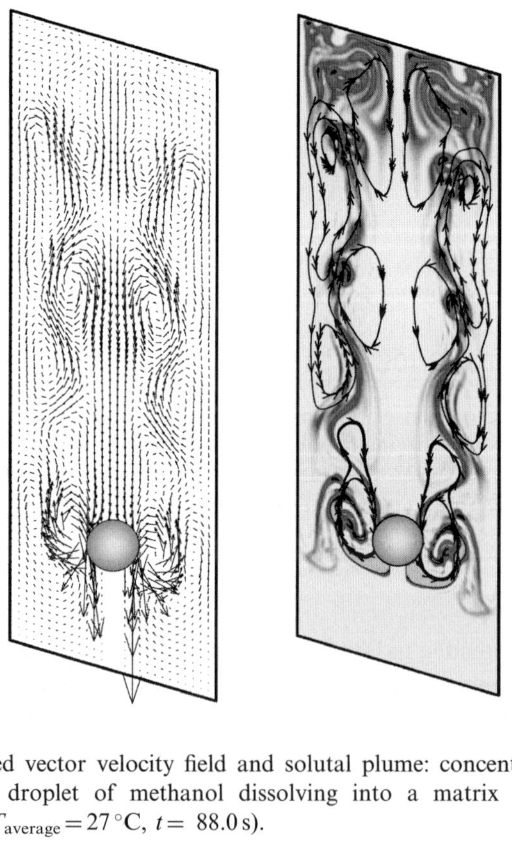

Fig. 3.31 Computed vector velocity field and solutal plume: concentration and flow-field patterns for a 5 μl droplet of methanol dissolving into a matrix of cyclohexane with $\partial T/\partial x = 5\,^\circ\mathrm{C\,cm^{-1}}$ ($T_{\mathrm{average}} = 27\,^\circ\mathrm{C}$, $t = 88.0$ s).

Table 3.4 Modes of convection as they appear in the generic meridian plane (5 μl droplet of methanol dissolving into the matrix of cyclohexane).

$\partial T/\partial x$ (°C cm^{-1})	Regime
< 1.2	Stable plume
1.5	Pulsating rising jet – single finger structure
5	Periodic shooting of solutal packets ≃ sinusoidally shaped branches
8	Quasiperiodic shooting of solutal packets ≃ sinusoidally shaped branches

Progression from spatial "order" to "disorder" versus $\partial T/\partial x$ is shown in Figs. 3.32, 3.33, and 3.34. Superimposed on these aspects is the evolutionary transition from the initial steady state to a final quasi-periodic fluctuating flow regime that appears if the phenomenon is considered from a temporal point of view.

When $\partial T/\partial x$ is sufficiently small a well-defined frequency (f_0) can be identified but other incommensurable frequencies superimpose on the principal one if $\partial T/\partial x$ is increased.

Table 3.4 summarizes the prominent spatial and temporal structure of the rising solutal jet as a function of the increasing $\partial T/\partial x$: "stable plume," "pulsating rising jet – single finger structure," "periodic shooting of solutal packets – sinusoidally shaped branches," "quasiperiodic shooting of solutal packets – sinusoidally shaped branches" are the different possible regimes of evolution of the system shown, respectively, in Figs. 3.27 and 3.32–3.34.

This time-dependent oscillatory convection is not only of inherent phenomenological significance, but it is also of prime practical interest to ground-based immiscible alloys processing. The oscillatory thermosolutal convection generated by the drop, in fact, could influence solute segregation and interfere with the solidification mechanisms in a significant manner.

3.11. Mixed buoyant–Marangoni instability of solutal jets

This section is devoted to the investigation of the intrinsic nature of the curious instabilities described in the earlier sections.

Along these lines, computations obtained "switching off" surface-tension effects, can be used to prove that the presence of the Marangoni effect is crucial in determining them. These simulations show, in fact, that in the absence of Marangoni effects the solutal jet is not broken into discrete events and the aforementioned characteristic curls do not appear.

It is well known that Marangoni forces drive liquid from hot regions towards cold regions. Accordingly, on the surface of the drop, Marangoni effects tend to drive liquid towards the bottom. On the contrary, solutal buoyancy effects act in the opposite direction. Therefore, if the test cell is heated from above and cooled from below, thermal Marangoni and solutal buoyancy flow counteract.

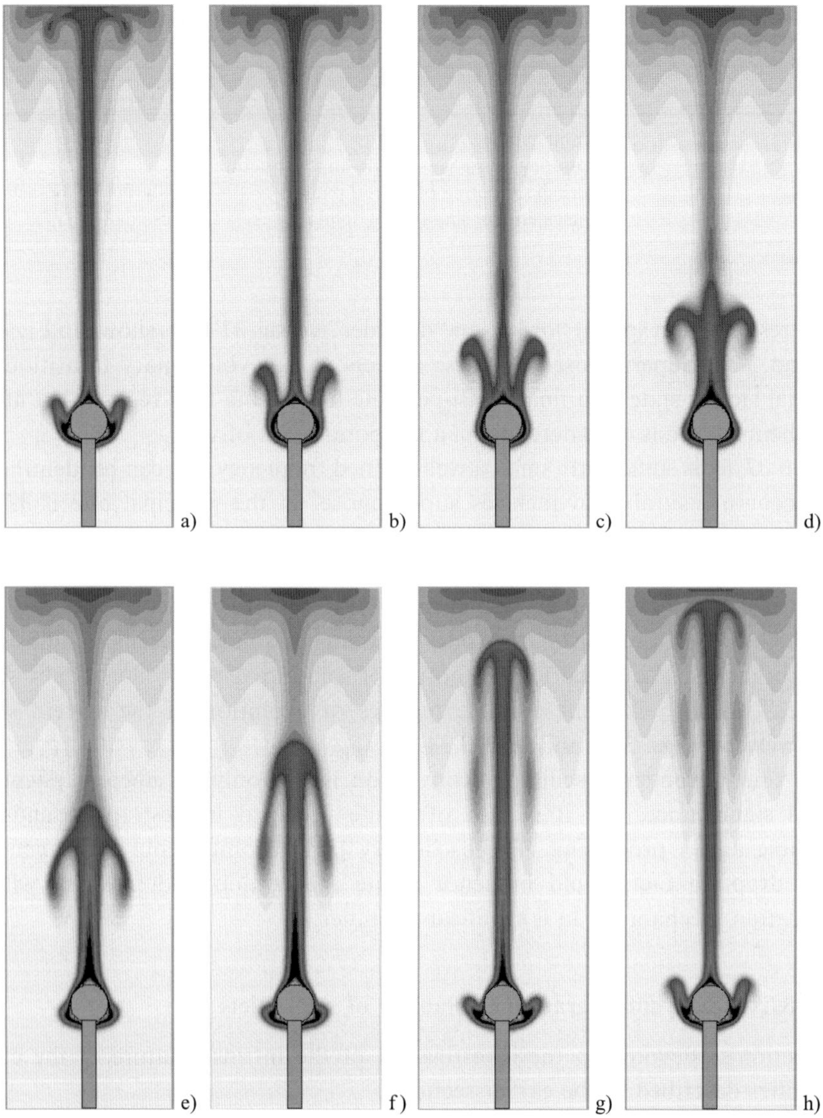

Fig. 3.32 Computed evolution of the unstable solutal plume: concentration and flow-field patterns for a 5 μl droplet of methanol dissolving into a matrix of cyclohexane with $\partial T/\partial x = 1.5\,^{\circ}\mathrm{C\,cm^{-1}}$ ($T_{\mathrm{average}} = 27\,^{\circ}\mathrm{C}$): one period of oscillation is shown by means of subsequent snapshots.

In view of these arguments and due to the fact that the instability does not appear if the surface-tension forces are artificially neglected in the simulations, a theory based on the delicate balance of the aforementioned counteracting phenomena could be proposed to explain the oscillatory behavior.

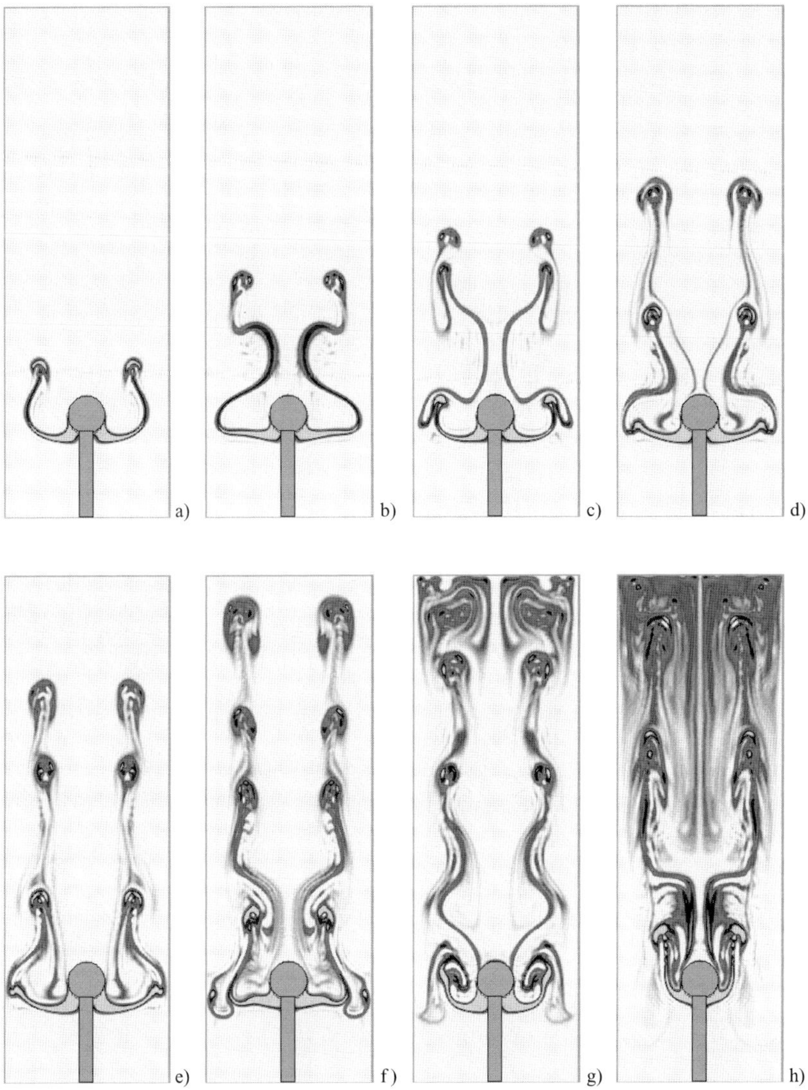

Fig. 3.33 Computed evolution of the unstable solutal plume: concentration and flow-field patterns for a 5 μl droplet of methanol dissolving into a matrix of cyclohexane with $\partial T/\partial x = 5\,^\circ\mathrm{C\,cm^{-1}}$ ($T_{\mathrm{average}} = 27\,^\circ\mathrm{C}$): (a) $t = 17.9$ s; (b) $t = 27.7$ s; (c) $t = 34.2$ s; (d) $t = 42.4$ s; (e) $t = 52.2$ s; (f) $t = 68.5$ s; (g) $t = 88.0$ s; (h) $t = 117.4$ s.

A particular procedure can be used to get insights into these aspects. The basic approach proposed herein for this purpose relies on the combination of the available experimental and numerical results with theoretical models. First, the trends in available data are used to postulate qualitatively mechanical principles for explaining the instability (for the case under investigation the aforementioned

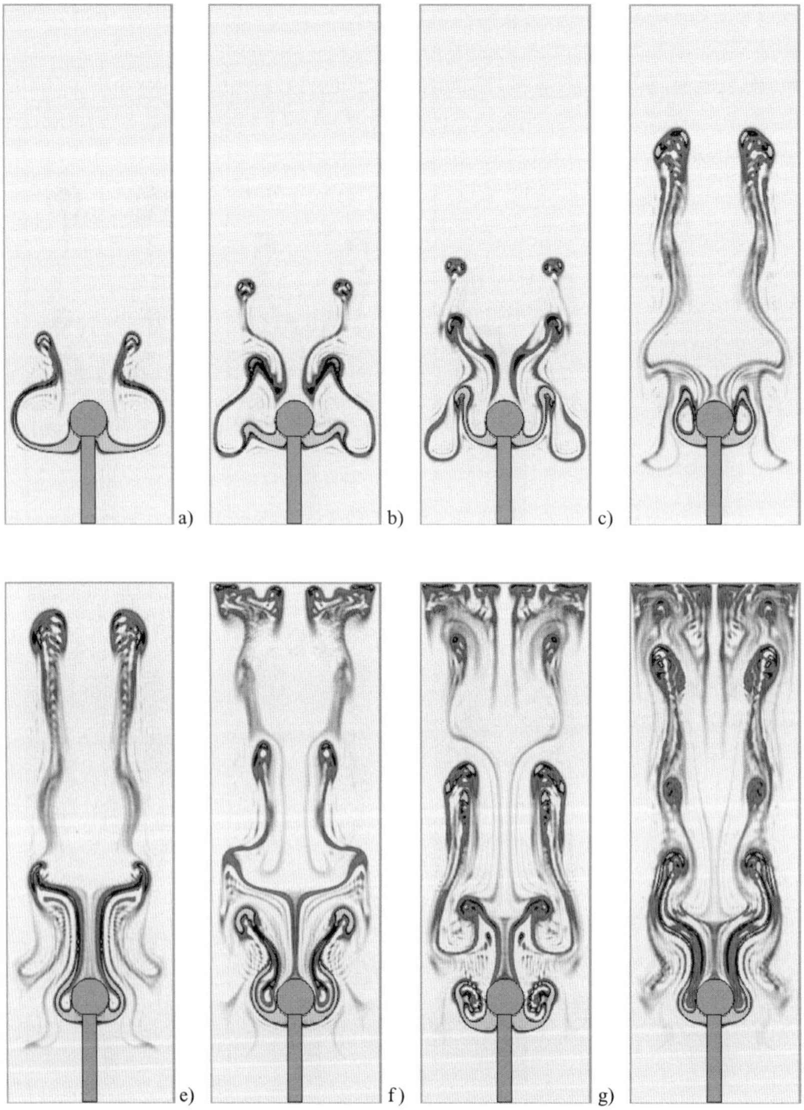

Fig. 3.34 Computed evolution of the unstable solutal plume: concentration and flow-field patterns for a 5 μl droplet of methanol dissolving into a matrix of cyclohexane with $\partial T/\partial x = 8\,^{\circ}\mathrm{C\,cm^{-1}}$ ($T_{\mathrm{average}} = 27\,^{\circ}\mathrm{C}$): (a) $t = 17.9$ s; (b) $t = 24.5$ s; (c) $t = 27.7$ s; (d) $t = 44.0$ s; (e) $t = 52.2$ s; (f) $t = 65.2$ s; (g) $t = 76.6$ s; (h) $t = 86.4$ s.

delicate balance between solutal buoyancy forces and Marangoni thermal stresses). Then, on the basis of the foregoing principles, explicit laws are proposed and tested for their ability to predict the experimentally determined patterns of the instability threshold.

In practice, this approach requires a very demanding (quite lengthy) parametric analysis about the effect of the initial volume of the droplet. A scaling analysis, in fact, must be carried out to discern the theoretical relationship between the critical temperature gradient and the initial radius of the droplet. See the detail in the discussion below:

If the instability resulted from a delicate counteracting behavior between solutal buoyancy and thermal surface driven effects, it would be possible to draw information about the critical threshold simply imposing that

Intensity of solutal buoyancy effects/intensity of Marangoni thermal effects = O(1).

The relative importance of these effects can be estimated by comparing the relevant characteristic velocities (see Chapter 2 for the relevant theoretical background).

The solutal buoyancy velocity reads:

$$V_{\mathrm{g}}^{\mathrm{C}} = \frac{g\beta_{\mathrm{c2}}\Delta C(2R)^2}{\nu} \tag{3.46}$$

where ΔC is the initial concentration difference between the drop surface and the top of the test cell (for the present case $\Delta C \cong C_{2(\mathrm{o})}$).

The thermal Marangoni velocity reads:

$$V_{Ma}^{\mathrm{T}} = \frac{\sigma_{\mathrm{T}}(\partial T/\partial x)\,2R}{\mu} \tag{3.47}$$

According to the foregoing arguments the conditions for the onset of the oscillatory instability should correspond simply to

$$\frac{V_{Ma}^{\mathrm{T}}}{V_{\mathrm{g}}^{\mathrm{C}}} = \mathrm{O}(1) \tag{3.48a}$$

$$\rightarrow \frac{\partial T}{\partial x} \propto \frac{g\beta_{\mathrm{c}}\rho\Delta C}{\sigma_{\mathrm{T}}} 2R = \frac{g\beta_{\mathrm{c2}}\rho\Delta C}{\sigma_{\mathrm{T}}} 2\sqrt[3]{\frac{3}{4\pi}\mathrm{Vol}} \tag{3.48b}$$

Equation (3.48b) is the aforementioned analytical relationship between the critical gradient and the droplet diameter provided by the theoretical scaling analysis carried out under the assumption that the instability is caused by the opposition of the surface Marangoni stress to the natural direction in which the fluid wants to move (upward, because of buoyancy). This explicit law shows that the critical $\partial T/\partial x$ is an increasing function of the initial volume of the droplet.

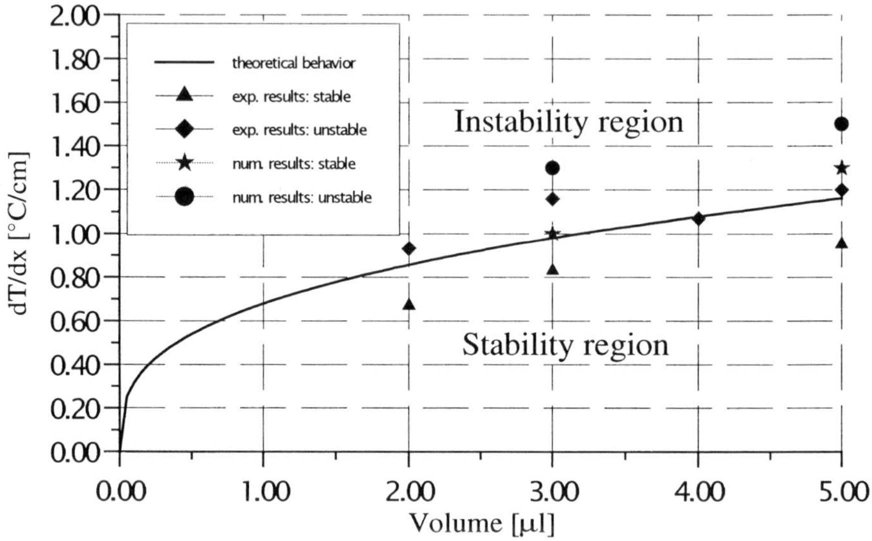

Fig. 3.35 Critical temperature gradient for the onset of oscillations as a function of the initial volume of the dissolving droplet (Lappa et al., 2004a).

To provide physical validation to these arguments, Eq. (3.48b) must be finally compared with the numerical and the experimental results. Such a comparison shows that the relationship presented herein fits the dependencies highlighted by the the experimental and the numerical results (Fig. 3.35). In turn, this good agreement provides validation for the proposed theory.

The consistency of model predictions with the experimental data suggests that rate-controlling steps have been taken into account, and that simplifications do not distort actual behavior. The proposed theory about the counteracting interplay of surface-tension and buoyancy forces exhibits quite exhaustive capabilities to predict and elucidate experimental observations (in particular with regard to the behavior of the system as a function of the initial droplet volume) and to identify cause-and-effect relationships, i.e. to give insights into the mechanisms driving the phenomena under investigation.

If the temperature gradient is sufficiently small, buoyancy effects prevail and the velocity along the drop surface is upward (i.e. directed from the capillary towards the point at the peak of the drop). In this case the release of solute occurs in the form of a continuous rising jet of lighter liquid starting at the peak of the methanol drop.

If the temperature gradient is increased, correspondingly, the relative importance of the surface Marangoni stress with respect to the local buoyancy forces increases.

The surface-tension-induced flow carries the hot fluid from the peak of the drop into the lower regions of the domain just below the drop base. The solutal buoyancy force opposes the thermocapillary flow and the stabilizing density gradients exert a restoring upward force on the downward flowing hot fluid. This restricts the flow to an area close to the drop.

The $\partial T/\partial x$ at which the aforementioned counteracting effects can be considered competitive in determining the direction of the surface flow, corresponds to the critical conditions for the onset of the instability. For temperature differences larger than the critical one, a "back and forth" mechanism occurs, i.e. an instability of the mixed buoyancy–Marangoni type.

The complex interplay among transport of heat, solute and momentum results in a pumping action first pushing down and then pulling up the fluid close to the drop surface. This results in a complex flow structure that affects the entire enclosure as depicted in the sequence in Figs. 3.29 and 3.30.

This explains why the instability occurs only if a certain initial value of the temperature gradient is exceeded, whereas the flow exhibits quite a stable solutal rising plumes in the case of small initial ΔT or isothermal conditions. The generation of curls close to the drop surface is strictly related to the counteracting behavior of buoyancy and surface-tension forces that leads to the production of vorticity.

The discussion above is quite an impressive example of the insights into the physics that can be obtained by comparison of experimental results with numerical simulations based on custom mathematical models. It is worthwhile to stress how the behaviors discussed above are heretofore unseen and represent an original contribution to the fluid dynamics of drops in liquid–liquid systems (fundamental research) as well as to the field of materials (with miscibility gap) science.

To conclude this chapter, it is interesting to point out how the oscillatory behaviors elucidated in the foregoing exhibit some remarkable analogies with other systems in nature.

For instance, typical instabilities of rising buoyant jets were investigated by Cetegen and coworkers for a number of different cases and fluid–fluid systems (e.g., candle flames and buoyant plumes).

Cetegen (1997) carried out experimental studies dealing with the instabilities and flow transitions of buoyant plume/jets of gas mixtures. The author investigated the case of a buoyant plume of helium or helium–air mixture originating from a large axisymmetric nozzle with low velocity. He found toroidal–vortex formation as a result of rapid buoyant acceleration of light plume fluid in heavier more or less quiescent surroundings.

It was found that these plumes may undergo periodic oscillations: as the buoyant fluid exits the nozzle, the plume boundary contracts towards the plume centerline as a result of buoyant acceleration due to the hydrostatic pressure field and the condition imposed by the conservation of mass; the plumes undergo periodic oscillations of the plume boundary starting in the immediate vicinity of the nozzle lip. This behavior might resemble that observed in the present chapter about the periodic release of curls close to the drop surface.

Moreover, in analogy with the present case of solutal plumes, it was found experimentally by Cetegen (1997) that not all buoyant plumes exhibit periodic oscillations. The onset of oscillations is, in general, a function of nozzle diameter, the nozzle exit velocity and the density ratio between the plume fluid and its surroundings (Cetegen, 1997) as for the present case, instability occurs only if a critical temperature gradient is exceeded.

Cetegen and Kasper (1996) provided a very interesting explanation of the observations: the mechanism leading to the periodic oscillatory state of the flow field is connected with the highly unstable (Rayleigh–Taylor) density stratification in the sharply contracting region of the flow just above the nozzle exit.

Important similarities can also be found about the behavior of pool fires. It is well known (Cetegen and Ahmed, 1993) that axisymmetric pool fires exhibit a periodic oscillatory motion close to their origin, often referred to as puffing. In a fire, these periodic oscillations result in the formation of large-scale (of the order of the burner diameter) flaming vortical structures at a short distance from the burner surface. These structures significantly modify the downstream flame behavior as they rise through the flame and finally burnout near the flame top.

Again, the behavior is very similar to that observed in the case of the periodic release of packets of fluid above the dissolving drop. The perturbation in the concentration field originated at the drop surface rises along the core of the jet and dies at the top.

It is crucial to point out how, despite the macroscopic analogies and apparent common features, the cause-and-effect relationships exhibit notable differences with respect to the present case of dissolving drops in nonisothermal conditions.

In the case of isothermal helium plumes and flames, the oscillatory behavior follows from an instability of the pure buoyant flow. In particular, for helium plumes the force driving the instability is the pressure gradient between the helium reservoir and the ambient. In the case of flames and fires, the driving force is the release of heat in the core of the flame due to the combustion process; puffing in pool fires, in fact, is typically stronger than buoyant nonreacting plumes due to the local heat release and the consequently maintained buoyancy. These instabilities are hydrodynamic in nature, i.e. bifurcation to oscillatory flow occurs when the velocity in the core of the rising plume exceeds a critical value. Despite the similarities on the macroscopic scale the underlying mechanisms for the case of a dissolving drop are very different. The related transition to oscillatory flow is not induced by the effect of gravity alone (like the instabilities investigated by Cetegen and coworkers) or by the action of surface-tension forces perpendicular to the interface (the aforementioned absolute instabilities in falling liquid jets, see Section 3.6); moreover, it does not follow from an instability of the rising jet; it is due to a delicate balance between two counteracting effects (buoyancy and surface-tension-driven stresses) at the surface of the dissolving drop (as proven by the numerical computations carried out by "switching off" the Marangoni effects).

These special features make this instability more similar to other categories of phenomena where free-fluid/fluid interfaces supporting Marangoni stresses are involved. Along these lines, in particular the oscillatory instability that occurs in the case of air bubbles surrounded by liquid and trapped by gravity below heated walls should be mentioned.

Experiments with transparent liquids indicate that, for a given radius, the flow and temperature fields generated by the bubble become oscillatory beyond a critical Marangoni number. From a kinematic point of view, two distinct classes of oscillatory flows were identified by Kassemi and Rashidnia (1999).

The initial or first class of oscillatory patterns are in the transverse direction and are mainly associated with azimuthally traveling waves in the flow field. In this case the phenomenon is fully three-dimensional and both symmetric and asymmetric oscillatory temperature and flow fields are possible in the generic meridian plane that can be related to the number (even or odd) of disturbance lobes formed by the hydrothermal-wave disturbance on the annular ring-shaped Marangoni roll around the bubble (with a mechanism somewhat similar to the one that will be illustrated in Section 4.12.3 in the case of high Prandtl number liquid bridges).

This first class of instabilities falls into the same category of instabilities characterizing the case of liquid bridges with $Pr \gg 1$, which have instigated much research in the literature (the related discussion is provided in Chapter 4).

The second class of oscillatory pattern has a higher frequency. It is symmetric in the axial direction, and is associated with an axially traveling wave generating a mainly up-and-down axial movement in the flow and temperature fields.

This second class, not present in the case of liquid bridges, seems to resemble the behavior previously described in the case of a dissolving drop.

The analogy between the two instabilities, however, also goes beyond the observed macroscopic oscillatory pattern. Some similarities can also be highlighted with respect to the underlying forces responsible for the time-dependent flow.

The available results for the bubble case clearly indicate that the axially-driven oscillatory state that occurs at high Marangoni numbers is induced by the dynamic interaction between the buoyancy and surface-tension effects and the resulting combined gravitational convection and Marangoni flow.

The mechanism is the following: At higher Marangoni numbers, the Marangoni convection is stronger and brings about a significant alteration of the temperature field around the bubble. An interesting feature of this temperature field is the existence of a distinct thermal structure in the form of a cold return flow intruding near the top of the bubble that is not present at lower *Ma* (hereafter referred to as "cold finger").

This cold finger plays a critical role in instigating the oscillatory response of the flow since the origin of the convective instability is closely related to an intricate temporal coupling between the large-scale thermal structures that develop in the fluid in the form of the cold finger and the temperature-sensitive surface of the bubble.

The cold finger can periodically reverse the local thermal gradient on the free surface leading to the onset of oscillations. Gravity enters the dynamics since it plays the important role of keeping the main Marangoni vortex near the bubble surface where all interaction occurs. In the absence of gravity the Marangoni vortex tends to grow unopposed until it nearly fills the entire enclosure; in such a situation the mechanism with which the cold finger affects the bubble surface is removed and the instability is suppressed.

In the light of these aspects and as also demonstrated in the analysis of Kassemi and Rashidnia (1999) by means of numerical simulation, this instability is basically hydrothermal in nature, like that described in the landmark analysis of Smith and Davis (1983) for the case of liquid layers (see also Chapter 2).

Therefore, despite all the analogies highlighted above, this type of instability is also not relevant for the present case related to dissolution of drops. In fact, visualization of the computed thermal field corresponding to Fig. 3.32 does not show the presence of any cold finger impinging on the fluid/fluid interface (Fig. 3.36). There is no reversion of the local thermal gradient on the free surface (Fig. 3.37). Moreover, the axisymmetric behavior does not belong to a secondary class of instabilities. It is a "primary" bifurcation of the flow.

As mentioned before, the instability investigated by Kassemi and Rashidnia (1999) is basically an instability of the Marangoni toroidal convection roll wrapped around the bubble very similar to that arising in the case of liquid bridges with $Pr \gg 1$. Gravity plays only a marginal role in the related mechanisms: as previously explained it makes possible the existence of a cold finger that can periodically perturb the surface-temperature distribution leading to a periodic reversion of the thermal gradient. For instance, the presence of gravity is not necessary in the case of liquid bridges since even in zero-g conditions a cold finger is always present near the hot disk, which is created by the return flow that brings cool fluid away from the cold wall along the symmetry axis (the cold liquid is carried towards the hot surface at a rather high position, see the discussion in Section 2.4.2).

The instability related to the dissolving drop is different as it follows from the direct opposition of counteracting forces (a correct denomination would be "thermocapillary-solutogravitational instability of a mixed type"). As a consequence of such a nature, in fact, the drop is not surrounded by a Marangoni vortex with fixed position and undergoing minute accelerations and decelerations in the axial direction (like in the bubble case). Rather, the drop continuously generates vorticity with the periodic (or quasi-periodic) release of rising toroidal rolls (the generation

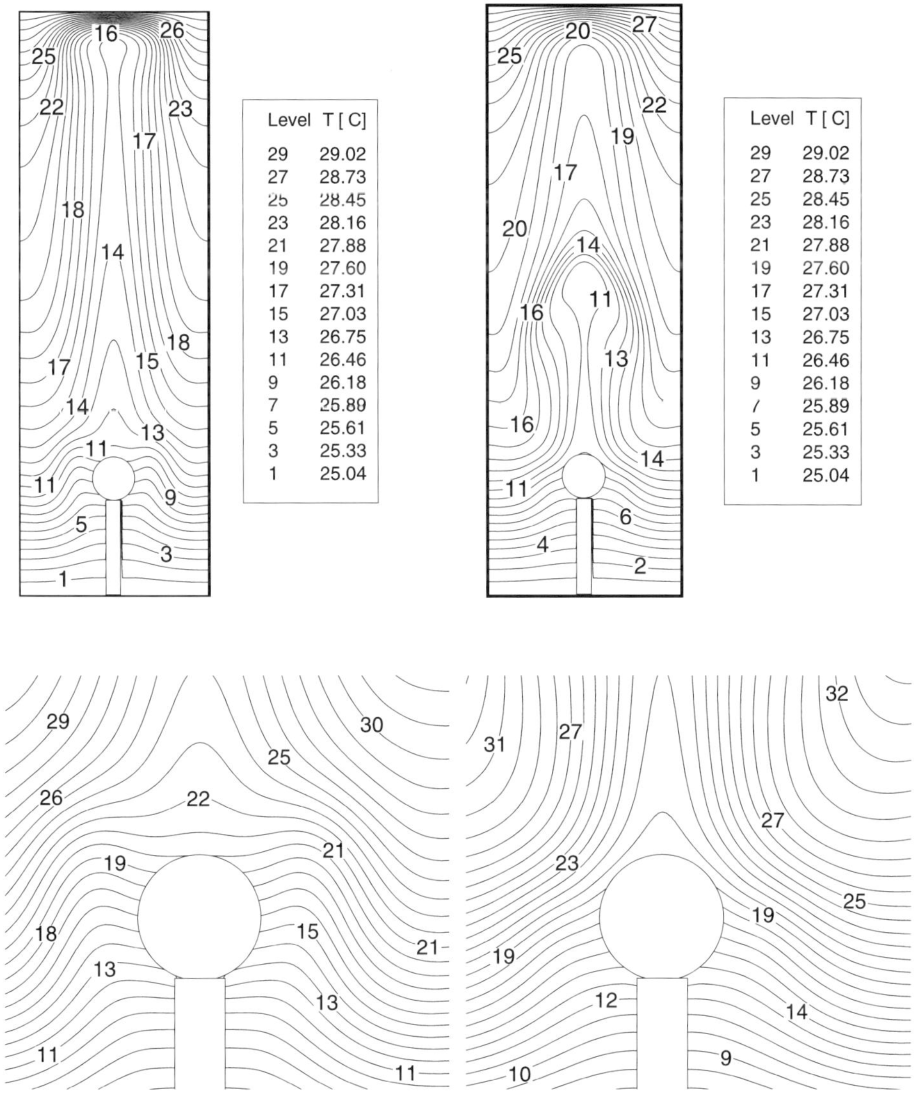

Fig. 3.36 Temperature distribution for a 5 μl droplet of methanol dissolving into a matrix of cyclohexane with $\partial T/\partial x = 1.5\,^{\circ}\mathrm{C\,cm^{-1}}$ ($T_{\mathrm{average}} = 27\,^{\circ}\mathrm{C}$): two snapshots with related enlargements are shown corresponding to $t = t_{\mathrm{o}}$ and $t = t_{\mathrm{o}} + \tau/2$ where τ is the period of oscillation.

of "curls" close to the drop surface is strictly related to the counteracting behavior of buoyancy and surface-tension forces that leads to the continuous production of vorticity).

A critical comparison of the results with other worthy and important studies concerning similar (from a macroscopic point of view) phenomena leads to the conclusion that the curious flow instability treated in the last part of this chapter,

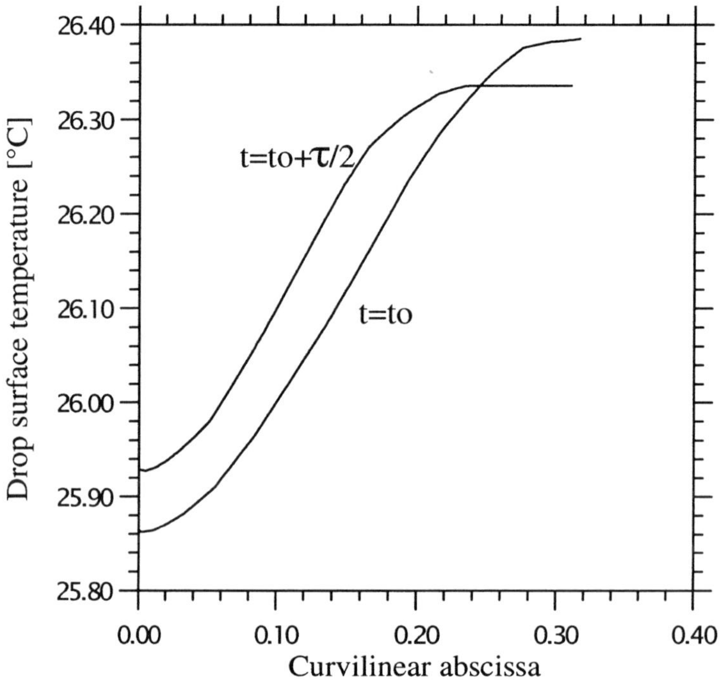

Fig. 3.37 Temperature distribution on the drop surface for the same conditions of Fig. 3.36 ($\partial T/\partial x = 1.5\,°\mathrm{C\,cm^{-1}}$).

following from the delicate balance between counteracting thermal Marangoni forces and solutal-buoyancy effects, is neither a hydrodynamic transition of the pure buoyant flow nor a hydrothermal bifurcation of the Marangoni convection, but represents a new and relevant class of instabilities driven by lack of thermodynamic equilibrium that can be present when processing liquid–liquid systems with a miscibility gap on the ground.

CHAPTER 4

Growth of semiconductors: the floating zone technique

4.1. Scientific rationale

In recent years increasing interest has been directed towards crystals of semiconductors and towards the crystallization process of special liquid metals and alloys. In fact, single crystals with superior purity and structural quality are needed to support evolutionary progress in materials science and computer technology.

Current experimentation along these lines is enabling the production of limited quantities of high-quality electronic materials and of materials that exhibit unique properties for use as benchmarks. In future, moreover, the critical knowledge gained from such pioneering research will be used for the design of the next-generation commercial crystal products (additional details about these branches of materials science research can be found in the introductory learning environment provided in Section 1.5.2).

Within this context, of particular relevance is the floating zone (FZ) technique that seems to be one of the most promising candidates for obtaining such advanced materials. Since it is capable of growing single crystals with a given orientation, as well as zone refining and leveling, this method is, in fact, ideally suited for the production of specimens with improved properties for semiconductor devices. For instance, most of the very pure silicon produced today is processed in this way.

By this containerless method, crystals can be grown with less contamination, more homogeneity and higher purity. Moreover, it is the most practical solution to problems dealing with the high chemical reactivity of these substances and their high melting point. The melt, in fact, is positioned and solidified without physically contacting the container's wall. As anticipated, this also minimizes container-induced contamination and heterogeneous nucleation. The liquid zone is produced by heating a short length of feed rod by a heat source such as a ring heater. The rod is slowly moved through the hot zone, and the crystal is obtained by resolidification of the liquid as it is moved away from the heat source (Fig. 4.1a). The polycrystalline material melts and resolidifies as the heat source passes. After one or several such passes the purity and the crystal structure of the rod is greatly improved, often leading to a single crystal.

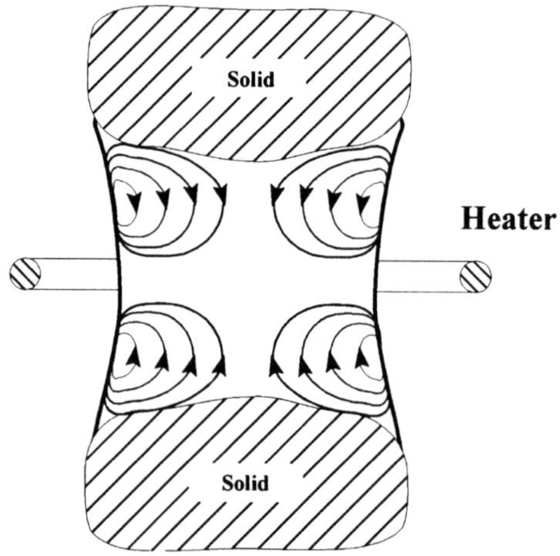

Fig. 4.1 The floating zone (FZ) technique.

This method has been used often in the terrestrial environment. However, the opportunity of accessing a low-gravity environment in recent years has given the possibility of growing larger crystals of higher quality. On Earth, in fact, the use of free melt zones is limited to small sizes for most materials due to the action of hydrostatic pressure. Moreover, buoyancy forces that are an important cause of convection and crystal imperfections (see also Section 2.2.4) are absent in space.

Besides the advantages offered in terms of size and purity (see also Muller, 1988; Benz, 1990) the use of this environment was also instrumental in unraveling and discerning processes that are interwoven or overshadowed in normal gravity. It has been understood, in fact, that even in the absence of gravity the presence of a free-melt gas interface at different temperatures induces another undesirable form of convection (the so-called Marangoni convection, see Section 2.3 for additional details dealing with the intrinsic nature of this type of flow) and that this convection can play an important role in determining crystal quality also on the ground.

Crystal growth by the floating zone technique under microgravity conditions was pioneered by Eyer et al. (1984, 1985). It was observed that the aforementioned Marangoni convection can be responsible for microscopic defects in the crystals.

Nearly all semiconductor crystals doped during growth (by the FZ technique) exhibit, in fact, microscopic variations of the dopant concentration. These striations are mostly caused by fluctuations of the growth rate or of the mixing of the melt. The fluctuations in turn result from time-dependent flows caused by an oscillatory instability of the Marangoni convection (see also Section 2.3.3 for additional theoretical background). The first experimental indication that surface-tension-driven convection could be the origin of striations in semiconductor crystals, was obtained in floating zone experiments in space by Eyer et al. (1984), in which no interacting buoyancy forces existed. Cröll et al. (1990) were the first to measure experimentally the critical temperature difference (ΔT) along the free-melt/gas surface at which the flow seems to change from steady to time-dependent.

For these reasons, the study of the structure and stability of the flow in "liquid bridges" and similar configurations with a wide free surface, introduced as models to capture the essential and relevant hydrodynamics of the FZ process, has become, over the years, a topic of great interest. Like the case of metal alloys (see Chapter 3), transparent organic liquids have enjoyed a widespread use also within the framework of these fundamental studies. Typically, in fact, liquid semiconductors are opaque and at high temperatures, so that it is hard to visualize and to study Marangoni flow and related transitions.

Experiments performed with these fluids (characterized by Prandtl numbers higher than those typical of liquid metals) showed that for sufficiently small values of the Marangoni number, the convection in liquid columns is laminar, steady and axisymmetrical, but when the Marangoni number exceeds certain critical values (the so-called critical Marangoni number Ma_c) depending on the Prandtl number of the liquid, on the geometry and on the boundary conditions, the liquid motion can undergo a transition to an oscillatory three-dimensional (3D) complex flow pattern.

The appearance of these instabilities in organic liquids was initially used by the investigators (in the 1980s) to provide a possible explanation for the undesired microscopic imperfections found in the final crystals obtained in microgravity conditions and even under normal-gravity conditions, in situations where the Marangoni flow is emphasized with respect to buoyancy forces (microscale experimentation).

Additional research carried out in the subsequent years pointed out, however, that the choice of the proper liquid model is a very delicate aspect of the problem. As highlighted by many linear stability analyses (Neitzel et al., 1992; Kuhlmann and Rath, 1993a; Chen et al., 1997a; Chen and Hu, 1998; Chen et al., 1999), in fact, the instability of Marangoni flows is initiated through different mechanisms according to the liquid used; by means of an energy analysis, in particular, Wanschura et al. (1995) elucidated that the first bifurcation of the flow is due to inertia instability of the axial shear layer below the free surface for small-Pr fluids (semiconductor melts and liquid metals) and hydrothermal wave for the large-Pr fluid (transparent organic liquids) case.

The instability is hydrothermal in nature for $Pr \gg 1$ (for this case it is strictly related to the coupling and interplay between the velocity and temperature fields) and hydrodynamic in nature for $Pr \ll 1$ (i.e. it does not depend on the behavior of the temperature field that in this case simply acts as a driving force for the velocity field).

These theoretical analyses also elucidated that for transparent organic liquids the transition to 3D flow is always associated with the onset of oscillatory flow (Hopf bifurcation) whereas for semiconductor melts the instability breaks the spatial axisymmetry (but the flow regime is still steady, stationary bifurcation) prior to the onset of a time-dependent flow field.

Camel et al. (1995) were the first to demonstrate that the aforementioned (first) nonoscillatory bifurcation of Marangoni flow can also have a significant effect on the solidification process. Their experiments were carried out during the D-2 Mission and dopant-concentration distributions inside the solidified specimen were measured by a posteriori spreading-resistance techniques. According to the fact that no oscillations were detected in the melt during the flight, on the ground analyses of the solidified samples did not show any striations. However, undesired macroscopic radial segregation with a nonaxisymmetric distribution was found. This effect was attributed to the presence of "weak nonaxisymmetric laminar convection" in the liquid phase (this aspect had been ignored for a long time by other investigators mainly involved in the study of the oscillatory flow and crystal defects on the microscopic scale length).

At this stage it is evident that the problem is very complex. The related analysis is a very challenging task considering that, in the light of the aforementioned arguments, transparent liquids cannot be used as surrogates for semiconductor melts and that the direct experimental investigation of Marangoni flow in opaque liquids offers special problems (the use of nonconventional and very expensive diagnostic techniques, substances with very high reactivity and temperature, etc.).

Therefore numerical simulation is of paramount importance for these subjects. It offers the means of yielding theoretical advances for conditions often unavailable experimentally or having a prohibitive cost.

Along these lines, the present chapter tries to shed new light on possible models (liquids, geometrical configurations, boundary conditions, etc.) of the floating zone process and the related possibility to use some different configurations for better numerical simulation of the FZ method under different conditions as well as for capturing the underlying relevant fluid dynamics in terms of transitions and bifurcations. Very recent results are discussed and used to draw some conclusions and to show some heretofore unseen physics and dynamics. A numerical technique for the simulation of the solidification process is introduced. Some effort is also provided to illustrate a parallel numerical strategy devoted to shorten the otherwise prohibitive CFD time that is required for the simulation of the 3D flow occurring in the liquid melt. Axisymmetric simulations alone, in fact, do not adequately address the role of disturbances in inducing the instability of Marangoni flow. These studies require fundamentally three-dimensional representations of the waves or modes.

4.2. Phase-change modeling theory: the enthalpy method

Enthalpy methods and similar techniques taking into account the release or absorption of latent heat have been successfully applied over recent years to the case of thermal phase-change problems characterized by the presence of moving solid/melt interfaces due to the heating or the cooling of the system under investigation. These techniques were pioneered by Voller and Prakash (1987), Bennon and Incropera (1987a, b), Brent et al. (1988) and have been strongly improved over the years for various systems and conditions (see, e.g., Beckermann et al., 1999; Knoll et al., 1999; and many others).

With regard to this aspect it is interesting to point out how the approaches to date undertaken by the different researchers seem to fall essentially into two categories: "adaptive methods" (see, e.g., the excellent works of Rao and Shyy, 1997; Lan and Liang, 1999; Lan and Liu, 2002) and "fixed-grid techniques" (related by a certain theoretical kinship to volume of fluid (VOF) and level-set methods). In the first case, numerical solutions are obtained through numerically generated orthogonal curvilinear coordinate system, automatically adjusted to fit the solidification front (boundary where the phase change takes place) at any instant. On the contrary, the so-called enthalpy formulation, which allows a fixed-grid solution to be undertaken, removes the need to explicitly satisfy conditions at the phase front and is therefore able to utilize standard solution procedures for the fluid flow and energy equations directly, without resorting to mathematical manipulations and transformations.

Owing to the simplicity and computational efficiency the second type of strategy has enjoyed a widespread use throughout the scientific community (see, e.g., Chen et al., 1997b; Kothe et al., 1997; Reddy et al., 1997; Udaykumur et al., 1999; Kim et al., 2000; Sekerka, 2004). These authors elaborated various casting simulation tools employing robust, high-resolution finite-volume algorithms for incompressible fluid flow, volume tracking of interfaces, and solidification physics on 2D and 3D structured and unstructured meshes.

In the light of these worthy examples, the present section relies on the second type of approach and the related fundamental equations are introduced and discussed in the framework of the floating zone technique.

The enthalpy methods simply account for the latent heat in the energy equations by assigning a nodal latent heat value to each computational cell according to the temperature of the cell. Upon changing in phase, the nodal latent heat content of the cell is adjusted to account for latent-heat release or absorption, this adjustment being reflected in the energy equation as either a source or sink (a distinctive advantage of this arrangement is that no explicit conditions for energy conservation at the solid/liquid interface need to be accounted for). Latent heat during a phase change is incorporated in the energy equation using for each phase the following definition of enthalpy:

$$h = C_P T + \Delta H \tag{4.1}$$

where $\Delta H = \phi L_m$, L_m is the latent heat, C_P the specific heat coefficient and ϕ the liquid fraction defined as

$$\begin{aligned} &\phi = 0 && \text{for } T < T_{sol} \\ &\phi = \frac{T - T_{sol}}{T_{liq} - T_{sol}} && \text{for } T_{sol} \leq T \leq T_{liq} \\ &\phi = 1 && \text{for } T > T_{liq} \end{aligned} \tag{4.2}$$

where $T_{sol} = T_m - \Delta T_m$, $T_{liq} = T_m + \Delta T_m$ and $2\Delta T_m$ is a temperature interval chosen to represent the range over which the phase change occurs (T_{liq} is the liquidus temperature at which solid formation commences and T_{sol} is the temperature at which full solidification is achieved, see Voller and Prakash, 1987). This means that an extended temperature range is considered for solidification. From a physical point of view this assumption is based on the idea (Voller and Prakash, 1987; Bennon and Incropera, 1987a), that in many cases (metal alloys and doped semiconductors) solid formation occurs as a permeable crystal-like matrix that coexists with the liquid phase (mushy region). From a numerical point of view a linear approximation (Eq. (4.2b)) must be considered for the liquid fraction since a step change in the liquid fraction could cause serious numerical instabilities.

During solidification, latent energy is released at the interfaces that separate the phases within the mushy region. Thus the energy equation can be written as:

$$\frac{\partial \rho h}{\partial t} + \frac{\partial \rho(\phi L_m)}{\partial t} = -\underline{\nabla} \cdot [\rho \underline{V} h] + \underline{\nabla} \cdot [\lambda \underline{\nabla} T] \tag{4.3}$$

where ρ is the density, λ is the thermal conductivity, $\underline{V}$ is the flow velocity; the unsteady latent heat content term on the left side of Eq. (4.3) is equal to zero for $T < T_{sol}$ and $T > T_{liq}$ (it influences the energy equation only within the mushy region).

Equation (4.3), together with the continuity and Navier–Stokes equations, can be posed in nondimensional form scaling the lengths by a reference distance L (see, e.g., Figs. 4.2 and 4.4) and the velocity by the energy diffusion velocity $V_\alpha = \alpha/L$ ($\alpha = \lambda/\rho C_P$ is the thermal diffusivity); the scales for time and pressure are, respectively, L^2/α and $\rho\alpha^2/L^2$.

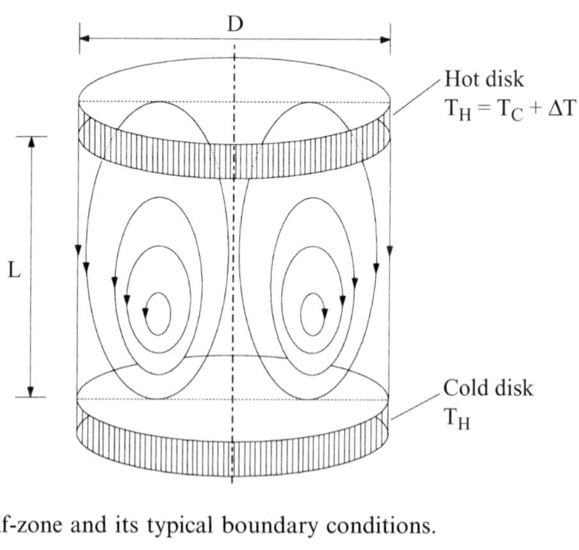

Fig. 4.2 The half-zone and its typical boundary conditions.

The nondimensional conservative form reads for a pure substance (e.g. pure silicon):

$$\underline{\nabla} \cdot \underline{V} = 0 \tag{4.4a}$$

$$\frac{\partial \underline{V}}{\partial t} = -\underline{\nabla}\pi - \underline{\nabla} \cdot [\underline{V}\,\underline{V}] + Pr\nabla^2 \underline{V} + Pr\,Ra_r(T - T_{ref})_g \quad Pr\frac{1}{\eta_p}\underline{V} \tag{4.4b}$$

$$\frac{\partial T}{\partial t} + \frac{L_m}{C_P}\frac{\partial \phi}{\partial t} = -\underline{\nabla} \cdot [\underline{V}T] + \nabla^2 T \tag{4.4c}$$

where V and π are the nondimensional velocity and pressure (in this chapter pressure is denoted by π), Pr is the Prandtl number, defined by $Pr = \nu/\alpha$ ($\nu = \mu/\rho$ is the kinematic viscosity), T is kept in its dimensional form (T_{ref} in Eq. (4.4b) is a reference temperature, for instance, the initial temperature T_o or T_m) and $Ra_r = g\beta_T L^3/\nu\alpha$, β_T being the thermal expansion coefficient (moreover, it is assumed that ρ, λ and C_P are constant).

An important problem with fixed-grid solution procedures, is accommodating the zero-velocity condition, which is required as a liquid region turns to solid. Various methods (see Brent et al., 1988) could be used in principle to "switch off" velocities in computational cells that are freezing (or "switch on" velocities in the case of melting). Velocities may simply be set to zero. Another possible approach is based on the viscosity. The viscosity of a cell can be driven to a very large value as the latent-heat content of the cell falls to zero. The increasing viscosity provides the necessary coupling between the physical state of the material in the cell and the momentum equations, thereby driving velocities to zero in cells that are solid. A third possible approach (enthalpy-porosity approach) requires that computational cells that are undergoing a phase change are modeled as a pseudoporous media, with the porosity η_p being a function of ϕ ranging between 1 (fully liquid cell) and 0 (fully solid cell). This approach is recommended herein. The term $-Pr\,V/\eta_p$ in Eq. (4.4b), in fact, is the Darcy term (see Bennon and Incropera, 1987b; Brent et al., 1988) added to the momentum equation to eliminate convection in the solid phase. The permeability can be assumed to vary according to the Carman–Kozeny equation:

$$\eta_p = \frac{\phi^3}{(1-\phi)^2} \tag{4.5}$$

In the pure solid ($\phi = 0$) and pure liquid ($\phi = 1$), Eq. (4.4b) reduces to the appropriate limits, namely $\eta_p = 0$ and $\eta_p = \infty$, respectively. In practice, the effect of η_p is as follows: in full liquid elements $1/\eta_p$ is zero and has no influence; in elements that are changing the phase, the value of $1/\eta_p$ will dominate over the transient, convective and diffusive components of the momentum equation, thereby forcing them to imitate the Carman–Kozeny law; in totally solid elements, the final large value of $1/\eta_p$ will swamp all terms in the governing equations and force any velocity predictions effectively to zero.

Since each of the continuum equations is valid throughout the entire domain, as anticipated, explicit consideration need not be given to boundaries between solid, multiphase and liquid regions.

A prototype application dealing with the tracking of the solid/liquid interface in the floating zone will be discussed in Section 4.11.

4.3. Modeling the floating zone: The half-zone and the full-zone

4.3.1 *The geometrical configurations*

One of the major difficulties in the experimental analysis of Marangoni flow in real floating zones is that, due to the aforementioned phase change related to melting and solidification of the material, the geometry of the boundary of the liquid volume is not known a priori. Numerical techniques such as the enthalpy method introduced in the previous section can help researchers in discerning (by means of computations) some important aspects of problem, however, irregular melting and freezing interfaces and thermal conditions associated with latent-heat effects make the analysis of the features of the Marangoni flow very difficult from an experimental point of view. These difficulties led the investigators, in the past, to create a new configuration, the so-called half-zone (usually referred to as a liquid bridge). This model of a float zone melt was introduced, in fact, in the mid-1970s as a vehicle for performing experiments in well-controlled conditions.

The half-zone model simulates half of a real floating zone (the liquid between one of the ends of the domain and the equatorial plane); it consists of a pair of coaxial, solid cylindrical disks with a bridge of liquid material suspended between them; the presence of a ring heater around the equatorial plane is simulated by heating one of the two supporting disks with respect to the other, whereas heat flow is neglected through the free surface. Surface-tension forces drive fast fluid motion in the free surface from the hot disk to the cold one giving rise to a single toroidal vortex (Fig. 4.2).

The Marangoni flow in the FZ process is characterized by two separated toroidal vortices in the upper and lower part of the zone (Fig. 4.1). Each of these parts corresponds approximately to a half-zone configuration. This model has distinct advantages for both the experimentalist and also for numerical investigators. In particular, the ends of the domain are isothermal, the interface is adiabatic and the applied temperature difference (difference of the temperature of the supporting disks) driving the Marangoni surface flow can be fixed a priori in the analyses.

Other models, however, have been introduced over the years. The ends of the full-zone, for instance, are plane and isothermal as in the case of the half-zone, but the supporting disks are posed at the same temperature (Fig. 4.3).

The presence of a ring heater around the equatorial plane of the zone is simulated in this case by imposing a specified heat-flux distribution on the free surface with a

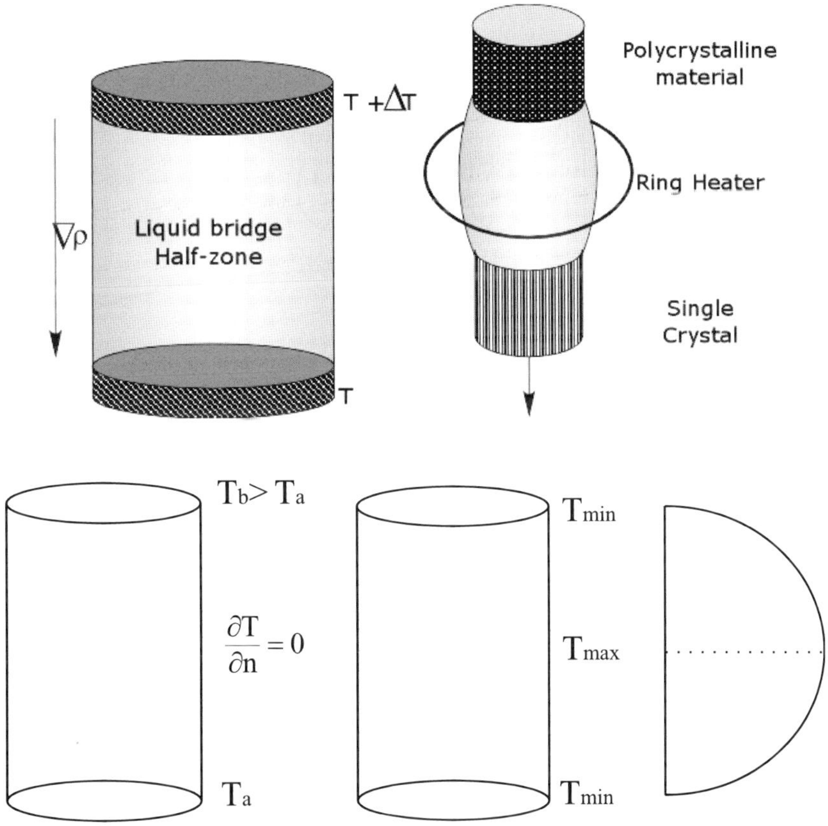

Fig. 4.3 Comparison between typical boundary conditions for the half-zone and the full-zone.

maximum of the flux in correspondence of the equatorial plane (in a half-zone model the end disks are held at the maximum and minimum temperatures, respectively, whereas in a full-zone model, the maximum temperature occurs between the disks). Marangoni flow directed from the equatorial region towards the supporting disks is generated due to the imposed flux distribution.

Typical flow configurations obtained for small Marangoni numbers consist of two separate axisymmetric toroidal vortices in the lower and upper half of the liquid zone with opposite senses of rotation (Fig. 4.4).

The geometrical aspect ratio of the liquid column of full-zone extent (A_F) is defined as $A_F = 2L/D$; the aspect ratio of the corresponding half-zone (half of the considered full-zone) is defined as $A_H = L/D = A_F/2$ (see Figs. 4.2 and 4.4, D is the diameter of the supporting disks).

Under normal-gravity conditions the upper half of the full-zone is influenced by the buoyancy forces so that the Marangoni effect can scarcely be recognized. On the

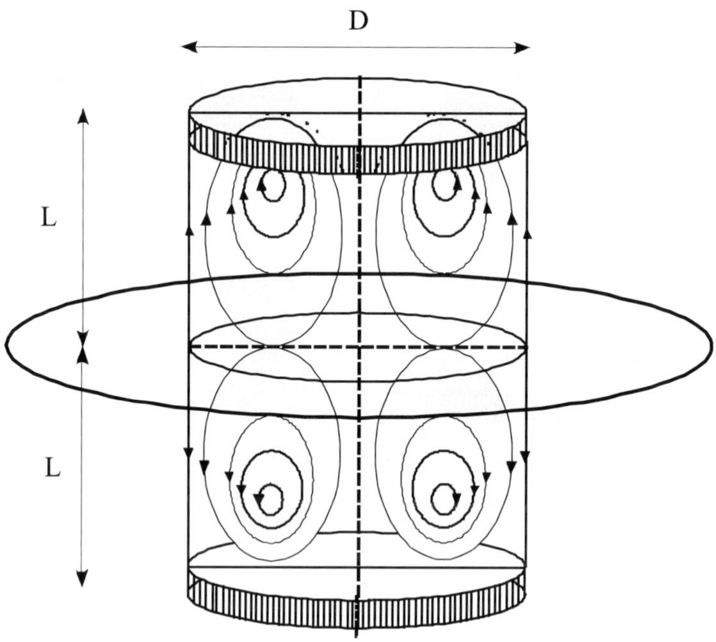

Fig. 4.4 The full-zone heated by an equatorial ring heater.

other hand, the lower half is less affected by buoyancy since the upper temperature is higher than the lower one. Consequently, the Marangoni effect dominates the convection in the lower half of the full-zone. For this reason, the half-zone (heated from above) configuration has been preferred in the on-the-ground fundamental research.

Moreover, the full-zone model has a disadvantage due to the fact that the temperature difference driving the surface flow between the ends of the domain and the equatorial cross-section cannot be fixed a priori since it is a consequence of the flow itself (i.e. ΔT is established by the convective and heat transport through the liquid zone). However, the thermal boundary conditions and the flow field are more realistic than those related to a half-zone configuration. The flow situation in the half-zone corresponds to that of half of a floating zone, but this correspondence is very approximate. In a "complete" floating zone, in fact, the equatorial plane is characterized by radial temperature gradients and radial flow whereas in the half-zone the boundary conditions for temperature and velocity on the hot disk, simulating the heat source around the equatorial plane of a real floating zone, require isothermal conditions and no-slip conditions, respectively.

In practice, the half-zone is supposed to model the flow in a half of the real floating zone under the constraint that the free surface is symmetric by reflection about the plane at midheight between the rods, and that the imposed no-slip condition on this plane does not significantly alter the dynamics.

In the on the ground environment, however, the mirror symmetry of the shape of a floating zone with respect to the equatorial plane does not occur. The gravitational acceleration, in fact, results in a surface shape that bulges out below the equatorial plane and necks in above it. This causes an asymmetry in the structure and intensity of the two aforementioned rolls.

Superimposed on these aspects is the fact that in a zero-gravity environment, even if the shape of the gas melt interface is symmetric with respect to the heating center, there is no physical constraint forcing the flow to maintain mirror symmetry with respect to the equatorial plane. The transition of Marangoni flow to a 3D pattern in the case of the half-zone flow may be not relevant for the real floating zone since, there, the mirror symmetry with respect to the equatorial plane could be broken.

In view of these arguments, even if the half-zone model has enjoyed widespread use over the years as a paradigm model for testing the ability of researchers to predict the behavior of surface-tension-driven flows, it is the opinion of the author that the full-zone should be used as a relevant testbed for applying simulations to the technologically important FZ problem (Lappa, 2004c).

However, each model, as pointed out in the foregoing discussion, has its own advantages and limitations and has been conceived in order to provide quite a simplified (easy to handle) configuration where well-established parameters can be fixed and the behavior of the system in response to changes in these parameters can be investigated.

For instance, of particular interest is the effect on the fluid flow of the geometrical aspect ratio since it is known that it can change during real applications. For this reason, usually the devices used for fundamental research about the flow within such suspended liquid zones are conceived by the researchers in order to easily allow changes of the distance between the supporting disks (Fig. 4.5).

Also, the effect of the shape (cylindrical, convex or concave under microgravity conditions, amphora-like on the ground) is expected to be important. Usually these aspects are studied in terms of the nondimensional parameter S (the so-called volume or shape factor) ratio of the volume held between the supporting disks and the volume of the corresponding straight configuration with a cylindrical liquid/air interface (for a fixed aspect ratio, the effective volume of liquid suspended between the supporting disks, in fact, determines the shape of the melt/gas interface, Fig. 4.6).

These configurations, as emphasized at the beginning of this section, have been expressly designed for static analyses, i.e. no phase change, fixed distance between the supporting disks and fixed volume and interface shape (for such conditions the latent-heat contribution $(L_m/C_P)(\partial\phi/\partial t)$ in Eq. (4.4c) is absent).

However, the half-zone has been used by some investigators also for the study of dynamical effects related to the advance of the solidification front (see Section 4.11 for further details).

4.3.2 *The Young–Laplace equation*

With regard to the computation of the free-surface shape, it is well known (e.g., by estimates of Chen and Roux, 1991; Mundrane and Zebib, 1995; Kuhlmann and Nienhüser, 2002) that in the limit of very small capillary number the shape of the free-surface is independent of the flow and temperature fields and depends only on the volume of liquid and the hydrostatic pressure difference.

The interface location is determined, in fact, by the hydrodynamic and the Laplace pressure. The relative importance of these quantities is given by the ratio of the relative magnitudes ($\sigma_T \Delta T/L$) and σ_o/L (see also Section 2.3). This ratio is the capillary number $Ca = \sigma_T \Delta T/\sigma_o$. If $Ca \to 0$ (i.e. $Ca \ll 1$) the dynamic surface deformation can be neglected. This aspect was demonstrated through numerical computations by Chen and Roux (1991), and Kuhlmann and Nienhüser (2002) for

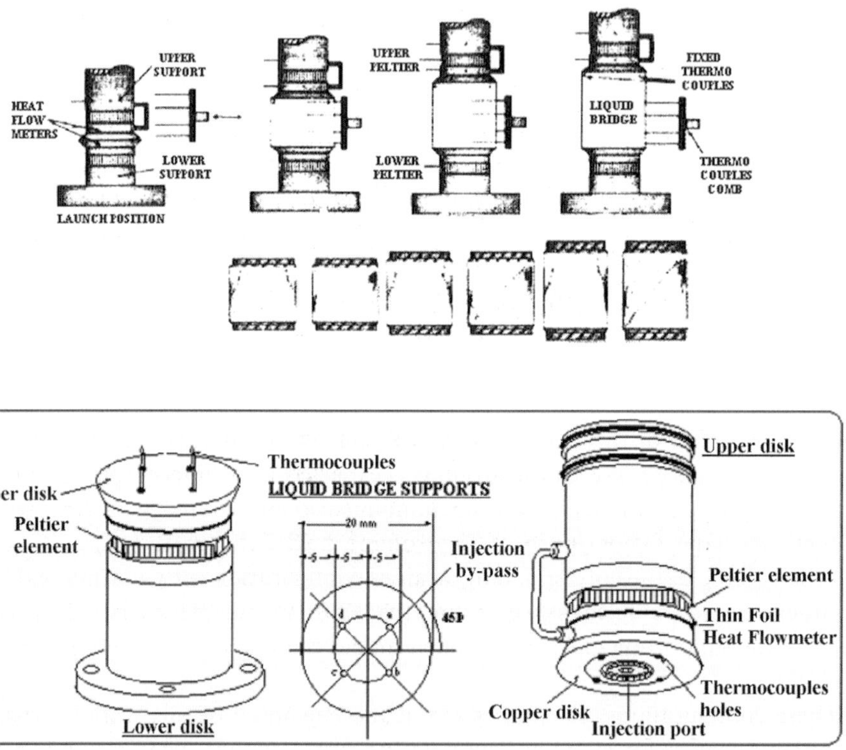

Fig. 4.5 Examples of experimental facilities conceived in order to allow changes of the distance between the supporting disks.

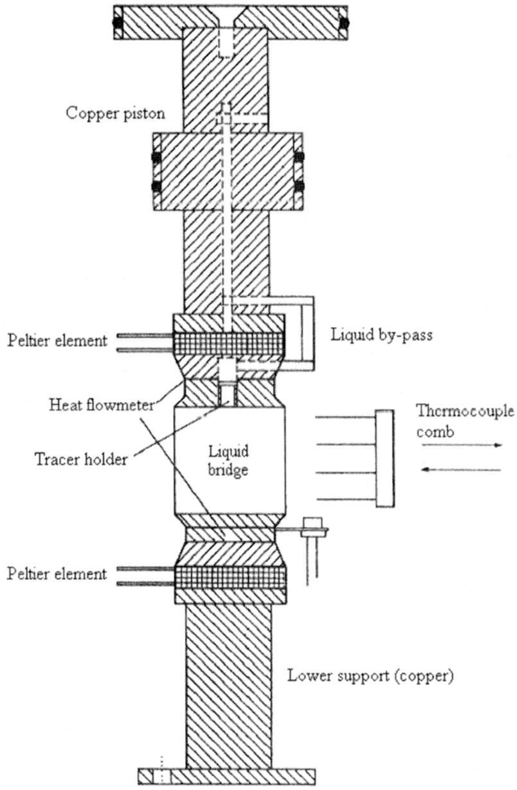

Fig. 4.5 Continued.

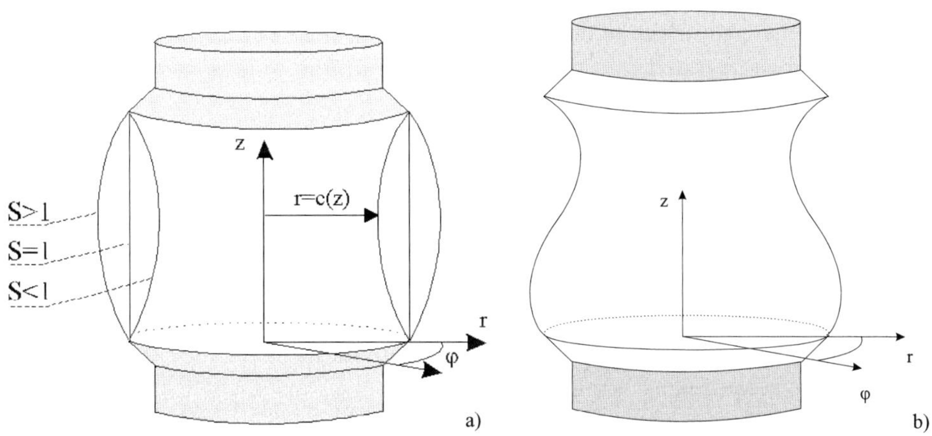

Fig. 4.6 Noncylindrical floating zones (a) microgravity (concave and convex surfaces), (b) on the ground shape.

values of the Prandtl number and conditions very close to that considered in the present book (silicon, $Pr = 0.01$). They provided quantitative information on the shape and amplitude of dynamic free-surface deformations for two- (Chen and Roux, 1991) and three-dimensional Marangoni flows (Kuhlmann and Nienhüser, 2002), respectively.

In the light of their results the interface can be assumed to be rigid and axisymmetric around the z-axis; its radial coordinate is a function only of the z variable ($r = c(z)$). Under such assumptions, the hydrostatic shape is obtained from the Young–Laplace equation, relating the local curvature of the surface to the pressure jump along the liquid/gas interface:

$$\Delta p + \rho g z = \sigma_o \left(\frac{1}{R_1} + \frac{1}{R_2} \right) \tag{4.6a}$$

where R_1 and R_2 are the principal radii of curvature at each point of the surface.

Equation (4.6) may be reformulated in dimensionless form (scaling the radial and axial coordinates by L) in the cylindrical coordinates by substituting the analytical expression of the principal radii of curvature in the axisymmetric geometry in terms of the surface equation $c(z)$:

$$\frac{1}{c(1 + c'^2)} - \frac{c''}{(1 + c'^2)^{3/2}} = \frac{L\Delta p}{\sigma_o} + \frac{\rho g L^2}{\sigma_o} z = k_1 + k_2 z \tag{4.6b}$$

The parameter $k_2 = \rho g L^2 / \sigma_o$, in practice, is the Bond number already introduced in Section 2.3. It takes into account the deformation of the shape under gravity conditions and depends on the value of the g-level, on the height of the floating zone and on the properties of the liquid under investigation (ρ and σ). Each value of the parameter $k_1 = L\Delta p / \sigma_o$ corresponds to a fixed volume of the liquid zone. The value of k_1 can be changed in Eq. (4.6b) in order to obtain the desired volume V and the equation can be integrated by a shooting method with the conditions that the liquid is attached to the solid supports:

$$\text{Half-zone:} \quad c(0) = c(1) = R/L \tag{4.7a}$$

$$\text{Full-zone:} \quad c(-1) = c(1) = R/L \tag{4.7b}$$

4.3.3 *Body-fitted curvilinear coordinates*

The nondimensional Navier–Stokes and energy equations in cylindrical coordinates and in the absence of phase change read:

$$\frac{\partial u}{\partial z} + \frac{\partial v}{\partial r} + \frac{v}{r} + \frac{1}{r} \frac{\partial V_\varphi}{\partial \varphi} = 0 \tag{4.8a}$$

$$\frac{\partial u}{\partial t} = -\frac{\partial \pi}{\partial z} - \left(\frac{\partial u^2}{\partial z} + \frac{\partial uv}{\partial r} + \frac{uv}{r} + \frac{1}{r}\frac{\partial uV_\varphi}{\partial \varphi}\right) + Pr\left(\frac{\partial^2 u}{\partial z^2} + \frac{\partial^2 u}{\partial r^2} + \frac{1}{r}\frac{\partial u}{\partial r} + \frac{1}{r^2}\frac{\partial^2 u}{\partial \varphi^2}\right) + PrRa_\mathrm{r}(T - T_\mathrm{ref}) \tag{4.8b}$$

$$\frac{\partial v}{\partial t} = -\frac{\partial \pi}{\partial r} - \left(\frac{\partial uv}{\partial z} + \frac{\partial v^2}{\partial r} + \frac{v^2}{r} + \frac{1}{r}\frac{\partial vV_\varphi}{\partial \varphi} - \frac{V_\varphi^2}{r}\right) + Pr\left(\frac{\partial^2 v}{\partial z^2} + \frac{\partial^2 v}{\partial r^2} - \frac{v}{r^2} + \frac{1}{r}\frac{\partial v}{\partial r} + \frac{1}{r^2}\frac{\partial^2 v}{\partial \varphi^2} - \frac{2}{r^2}\frac{\partial V_\varphi}{\partial \varphi}\right) \tag{4.8c}$$

$$\frac{\partial V_\varphi}{\partial t} = -\frac{1}{r}\frac{\partial \pi}{\partial \varphi} - \left(\frac{\partial uV_\varphi}{\partial z} + \frac{\partial vV_\varphi}{\partial r} + \frac{2vV_\varphi}{r} + \frac{1}{r}\frac{\partial V_\varphi^2}{\partial \varphi}\right) + Pr\left(\frac{\partial^2 V_\varphi}{\partial z^2} + \frac{\partial^2 V_\varphi}{\partial r^2} + \frac{1}{r}\frac{\partial V_\varphi}{\partial r} - \frac{V_\varphi}{r^2} + \frac{1}{r^2}\frac{\partial^2 V_\varphi}{\partial \varphi^2} + \frac{2}{r^2}\frac{\partial v}{\partial \varphi}\right) \tag{4.8d}$$

$$\frac{\partial T}{\partial t} = -\left(\frac{\partial uT}{\partial z} + \frac{\partial vT}{\partial r} + \frac{vT}{r} + \frac{1}{r}\frac{\partial V_\phi T}{\partial \phi}\right) + \left(\frac{\partial^2 T}{\partial z^2} + \frac{\partial^2 T}{\partial r^2} + \frac{1}{r}\frac{\partial T}{\partial r} + \frac{1}{r^2}\frac{\partial^2 T}{\partial \varphi^2}\right) \tag{4.8e}$$

where u, v and V_φ are the velocity components along z, r and φ, respectively.

To adequately take into account the deformation of the free surface due to $S \neq 1$ and/or to gravity, the body-fitted curvilinear coordinates can be adopted. The noncylindrical original physical domain in the (r, z, φ) space is transformed into a cylindrical computational domain in the (ξ, η, φ) space by

$$\begin{matrix} z = \xi & & \xi = z \\ & \rightarrow & \\ r = \eta c(\xi) & & \eta = r/c(\xi) \end{matrix} \tag{4.9a}$$

the radial coordinate r ranges from $\eta = 0$ (at the symmetry axis) up to $\eta = 1$ at the free surface; the axial coordinate varies from $\xi = 0$ (half-zone) or $\xi = -1$ (full-zone) at the bottom disk up to $\xi = 1$ at the top disk.

The first-order spatial derivatives read

$$\begin{aligned} \frac{\partial f}{\partial z} &= \frac{\partial f}{\partial \xi}\frac{\partial \xi}{\partial z} + \frac{\partial f}{\partial \eta}\frac{\partial \eta}{\partial z} \quad \longrightarrow \quad \frac{\partial f}{\partial z} = \frac{\partial f}{\partial \xi} - \frac{c'}{c}\eta\frac{\partial f}{\partial \eta} \\ \frac{\partial f}{\partial r} &= \frac{\partial f}{\partial \xi}\frac{\partial \xi}{\partial r} + \frac{\partial f}{\partial \eta}\frac{\partial \eta}{\partial r} \quad \longrightarrow \quad \frac{\partial f}{\partial r} = \frac{1}{c}\frac{\partial f}{\partial \eta} \end{aligned} \tag{4.9b}$$

The second-order spatial derivatives read

$$\begin{aligned}
&\frac{\partial^2 f}{\partial z^2}=\frac{\partial^2 f}{\partial \xi^2}+\left(\frac{c'}{c}\right)^2\eta^2\frac{\partial^2 f}{\partial \eta^2}-\frac{2c'}{c}\eta\frac{\partial^2 f}{\partial \xi\partial \eta}-\frac{c''}{c}\eta\frac{\partial f}{\partial \eta}\\
&\frac{\partial^2 f}{\partial r^2}=\frac{1}{c^2}\frac{\partial^2 f}{\partial \eta^2}
\end{aligned} \tag{4.9c}$$

Thus the transformed equations are:

$$\frac{\partial u}{\partial \xi}-\frac{c'}{c}\eta\frac{\partial u}{\partial \eta}+\frac{1}{c}\frac{\partial v}{\partial \eta}+\frac{v}{\eta c}+\frac{1}{\eta c}\frac{\partial V_\varphi}{\partial \varphi}=0 \tag{4.10a}$$

$$\begin{aligned}
\frac{\partial u}{\partial t}=&-\left(\frac{\partial \pi}{\partial \xi}-\frac{c'}{c}\eta\frac{\partial \pi}{\partial \eta}\right)-\left(\frac{\partial u^2}{\partial \xi}-\frac{c'}{c}\eta\frac{\partial u^2}{\partial \eta}+\frac{1}{c}\frac{\partial uv}{\partial \eta}+\frac{uv}{\eta c}+\frac{1}{\eta c}\frac{\partial uV_\varphi}{\partial \varphi}\right)\\
&+Pr\left[\frac{\partial^2 u}{\partial \xi^2}+\left(\left(\frac{c'}{c}\eta\right)^2+\frac{1}{c^2}\right)\frac{\partial^2 u}{\partial \eta^2}-2\frac{c'}{c}\eta\frac{\partial^2 u}{\partial \xi\partial \eta}+\left(\frac{1}{\eta c^2}-\frac{c''}{c}\eta\right)\frac{\partial u}{\partial \eta}+\frac{1}{\eta^2c^2}\frac{\partial^2 u}{\partial \varphi^2}\right]\\
&+PrRa_{\mathrm{r}}(T-T_{\mathrm{ref}})
\end{aligned} \tag{4.10b}$$

$$\begin{aligned}
\frac{\partial v}{\partial t}=&-\left(\frac{1}{c}\frac{\partial \pi}{\partial \eta}\right)-\left(\frac{\partial uv}{\partial \xi}-\frac{c'}{c}\eta\frac{\partial uv}{\partial \eta}+\frac{1}{c}\frac{\partial v^2}{\partial \eta}+\frac{v^2}{\eta c}-\frac{V_\phi^2}{\eta c}+\frac{1}{\eta c}\frac{\partial vV_\varphi}{\partial \varphi}\right)\\
&+Pr\left[\frac{\partial^2 v}{\partial \xi^2}+\left(\left(\frac{c'}{c}\eta\right)^2+\frac{1}{c^2}\right)\frac{\partial^2 v}{\partial \eta^2}-2\frac{c'}{c}\eta\frac{\partial^2 v}{\partial \xi\partial \eta}\right.\\
&\left.+\left(\frac{1}{\eta c^2}+2\left(\frac{c'}{c}\right)^2-\frac{c''}{c}\eta\right)\frac{\partial v}{\partial \eta}-\frac{v}{\eta^2c^2}+\frac{1}{\eta^2c^2}\frac{\partial^2 v}{\partial \varphi^2}-\frac{2}{\eta^2c^2}\frac{\partial V_\varphi}{\partial \varphi}\right]
\end{aligned} \tag{4.10c}$$

$$\begin{aligned}
\frac{\partial V_\varphi}{\partial t}=&-\left(\frac{1}{c\eta}\frac{\partial \pi}{\partial \varphi}\right)-\left(\frac{\partial uV_\varphi}{\partial \xi}-\frac{c'}{c}\eta\frac{\partial uV_\varphi}{\partial \eta}+\frac{1}{c}\frac{\partial vV_\varphi}{\partial \eta}+\frac{2vV_\varphi}{\eta c}+\frac{1}{\eta c}\frac{\partial V_\varphi^2}{\partial \varphi}\right)\\
&+Pr\left[\frac{\partial^2 V_\varphi}{\partial \xi^2}+\left(\left(\frac{c'}{c}\eta\right)^2+\frac{1}{c^2}\right)\frac{\partial^2 V_\varphi}{\partial \eta^2}-2\frac{c'}{c}\eta\frac{\partial^2 V_\varphi}{\partial \xi\partial \eta}\right.\\
&\left.+\left(\frac{1}{\eta c^2}+2\left(\frac{c'}{c}\right)^2-\frac{c''}{c}\eta\right)\frac{\partial V_\varphi}{\partial \eta}-\frac{V_\varphi}{\eta^2c^2}+\frac{1}{\eta^2c^2}\frac{\partial^2 V_\varphi}{\partial \varphi^2}+\frac{2}{\eta^2c^2}\frac{\partial v}{\partial \varphi}\right]
\end{aligned} \tag{4.10d}$$

$$\begin{aligned}
\frac{\partial T}{\partial t}=&-\left(\frac{\partial uT}{\partial \xi}-\frac{c'}{c}\eta\frac{\partial uT}{\partial \eta}+\frac{1}{c}\frac{\partial vT}{\partial \eta}+\frac{vT}{\eta c}+\frac{1}{\eta c}\frac{\partial V_\varphi T}{\partial \varphi}\right)\\
&+\left[\frac{\partial^2 T}{\partial \xi^2}+\left(\left(\frac{c'}{c}\eta\right)^2+\frac{1}{c^2}\right)\frac{\partial^2 T}{\partial \eta^2}-2\frac{c'}{c}\eta\frac{\partial^2 T}{\partial \xi\partial \eta}\right.\\
&\left.+\left(\frac{1}{\eta c^2}+2\left(\frac{c'}{c}\right)^2-\frac{c''}{c}\eta\right)\frac{\partial T}{\partial \eta}+\frac{1}{\eta^2c^2}\frac{\partial^2 T}{\partial \varphi^2}\right]
\end{aligned} \tag{4.10e}$$

For the sake of brevity the details of the transformation are not shown (they would require many pages, see, for instance, Sim and Zebib (2002). At the interface the following boundary conditions must be imposed:

$$\left(\frac{1+c'^2}{c}\right)\frac{\partial u}{\partial \eta}-2c'\frac{\partial u}{\partial \xi}+\left(\frac{c'+c'^3}{c}\right)\frac{\partial v}{\partial \eta}+(1-c'^2)\frac{\partial v}{\partial \xi}=-Ma_{\mathrm{r}}(1+c'^2)^{1/2}\frac{\partial T}{\partial \xi} \tag{4.11a}$$

$$\frac{\partial V_\phi}{\partial r}-\frac{V_\phi}{r}+\frac{1}{r}\frac{\partial v}{\partial \varphi}-c'\left(\frac{\partial V_\varphi}{\partial z}+\frac{1}{r}\frac{\partial u}{\partial \varphi}\right)=-Ma_{\mathrm{r}}(1+c'^2)^{1/2}\frac{1}{r}\frac{\partial T}{\partial \varphi} \tag{4.11b}$$

$$\left(\frac{1}{c}+\frac{c'^2}{c}\right)\frac{\partial T}{\partial \eta}-c'\frac{\partial T}{\partial \xi}=-(1+c'^2)^{1/2}J(\xi) \tag{4.11c}$$

where $Ma_{\mathrm{r}}=\sigma_{\mathrm{T}}L/\mu\alpha$ ($Ma=Ma_{\mathrm{r}}\cdot\Delta T$) is the reference Marangoni number, $J(\xi)$ is the heat flux at the free interface ($J(\xi)=0$ for the adiabatic half-zone case). No-slip conditions must be imposed on the supporting disks.

4.4. Numerical simulation and parallel strategy

Sections 4.2 and 4.3 can be regarded as a significant effort provided to approach the FZ problem in terms of fundamental concepts, geometrical models and governing equations. The next natural step is the introduction of reliable and robust numerical techniques.

As anticipated in the introduction of this chapter, numerical simulations dealing with both the aforementioned models of the FZ technique, are a decisive tool to reduce the number of expensive experiments, to plan and improve the experimental setups and therefore to optimize the FZ production technique.

In the literature, Marangoni convection in axisymmetric floating zones has often been investigated numerically under the constraint of axial symmetry of the flow (see, e.g., Kazarinoff and Wilkowski, 1990; Griaznov, 1994; Chen and Chin, 1995; Tang et al., 1995; Okano et al., 1997; Shevtsova and Legros, 1997; Chenier et al., 1999; Bazzi et al., 2000; Kasperski et al., 2000). Two-dimensional methods are, in fact, convenient to use due to their efficiency in memory, computing time and costs. They can be reliable for symmetrical geometries and boundary conditions if the flow is in subcritical conditions, or if the third dimension is assumed to be infinitely extended. However, if the flow is actually three-dimensional, vital information is lost. In particular, as discussed in Section 4.1, real experiments on Marangoni flow instability show that the flow and temperature fields are three-dimensional, and this peculiar information is lost in 2D models.

Linear or energy stability theories provided important information on the critical conditions for the onset of instability and on the flow-field organization in the neighborhood of the transition point from axisymmetric to three-dimensional flow;

unfortunately these theories cannot predict the nonlinear behavior of the flow far from the bifurcation point (see also Section 1.4.1).

In recent years the availability of supercomputers and the development of efficient numerical methods has enhanced the possibility of studying the problem by direct numerical solution of the three-dimensional Navier–Stokes equations. However, only a few studies have been published on this subject due to the fact that enormous speed and memory requirements are necessary. This is particularly true for large Marangoni numbers, due to the presence of very thin boundary layers (see Section 2.8.3) that require a large number of computational grid points and consequently a large number of unknown variables. In fact, coarse grids cannot be used because a low resolution does not permit the capture of instabilities of the Marangoni flow. For these reasons, grid quality and computation speed are often critical factors for the study of these problems.

Two possibilities exist for achieving high computational speeds: advanced computers with their intrinsic physical limitations, and parallel computations, which increase the computational speed as the number of processors is increased.

The main objective of this section is to introduce a parallel and scalable flow-solver algorithm to solve a variety of nonsmooth problems related to the instability of Marangoni flow in floating zones and related model configurations.

The Navier–Stokes algorithm must be numerically stable, physically robust and computationally efficient. Previous results (see, e.g., Rupp et al., 1989; Levenstam and Amberg, 1995) indicate that a Simplified Marker and Cell method (SMAC) can be successfully utilized for simulations of complex Marangoni flows (see Chapter 2 for further details on this technique).

The parallelization of this algorithm can only be performed by an articulate numerical solution procedure, in order to optimize simultaneously the numerical scheme, the grid generation and the domain decomposition related to the parallel method (Lappa, 1997; Lappa and Savino, 1999).

4.4.1 *Domain decomposition and data mapping*

True parallel processors have the capability of executing different instruction streams (tasks) in parallel, and that of permitting communication between tasks.

In these computers (the present book refers to distributed memory systems NORMA: No Remote Memory Access) the global algorithm has to be split into several subkernels (Häuser and Williams, 1992; Keyes et al., 1992; Lou and Ferraro, 1996; Smith et al., 1996). Each processor solves a task and cannot 'see' directly the variables stored in the local memory of the other processors solving the other tasks. The different nodes have to communicate via special messages using an 'internode communication network' (message-passing philosophy).

This section is dedicated to the description of the parallel algorithm to be used for the solution of the parabolic equations (momentum and energy equations) on a parallel supercomputer. For the sake of simplicity, the case of a floating zone having a cylindrical interface and of a uniform grid is considered.

It is worthwhile to stress how the data distribution among the various processors of a parallel machine is a key factor in the efficiency of a parallel implementation. For many scientific applications, the optimal data distribution is not obvious and requires experimentation.

The parallelization approach proposed in the present section is based on domain decomposition. The computational domain can be split into p subblocks (or subdomains) without overlapping neighboring parts in a manner that provides a complete partition (Keyes et al., 1992). Each subdomain is topologically equivalent to a simple domain such as a cylinder: This splitting comes from the grid-generation process.

Hereafter, p denotes the number of processors available on the parallel machine, and N_z, N_r, N_φ denote the number of data items (i.e. grid points) in the axial, radial and azimuthal directions.

The three-dimensional data are partitioned into N_z two-dimensional planes of $N_r N_\varphi$ data items each. The first N_z/p planes are mapped to processor P_0, the next N_z/p planes are mapped to processor P_1 and so on. An example of this mapping is shown in Fig. 4.7 (for $p=4$).

In this way the solution domain is split into a connected set of axisymmetric subblocks in three-dimensional space. Each block has at most two surrounding blocks, although some at physical boundaries have fewer neighbors. In fact, interior blocks have two neighboring blocks, whereas those at the physical boundary only have one neighbor.

Since the domain is partitioned in equal subdomains, each processor has exactly the same number of grid points. The parabolic equations (4.4b) and (4.4c) can be solved within each subdomain. Since an explicit, in time, scheme is considered (as anticipated in Section 2.8 the use of explicit schemes facilitates the parallel implementation of the code), at each time step each subdomain is completely independent of its surrounding (the variables computed at the step $n+1$ are completely uncoupled since they depend exclusively on the variables at the previous time step), parallelism simply being achieved by introducing a new boundary condition, denoted as interblock or interdomain boundary condition. Each processor works with one of these subblocks, and exchanges data across the boundaries of the subblocks. The entire application can be imagined as a number of kernels with different data-distribution requirements, which necessitates data movement between the different kernels.

4.4.2 *Interprocessor communication and synchronization*

To ensure a correct implementation of the numerical algorithm on the partitioned grid, data exchange is necessary close to the "partition boundaries" of each subgrid local to a certain processor.

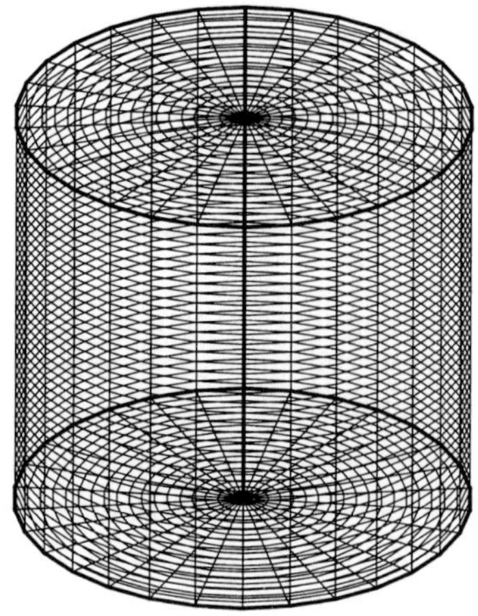

a)

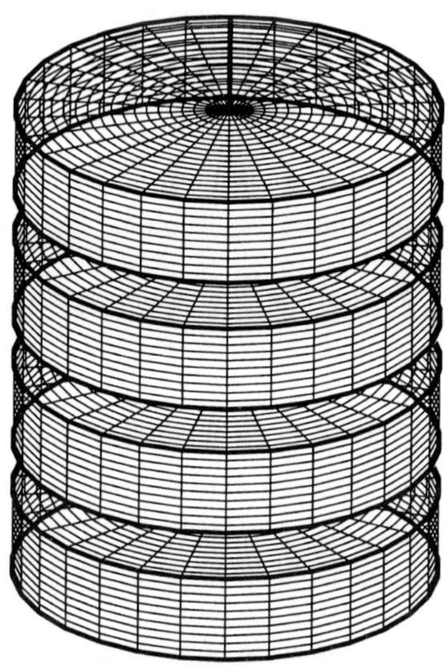

b)

Fig. 4.7 Computational domain (a) and grid partition (b) for the floating zone problem.

Each processor contains a cylindrical subgrid surrounded by some "ghost grid points" that are duplicates of grid points contained in other processors, as depicted in Fig. 4.8, where the computational mesh in a generic meridian plane is shown. In addition to the block of data owned by each processor, there are two surrounding layers of "ghost" points. These ghost points must be updated by the message-passing phase of the calculation, and provide data continuity from the neighboring subblock for the computation.

An example of the parallel numerical procedure and of the utilization of the ghost points is shown in Fig. 4.8. Suppose we wish to compute any variable

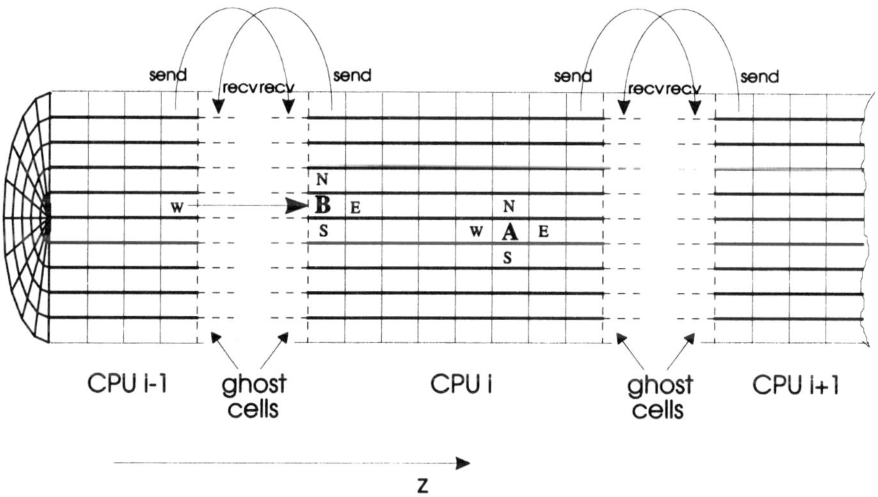

Fig. 4.8 Data exchanges between neighboring processors. The data lying in partition boundary layers in each processor are sent out and are stored in the blocks for ghost grid points in the neighboring processors.

(velocity component or temperature) at the time step t^{n+1} in the "internal" point A as a function of the solution at the previous step t^n.

The processor (i) is able to compute the diffusive and convective fluxes across the faces of the control volume containing the grid point A, since the values related to the grid points adjacent to this control volume are stored in its local memory.

Similarly, at the "boundary" point B, the processor is able to compute the diffusive and the convective fluxes across the faces N, S and E but it "cannot" compute directly these fluxes across the face W, since external values are not stored in its local memory, but they are stored in the local memory of the processor $(i-1)$. Therefore the processor $(i-1)$ has to "communicate" these values to (i) to complete its local computations. These values are "received" by the processor (i) and "located" in a "geometrical extension" of its subdomain represented by the ghost cells.

The number of ghost points on each partition boundary of the subgrid depends on the numerical algorithm. For the SMAC method using a standard Laplacian stencil (see Section 2.8 and Fletcher, 1991), one ghost grid point on each side is needed for the local subgrid at each level. Therefore, in the Navier–Stokes flow solver, it is necessary to allocate storage for one ghost grid point for each subgrid on each partition boundary.

The code can be implemented with an extremely simple message-passing model, consisting of *send* and *receive* only (see Fig. 4.8). The simplicity of this model implies easy portability. The message-passing can be written using the Message Passing Interface (for further details see Gropp et al., 1994; and the Message Passing Toolkit, Cray publication 1996), ensuring portability.

The sending data contains boundary information that is placed into the ghost locations of the neighboring processors. After data exchanges are performed, a new computation step can start. Each processor, therefore, only needs to know its nearest surrounding processors on the logical processor mesh.

Each processor computes the data of its subblock independently of the others, then exchanges data with its neighboring subdomains. The exchange consists of sending messages to each of the neighboring blocks, where the size of each message is proportional to the number of grid points on the common face. The processor then waits to receive the corresponding messages from the neighbor processors, and when these are received, the cycle starts again with the computation phase.

This means that the code runs in loosely synchronous cycles of computation followed by data exchange across the faces of the subblocks.

In order to optimize the synchronization, the subroutines for the management of parallel events, provided by the MPI FORTRAN library (mpi_barrier), for instance, can be used. Hereafter, the typical sequence of FORTRAN instructions for the initialization of the Message Passing Interface and for the data exchange (A is the generic array storing the fluid-dynamic variables, e.g., the temperature, the pressure, the velocity components, etc.) are shown.

```
  real*4, target, allocatable :: A(:,:,:)

C MESSAGE PASSING INTERFACE INITIALIZATION

  call mpi_init(IERROR)
c if (IERROR.ne.0) STOP 'BAD mpi_init'

C Number of available processors (nb_proc) is given by the next
  instruction

  call mpi_comm_size(MPI_COMM_WORLD,Nb_proc,IERROR)

c if (IERROR.ne.0) STOP 'BAD mpi_comm_size'

c if (nb_proc.eq.1) stop

  IF(mod((Nz-2), Nb_proc).ne.0) STOP 'grid points number not
  divisible'

C size of each subdomain
C Nparz is the number of grid points along z for each subdomain
  Nparz=(Nz-2)/Nb_proc

C Nloc takes into account the GHOST CELLS
C on the left and on the right of each subdomain
  Nloc=Nparz+2

C The next instruction determines the number used to identify the
  current processor
```

```
   call mpi_comm_rank(MPI_COMM_WORLD, pe,IERROR)
   if(IERROR.ne.0) STOP 'BAD mpi_comm_rank'
C  ip is CPU identification number plus 1 (the initial value of pe is 0)
   ip=pe+1
C  pes is the identification number of the processor used
C  for the subdomain on the left of the current one
   pes=pe-1
C  ped is the identification number of the processor used
C  for the subdomain on the right of the current one
   ped=pe+1
C  Each CPU stores in its dedicated memory only the values related to
   its subdomain
   allocate (A(Ny, Nf, Nloc))
C  TYPICAL SEQUENCE OF INSTRUCTIONS FOR DATA EXCHANGE
   IF(ip.eq.1) THEN
   CALL mpi_recv(A(1,1,Nloc), Ny*Nf, mpi_real, ped, pe,
   * MPI_COMM_WORLD, status, IERROR)
c  if (IERROR.ne.0) STOP 'BAD MPI_RECV'
c  write(*,*) 'right receive OK ',ip
   END IF
   IF(ip.Eq.Nb_proc) THEN
   CALL mpi_send(A(1,1,2), Ny*Nf, mpi_real, pes, pes,
   * MPI_COMM_WORLD, IERROR)
c  if (IERROR.ne.0) STOP 'BAD MPI_SEND'
c  write(*,*) 'left send OK',ip
   END IF
   IF(ip.gt.1.and.ip.lt.Nb_proc) THEN
   CALL mpi_send(A(1,1,2), Ny*Nf, mpi_real, pes, pes,
   * MPI_COMM_WORLD, IERROR)
c  if (IERROR.ne.0) STOP 'BAD MPI_SEND'
c  write(*,*) 'left send OK ',ip
   CALL mpi_recv(A(1,1,Nloc), Ny*Nf, mpi_real, ped, pe,
   * MPI_COMM_WORLD, status, IERROR)
c  if (IERROR.ne.0) STOP 'BAD MPI_RECV'
c  write(*,*) 'right receive OK',ip
   END IF
```

```
      IF(ip.eq.1) THEN

      CALL mpi_send(A(1,1,Nloc-1), Ny*Nf, mpi_real, ped, ped,
      * MPI_COMM_WORLD, IERROR)
c     if (IERROR.ne.0) STOP 'BAD MPI_SEND'
c     write(*,*) 'right send OK',ip

      END IF

      IF(ip.eq.Nb_proc) THEN

      CALL mpi_recv(A(1,1,1), Ny*Nf, mpi_real, pes, pe,
      * MPI_COMM_WORLD, status, IERROR)
c     if (IERROR.ne.0) STOP 'BAD MPI_RECV'
c     write(*,*) 'left receive OK',ip

      END IF

      IF(ip.gt.1.and.ip.lt.Nb_proc) THEN

      CALL mpi_send(A(1,1,Nloc-1), Ny*Nf, mpi_real, ped, ped,
      * MPI_COMM_WORLD, IERROR)
c     if (IERROR.ne.0) STOP 'BAD MPI_SEND'
c     write(*,*) 'right send OK',ip

      CALL mpi_recv(A(1,1,1), Ny*Nf, mpi_real, pes, pe,
      * MPI_COMM_WORLD, status, IERROR)
c     if (IERROR.ne.0) STOP 'BAD MPI_RECV'
c     write(*,*) 'left receive OK',ip

      END IF
```

4.4.3 *The elliptic equation and the multisplitting technique*

When the grid-partition strategy is employed, the code consists of a number of computationally intensive kernels. Depending upon the equations (parabolic or elliptic) some of these kernels are executed locally on a single processor, and the rest are executed globally across the processors. The kernels that are executed globally require strong communication between the processors.

The efficiency of the numerical solution of the pseudopressure equation (Eq. (2.82)) introduced by the SMAC method is of paramount importance in unsteady three-dimensional, incompressible flow simulations. The pressure solution demands, in fact, at least 50–70% of the CPU time depending on the grid size.

In the "serial" version of the code usually an iterative SOR method is adopted for the solution of the linear elliptic equation. The disadvantage is that a fully SOR or (Gauss–Seidel) method involves the solution of a linear system of equations over the whole solution domain and is therefore, on a parallel machine, much more computationally intensive than explicit steps for parabolic equations.

Over the years, many solution methods have been proposed for linear problems. In particular, much attention has recently been paid to iterative "splitting" methods that generalize the matrix-splitting methods (e.g., Jacobi, Gauss–Seidel and SOR) for solving systems of linear equations. O'Leary and White (1985) have proposed a parallel "multisplitting" iterative method using different splittings of the coefficient matrix. The results obtained from the splitting iterations are combined to define the multisplitting cycles. Thus, the method may be effectively implemented on multi-processors.

This technique is recommended to solve the present problem; the system matrix can be decomposed by columns into p blocks. The global problem is then replaced by a sequence of local problems to be solved in parallel.

This approach has a great potential for parallelism (it runs as a convergence accelerator for the parallel code), is essentially architecture independent and uses a much serial expertise. In fact, the multisplitting philosophy can be regarded as a domain-decomposition algorithm, in which each CPU only provides an updated solution for a portion of the domain.

Specifically, the pressure-variable domain $\pi \in R^k$ (where k is the number of grid points, i.e. $k = N_z N_r N_\varphi$) is partitioned according to $\pi = (\pi_1, \pi_2, \ldots, \pi_p)$ where each subdomain has $\pi_i \in R^{k_i}$ and $\sum_{i=1}^{p} k_i = k$ without overlap of subdomains (Renaut and Mittelmann, 1995).

The solution π_i is then the solution with updated values only on the i portion of the partition. The basic algorithm using p processors follows for all processes i $(1 \leq i \leq p)$:

1. Communicate values near partition boundaries to the neighboring processors
2. Calculate the pseudopressure π over each subdomain using the SOR method
3. Synchronization
4. Test for convergence locally
5. Communicate the convergence result to all processors

The method can be seen as a Jacobi method from a global point of view (each processor does not "see" the values computed by the others processors until each iterative step has been completed) and as an SOR method from a local point of view (each processor uses a SOR method over the variables stored in its local memory). Within the local iterations the system of linear equations is solved neglecting coupling matrix elements related to grid points in different blocks. This means that each iterative step is performed within a single processor with no communication to other processors. Only after the iterative step has been completed is there communication with other processors.

Testing for convergence can be carried out either locally or globally. In the former case it is only necessary to share a logical variable with all the other processes. Otherwise the array must be accumulated and a global check performed. In the framework of the present book, testing globally is recommended. The pseudopressure array is accumulated and the convergence check is performed by a single CPU (the master) that then communicates the result of the test to the other processors (slaves). A master–slave model is suggested herein because it requires minor modification of the original test employed in the serial version of the code for convergence.

Convergence results show that extra iterations are needed by the multisplitting scheme with respect to the serial version of the SOR method. The number of iterations required by the parallel SOR method described above is, in fact, larger than that required by the serial version; however, it is always smaller than that required by the serial Jacobi method (according to this result and to the fact that the multisplitting method can be seen as a Jacobi method from a global point of view and as an SOR method from a local point of view, it follows that this method can be regarded as a hybrid technique).

4.4.4 *Program organization*

The code can be written entirely in FORTRAN 90, making full use of the advanced concepts provided by the language, such as dynamic data structures and pointers, thus obtaining a much more compact, well-structured, and maintainable code.

Storages for all grid variables can be allocated at run time. This strategy of dynamic memory allocation offers a greater flexibility in data-structure manipulations and more efficient use of memory than a static memory allocation, and the user is also alleviated of the burden of calculating storage requirements.

After initialization of the problem to be solved and some algorithm parameters, a preprocessing routine is called. The preprocessing routine constructs the set of nested grids and the corresponding set of logical processor meshes.

The numerical solution proceeds in three major stages:

1. Topology: Perform domain decomposition of the solution domain.
2. Grid generation: Create a high-quality grid within each domain. Spatial discretization reduces the Navier–Stokes equations to a set of partial differential equations.
3. Explicit solution: Advance explicitly in time by using SMAC method.

   For stage 3, the combined use of locks and events permits creation of complex communication and synchronization patterns:

   3a. Solution of the parabolic momentum equations neglecting the pressure gradient (Eq. (2.81))
   3b. Synchronization
   3c. Solution of the elliptic equation (see Eq. (2.82))
   3d. Synchronization
   3e. Update using the computed pressure field to account for continuity (Eq. (2.83)).
   3f. Synchronization
   3g. Solution of the parabolic energy equations
   3h. Synchronization

Parallelization can be performed using FORTRAN extension and special library subroutines (mpi_send, mpi_recv, mpi_barrier).

When making the effort to parallelize the code, it is of great interest to reduce the inefficiency, which is defined as $1-p(T(1)/T(p))$ where p is the number of processors, $T(p)$ is the time taken to run on the p processors, and $T(1)$ is the time taken to run on a single processor (serial version of the code).

The parallel code developed from a theoretical point of view in this chapter has an inner loop structure of loosely synchronous cycles of computation and communication between processors whose physical domains are adjacent. In this case, there are usually two contributions to inefficiency: load-balance inefficiency and communication inefficiency. The load-balance inefficiency could occur because the processors have different amounts of computation to perform, which for the code means different numbers of grid points. In this case the algorithm, in fact, runs at the speed of the processor with the largest number of grid points (the minimum speed). This cause of inefficiency, however, can be eliminated by splitting the grid into equal subblocks.

A communication inefficiency arises from the time taken for messages to pass between the processors. This cause of inefficiency can be reduced in the grid-generation phase. For instance, for the present case of cylindrical subdomains, the subblocks have reasonably low surface area to volume ratios, so that the communication inefficiency tends to be low.

4.4.5 *Parallel performances*

As an example, the Cray T3E supercomputer is considered for comparison. It has 128 DEC Alpha processors (model 21164). The network topology is 3D Torus (480 Mbyte/sec payload bandwidth through each one of the 6 torus directions). The clock cycle is 3.33 ns. The memory is 16 Mword (64 bits) per processor (2 Gword global memory) and the peak performance is 76.8 GigaFLOPS.

The parallel performance of an application code is usually judged by two measurements: Speed-Up and Scaling. The Speed-Up is measured by fixing the problem size (or grid size in our case) and increasing the number of processors used. The scaling is measured by fixing the local problem size in each processor and increasing the number of processors used.

The Speed-Up parameter is defined as

$$S_{\mathrm{p}}=\frac{T(1)}{T(p)} \tag{4.12}$$

where $T(1)$ is the time required for the serial computation (single CPU) and $T(p)$ is the time required by the algorithm to perform computations using p parallel processors.

The serial version of the algorithm does not correspond to the parallel one processor execution of the code since the latter requires additional time for the initialization of the tools introduced by the compiler for the management of the multiprocessing environment.

$T(p)$ may be expressed as the sum of the time required for the serial operations (operations that cannot be executed in parallel mode, e.g., input/output operations) plus the time required for the "true" computations on each processor T_p and the communication overhead T_o:

$$T(p) = T_s + T_\mathrm{p} + T_\mathrm{o} \tag{4.13}$$

Typically, T_p and T_o are combined in a single nondimensional parameter known as the communication cost:

$$c = \frac{T_\mathrm{o}}{T_\mathrm{p}} \tag{4.14}$$

The efficiency of the algorithm is defined as:

$$E_\mathrm{p} = \frac{S_\mathrm{p}}{p} \tag{4.15}$$

The Speed-Up function is actually a comparison between the performances of the serial version of the code and of the parallel program running on p processors. The parallel version takes into account the additional source code that the programmer may have to insert (typically for synchronizing different tasks and for implementing the required communications between them) in the serial code in order to obtain a parallel operation. Moreover, it takes into account the additional object code that the compiler may have to insert for managing the multiprocessing environment (parallel overhead). For these reasons the Speed-Up obtained by parallelizing a given code is always lower than p (and consequently the efficiency will be lower than 1). On the other hand, it can be made very close to p by taking all limiting factors into account during the design of the code.

Figure 4.9 shows the speed-up function S_p related to the present algorithm running on the Cray T3E for a typical case dealing with the 3D simulation of flow in floating zones. For a fixed problem size (i.e. for a fixed grid) S_p increases with the number of processors employed and for a fixed number of CPUs it increases with the dimension of the problem. Figure 4.10 shows the corresponding parallel efficiency E_p. Figures 4.9 and 4.10 show that the advantages obtained by parallelism diminish rapidly as the number of processors exceeds some threshold. This behavior is explained by the fact that the parallel efficiency E_p is largely determined by the ratio of local computations over interprocessor communications. As the number of processors increases, for a fixed problem size, the local communication cost and the cost for global operations will eventually become dominant over the local computation cost after a certain stage.

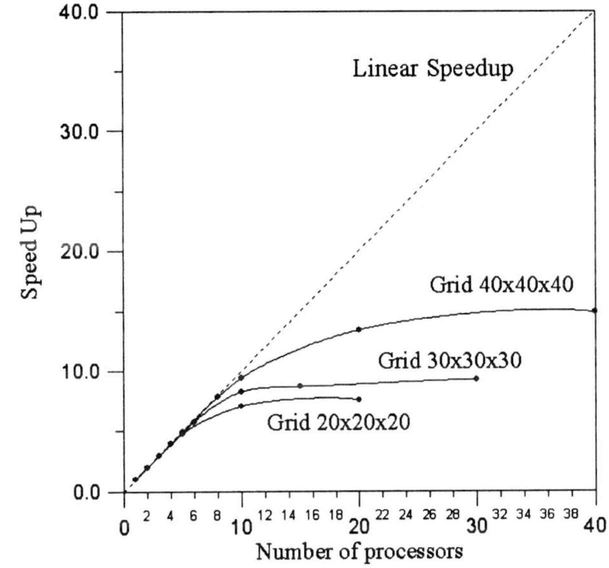

Fig. 4.9 Speed-Up function obtained on the Cray T3E (Lappa, 1997).

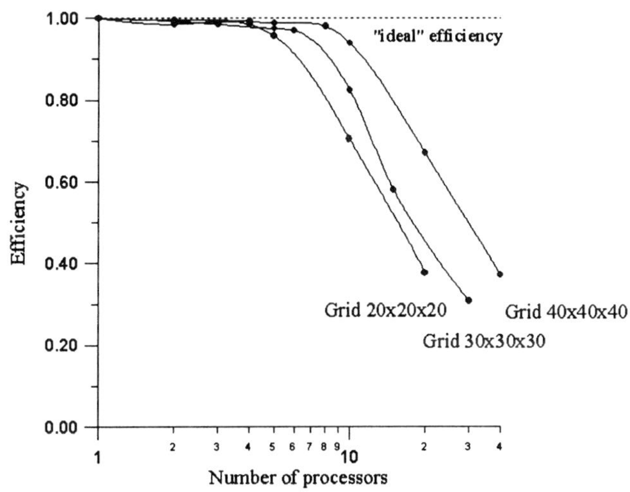

Fig. 4.10 Efficiency obtained on the Cray T3E (Lappa, 1997).

This high ratio of communication to computation makes the influence of a further reduction in the local computation very small on the overall cost of running the application.

For the present algorithm, the best parallel efficiency is achieved on the finest grid (see Fig. 4.10), where the communication overhead is less important than the large number of computations.

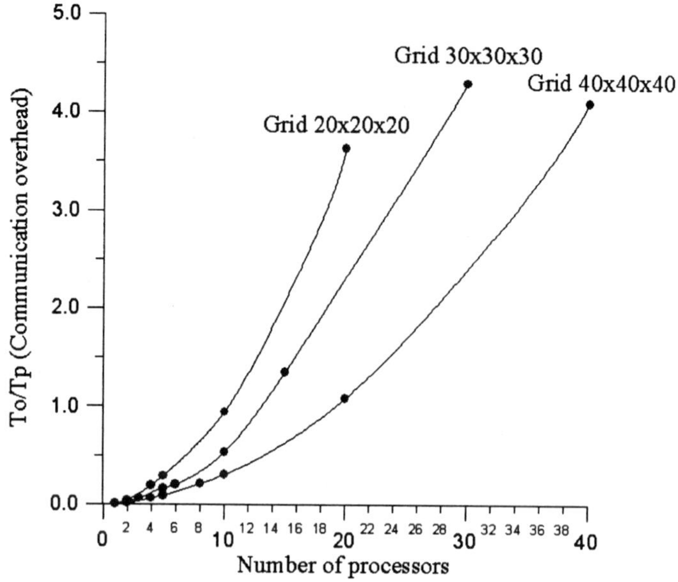

Fig. 4.11 Communication cost (Lappa, 1997).

Indeed, the computational cost is proportional to the number of grid points, and the communication overhead is proportional to the number of grid points associated with the boundary of the subdomains. Thus the communication cost $c = T_o/T_p$ of the code is proportional to the surface area to volume ratio for a processor; and also proportional to the ratio of the amount of communication per boundary grid point, to the amount of work (flops) per internal grid point.

Therefore, the best performances obtained using the finest grids (Figs. 4.10 and 4.11) can be explained by the fact that, for the present algorithm, the computation cost T_p scales as $O(N_r N_\varphi N_z/p)$; whereas the communication overhead T_o scales as $O(N_r N\varphi)$. In conclusion, the communication cost $c = T_o/T_p$ scales as $O(p/N_z)$ as is well illustrated in Fig. 4.11.

4.5. Numerical simulations and theory of bifurcation

Before starting to deal with the detailed discussion of the fluid-dynamics of the FZ and related instabilities some clarification is necessary with regard to the effect of initial guesses and disturbances introduced by the numerical simulations as well as to *the picture of the nonlinear behavior*. This problem, in fact, is very delicate (see also Section 1.4.1 for additional introductory theoretical background).

The numerical procedure proposed herein for the investigation of these topics requires that the steady 2D (axisymmetric) solution of Marangoni flow is used as the "basic state" (i.e. initial condition) for the 3D nonlinear simulations (based on the solution of the 3D and time-dependent Navier–Stokes equations). The simulations lead to the amplification (if $Ma \geq Ma_c$) of initial very small disturbances superimposed on the basic state. This means that it is investigated *if the system is "conditionally" stable (or unstable)*. Conditionally (un)stable means, in fact, that the system is (un)stable to small (infinitesimal) disturbances (Joseph, 1976). For this case, the properties of the bifurcation and of the final state after the bifurcation (*critical Marangoni number, three-dimensional structure of the flow field, final amplitude of the saturated disturbances*) *do not depend on the initial (infinitesimal) disturbances*: if arbitrary very small disturbances are added to the solution, noncritical infinitesimal disturbances decay, whereas the dangerous part (i.e. with the most dangerous value of the azimuthal wave number) is amplified. This behavior, for instance, was clearly observed in the very interesting work of Leypoldt et al. (2000) (half-zone): very small initial sinusoidal disturbances with a priori fixed shape (mode) were superimposed on the basic state; only the disturbances with the mode predicted by the linear stability analysis were observed to be amplified by the nonlinear simulation; on the contrary noncritical disturbances were observed to decay or to be amplified only for larger values of the Marangoni number ($Ma > Ma_c$). This means that the 3D structure of the final state and associated critical Marangoni number are not affected by the initial guesses.

According to the above arguments in the present chapter only the conditional stability is discussed; it is shown, however, that no a priori known small azimuthal disturbances have to be superimposed on the solution to facilitate the transition. It is very worthwhile to highlight how disturbances, in fact, can be freely produced in a spontaneous way by the computations. For the computations discussed in the next sections the infinitesimal disturbances perturbing the system are not a priori fixed but follow directly from the truncation error associated with the finite length of the computer registers storing the flow and temperature values (for more details about this aspect see also Lappa and Savino, 1999; the same procedure has also been used by Imaishi and coworkers and by many others).

In practice, the role of the infinitesimal disturbances is played by the numerical noise introduced in the computation by the round-off errors of the memory registers of the computer used for the simulation (for instance, if double precision is used the initial disturbances are of the order $O(10^{-16})$).

It is interesting to point out how the procedure considered in the present book should provide the same information (in terms of critical Marangoni number and critical mode) that usually is obtained by the linear stability analysis. The latter relies on a spectral problem where the stability properties of the basic state are gathered from the behavior of the eigenvalues of the Jacobian of the linearized equations obtained by adding infinitesimal disturbances to the basic state. The present nonlinear computations provide the critical values of the *Ma* and of the mode that would be predicted by the linear stability theories.

Accordingly, the critical Marangoni number must be defined as the threshold value of the Marangoni number at which the growth rate of the disturbances is zero. It can be computed using an extrapolation of the growth rate to zero. The growth rate in turn is computed by analyzing the behavior of the time-dependent history of the maximum of the azimuthal (3D disturbance) velocity on the free surface of the floating zone.

When the critical Marangoni number Ma_c is approached from within the unstable regime (i.e. $Ma > Ma_c$) the maximum azimuthal surface velocity decreases, becoming zero at the critical point. Thus a scheme to predict the critical Marangoni number can be constructed by extrapolation of $(Ma, V_{\varphi max}) \rightarrow (Ma_c, 0)$.

The same criterion has been used by Leypoldt et al. (2000) and in the works of Imaishi and coworkers. They obtained a very good agreement between the critical Marangoni numbers computed according to this technique and those evaluated by the linear stability analyses.

It must be emphasized, however, that nonlinear simulations also provide further information about the saturated (final) amplitude of the azimuthal disturbances in the final 3D state (i.e. the final configuration of the flow) as well as about the other nonlinear phenomena; this information cannot be obtained on the basis of linearized theories that do not provide quantitative data about the amplitude of the disturbances in the supercritical state and give only qualitative (nevertheless important) information about their structure (i.e. the azimuthal mode).

The foregoing discussion clarifies the differences as well as the analogies between the approach proposed in the present book (relying on the direct solution of the complete and nonlinear model equations) and other available linear stability techniques, while providing a rigorous framework for a deeper understanding of the role played by initial guesses and the related link with the theory of bifurcation.

Finally, it should be pointed out that for the particular case under investigation, the existence of multiple states after any bifurcation is possible in theory: the governing equations with azimuthally uniform boundary conditions (i.e. the case of the FZ process and its models) admit, in fact, rotational transformation on any azimuthal angle (see Eqs. (4.8) and (4.10)); this means that possible solutions can be azimuthally shifted with respect to each other, but being absolutely identical.

The system may tend indifferently to one or the other solution according to the symmetry of the initial infinitesimal disturbances (in practice according to the registers of the computer used, see Lappa and Savino, 1999) even if, as pointed out in the foregoing discussion, the features of the final state (mode, amplitude of the saturated disturbances) are the same.

Figure 4.12 shows, for instance, two final possible steady states (two different branches): one is the mirror image of the other; the mirror symmetry of one state with respect to the other occurs (for the case under investigation) with respect to a generic meridian plane cutting the whole floating zone (i.e. the two states become identical for a rotation of 180°). In practice, for the problem

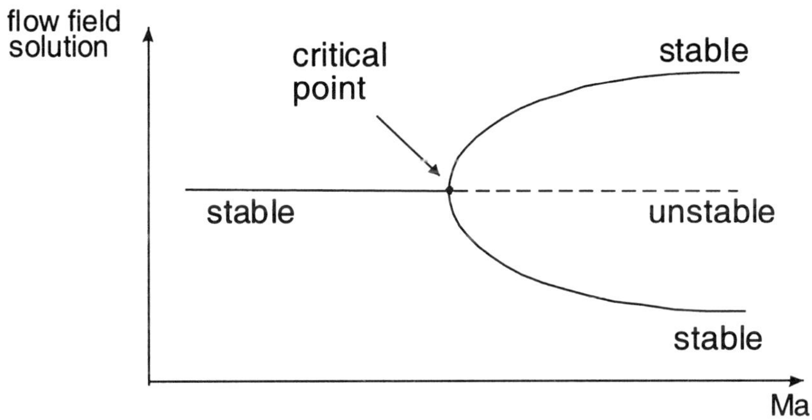

Fig. 4.12 Bifurcation diagram.

under investigation the complete bifurcation diagram would correspond to a 3D paraboloid (as previously pointed out, any possible solution can be azimuthally shifted with respect to each other).

4.6. The half-zone: historical perspective

The half-zone has become, over the years, a typical model for the study of Marangoni flows, their stability, and their bifurcations. Due to the complexity of Marangoni flow and of the related sequence of bifurcations (up to the onset of full turbulence), besides possible application to the FZ technological problem, this model has also been used by many investigators simply to test their ability to predict the behavior of nonlinear systems. Consequently, a rich variety of interesting half-zone-based results has been obtained. Among other things, these studies were the first to show that for liquid metals, the first bifurcation of the Marangoni flow is stationary (i.e. the supercritical state is three-dimensional and steady) and that the oscillations (known to be the source of the aforementioned crystal striations) appear only if the Marangoni number is further increased (secondary bifurcation) (Rupp et al., 1989; Levenstan and Amberg, 1995; Leypoldt et al., 2000).

Imaishi et al. (1999, 2000, 2001) and Yasushiro et al. (2000) depicted in detail the complex spatiotemporal evolution of the flow field that occurs after the secondary (oscillatory) bifurcation of the Marangoni flow for different values of the aspect ratio A_{II} and of the Prandtl number ($0 \leq Pr \leq 0.02$), elucidating different (referred to as "torsional") oscillatory behaviors.

The next sections are devoted to an exhaustive analysis of the influence that different geometrical parameters and environments (0g and 1g) can have on the structure of the 3D flow. The state-of-the-art about these subjects is illustrated.

For the sake of clarity and brevity, however, the discussion is limited to the primary bifurcation of Marangoni convection (Ma_{c1}).

4.7. Structure of the 3D steady flow

As discussed in Section 4.3, in the axisymmetric state (prior to the transition) the fluid-dynamic field inside the half-zone corresponds to a single axisymmetric toroidal vortex. The primary bifurcation of the basic flow described above breaks the spatial axial symmetry of the basic solution and replaces this solution of maximum symmetry with another steady flow having a different pattern of symmetry.

The 3D simulations show that the disturbances grow exponentially with time, i.e. $V_\varphi \propto \exp(\gamma t)\sin(m\varphi)$ (where V_φ is the azimuthal velocity component induced by the instability and γ is the so-called growth rate), and after a long incubation time macroscopic 3D steady velocity and temperature fields are established (stationary bifurcation).

Investigators usually characterize this pattern by specifying the azimuthal wave number m (mode) of the instability. After the bifurcation the flow field, in fact, can be interpreted as the superposition of a steady sinusoidal azimuthal disturbance (characterized by the appropriate wave number) to a reference axisymmetric state, i.e. the generic flow field variable $F(r, z, \varphi)$ can be expressed as:

$$\mathrm{F}(r, z, \varphi) = F_{\mathrm{o}}(r, z) + \tilde{f}(r, z)\sin(m\varphi + G) \tag{4.16}$$

where the subscript (o) refers to the axisymmetric field, m is the aforementioned azimuthal wave number (from a physical point of view m represents the number of sinusoidal distortions in the azimuthal direction), $\tilde{f}$ is the perturbation amplitude and G is a constant phase shift due to the fact that the azimuthal position of the disturbances can be random, as pointed out in the last part of Section 4.5 (possible solutions may be azimuthally shifted with respect to each other).

Hereafter, the temperature fields obtained as differences between the numerical three-dimensional solutions and the corresponding azimuthally averaged fields will be referred to as temperature disturbances.

When m increases (see the next section for possible reasons that can induce m changes) multicellular structures are formed. In the generic section orthogonal to the z axis (symmetry axis of the half-zone) $2m$ convective cells are present and on the bridge-free surface, $2m$ temperature spots (m cold and m hot) appear (each surface spot having an angular extension of $360/2m°$).

The radial position of the nodal lines of the temperature perturbations in the cross-sections perpendicular to the liquid-zone axis corresponds to the position of the vortex centerline after the bifurcation. The position of the vortex core is deformed and, in particular, displaced sinusoidally along the perimeter of the toroidal convection roll. This is due to the occurrence of the aforementioned

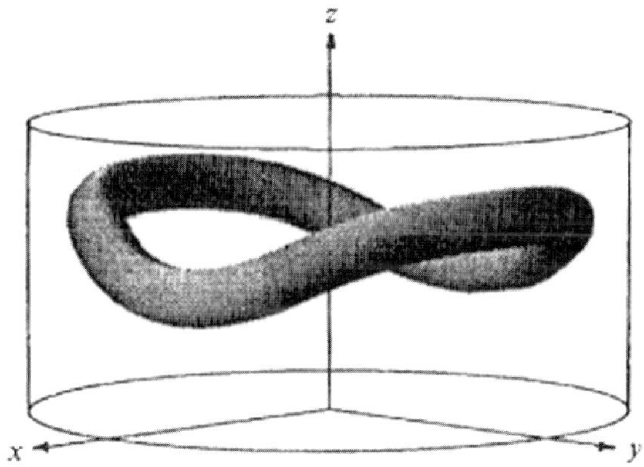

Fig. 4.13 Toroidal vortex in the 3D state.

additional convective cells in the sections orthogonal to the bridge axis. The position of the vortex centerline also describes in space a sine curve having m maxima and m minima in the z-direction (Fig. 4.13). According to the numerical simulations the part of the toroidal vortex that is displaced downwards is also displaced away from the center of the zone and the part that is displaced towards the center is, at the same time, displaced upwards (the displacement up and down is coupled to displacement in and out, respectively). For this reason each plane orthogonal to the axis of symmetry shows temperature nodes that correspond to the intersection of the distorted vortex centerline with the considered plane.

Critical wave numbers $m=1$, 3, 5, 7, etc. belong to the class of asymmetrical modes; $m=2$, 4, 6, etc. are instead symmetric modes. More generally, when the critical disturbance number (m) is odd, each meridian plane of the half-zone is characterized by two asymmetrical vortex cells (for $m=1$, in particular, one of the two vortex cells in the section can prevail over the other and can be extended along the whole axial plane of the bridge). For even critical wave numbers, the flow-field structure is, on the whole, three-dimensional and depends on the azimuthal coordinate, but in each meridian plane the velocity and temperature are symmetric.

Some examples of these behaviors are illustrated in the next section where the effect of some geometrical parameters under microgravity conditions is elucidated.

Hereafter, all the simulations deal with the case of a silicon melt ($Pr=0.01$, see Table 4.1 for the physical properties) or liquid gallium.

4.8. Effect of geometrical parameters

The critical azimuthal wave number increases when the geometrical aspect ratio of the bridge is decreased and for a fixed aspect ratio, it can be shifted to higher values by increasing the volume or to lower values by decreasing the volume.

Table 4.1 Physical properties of silicon melt.

Melting point temperature T_m (K)	1685
Density ρ (g/cm^3)	2.42
Thermal diffusivity α (cm^2/s)	2.44×10^{-1}
Kinematic viscosity ν (cm^2/s)	2.45×10^{-3}
Prandtl number	0.01
Thermal conductivity λ (W/cm K)	0.64
Thermal expansion coefficient β_T (1/K)	1.4×10^{-4}
Surface tension σ (dyne/cm)	7.33×10^2
σ_T (dyne/cm K)	1×10^{-1}
Emissivity (–)	0.3

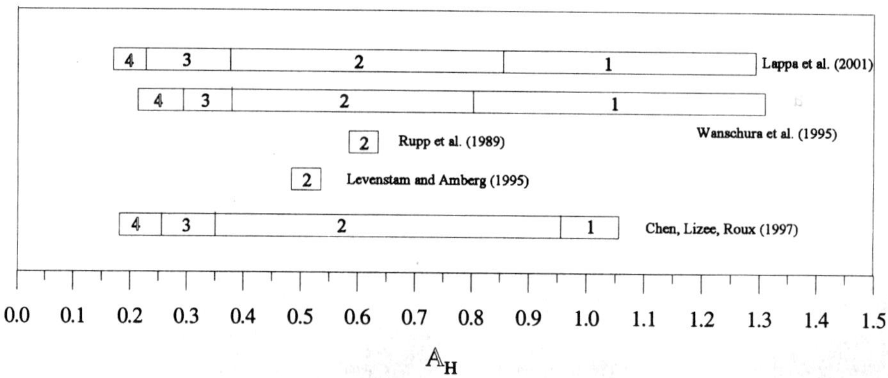

Fig. 4.14 Azimuthal wave number versus the aspect ratio A_H of the half-zone; comparison among different results for liquid metals.

The current state-of-the-art for $S=1$ and microgravity conditions (cylindrical interface) in particular, suggests an empirical correlation between the geometrical aspect ratio and the critical azimuthal wave number of the instability.

Rupp et al. (1989) found $m=2$ for $A_H=0.6$ and for Pr ranging from 0.007 to 0.16. Levenstam and Amberg (1995) found $m=2$ for $A_H=0.5$ and $Pr=0.01$. Lappa and Savino (1999) found $m=1$ and $m=2$ for $A_H=1.0$ and $A_H=0.6$, respectively ($Pr=0.04$). Wanschura et al. (1995) gave the dependence of the most dangerous azimuthal wave number on the aspect ratio for $Pr=0.02$. In their results the most dangerous mode is $m=1$ for $1.5>A_H>0.8$, $m=2$ for $0.75>A_H>0.4$, $m=3$ for $0.35>A_H>0.28$ and $m=4$ for $0.28>A_H>0.25$. Imaishi and coworkers (2000) found $m=2$ for $0.7 \geq A_H \geq 0.5$ and $m=1$ for $A_H=0.9$ for $Pr=0.01$. Lappa et al. (2001a) found $m=1$ for $A_H \geq 0.9$, $m=2$ for $0.85 \geq A_H \geq 0.4$, $m=3$ for $0.35 \geq A_H \geq 0.25$, $m=4$ for $A_H=0.2$.

In the light of these results the aforementioned correlation is

$$mA_H \cong 1 \tag{4.17}$$

Figure 4.14 summarizes the results in a single frame. If the shape is not cylindrical the effect of the parameter S must also be taken into account for a correct simulation of the problem (Lappa et al., 2001a) as shown in Figs. 4.15–4.32 for the case $Pr = 0.01$. In these figures, frames (a) and (b) are disposed on the left and right of the pages and refer to the case of straight and curved configurations, respectively; such an arrangement is adopted for the convenience of the reader and for emphasizing, by close comparison, the effect of the noncylindrical volume on the 3D structure with respect to $S = 1$.

In particular, for $A_H = 1$ and a cylindrical shape ($S = 1$) the wave number is $m = 1$. Two thermal spots are present on the liquid-bridge surface (Fig. 4.15a). Moreover, in the generic cross-section orthogonal to the liquid-bridge axis there are two azimuthal convective cells (see Fig. 4.19a) and two thermal spots (Fig. 4.20a). For $S = 1.22$ (fat liquid bridge) the wave number increases. A comparison of frames (a) and (b) shows, in fact, that in this case the convective cells and the temperature spots become four

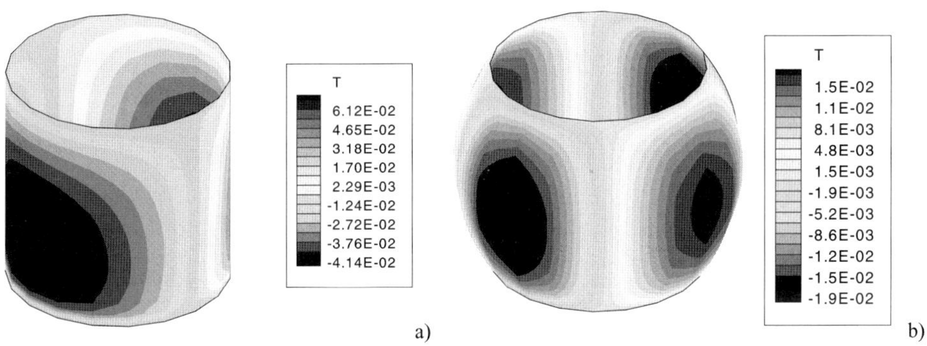

Fig. 4.15 Temperature disturbances on the liquid-bridge surface for $A_H = 1$ (half-zone) and zero-g: (a) $S = 1$, (b) $S = 1.22$.

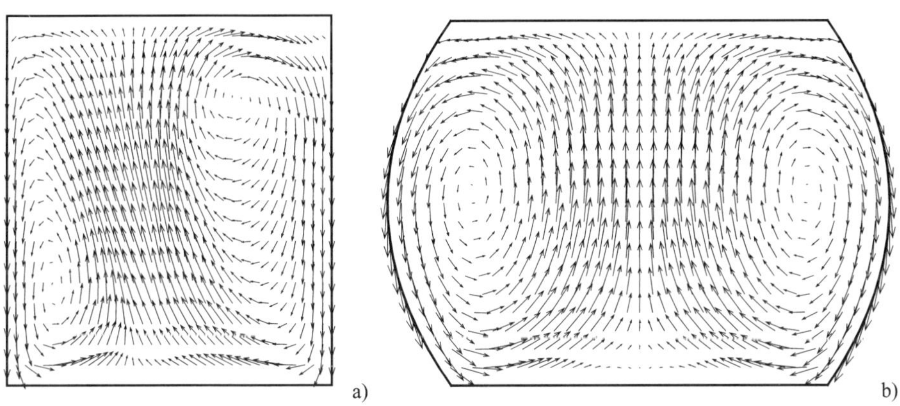

Fig. 4.16 Velocity field in the meridian plane $\varphi = 0$ for $A_H = 1$ (half-zone) and zero-g: (a) $S = 1$, (b) $S = 1.22$.

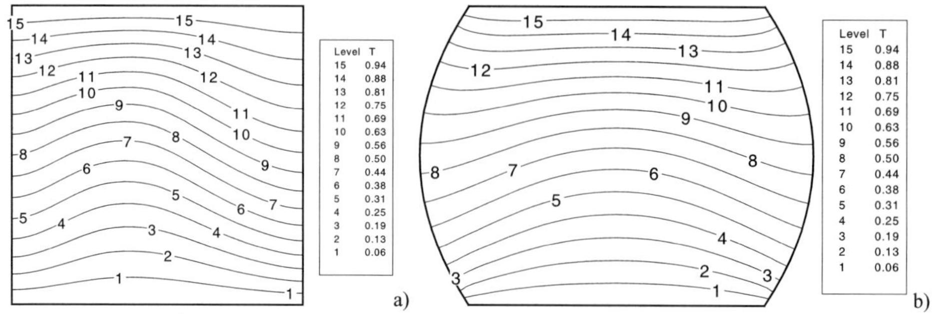

Fig. 4.17 Nondimensional temperature field in the meridian plane $\varphi = 0$ for $A_H = 1$ (half-zone) and zero-g: (a) $S = 1$, (b) $S = 1.22$.

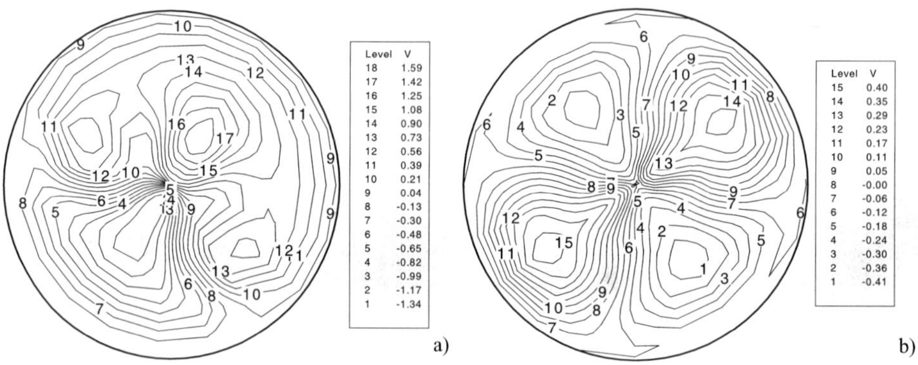

Fig. 4.18 Radial velocity component in the section $z = 0.5$ for $A_H = 1$ (half-zone) and zero-g: (a) $S = 1$, (b) $S = 1.22$.

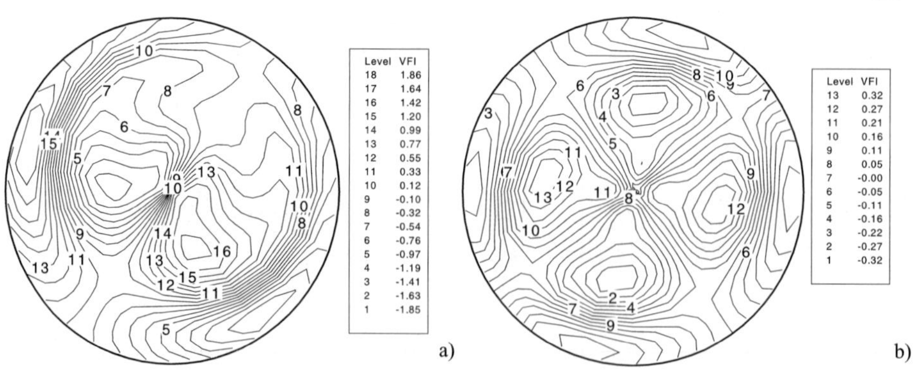

Fig. 4.19 Azimuthal velocity component in the section $z = 0.5$ for $A_H = 1$ (half-zone) and zero-g: (a) $S = 1$, (b) $S = 1.22$.

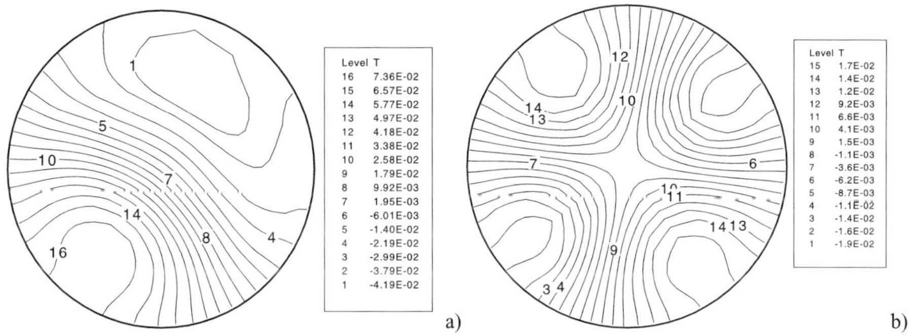

Fig. 4.20 Temperature disturbances in the section $z = 0.5$ for $A_{\mathrm{H}} = 1$ (half-zone) and zero-g: (a) $S = 1$, (b) $S = 1.22$.

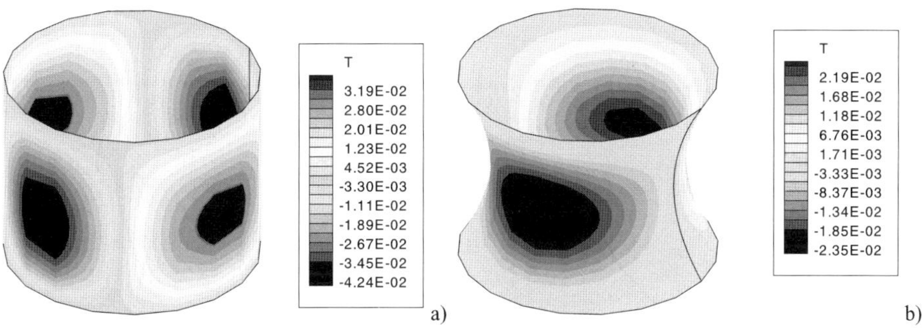

Fig. 4.21 Temperature disturbances on the liquid-bridge surface for $A_{\mathrm{H}} = 0.75$ (half-zone) and zero-g: (a) $S = 1$, (b) $S = 0.83$.

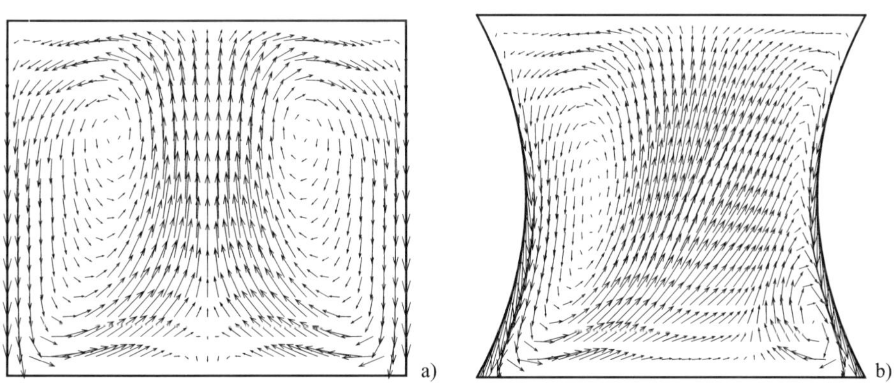

Fig. 4.22 Velocity field in the meridian plane $\varphi = 0$ for $A_{\mathrm{H}} = 0.75$ (half-zone) and zero-g: (a) $S = 1$, (b) $S = 0.83$.

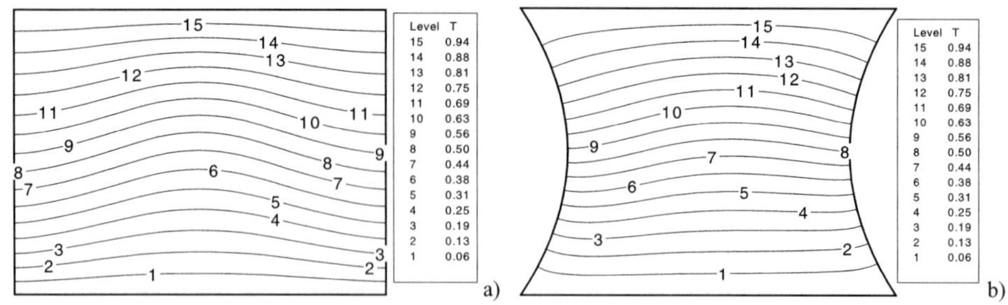

Fig. 4.23 Temperature field in the meridian plane $\varphi=0$ for $A_{\mathrm{H}}=0.75$ (half-zone) and zero-g: (a) $S=1$, (b) $S=0.83$.

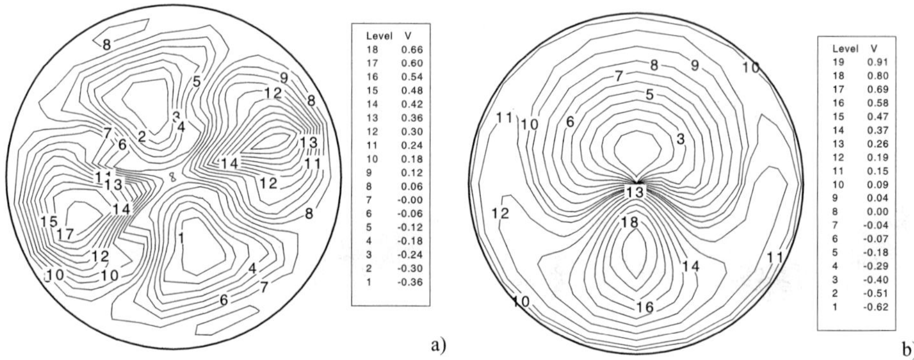

Fig. 4.24 Radial-velocity component in the section $z=0.5$ for $A_{\mathrm{H}}=0.75$ (half-zone) and zero-g: (a) $S=1$, (b) $S=0.83$.

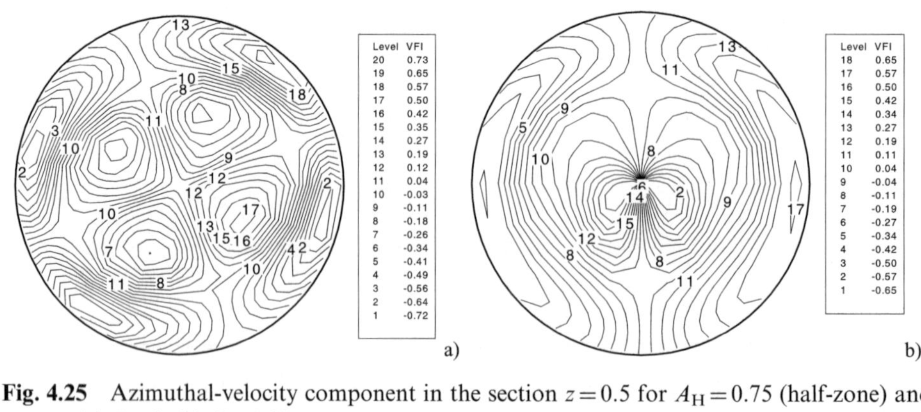

Fig. 4.25 Azimuthal-velocity component in the section $z=0.5$ for $A_{\mathrm{H}}=0.75$ (half-zone) and zero-g: (a) $S=1$, (b) $S=0.83$.

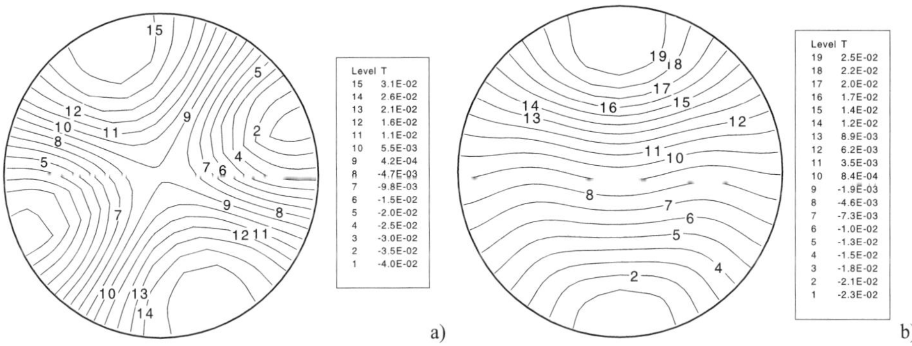

Fig. 4.26 Temperature disturbances in the section $z=0.5$ for $A_{\mathrm{H}}=0.75$ (half-zone) and zero-g: (a) $S=1$, (b) $S=0.83$.

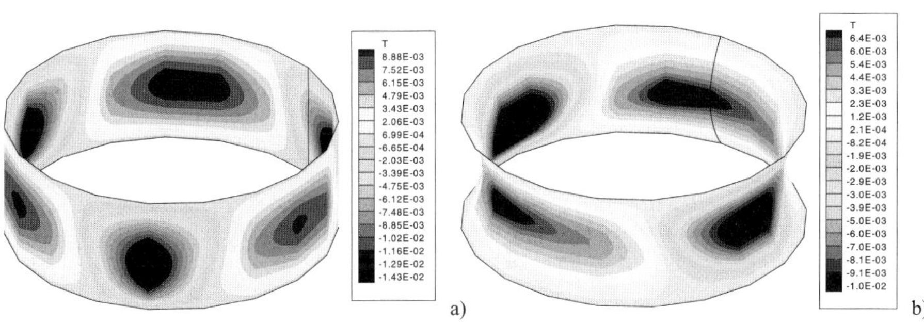

Fig. 4.27 Temperature disturbances on the liquid-bridge surface for $A_{\mathrm{H}}=0.35$ (half-zone) and zero-g: (a) $S=1$, (b) $S=0.915$.

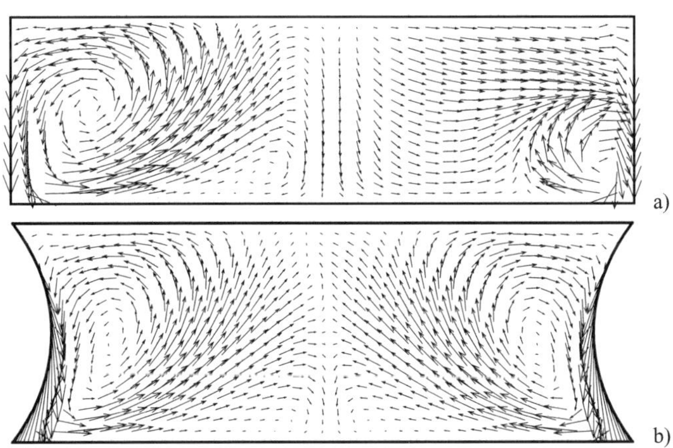

Fig. 4.28 Velocity field in the meridian plane $\varphi=0$ for $A_{\mathrm{H}}=0.35$ (half-zone) and zero-g: (a) $S=1$, (b) $S=0.915$.

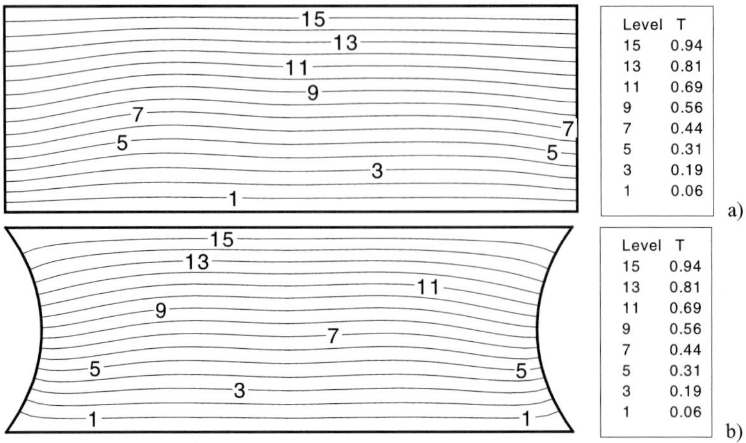

Fig. 4.29 Temperature field in the meridian plane $\varphi=0$ for $A_{\mathrm{H}}=0.35$ (half-zone) and zero-g: (a) $S=1$, (b) $S=0.915$.

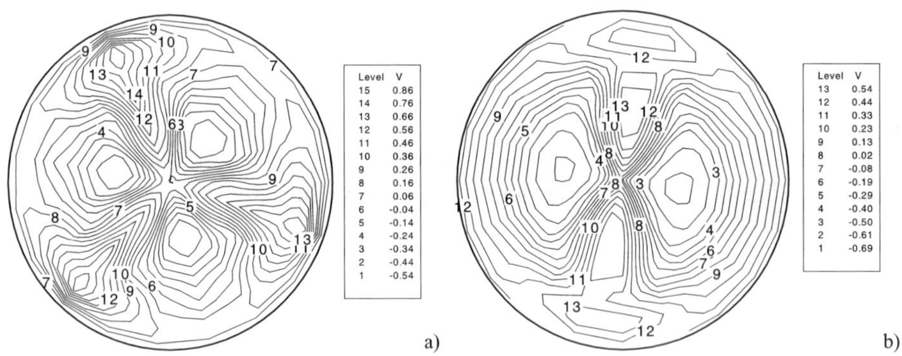

Fig. 4.30 Radial-velocity component in the section $z=0.5$ for $A_{\mathrm{H}}=0.35$ (half-zone) and zero-g: (a) $S=1$, (b) $S=0.915$.

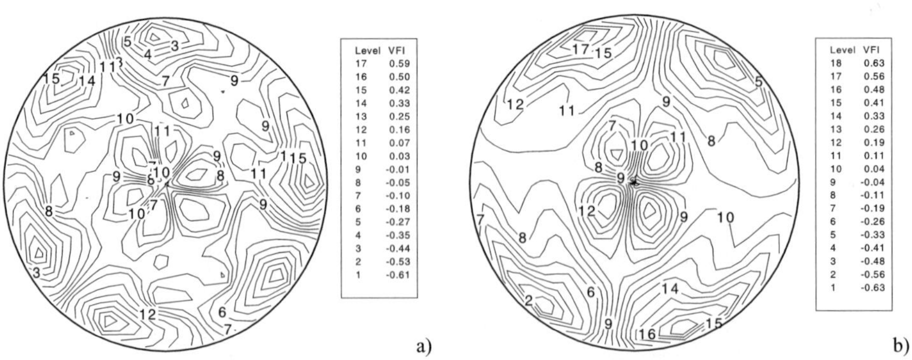

Fig. 4.31 Azimuthal-velocity component in the section $z=0.5$ for $A_{\mathrm{H}}=0.35$ (half-zone) and zero-g: (a) $S=1$, (b) $S=0.915$.

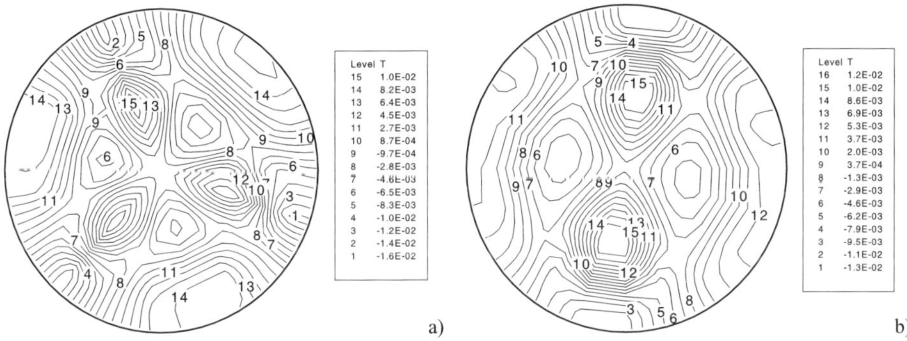

Fig. 4.32 Temperature disturbances in the section $z=0.5$ for $A_{\mathrm{H}}=0.35$ (half-zone) and zero-g: (a) $S=1$, (b) $S=0.915$.

(Figs. 4.15b, 4.19b, 4.20b). Moreover, comparing Fig. 4.16a with Figs. 4.16b and 4.17a with Fig. 4.17b, it is evident how in the first case the flow pattern in the generic meridian plane is asymmetric (due to an odd mode of convection) whereas in the latter it is symmetric (due to an even mode of convection).

For $A_{\mathrm{H}}=0.75$ and cylindrical shape the critical wave number is $m=2$ (Figs. 4.21a–4.26a). In this case, in contrast to the result found for $A_{\mathrm{H}}=1$, the critical mode number does not change if the volume is increased but is shifted to a lower value ($m=1$) if the volume is reduced (concave liquid bridge $S=0.83$). Comparing Fig. 4.22a and b it is evident, in fact, how in the former the flow pattern is symmetric, whereas in the latter it is asymmetric.

For $A_{\mathrm{H}}=0.35$ the critical wave number becomes $m=3$ for the case of a cylindrical shape and is shifted to $m=2$ if S is reduced. Consequently, in the generic cross-section orthogonal to the liquid-bridge axis there are six azimuthal convective cells and six temperature spots, and six thermal spots on the liquid-bridge surface for $S=1$ (Figs. 4.27a, 4.31a, 4.32a) but for $S=0.915$ the convective cells and the temperature spots are four (Figs. 4.31b, 4.32b) and there are four thermal spots on the free surface (two hot and two cold, Fig. 4.27b). Moreover comparing Fig. 4.28a and b it is evident how in the former the flow pattern is asymmetric (owing to a critical wave number $m=3$) whereas in the latter it is symmetric (due to a critical wave number $m=2$).

With regard to the instability threshold (see Table 4.2a), it is well known that the Ma_{c} increases if the aspect ratio is reduced (see, e.g., Imaishi et al., 1999, 2000 in the case of a cylindrical interface) and is characterized by a minimum value around $S=1$ if the effect of the volume is taken into account ($S\neq 1$) (Chen et al. 1999).

The foregoing arguments, supported by the numerical examples shown in the figures, summarize in a single brief and focused overview both the effects of the aspect ratio and the volume factor under microgravity conditions.

In the light shed on the problem by these studies, most of the scientific community is currently aware of the importance that the control of the boundary conditions and

Table 4.2a Critical Marangoni number versus the geometrical aspect ratio, $Pr=0.01$, $S=1$, microgravity conditions (results of Imaishi et al., 1999, 2000; Chen et al., 1999).

A_H	Ma_{c1}
0.4	12.03
0.45	11.33
0.5	12.41
0.55	14.91
0.6	15.09
0.65	11.35
0.75	10.15
0.9	10.36
1	9.923

Table 4.2b Critical Marangoni number for the half-zone under normal-gravity conditions ($L=1$ cm, $A_H=1$, $S=1$, $Pr=0.01$) after Lappa et al. (2003).

	Cylinder 1g heating from above	Cylinder 0g	Cylinder 1g heating from below	1g shape heating from above	1g shape heating from below
Ma_{c1}	15.36	15.24	14.92	16.62	18.78

the geometry of the liquid zone may have on the threshold values and observed flow patterns.

These studies have also provided some interesting insights into the intrinsic nature of the instability. For instance, a comparison between Figs. 4.19 and 4.20 shows that on the free surface the azimuthal flow is directed from the cold spots towards the hot spots. This is an indirect evidence of the fact that the mechanisms of the instability are hydrodynamic. The azimuthal convective cells are not driven by the surface temperature gradients. On the contrary, in the hydrothermal case, the interplay between such a flow and surface thermal spots in the azimuthal direction is of paramount importance (see the discussion on the role played by the surface-temperature disturbances in the case of high Prandtl number liquids in Section 4.12).

4.9. Gravity effects and heating direction

In order to estimate the gravity effects, both the case of a cylindrical shape (simplified model) or of a melt/air interface deformed by the effect of gravity (real conditions) must be considered from a theoretical point of view (see Table 4.2b for $Pr=0.01$). In practice, comparison among these situations is a theoretical craft that gives insight into the separate (gravity) influences of buoyancy forces and of the free-surface deviation with respect to the straight (cylindrical) configuration. It is also worthwhile to stress how within the context of these studies the heating direction (bridge heated from above or bridge heated from below) plays a critical role in the dynamics under investigation (Yasushiro et al., 2001; Lappa et al., 2003a).

For instance, Fig. 4.33a shows the values of the growth-rate constants (γ) obtained for $A_H = 1$, $S = 1$ and different values of the Marangoni number and for different conditions (i.e. zero g conditions ($Ra = 0$), on the ground with heating from above ($Ra < 0$) and heating from below ($Ra > 0$)) in the case of a cylindrical interface. The growth rates obtained in the case of normal-gravity conditions and heating from above ($Ra < 0$) lie below the corresponding ones obtained in the case of microgravity conditions ($Ra = 0$). Vice versa the growth rates obtained in the case heating from below are above those corresponding to zero-g conditions.

For these different cases the wave number is always equal to one ($m = 1$) thus showing that in the case of a silicon liquid bridge having a length of 1 cm, $S = 1$ and

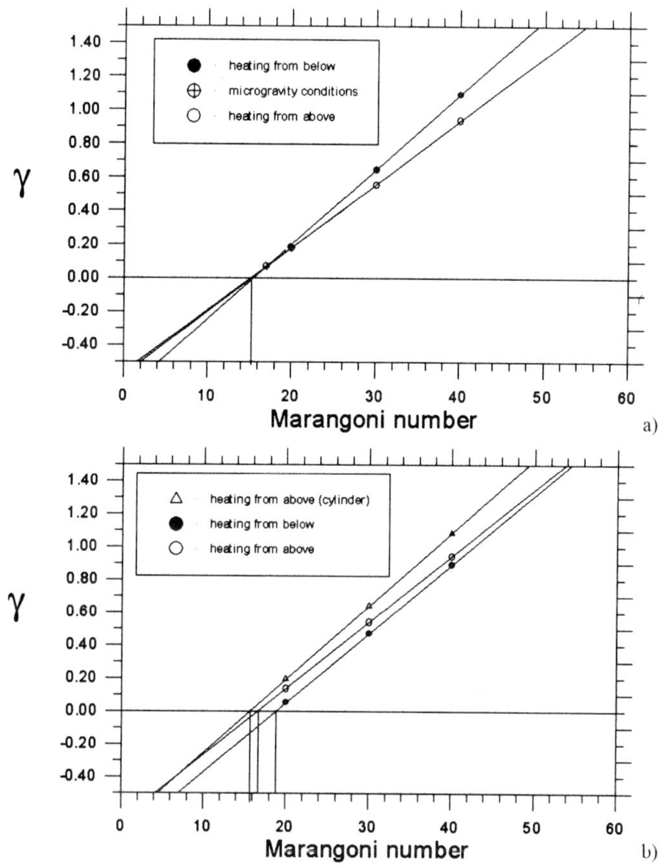

Fig. 4.33 Growth rates as function of the Marangoni number and determination of the first critical Marangoni number (silicon, $Pr = 0.01$, $A_H = 1$ $S = 1$ cm, results after Lappa et al., 2003a): (a) cylindrical shape, (b) gravity-deformed shape.

$A_H = 1$, the azimuthal wave number is not affected by the buoyancy forces. Nevertheless, according to the behavior of the evaluated growth rates, the on the ground critical Marangoni numbers in the case of heating from above ($Ma_{c1} = 15.36$) and heating from below ($Ma_{c1} = 14.92$) are increased and decreased, respectively, with respect to the case of microgravity conditions ($Ma_{c1} = 15.24$).

These behaviors are in agreement with the theoretical results of the linear stability analysis of Wanschura et al. (1995). They found, in fact, that the efficiency of the process of amplification of the azimuthal disturbances responsible for the inertia instability (that characterizes semiconductor melts and liquid metals) is proportional to the radial gradient of the basic state axial velocity, i.e. the axial shear. The Marangoni forces driving the basic flow field induce a strong axial shear flow below the free melt/air interface. The higher the intensity of the axial share, the lower should be the critical Marangoni number.

On the basis of this theory, the stabilization of the flow field in the case of heating from above simply follows from the decrease of the axial share intensity that in this case is reduced since buoyancy and Marangoni forces counteract in the liquid.

As a second step, in order to elucidate the overall effect of the gravity field (i.e. buoyancy forces + surface deviation from the straight configuration), the deformation of the shape must be taken into account. According to the results in Fig. 4.33b, the growth rates obtained in the case of heating from above and deformed shape lie below the corresponding ones obtained in the case of a cylindrical shape thus giving rise to a stabilization of the flow field ($Ma_{c1} = 16.62$). Despite the different value of the critical Marangoni number ($Ma_{c1} = 16.62$ instead of $Ma_{c1} = 15.36$) this trend is similar to that already discussed for the cylindrical case. However, contrary to the case of a straight surface, if the surface deformation is not ignored, the flow exhibits further stabilization for the heating from below condition. The corresponding values of the growth rate, in fact, are below those obtained in the opposite case ($Ma_{c1} = 18.78$).

Thus there is a significant effect of the shape on the instability boundaries. If the liquid surface of the half-zone is allowed to deform, in fact, gravity always acts to stabilize the Marangoni flow regardless of its direction (parallel or antiparallel to the z axis).

The combined effect of the shape and of the heating direction is also crucial in determining the azimuthal organization. The structures of the flow and temperature fields for the case under investigation ($A_H = 1$, $S = 1$), in fact, are characterized by $m = 1$ in the case of heating from below and $m = 2$ if the direction of gravity is reversed. The different azimuthal structures of the thermo-fluid-dynamic field at the steady state, according to the gravity level and according to the heating condition (from above and from below) are shown in Figs. 4.34 and 4.35.

To summarize, the liquid bridge exhibits different behaviors according to the allowed bridge shape. If the shape is forced to be cylindrical, the flow field is stabilized in the case of heating from above and destabilized if gravity is reversed.

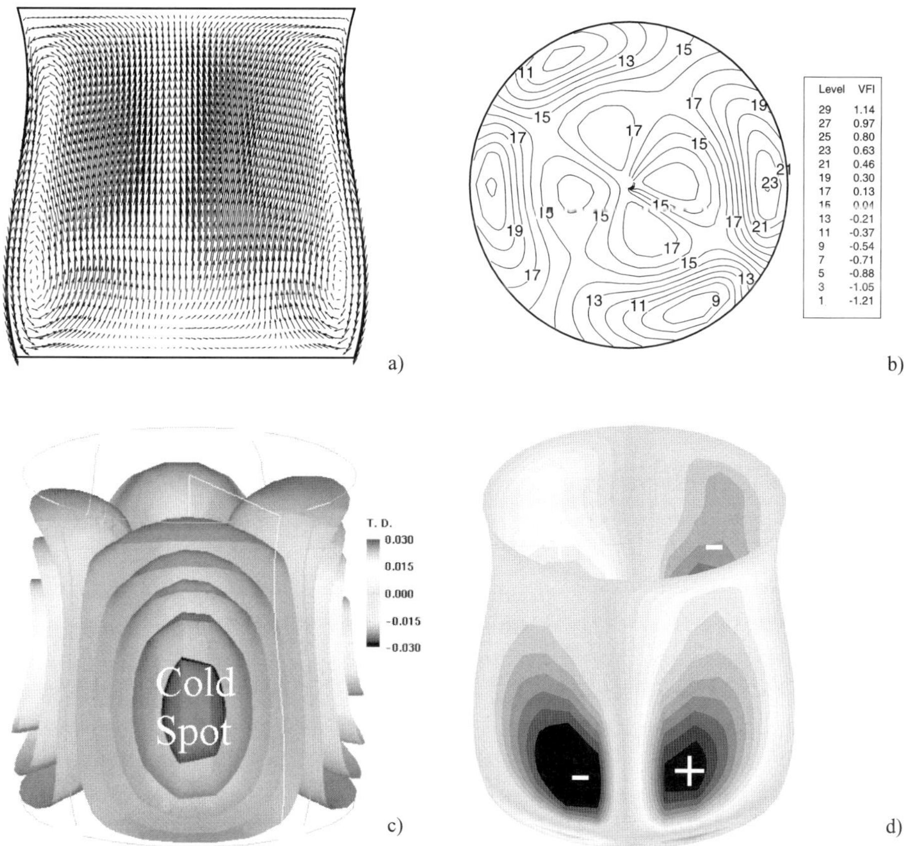

Fig. 4.34 Structure of 3D Marangoni flow with $m = 2$ in the case of a silicon liquid bridge, $L = 1\,\text{cm}$, $A_{\text{H}} = 1$, $S = 1$, $Ma = 40$ at the steady state, 1g, bridge heated from above. (a) Velocity field in the meridian plane $\varphi = 3\pi/4$, (b) azimuthal velocity in the cross section $z = 0.5$, (c) 3D view of the isosurfaces of the temperature disturbance, (d) azimuthal velocity on the free surface.

If the deformation is taken into account, gravity always stabilizes the Marangoni flow regardless of its direction (parallel or antiparallel to the axis) and the three-dimensional flow structure is different according to the heating condition (from above or from below). In the latter case, the critical Marangoni number is larger and the critical wave number is smaller, compared with the opposite condition.

Despite some remarkable analogies existing with the corresponding case of high Prandtl number liquids that have been briefly described in Chapter 2, the case of a small Prandtl number fluid treated herein has nothing to do with the case of silicone oils. For $Pr \gg 1$ the instability of the Marangoni flow is hydrothermal in nature like that disclosed by Smith and Davis (1983) in the case of liquid layers of infinite extent (some additional revealing insights about this type of transition in liquid bridges will

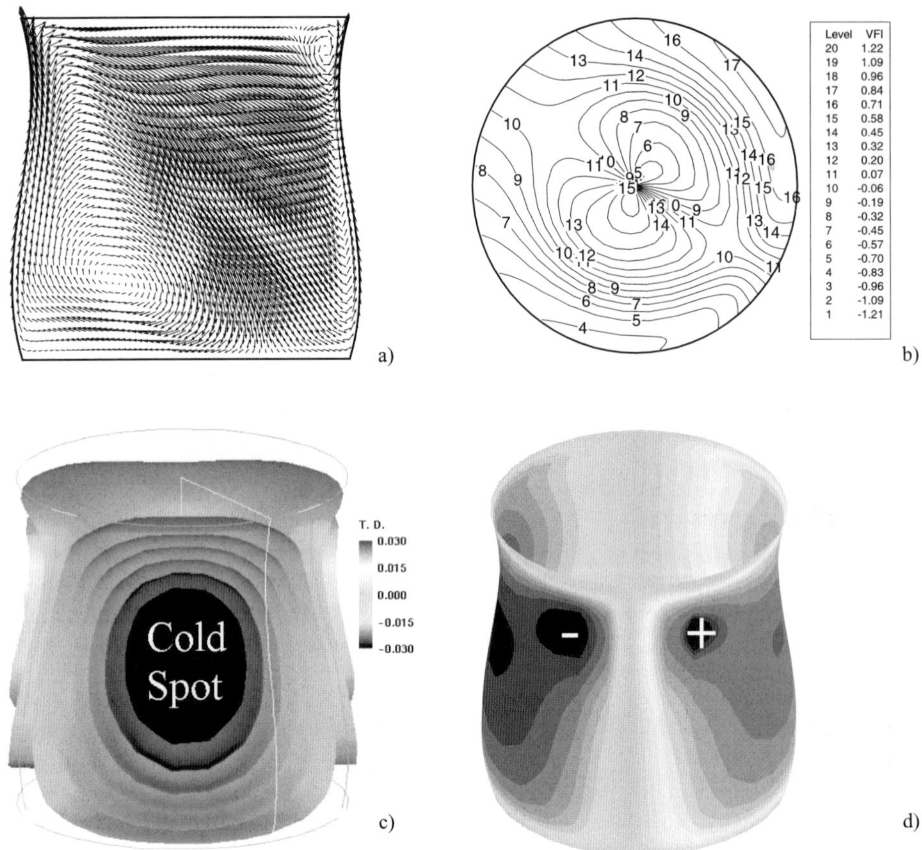

Fig. 4.35 Structure of 3D Marangoni flow with $m=1$ in the case of a silicon liquid bridge, $L=1$ cm, $A_H=1$, $S=1$, $Ma=40$ at the steady state, 1g, bridge heated from below. (a) Velocity field in the meridian plane $\varphi=3\pi/4$, (b) azimuthal velocity in the cross section $z=0.5$, (c) 3D view of the isosurfaces of the temperature disturbance, (d) azimuthal velocity on the free surface.

be provided in Section 4.12). Semiconductor melts and transparent silicone oils (as well as other organic liquids) and semiconductor melts are characterized by primary instabilities that exhibit a very different nature, as explained in Section 4.1.

For the first case the different structure of the flow is determined by the delicate interplay of shape deformation and heating direction in inducing instability of the shear layer below the free surface; for the latter the change in the azimuthal structure can occur even if the interface is maintained cylindrical since it is due to the effect of aiding or counteracting buoyancy forces on the dominating Marangoni convection. On the contrary, in the case of semiconductor melts gravity plays its role basically through the deformation induced in the shape of the interface.

Owing to this feature, the aforementioned stabilization of the flow induced by the deformation of the free surface dramatically changes according to the substance under consideration. In fact, melts exhibiting different densities and surface tensions tend to be characterized by very different slopes of the amphora-like shape.

As an example, Fig. 4.36 shows the computed free-melt/air interface in the case of silicon and gallium ($L=1$ cm, $A_H=1.0$, $S=1$). The shape of the liquid bridge of gallium is more considerably deformed than that of silicon because of the higher density of Ga and its lower surface tension (for Si and Ga the Bond number $\rho g L^2/\sigma_o$ is 3.386 and 8.3344, respectively, see Table 4.3 for the physical properties of gallium).

Along these lines, Figs. 4.37 and 4.38 are very important since they show how the influence of the deformation of the interface on the surface temperature and velocity distributions becomes more significant in the case of gallium.

In the case of heating from above the most dangerous azimuthal wave number is $m=2$. The value of the critical Marangoni number ($Ma_{c1}=89.51$) is significantly increased with respect to the cases where the straight surface is assumed ($m=1$ and $Ma_{c1}=36.19$).

This behavior is similar from a qualitative point of view to that already discussed for the case of silicon. If the direction of gravity is reversed (heating from below) however, a completely different scenario arises: The value of the critical Marangoni number is increased up to $Ma_c=242.2$; no steady bifurcation occurs in this case and the instability threshold value is much higher than the second critical Marangoni number for a cylindrical liquid bridge of $Pr=0.02$ fluid ($Ma_{c2}=131.8$ for $g=0$).

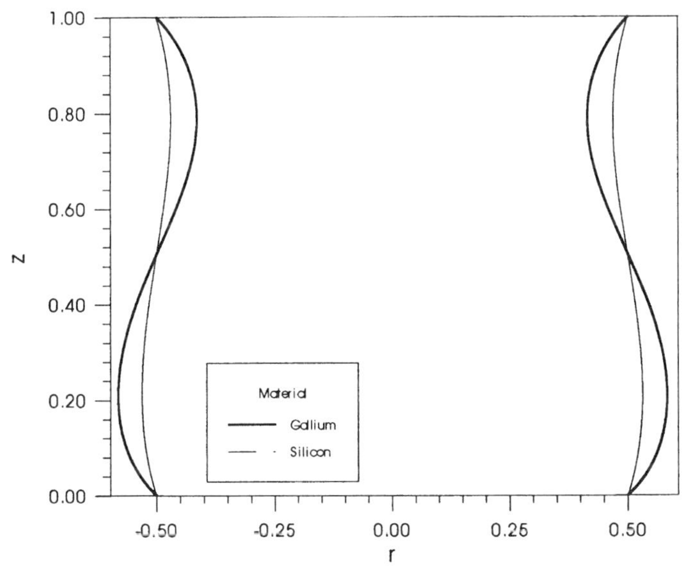

Fig. 4.36 Computed free-surface shapes in the case of silicon and gallium ($L=1$ cm, $A_H=1.0$, $S=1$).

Table 4.3 Physical properties of gallium.

Melting point temperature T_m (K)	302.91
Density ρ (g/cm^3)	6.10
Thermal diffusivity α (cm^2/s)	1.66×10^{-1}
Kinematic viscosity ν (cm^2/s)	3.44×10^{-3}
Dynamic viscosity μ (g/cm s)	2.12×10^{-2}
Prandtl number	0.0207
Thermal expansion coefficient β_T (1/K)	1.26×10^{-4}
Surface tension σ (dyne/cm)	7.18×10^{2}
σ_T (dyne/cm K)	1×10^{-1}

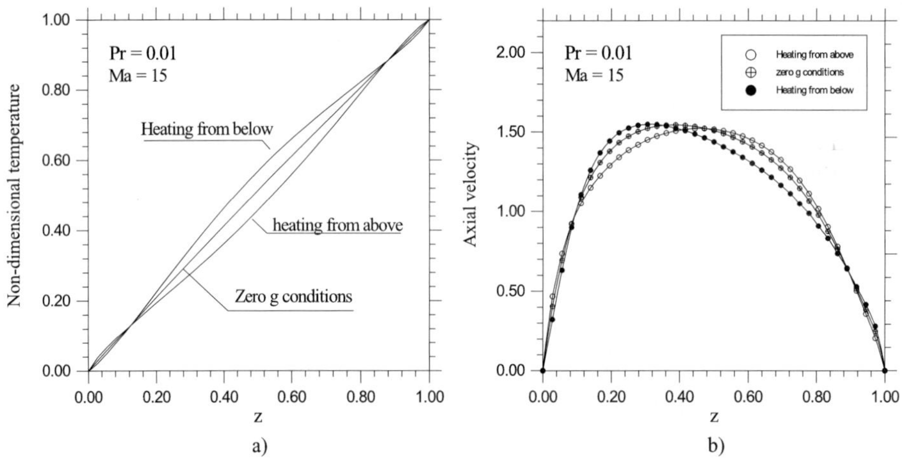

Fig. 4.37 Basic state in the case of a silicon liquid bridge (deformed shape, $A_H = 1$, $S = 1$): surface-temperature distribution (a) and surface velocity profile (b).

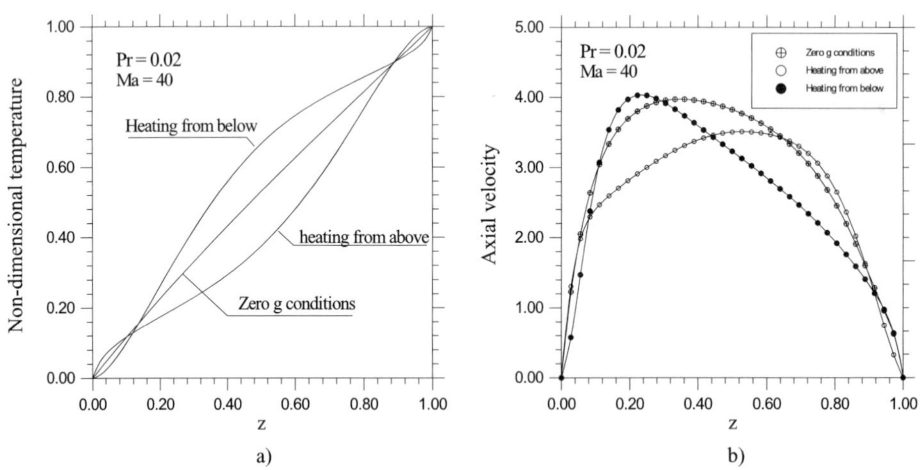

Fig. 4.38 Basic state in the case of a gallium liquid bridge (deformed shape, $A_H = 1$, $S = 1$): surface-temperature distribution (a) and surface velocity profile (b).

The flow becomes unstable against the 3D disturbances directly through a Hopf bifurcation characterized by a value of the azimuthal wave number $m = 1$ and a critical frequency $f = 0.31$ Hz (obtained by an extrapolation). Such a behavior has been reported for the first time by Lappa et al. (2003a).

The numerical simulations also highlight that the spatiotemporal behavior is quite different with respect to the case of the second bifurcation of flow in a liquid bridge of low Prandtl number fluid already depicted in detail in previous studies (Levenstan and Amberg (1995); Leypoldt et al. (2000); Imaishi et al. (1999, 2000)). In those cases, the oscillatory flow is superimposed on a steady 3D field established through the first flow transition.

The present case uncovers a new, heretofore unseen, transition directly from steady to oscillatory flow without the intermediate state of steady, asymmetric flow traditionally computed for the low-*Pr* case.

At an early stage, during a transient unsteady phase the spatiotemporal behavior is "rotating." The temperature disturbances on the interface (surface temperature spots) rotate around the perimeter of the liquid bridge (in Fig. 4.39, the period τ has been divided into 4 parts and the temperature field is shown in figures a–d, corresponding to $t = 0$, $\tau/4$, $\tau/2$, $3\tau/4$). The time-dependent behavior of the

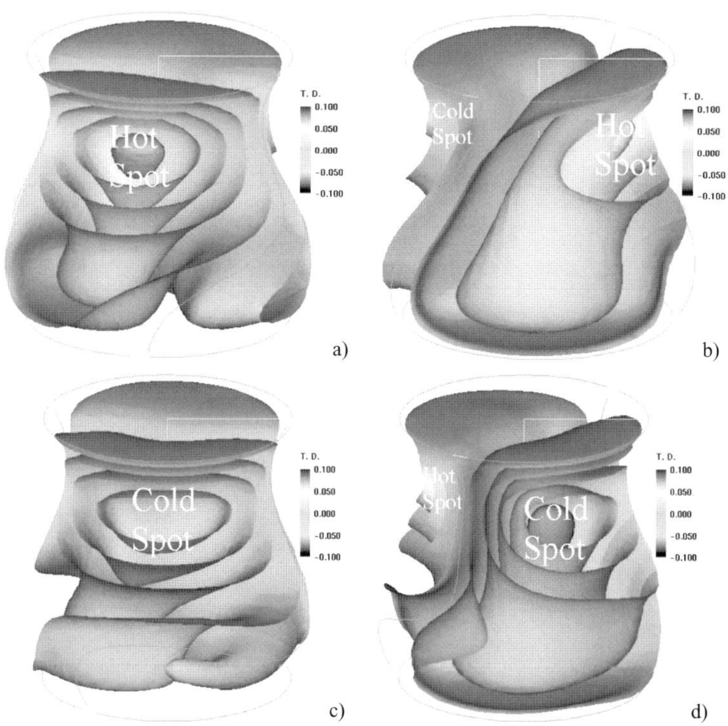

Fig. 4.39 Snapshots of temperature-disturbance isosurfaces over a period of oscillation ($Ma = 300$, gallium liquid bridge, 1g, heating from below, rotating regime).

temperature field is simply characterized by a full rotation of the entire flow pattern in the azimuthal direction (Fig. 4.40).

After a transient phase, the early rotating regime is taken over by a new spatiotemporal mode of azimuthal convection. During the rotating regime, the frequency is $f=0.449$ Hz but after a transient time the frequency becomes $f=0.386$ Hz. In this case the behavior is "pulsating." The pulsating temperature spots on the surface of the bridge are shown in Fig. 4.41 (the period τ has been divided into 4 parts and the fields are shown in figures a–d corresponding to $t=0$, $\tau/4$, $\tau/2$, $3\tau/4$). These temperature spots "pulsate," i.e. the cold spot grows in the axial direction during the shrinking of the hot spot and vice versa, but the azimuthal positions of these extrema do not change (see Fig. 4.42).

Though the similarity of this oscillatory flow with Marangoni-flow instabilities in the case of high Prandtl number liquids (standing waves and traveling waves, see Section 4.12 for further details), the mechanism of the instability observed in these results is still hydrodynamic. In fact, the azimuthal flow is not driven by the surface-temperature distribution. On the free surface, the azimuthal flow is directed from the cold spot towards the hot spot (azimuthal flow due to the instability and azimuthal

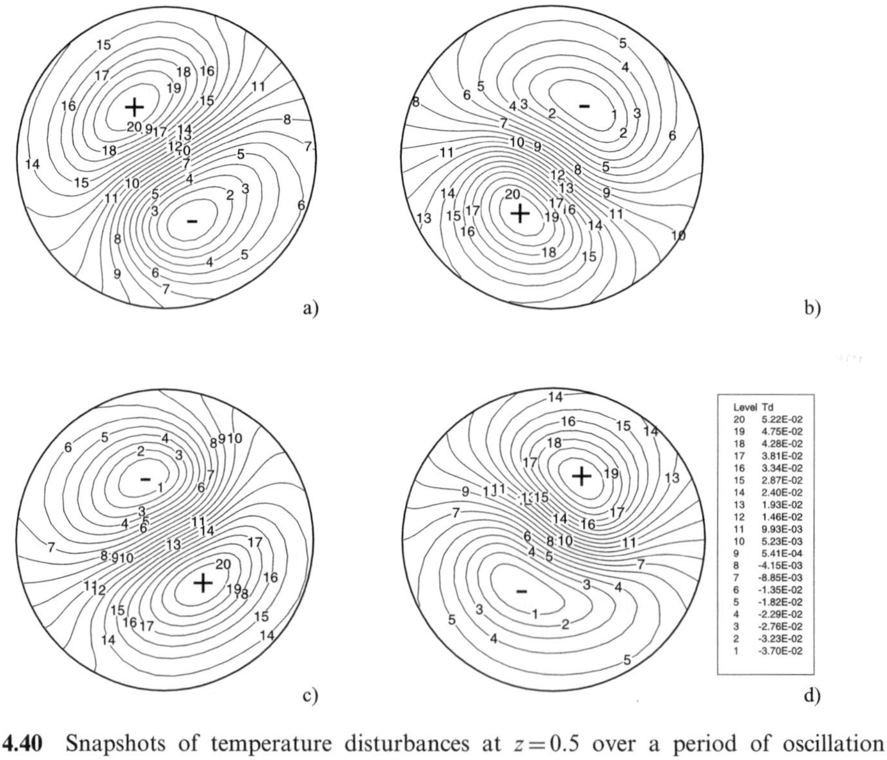

Fig. 4.40 Snapshots of temperature disturbances at $z=0.5$ over a period of oscillation (gallium liquid bridge, 1g, $Ma=300$, heating from below, rotating regime).

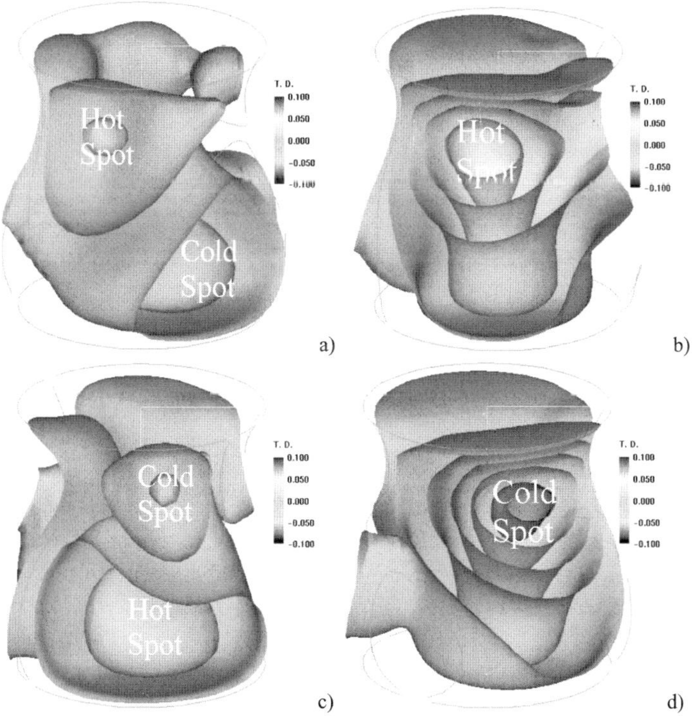

Fig. 4.41 Snapshots of temperature-disturbance isosurfaces over a period of oscillation (gallium liquid bridge, 1g, $Ma = 300$, heating from below, pulsating regime).

Marangoni effect "counteract" on the free surface), thus proving that in this case the instability is not hydrothermal.

Finally, Fig. 4.43 show the oscillatory behavior of the temperature field for the case of the cylindrical liquid bridge heated from below ($Ra > 0$) at $Ma = 160$, slightly above the second critical Marangoni number under microgravity conditions ($Ma_{c2} = 131.8$). In this case a "torsional" oscillatory motion with $m = 1$ is dominant. The wave number is the same as that for the deformed liquid bridge, but the oscillatory behavior is completely different, being characterized by a periodic twist (back and forth motion) in the azimuthal direction of the 3D $m = 1$ disturbance.

4.10. A generalized theory for the azimuthal wave number

It is known that discrete wave numbers of disturbances are selected out of the full spectrum of disturbances because the convection roll is closed in a special zone geometry. Since the instability in the case of liquid metals is hydrodynamic in nature (Levenstam and Amberg, 1995), i.e. it does not depend on the behavior of the temperature field (for this instability the temperature field simply acts as a driving force for the velocity field), the selection rule is given simply by the constraint that the

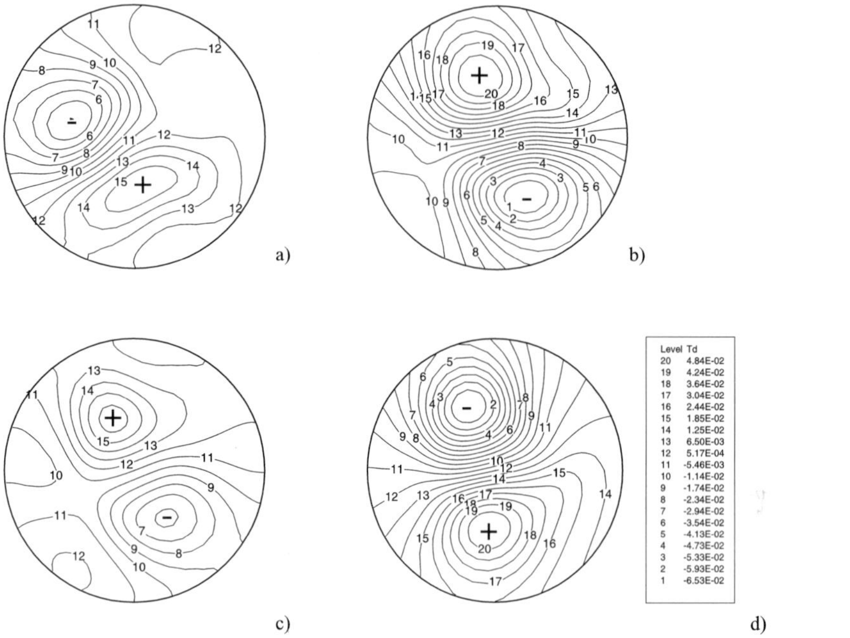

Fig. 4.42 Snapshots of temperature disturbances at $z=0.5$ over a period of oscillation (gallium liquid bridge, 1g, $Ma=300$, deformed shape, heating from below, pulsating regime).

azimuthal wavelength must be an aliquot of the toroidal vortex core circumference (as stated by Chun and Wuest, 1979) and by the fact that the convection roll is axially limited (as stated by Xu and Davis, 1984). Accordingly, the critical wave number has to be related to the axial extension L_V and to the diameter D_V of the centerline of the convection roll. In practice it scales with the parameter $A_V=L_V/D_V$.

The arguments below explain this dependence in detail:

Preisser et al. (1983) pointed out that in the case of half-zones with $A_H=O(1)$, the radial penetration depth of the convection in the bulk of the liquid is approximately given by the extension along z of the toroidal convection roll (L_V); since the size of the azimuthal convective cells induced by the flow instability in the cross-sections perpendicular to the liquid zone axis is equal to the radial extension (L_V) of the toroidal vortex (they are, in fact, superimposed on the basic convection roll and gain energy at its expense), the $2m$ convective cells cover a circumference $2mL_V$; the circumference of the toroidal vortex can be also computed as πD_V, therefore:

$$mA_V \cong \frac{\pi}{2} \tag{4.18}$$

Equation (4.18) provides an analytical relationship between the azimuthal wave number and the "aspect ratio" of the toroidal vortex supported by precise theoretical arguments. Along these lines, it is possible to link the behaviors described in the foregoing sections to the geometrical aspect ratio in the case of a straight surface or to

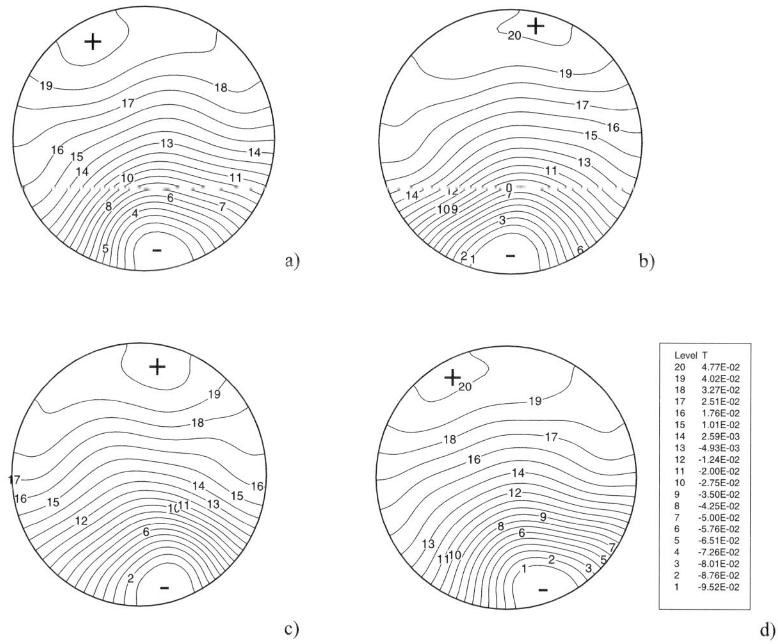

Fig. 4.43 Snapshots of temperature disturbances at $z=0.5$ over a period of oscillation (gallium liquid bridge, 1g, cylindrical interface, $Ma=160$, heating from below, secondary instability of Marangoni flow, "torsional" regime).

the effective aspect ratio of the toroidal convection rolls in the case of a noncylindrical volume. Such arguments also provide a theoretical justification to Eq. (4.17):

In the case of a cylindrical half-zone with an aspect ratio of O(1), since $D \propto D_{\mathrm{V}}$ and $L \propto L_{\mathrm{V}}$ Eq. (4.18) reads:

$$mA_{\mathrm{H}} \cong \text{const.}$$

Since in the case of the convex or the concave surface the diameter of the toroidal convection roll scales with the maximum diameter of the floating zone (Lappa et al., 2001a) and consequently according to the volume factor S:

$$L_{\mathrm{V}} \cong L \quad \text{and} \quad D_{\mathrm{V}} \propto D \cdot S \rightarrow A_{\mathrm{V}} \propto \frac{L}{DS} = \frac{A_{\mathrm{H}}}{S} \tag{4.19a}$$

Therefore, Eq. (4.18) reads:

$$mA_{\mathrm{H}} \propto S \tag{4.19b}$$

The foregoing relationships elucidate why under microgravity conditions the azimuthal wave number of the half-zone m increases if the aspect ratio is reduced (and vice versa) and scales with the volume.

Finally, it should be highlighted that the theory relying on Eq. (4.18) also provides a theoretical justification for some effects dealing with the dynamical change of the axial extension of the liquid zone that occurs during the half-zone-based solidification processes (i.e. it can also take into account effects related to variation of L_V in addition to the ones related to D_V, see the next section).

4.11. 3D analysis of crystal/melt interface shape in the half-zone

Landmark experiments dealing with the half-zone configuration in the case of liquid semiconductors and the related solidification process were carried out by Nakamura et al. (1995, 1998) and Hibiya et al. (1998a, b). These researchers studied the Marangoni flow azimuthal organization in a half-zone of pure silicon (diameter $D = 1\,\text{cm}$) under highly supercritical conditions using very sophisticated and nonconventional diagnostic devices (X-ray radiography with zirconium-core tracers). During a sounding rocket mission and other parabolic flights they observed that when the column height was short (column-formation period), the tracer particles preferred to stay at positions 90° away from each other, while after a silicon column with full height was formed, tracer particles preferred to stay at positions of 180° apart from each other. This behavior was interpreted as a change of the azimuthal wave number during the melting process.

This can be regarded as a typical example of a problem where the dynamical change of the aspect ratio influences the flow field directly.

The present section deals with the introduction of appropriate models and suitable techniques for the simulation of these aspects. A prototype investigation is discussed in the framework of the enthalpy method introduced in Section 4.2.

To better analyze the above phenomena an ideal solidification process of pure gallium (Table 4.3) in half-zone liquid bridges in zero-g conditions is considered as the reference and relevant case. Gallium is chosen for the prototype simulation owing to the low melting point that makes it suitable for future experimental investigations at relatively small temperatures (close to the ambient temperature).

The results presented in the next sections represent a first contribution to the modeling and understanding of complex phenomena caused by fluid flow and simultaneous solidification during half-zone crystal growth. Particular care is devoted to some critical issues like the definition of the phase-change zone thickness and the degree of sensitivity of the numerical solution to such a factor.

4.11.1 *Half-zone basic assumptions*

The geometry of the problem is shown in Fig. 4.44a. At the instant $t = 0$, a cylindrical liquid bridge of length L and diameter D is held between two coaxial disks at different temperatures under zero-g conditions (cylindrical interface). The upper disk

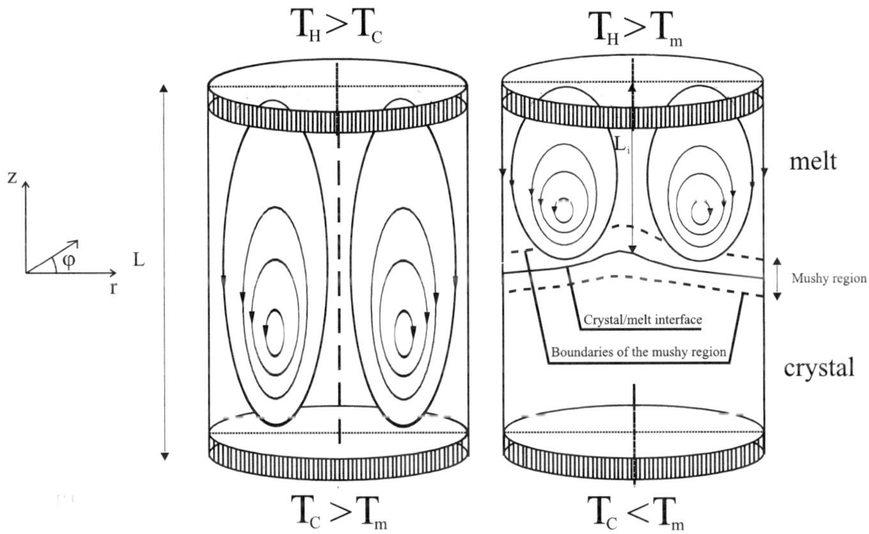

Fig. 4.44 Sketch of the half-zone-based solidification process.

in Fig. 4.44a is kept at the temperature T_H higher than the temperature T_c of the lower cold disk.

The solidification takes place as the temperature T_c is decreased to a value that is less than the melting temperature T_m. The instantaneous distance from the disk at temperature T_H of the point given by the intersection of the crystal/melt interface and of the liquid-bridge symmetry axis is denoted as L_i (Fig. 4.44b). The ratio $A_i = L_i/D$ is denoted as the melt aspect ratio. The temperature difference imposed at $t = 0$ is denoted by ΔT.

4.11.2 *Initial and boundary conditions for the solidification process*

The initial conditions are:

$$t = 0: \quad \underline{V}(z, r, \varphi) = 0, \qquad T(z, r, \varphi) = T_{\text{liq}} + z\Delta T \tag{4.20}$$

i.e. the liquid is motionless and the temperature is $T_c = T_{\text{liq}}$ on the cold disk and $T_H = T_{\text{liq}} + \Delta T$ on the hot disk.

For $t > 0$, no-slip conditions are considered for both the supporting disks. When, for $t > t^*$ the solidification process starts (t^* is the instant at which solidification takes place), the thermal boundary conditions on the disks become (see also Lappa and Savino, 2002):

$$\text{on the cold disk} \qquad \begin{array}{l} 0 \le r \le R/L; \quad 0 \le \varphi \le 2\pi \\ T_c = T_{\text{liq}} - c(t - t^*) \end{array} \tag{4.21a}$$

where c is the ramping rate used for the solidification process.

$$\text{on the hot disk} \quad 0 \le r \le R/L; \quad 0 \le \varphi \le 2\pi \qquad T_H = T_{liq} + \Delta T \tag{4.21b}$$

the kinematic condition of the stream surface (zero normal velocity), the Marangoni conditions (shear stress balance) and the adiabatic condition must be applied on the cylindrical interface:

$$\text{on the cylindrical free surface} \quad 0 \le z \le 1; \quad 0 \le \varphi \le 2\pi \qquad v(z, r = R/L, \varphi, t) = 0 \tag{4.22a}$$

$$\frac{\partial u}{\partial r}(z, r = R/L, \varphi, t) = -Ma_r \frac{\partial T}{\partial z}(z, r = R/L, \varphi, t)\, \phi(z, r = R/L, \varphi, t) \tag{4.22b}$$

$$r\frac{\partial V_\varphi}{\partial r}(z, r = R/L, \varphi, t) - V_\varphi\,(z, r = R/L, \varphi, t) = -Ma_r\, \phi(z, r = R/L, \varphi, t)\frac{\partial T}{\partial \varphi}(z, r = R/L, \varphi, t) \tag{4.22c}$$

$$\frac{\partial T}{\partial r}(z, r = R/L, \varphi, t) = 0 \tag{4.22d}$$

where the reference Marangoni number Ma_r is defined as $Ma_r = \sigma_T L/\mu\alpha$.

The Marangoni number corresponding to the applied temperature difference can be computed as $Ma = \Delta T Ma_r = \Delta T \sigma_T L/\mu\alpha$; in addition to the Marangoni number, the instantaneous Marangoni number is defined as $Ma_i = \Delta T \sigma_T L_i/\mu\alpha$.

The liquid fraction ϕ directly enters the interface boundary conditions where it is somehow used for weak regularization of the Marangoni stress condition close to the solid phase (see also the discussion in Section 2.8).

4.11.3 *Definition of the phase-change zone thickness*

According to Brent et al. (1988), in the case of a pure substance, the flow inside the computational cells that are melting or freezing does not follow exactly the governing laws for flow in a mushy zone. For a pure semiconductor, the phase front is a well-defined line (in reality there will be a boundary layer on the solidification surface) and has no band or width associated with it as in the case of a mushy phase change.

Even if, in the case of pure liquid metals, the mushy zone from a physical point of view is reduced to a very thin boundary layer, in a discretized calculation domain, anyway, a fixed-grid-solution technique cannot capture this layer in detail.

The numerical technique, in fact, cannot handle thin boundary layers since, as pointed out by Brent et al. (1988), the phase front will have a finite width associated with it, this width being at least that of a single computational cell. When the fixed-grid-enthalpy approach is employed, the only way to capture the physics of the boundary layer on the solidification surface would be to reduce the width of the computational cells up to the real thickness of the layer where phase change occurs (few μm). This is not possible for many situations. For the case under investigation, for instance, (liquid column having initial length $L = 1\,\text{cm}$) this would require a prohibitive number of computational points along the axis of the liquid zone (of the order of 10^3–10^4). Due to the fact that enormous speed and memory requirements are necessary for the direct numerical solution of the three-dimensional Navier–Stokes equations, a maximum of 100 points may be used in the axial direction. The thickness of the mushy zone is inevitably related to this number since its width corresponds to the Δz arising from the discretization of the liquid zone in axial direction.

However, as pointed out by Voller and Prakash (1987), Bennon and Incropera (1987a, b) and Brent et al. (1988) often this nonzero thickness is an advantage rather than a problem. It is desirable, in fact, that the method chosen allows a smooth, gradual transition rather than a step change (there is a certain theoretical kinship with the way in which the surface-tension force is taken into account by the VOF method in Chapter 3).

In the light of these arguments, the amplitude of the temperature range over which phase change occurs, has to be chosen in order to avoid step changes in the temperature fields (step changes in the momentum and energy equations source terms tend to retard convergence of their numerical solution, and sometimes lead to oscillations that may result in divergence), and at the same time in order to be consistent with the physics of the phenomenon under investigation: for instance, for the present prototype example (phase-change zone reduced to a thin boundary layer and 40 computational cells distributed in axial direction) a temperature range of 0.5 K must be used; this choice, in fact, is a good compromise between the need to avoid step changes and the need to represent the phase-change boundary layer with a number of computational cells in the axial direction as low as possible.

Usually, such a choice requires preliminary computations. These computations must be repeated to find a temperature range ΔT_{m} that guarantees numerical stability and makes the shape of the solidification front independent of it.

This is a very difficult task in the case of high Prandtl number liquids. In these cases, in fact, due to the low thermal diffusivity of the material (in the case of transparent substances usually $\alpha = \mathrm{O}(10^{-4})\,\text{cm}^2/\text{s}$), in many situations the thickness of the mushy zone is very large. Owing to this large thickness, the amount of convection protruding in the mushy zone is great and the flow inside this zone largely influences its shape. Fortunately, this, however, is not the case of a liquid metal ($Pr \ll 1$, $\alpha = \mathrm{O}(10^{-2})\,\text{cm}^2/\text{s}$). As previously explained, in the case of a low Prandtl number liquid, the mushy zone is reduced to a boundary layer (whose thickness corresponds to one/two layers of computational cells) and due to the "switch off" condition that drives velocities to zero in such a thin zone, the effect of the convection on the shape is very weak.

A few tests with a different size mushy region, in principle, must be carried out for $Pr \ll 1$ to assess that the shape of this zone is not influenced by the temperature range chosen for the phase change. For instance, since one may suspect that differences with respect to the computations carried out for an amplitude of the temperature range equal to 0.5 K may arise when further reducing the size of the mushy region down to zero, the amplitude of the temperature range has been reduced from 0.5 K to 0.25 K for the prototype application. The numerical simulations have shown that the influence of the amplitude of the temperature range on the shape of the mushy zone is very weak. The size is reduced but the crystal/melt interface defined as the isoline at $T = T_m$ does not change its shape.

Regarding the effect of the thickness and curvature of the mushy zone on the flow, it should be pointed out, moreover, that the Marangoni flow is a large-scale flow whose properties do not depend upon the local curvature of the boundary of the mushy zone and/or upon its thickness.

If at any instant the axial boundaries of the mushy region computed according to the temperature distribution are replaced with average axial values obtained averaging the z-coordinate of the boundaries over their entire surface (thus making the mushy zone a horizontal band), the structure of the field in the meridian plane and the azimuthal wave number of the instability and their modification as time passes do not change.

4.11.4 *Azimuthal structure evolution*

Liquid bridges of gallium (see Table 4.3 for the physical properties), having an initial length $L = 1$ cm and two different initial aspect ratios ($A_H = 1.0$ and $A_H = 0.5$), are considered for the prototype application. For both the aspect ratios, a temperature difference of 4 K is initially imposed (corresponding to a Marangoni number $Ma = \Delta T \sigma_T L/\mu\alpha \cong 110$).

For this value of the Marangoni number, the axisymmetric flow field is unstable and for both the aspect ratios a steady, three-dimensional flow regime is established after a certain time denoted by t^*.

For an initial aspect ratio $A_H = 1.0$, the steady and asymmetric Marangoni flow after the bifurcation is characterized by the azimuthal wave number $m = 1$. Figures 4.45b and c show that in the generic cross-section orthogonal to the liquid-bridge axis there are two azimuthal convective cells and two thermal spots on the liquid-bridge surface (one cold and the other hot). The temperature contours in Fig. 4.45d show a lower-temperature inner region whose shape is circular but whose position is

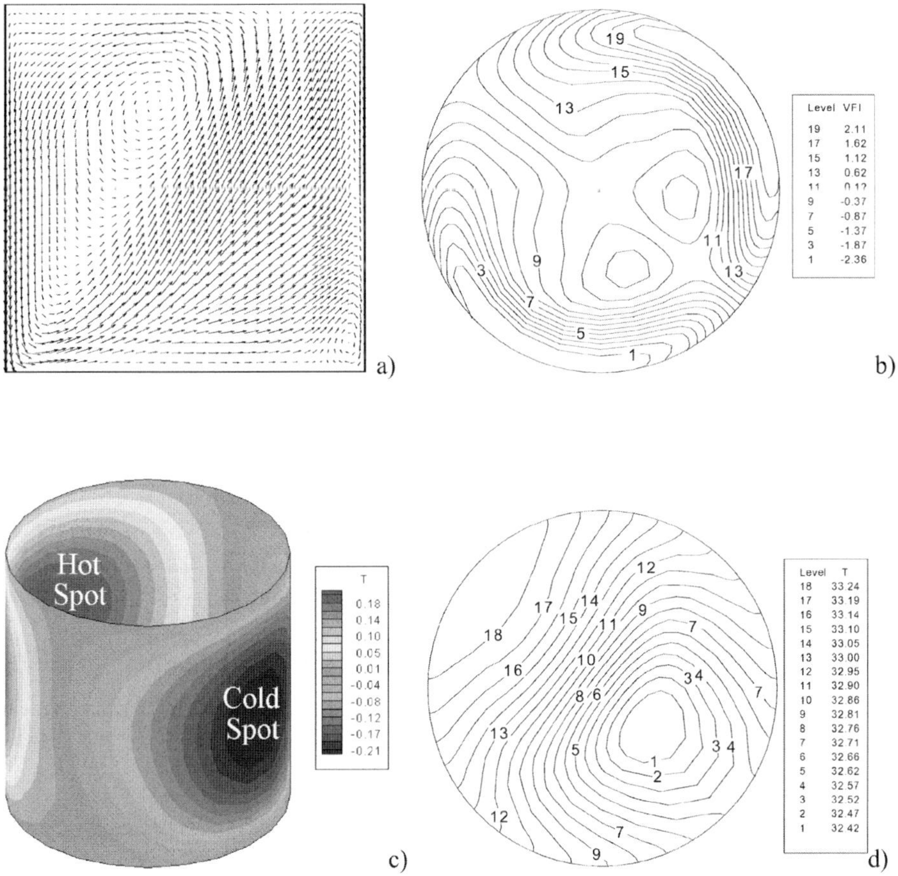

Fig. 4.45 ($A_H = 1.0$, $Ma = 110$, $t = t^*$): (a) Flow field in the meridian plane $\varphi = -\pi/6$, (b) azimuthal-velocity distribution in the section $z = 0.75$, (c) temperature disturbances $r = 1/2A$, (d) temperature distribution $z = 0.75$.

eccentric to the geometrical axis of symmetry of the liquid bridge. In this case the 3D flow appears as an inclined toroidal vortex. In each meridian plane of the liquid bridge there are two asymmetrical vortex cells (one of the two vortex cells in the section prevails over the other and is extended along the whole axial plane of the bridge, for instance see Fig. 4.45a). These behaviors have already been described in Sections 4.7 and 4.8.

For $t > t^*$, the temperature on the cold disk is reduced with a ramp of 0.1 K/s and the crystallization front advances, as shown in Figs. 4.46–4.50.

Owing to the azimuthal distortion of the thermo-fluid-dynamic field, the shape of the crystal is not axisymmetric. This is shown clearly in Fig. 4.46a and in Fig. 4.51a for $A_i = 0.885$ ($Ma_i = 100$). In Fig. 4.46a, the distortion of the boundaries of the

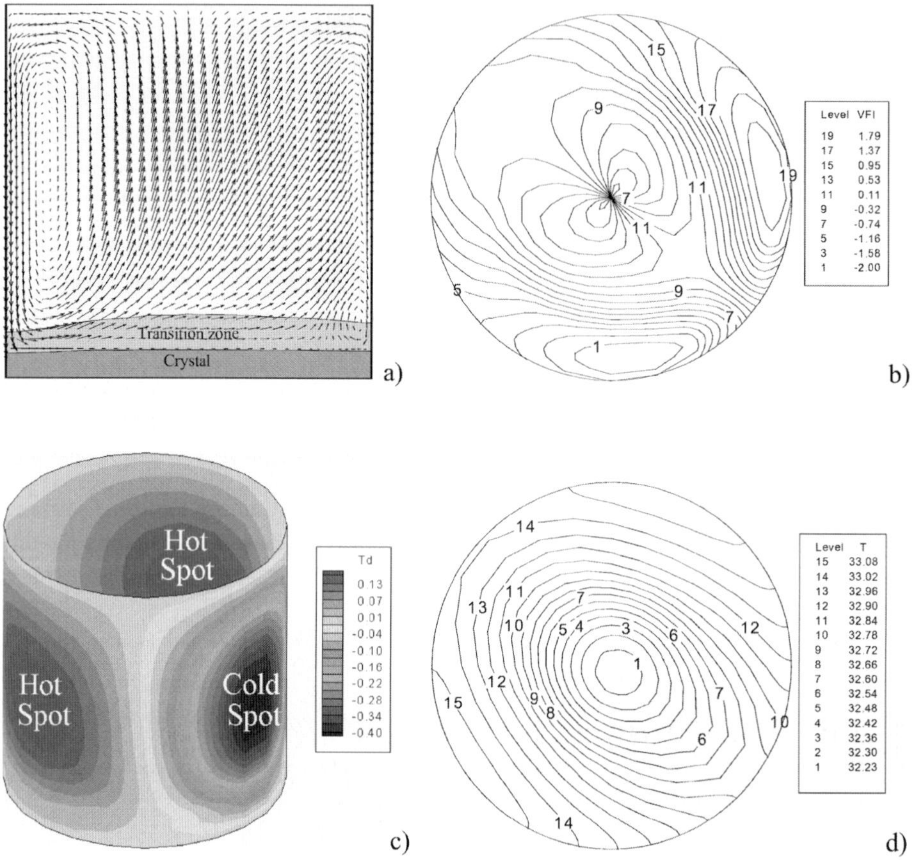

Fig. 4.46 ($A_i=0.885$, $Ma_i=100$, $t-t^*=2.6$): (a) flow field in the meridian plane $\varphi=-\pi/6$, (b) azimuthal-velocity distribution in the section $z=0.75$, (c) Temperature disturbances $r=1/2A$, (d) temperature distribution $z=0.75$.

transition zone is shown in relationship to the asymmetric flow-field structure in the meridian plane $\varphi=-\pi/6$.

The distortion of the boundaries of the mushy zone is more pronounced on the left side of the liquid zone where the strongest vortex cell is present. Due to this distortion, the thickness of the transition zone is not uniform in the meridian plane; the thickness is larger in the right side. In Fig. 4.51a the temperature isosurface corresponding to $T=T_m$ (average crystal/melt interface) is shown. The shape of the surface is convex. Due to the steady Marangoni instability, it is azimuthally distorted and has an axial minimum at $\varphi=-\pi/6$. The corresponding surface temperature disturbance distribution (Fig. 4.46c) shows that the azimuthal wave number is still $m=1$ since only two temperature spots are present. However, the hot spot is more extended in the azimuthal direction with respect to the cold one. This situation is different compared to that observed at $t = t^*$, where the two spots have exactly the same azimuthal extension. This behavior suggests that the azimuthal organization of the flow has been influenced by the presence of the crystal.

For $A_\mathrm{i} = 0.72$ ($Ma_\mathrm{i} = 82$, Fig. 4.47c), the spots on the free surface become four (two hot and two cold). This indicates that the azimuthal wave number is changed from $m = 1$ to $m = 2$, due to the solidification process. Four convective cells appear in the section as shown in Fig. 4.47b; the inner region in this case is not circular but approximately elliptic with center corresponding to the geometrical axis of symmetry of the liquid bridge (see Fig. 4.47d).

As previously pointed out (see Section 4.7), for even azimuthal wave numbers, the flow-field structure is, on the whole, three-dimensional and depends on the azimuthal coordinate, but in each meridian plane the velocity and the temperature are

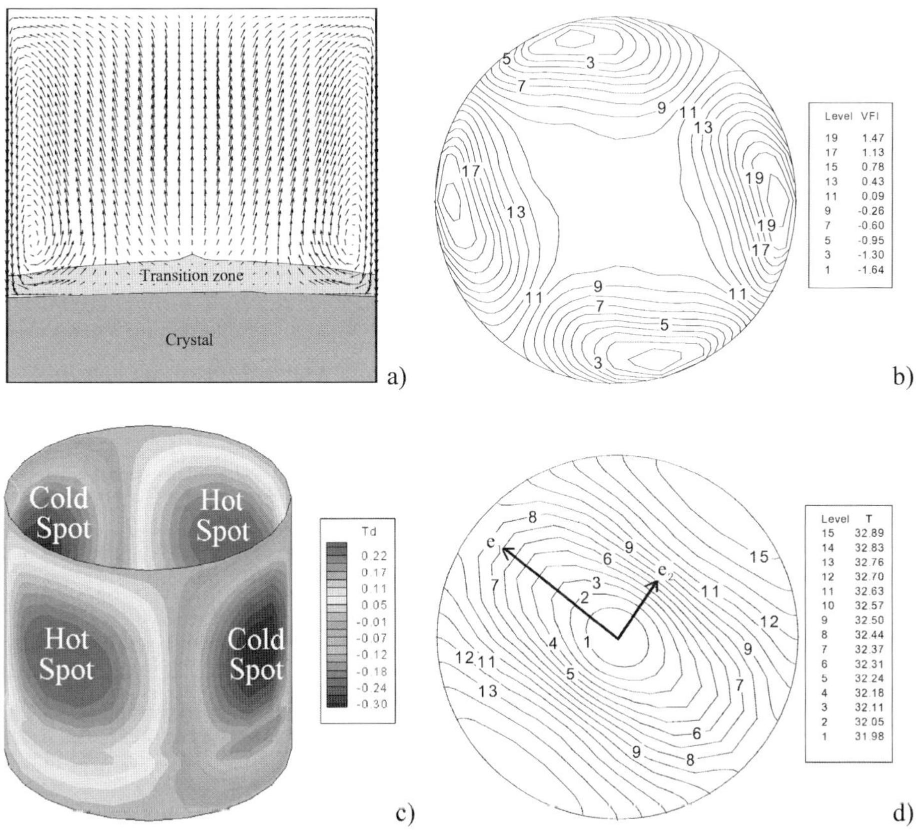

Fig. 4.47 ($A_\mathrm{i} = 0.72$, $Ma_\mathrm{i} = 82$, $t - t^* = 6.2$): (a) flow field in the meridian plane $\varphi = -\pi/6$, (b) azimuthal-velocity distribution in the section $z = 0.75$, (c) temperature disturbances $r = 1/2A$, (d) temperature distribution $z = 0.75$.

symmetric (for instance, see Fig. 4.47a). For this reason the azimuthal distortion of the mushy region boundaries due to the Marangoni flow instability cannot be highlighted in the generic meridian plane. Figure 4.51b shows that the shape of the crystal/melt interface is azimuthally distorted and has two axial minima ($\varphi = 5\pi/6$ and $\varphi = -\pi/6$) and two maxima ($\varphi = \pi/3$ and $\varphi = 4\pi/3$) in the azimuthal direction.

For $A_i = 0.5$ ($Ma_i = 57$, Fig. 4.48a–d), the azimuthal wave number is $m = 2$, as for $A_i = 0.72$; Fig. 4.48c shows that in this case the amplitude of the temperature disturbances on the free surface is very small.

When the melt aspect ratio is decreased further ($A_i < 0.5$) due to the solidification process, the thickness of the mushy region decreases (see Figs. 4.49a and 4.50a) with a linear dependence on A_i. The Marangoni flow becomes more and more confined near the free surface (the radial extension of the vortex cells in the meridian plane becomes approximately equal to the axial extension of the liquid zone held between

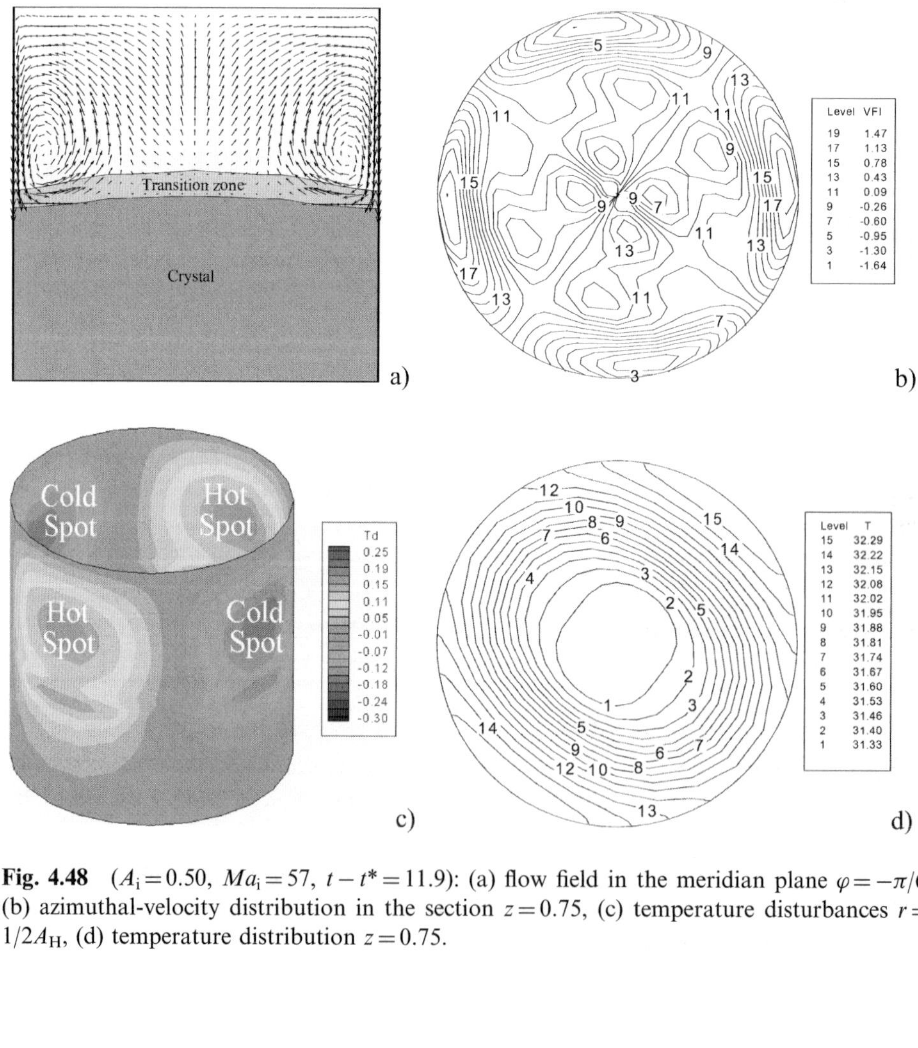

Fig. 4.48 ($A_i = 0.50$, $Ma_i = 57$, $t - t^* = 11.9$): (a) flow field in the meridian plane $\varphi = -\pi/6$, (b) azimuthal-velocity distribution in the section $z = 0.75$, (c) temperature disturbances $r = 1/2A_H$, (d) temperature distribution $z = 0.75$.

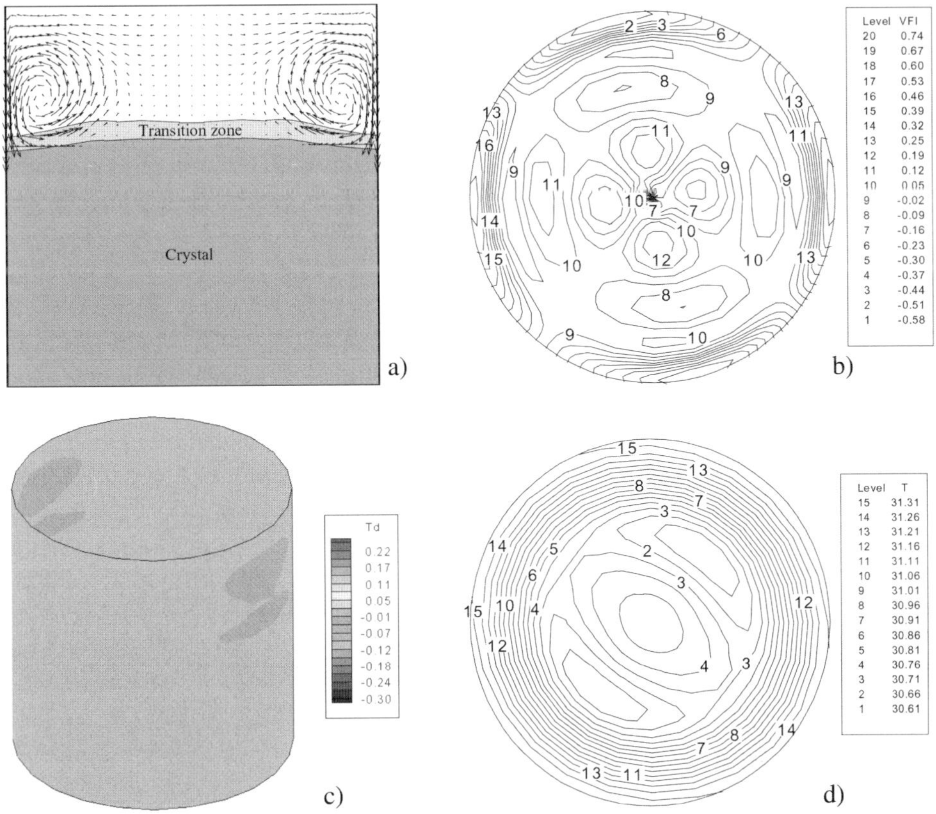

Fig. 4.49 ($A_i = 0.33$, $Ma_i = 37$, $t - t^* = 17.5$): (a) flow field in the meridian plane $\varphi = -\pi/6$, (b) azimuthal-velocity distribution in the section $z = 0.75$, (c) temperature disturbances $r = 1/2A_H$, (d) temperature distribution $z = 0.75$.

the crystal surface and the hot supporting disk, see Figs. 4.49a and 4.50a). For this reason the isolines of the temperature field in the inner part (near the axis) of the bridge are no longer convex and become concave. Correspondingly, the isosurface $T = T_m$ for $A_i = 0.33$ (Fig. 4.51d) exhibits a concave shape in the inner part (near the axis) and a convex shape near the free surface. The azimuthal distortion of the shape is more pronounced in the inner part, whereas an axisymmetric behavior can be observed far from the axis.

This tendency is confirmed by the fact that the temperature disturbances are negligible on the free surface (Fig. 4.49c) whereas the temperature distribution is not symmetric in the inner part of the generic cross-section orthogonal to the axis (Fig. 4.49d).

For $A_i = 0.28$ ($Ma_i = 32$) a full stabilization of the flow field is reached (subcritical bifurcation). The azimuthal disturbances vanish (Fig. 4.50b and c) and the shape of the crystal is axisymmetric (Fig. 4.51e).

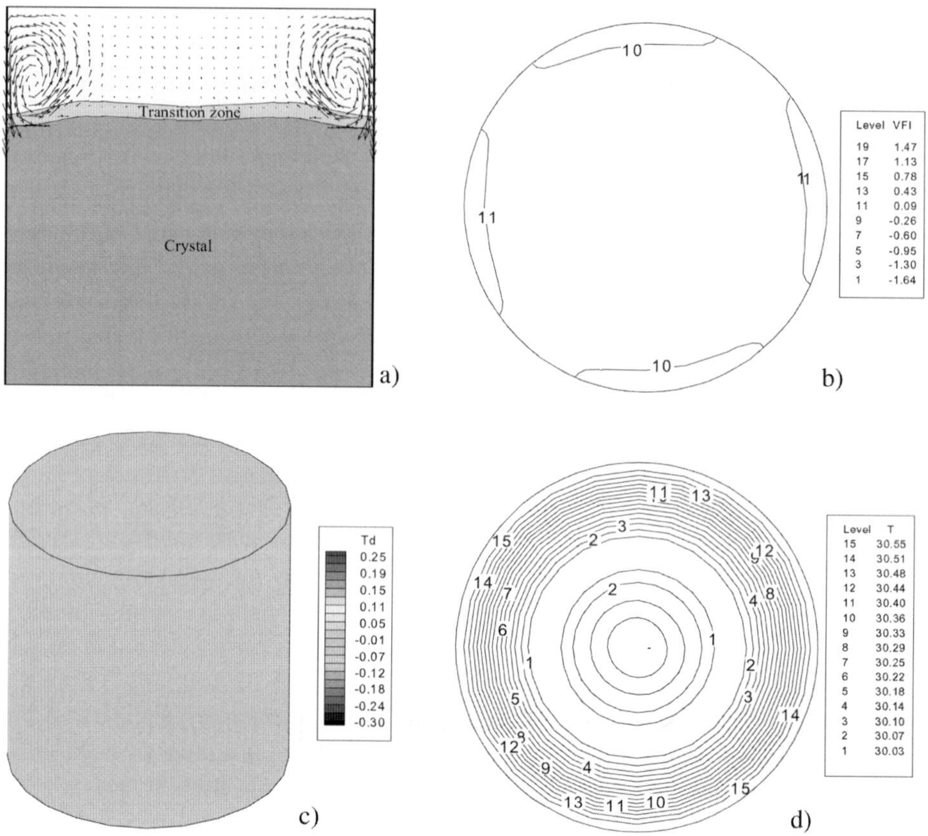

Fig. 4.50 ($A_i = 0.28$, $Ma_i = 32$, $t - t^* = 20.6$): (a) Flow field in the meridian plane $\varphi = -\pi/6$, (b) azimuthal-velocity distribution in the section $z = 0.75$, (c) temperature disturbances $r = 1/2A_H$, (d) temperature distribution $z = 0.75$.

For an initial aspect ratio $A_H = 0.5$, at the initial time $t = t^*$ the supercritical Marangoni flow is still steady and asymmetric (as for $A_H = 1.0$) but it is characterized by an azimuthal wave number $m = 2$. There are four convective cells in the cross-section orthogonal to the axis and four temperature spots on the free surface (two hot and two cold, see Fig. 4.52). In the generic meridian plane, the flow field is symmetric but with different radial and axial organization with respect to the other planes corresponding to other values of the azimuthal coordinate φ (Fig. 4.52a and b).

When solidification occurs (ramp of 0.1 K/s), for $0.24 < A_i \leq 0.5$, the azimuthal wave number $m = 2$ is stable. For $A_i = 0.24$ ($Ma_i = 55$, see Fig. 4.53), however, the hot spots on the free surface show a doubling mechanism similar to that described in the case $m = 1$ ($A_H = 1.0$). Each hot spot seems to double, increasing its azimuthal extension (see Fig. 4.53c). The crystal/melt interface is concave near the axis and convex near the free surface and the azimuthal distortion is more pronounced in the inner part (Fig. 4.55a) as described for the cases $A_H = 1.0$ and $A_i < 0.5$.

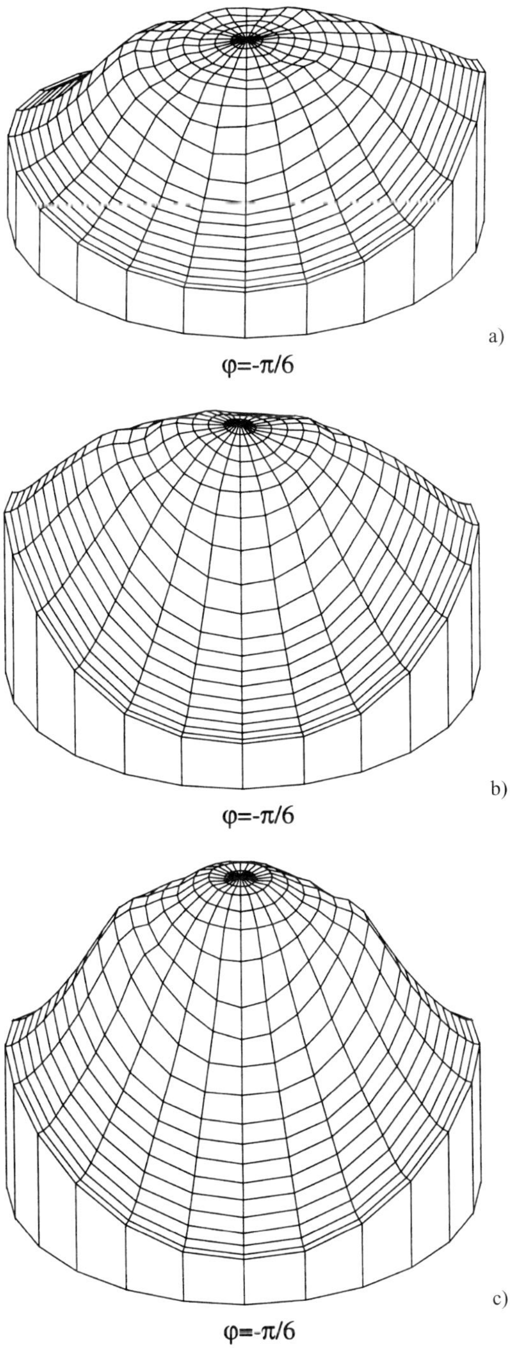

Fig. 4.51 Crystal/melt interface: (a) $A_i = 0.885$, $Ma_i = 100$, $t - t^* = 2.6$, (b) $A_i = 0.72$, $Ma_i = 82$, $t - t^* = 6.2$, (c) $A_i = 0.50$, $Ma_i = 57$, $t - t^* = 11.9$, (d) $A_i = 0.33$, $Ma_i = 37$, $t - t^* = 17.5$, (e) $A_i = 0.28$, $Ma_i = 32$, $t - t^* = 20.6$.

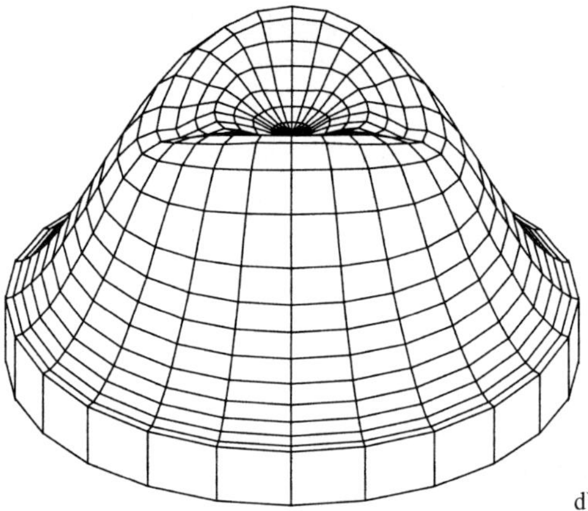

d)

φ=-π/6

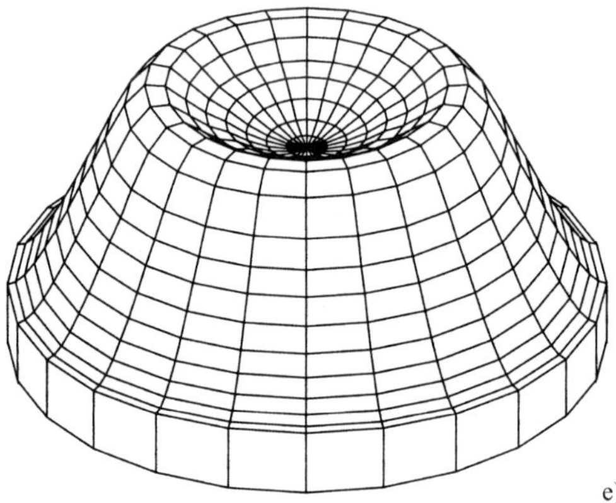

e)

φ=-π/6

Fig. 4.51 Continued.

For $A_i = 0.18$ ($Ma_i = 41$), a mode $m = 4$ appears. There are, respectively, eight vortex cells and surface spots, as shown in Fig. 4.54b and c. Moreover a more complicated noneccentric pattern appears in the temperature field of the cross-section. For $m = 4$ the inner region has the form of a quadrangle (Fig. 4.54d). Correspondingly, the crystal/melt interface shape exhibits four axial maxima and four minima ($\varphi = 3\pi/7$, $\varphi = \pi - \pi/14$, $\varphi = 10\pi/7$, $\varphi = -\pi/14$) in the azimuthal direction (Fig. 4.55b).

For $A_i = 0.135$ ($Ma_i = 31$) full stabilization of the thermo-fluid-dynamic field occurs (see Fig. 4.55c).

4.11.5 *Some insights into the physics*

It is known, in principle, that the crystallization process is favored if conditions making the melt/crystal interface flat during the resolidification are attained.

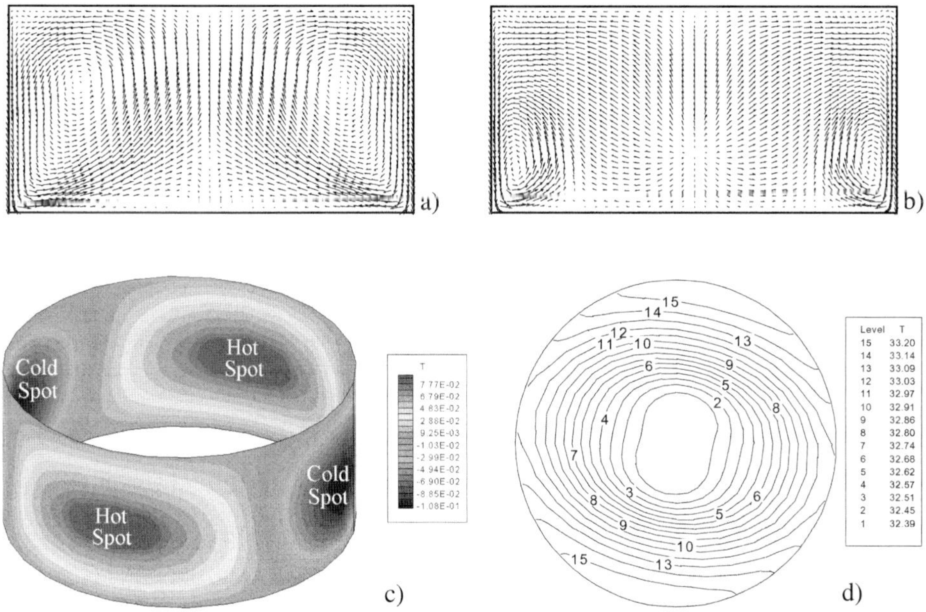

Fig. 4.52 Flow field in the meridian plane: ($A_H = 0.50$, $Ma = 110$, $t = t^*$): (a) $\varphi = 3\pi/7$, (b) $\varphi = 10\pi/7$, (c) Temperature disturbances $r = 1/2A_H$, (d) temperature distribution $z = 0.75$.

Therefore, from a theoretical point of view, the quality of crystals is highly dependent upon the uniformity of the thermal field at the solidifying interface and hence on the convection fields there. Accordingly, the azimuthal distortion of the shape of the crystal/melt interface can be explained on the basis of the toroidal structure of the supercritical flow field and of the temperature distribution.

The temperature disturbances have m maxima (minima) in the azimuthal direction (as shown by Figs. 4.45c–4.49c). As pointed out earlier (Section 4.7) the nodes of the geometrical patterns given by the isolines of the temperature field in the cross-section orthogonal to the axis simply correspond to these extrema in the temperature field. For $m = 1$ the geometrical pattern is an eccentric circle (Fig. 4.45d) since there is a minimum and a maximum in the temperature field (the circle corresponds to the minimum). For $m = 2$ the geometrical pattern is an ellipse since there are two minima and two maxima (in Fig. 4.47d the main axes of the ellipse connect, respectively, the axis of symmetry of the bridge with the positions corresponding to a minimum and to a maximum of the temperature-disturbance distribution). For $m = 4$ (see, e.g., Fig. 4.54d) the geometrical pattern is a quadrangle since the temperature disturbances have four maxima.

The results dealing with the crystal/melt interface shape show that this surface describes, in the azimuthal direction, a sine curve having m maxima and m minima in the z-direction. This distortion can be simply related to the sinusoidal spatial

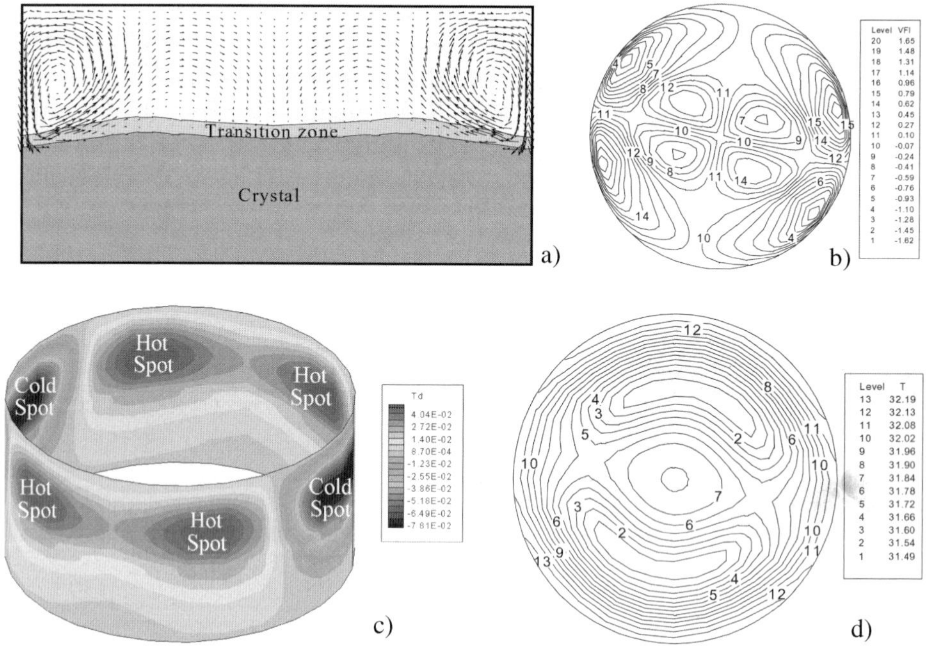

Fig. 4.53 ($A_i = 0.24$, $Ma_i = 55$, $t - t^* = 23$): (a) Flow field in the meridian plane $\varphi = 3\pi/7$, (b) azimuthal-velocity distribution $z = 0.75$, (c) Temperature disturbances $r = 1/2A_H$, (d) temperature distribution $z = 0.75$.

deformation of the temperature field, due to the distorted spatial displacement of the vortex core in the supercritical state (Section 4.7 and Fig. 4.13).

Regarding the occurrence of different values of the azimuthal wave numbers during the solidification process ($m = 1$ and $m = 2$ for $A_H = 1.0$, and $m = 2$ and $m = 4$ for $A_H = 0.5$), the theoretical justification for this behavior can be found in the relation between the three-dimensional disturbances and the geometry of the liquid zone elucidated in Section 4.10.

According to this theory, for the problem under investigation, the critical wave number must be simply related to the axial length (distance between the hot disk and the crystal/melt interface) and to the diameter of the liquid zone, i.e. it should scale with the parameter $A_i = L_i/D$ (D being constant). In fact, according to the results provided by direct numerical simulation in Figs. 4.45–4.55 the following relationship holds:

$$mA_i \cong 1 \qquad (4.23)$$

Such a finding provides additional validation for the theory in Section 4.10.

The stabilization of the flow field, observed for both cases considered herein ($A_H = 1.0$ and $A_H = 0.5$) when the melt aspect ratio becomes less than a certain value,

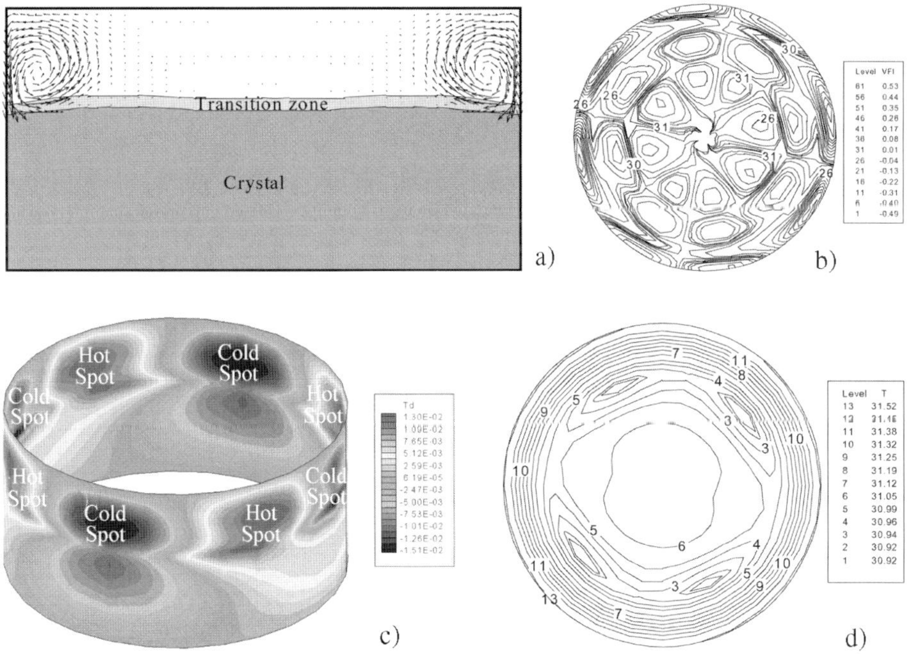

Fig. 4.54 ($A_i = 0.18$, $Ma_i = 41$, $t - t^* = 27.7$): (a) Flow field in the meridian plane $\varphi = 3\pi/7$, (b) azimuthal-velocity distribution $z = 0.75$, (c) Temperature disturbances $r = 1/2A_H$, (d) temperature distribution $z = 0.75$.

can be explained by the simultaneous reduction of the instantaneous Marangoni number and of the melt aspect ratio. There are, in fact, two different effects leading to the stabilization of the flow field. For a fixed aspect ratio, decreasing the Marangoni number, subcritical conditions prevail. On the other hand, as discussed in Wanschura et al. (1995), for very small values of A_H, the lower the aspect ratio, the higher should be the critical Marangoni number.

The stabilization of the flow field can be explained by also focusing directly on the nature of the instability. As highlighted by Wanschura et al. (1995), for low Prandtl number liquids, the bifurcation of the flow field to a steady three-dimensional configuration is directly related to the inertial instability of the shear layer below the free surface; the instability takes energy from the radial gradient of the axial velocity, i.e. the axial share. Marangoni forces driving the flow field induce a strong axial shear flow. For this reason, the bifurcation is favored by increasing the Marangoni number (driving action) and opposed by the presence of walls (where no-slip prevails). For the specific problem under investigation, when the aspect ratio decreases, the ratio of the free surface area $S_l = \pi D L_i$ (where the driving force is present) divided by the end-walls area $S_W = \pi D^2/2$ (where no-slip prevails), $S_l/S_W = 2A_i$ also decreases, increasing viscous effects; moreover, the decrease of

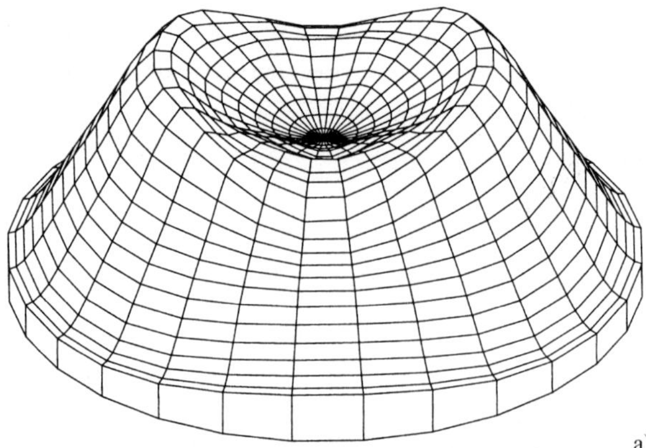

a)

φ=-π/14

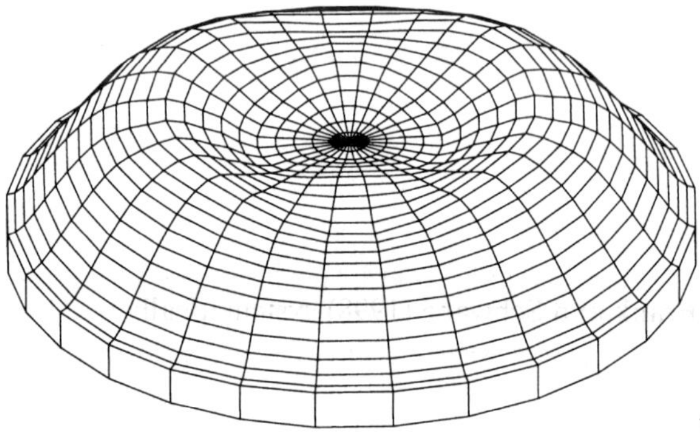

b)

φ=-π/14

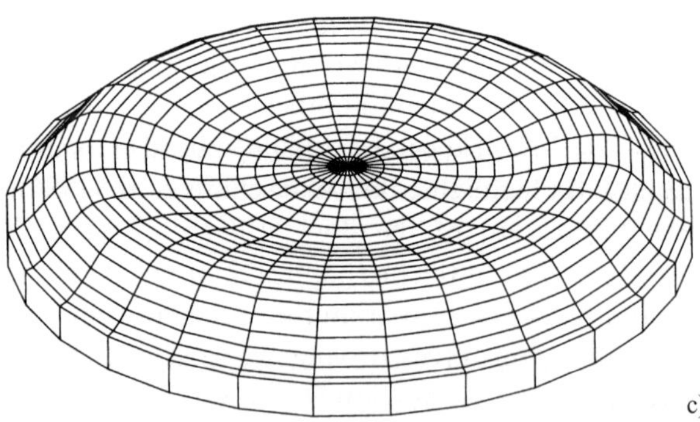

c)

φ=-π/14

Fig. 4.55 Crystal/melt interface: (a) $A_i=0.24$, $Ma_i=55$, $t-t^*=23$, (b) $A_i=0.18$, $Ma_i=41$, $t-t^*=27.7$, (c) $A_i=0.135$, $Ma_i=31$, $t-t^*=33.2$.

the instantaneous value of the Marangoni number decreases the Marangoni stress on the free-liquid surface. The simultaneous increase of the viscous effects and decrease of the Marangoni stress stabilize the flow field.

4.12. High Prandtl number liquids

4.12.1 *Transparent liquids and oscillatory behaviors*

Within the context of the present book, the case (half-zone) of transparent high Prandtl number liquids is discussed only for the sake of completeness. It has been widely used as a model for direct visualization of the flow structure but in the opinion of the author it is not relevant for the FZ technique as elucidated in Section 4.1 (even if the possible existence of steady hydrodynamic bifurcations also for transparent liquids has been recently pointed out for fat (convex) liquid bridges in the interesting analysis of Hu and Tang (2003)).

Since the end of 1970s (Chun and Wuest, 1979; Chun, 1980), on the ground experiments with small-size half-zones and high Prandtl number liquids ($Pr \gg 1$) pointed out that the Marangoni convection in liquid bridges heated from above or from below shows a direct transition from a steady axisymmetric toroidal flow to a three-dimensional oscillatory flow, when a critical temperature difference between the liquid-bridge supports is exceeded. Schwabe et al. (1982), Preisser et al. (1983), Velten et al. (1991) and Frank and Schwabe (1998) systematically performed on the ground research over a large range of experimental conditions. They found that the oscillatory Marangoni flow exhibits different behavior depending on the geometrical aspect ratio of the zone A_H.

The effect of volume was studied on the ground by Hu et al. (1995) and Hirata et al. (1997a–c). It was found that an overstability gap (i.e. very high values of the critical Marangoni number) exists for intermediate values of the volume ($S \cong 1$) whereas the flow is destabilized for large or small volumes (for $Pr \ll 1$ the behavior is the opposite, see Section 4.8).

The onset of instability of Marangoni convection was investigated in space using large transparent liquid bridges with a length of the order of several centimeters and in the absence of buoyancy effects (see, e.g., Monti, 1987 and Monti et al., 1995). Additional experimental studies have been performed recently on the ground by Muehlner et al. (1997), Schwabe et al. (1996).

Stability analyses were carried out by several authors. Xu and Davis (1984) investigated the case of half zones with infinite aspect ratio. They found that there is a critical value Pr^* ($\cong 50$) of the Prandtl number such that if $Pr < Pr^*$, the mode $m-1$ is preferred while if $Pr > Pr^*$ the mode $m=0$ (two-dimensional) is preferred.

The effect of axial boundaries was introduced later by Shen et al. (1990), Neitzel et al. (1991, 1992), Kuhlmann (1992), Kuhlmann and Rath, (1993a, b), Wanschura et al. (1995) to define, in the nondimensional parameter space, sufficient conditions for stability and instability with respect to two-dimensional and three-dimensional disturbances. In agreement with the experimental evidence it was shown that for all cases the three-dimensional disturbances ($m \neq 0$) are the most dangerous.

Chen and Hu (1998) and Ermakov and Ermakova (2004) extended the previous studies accounting for the effect of a noncylindrical interface in zero-g conditions and on Earth, respectively. They found that for high Prandtl liquids the critical values of the Marangoni number are strongly influenced by the liquid bridge volume (e.g., for $Pr = 10$ a variation of 20% of the volume of the liquid bridge can lead to a variation of 100% of the critical Marangoni number).

Landmark results on the subject are those reported in the linear stability analyses of Kuhlmann and Rath (1993a) and Wanschura et al. (1995). The basic axisymmetric Marangoni flow was obtained by numerical solutions of the Navier–Stokes equations, together with the appropriate boundary and symmetry conditions, and the eigenvalue problem for the three-dimensional disturbances was solved over a range of Prandtl numbers and for aspect ratios close to unity. These results predicted the critical Marangoni numbers and the form of the most dangerous disturbances, characterized by the appropriate value of the critical wave number, in the neighborhood of the neutral stability point (i.e. at the onset).

They found that the 3D supercritical state after the Hopf bifurcation point, is given by superposition of two counterpropagating hydrothermal waves, with axial and azimuthal components, similar to those already disclosed in the milestone analysis of Smith and Davis (1983) in the case of infinite horizontal liquid layers (see also Section 2.3).

In principle, if the two waves have equal amplitude the resulting disturbance is a "standing wave," with the minimum and maximum disturbances pulsating at fixed azimuthal positions; while the superposition of waves with different amplitudes gives rise to a "traveling wave," with the minimum and maximum disturbances traveling in the azimuthal direction. Unfortunately, nothing can be predicted by the linear stability analyses about the amplitude of these disturbances.

The experiments performed on the ground and in microgravity conditions revealed one or other instability mechanism. For instance, the traveling wave was observed by Chun and Wuest (1979), Preisser et al. (1983), Schwabe et al. (1996), Monti (1987), Muehlner et al. (1997) and Frank and Schwabe (1998). The standing wave, instead, was observed during experimental ground-based activities (Velten et al., 1991; Frank and Schwabe, 1998).

A number of diagnostic techniques were used for the experimental investigation obtaining different glimpses of various aspects of the problem.

Velten et al. (1991), measuring the temperature near the free surface by three thermocouples positioned at the same radial coordinate but at different azimuthal positions, detected two different nonaxisymmetric spatial structures of Marangoni convection: (1) running waves with an azimuthal component (corresponding to the aforementioned traveling wave regime); (2) axially running waves with deformed wave front. The latter regime corresponds to the standing wave, as demonstrated by Kuhlmann and Rath (1993b) in a subsequent analysis.

Schwabe et al. (1996) used a stereo microscope (rather than the usual light-cut-technique-based visualization following tracers illuminated by a laser sheet in the generic meridian plane) and observed in the bulk fluid dark inclined stripes moving in a horizontal direction from one side of the floating zone to the other. These stripes (moving horizontally) were interpreted as a direct manifestation of azimuthally traveling thermal waves since they were generated by densely packed tracer particles bunched in a line wound around the toroidal convection roll moving azimuthally.

Muehlner et al. (1997) investigated the time-dependent features of the instability on Earth using a thermographic system to visualize directly the temperature field. They employed an infrared thermocamera with wavelengths centered at 4.61 μm. Since in the wavelength band centered at 4.61 μm, the substance they considered (tetradecamethylhexasiloxane with $Pr = 35$) is partially transparent, these authors observed an average temperature distribution over a depth of about 70% of the radius of the liquid bridge. A thermographic diagnostic system able to provide direct visualization of the surface temperature spots induced by the Marangoni flow instability on the microzone surface was used by Lappa et al. (2000); a thermocamera with wavelengths 8–12 μm, in fact, was employed that does not make the silicone oil transparent providing thus an quite clear picture of the surface temperature field (see, e.g., Figs. 2.24–2.27 in Chapter 2).

Recently Frank and Schwabe (1998) by means of a light cut technique clearly observed standing and traveling waves. For this purpose a nonconventional setup with a transparent (quartz glass) upper disk heated by an electrical resistance was used for direct visualization of the azimuthal motion of tracers in cross-sections perpendicular to the axis of the bridge. Similar approaches are currently used by the Japanese group of Kawamura and coworkers (see, e.g., Kawamura and Harada, 1998; Ueno et al., 2000).

These modes of convection were also studied by means of direct numerical solution of the time-dependent and the nonlinear Navier–Stokes equations (Yasushiro et al., 1997, 1999; Tang et al., 1997; Zeng et al., 1999; Leypoldt et al., 2000; Shevtsova et al., 2001). Along these lines, recent numerical results obtained by Lappa et al. (2001b), have elucidated that the standing wave and the traveling wave models correspond to two consecutive transitions of the Marangoni flow. In other words, the flow exhibits a first transition from the axisymmetric, steady to the three-dimensional oscillatory state, characterized by the standing wave and, after a certain time, a second transition from the standing wave to the traveling wave. The numerical results also pointed out that the transition from one mode of oscillatory convection to the other depends on the aspect ratio of the liquid bridge.

The azimuthal standing-wave regimes tend to be stabilized (they become long-lasting) when the aspect ratio is reduced (Lappa et al., 2001b).

4.12.2 *Standing waves and traveling waves*

The three-dimensional temperature disturbance for the azimuthally standing wave consists of a number *m* of couples of disturbances (hot and cold) pulsating at the same azimuthal positions along the interface and in the bulk (minimum and maximum disturbances at fixed azimuthal positions). These spots are responsible for thermocapillary effects in the azimuthal direction, causing 2*m* counter-rotating vortex cells in the transversal section.

After the aforementioned second transition, the instability mechanism is taken over by a traveling wave and the temperature spots rotate around the axis of the liquid bridge.

The computed temperature and the velocity fields in a transversal cross-section for the standing-wave regime and $m=2$ ($A_H=0.4$, $Ma=3.6\times10^4$) are illustrated in Figs. 4.56 and 4.57. The pulsating temperature spots on the surface of the bridge are

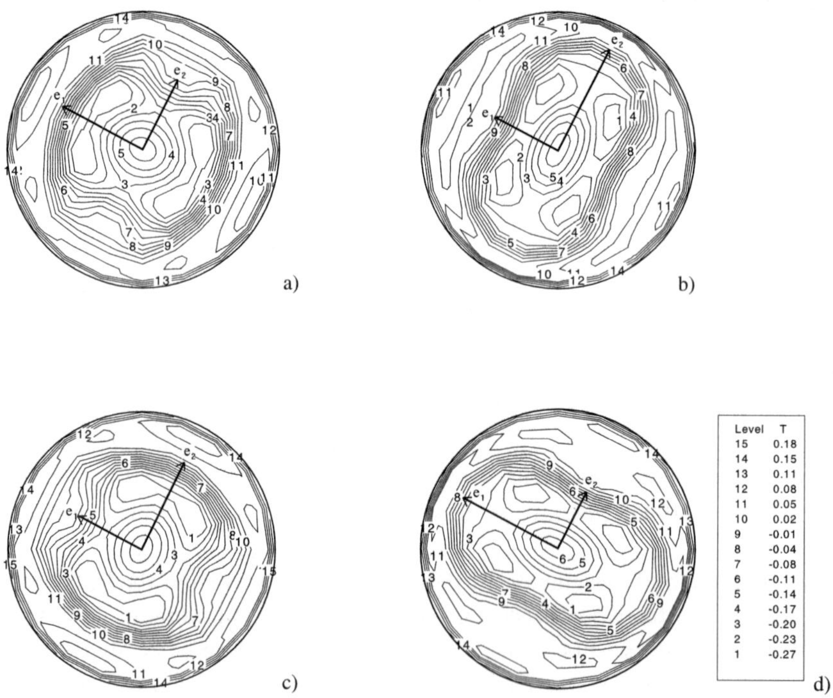

Fig. 4.56 Temperature distribution in the section $z=0.75$ for $A_H=0.4$ ($Pr=30$, $Ma=3.6\times10^4$, zero-g) and standing-wave regime (the field is shown in figures a–d corresponding to $t=0$, $\tau/4$, $\tau/2$, $3\tau/4$).

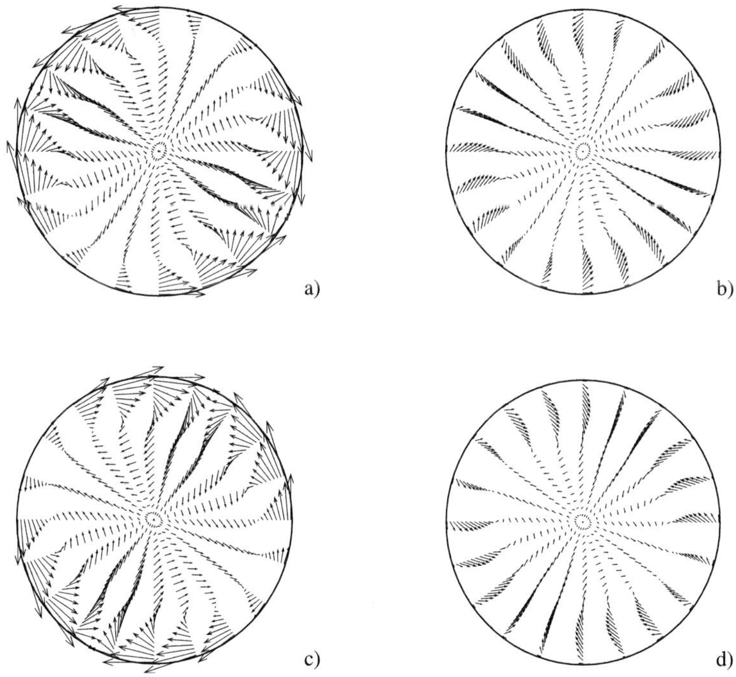

Fig. 4.57 Velocity field in the section $z=0.5$ for $A_H=0.4$ ($Pr=30$, $Ma=3.6\times10^4$, zero-g) and standing-wave regime (the field is shown in figures a–d corresponding to $t=0$, $\tau/4$, $\tau/2$, $3\tau/4$).

shown in Fig. 4.58 (the period τ has been divided into 4 parts and the fields are shown in figures a–d corresponding to $t=0$, $\tau/4$, $\tau/2$, $3\tau/4$). These temperature spots pulsate, i.e. the two cold spots grow in the axial direction during the shrinking of the two hot spots and vice versa, but the azimuthal positions of these extrema do not change. Accordingly, the convective cells in Fig. 4.57 change their sense of rotation periodically. Their intensity is not constant in time; if the considered convective cell is clockwise oriented during the first half-period, then during the second half of the period it vanishes and then reappears in the same azimuthal position anticlockwise oriented.

The behavior of the surface spots is also directly related to the motion of the toroidal vortex centerline. As explained in Section 4.7 the position of the vortex centerline describes in space a sine curve having 2 maxima and 2 minima in the z-direction if $m=2$. During the half-period the maxima travel in the z-direction toward the cold disk and the minima travel in the z-direction towards the hot disk; in the second half the minima become maxima and vice versa. Since the displacement up and down of the toroidal convection roll is coupled to displacement out and in, respectively, with respect to the axis of symmetry of the zone, this gives rise to an alternate expansion and contraction of the axes e_1 and e_2 of the elliptic inner region

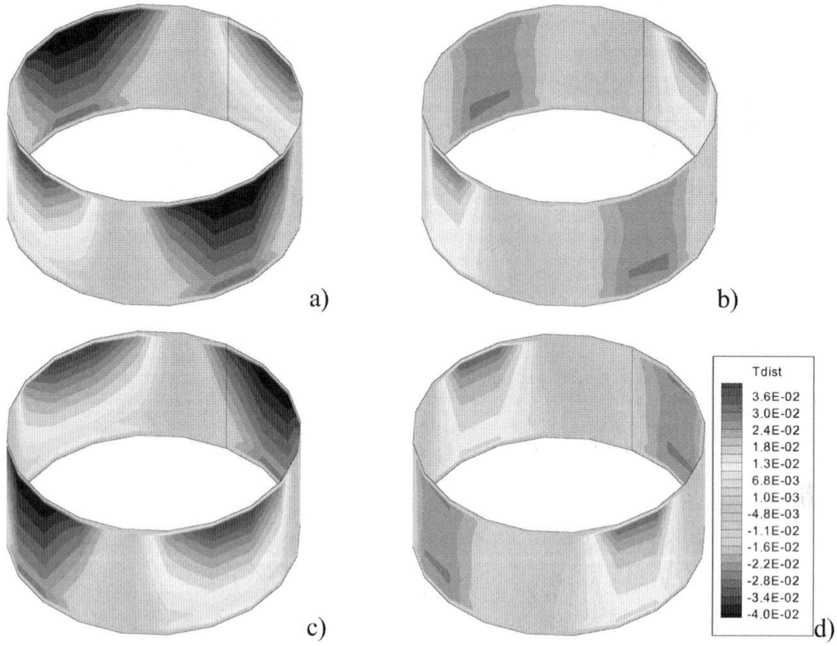

Fig. 4.58 Temperature disturbance on the bridge surface for $A_H = 0.4$ ($Pr = 30$, $Ma = 3.6 \times 10^4$, zero-g) and standing-wave regime (the field is shown in figures a–d corresponding to $t = 0$, $\tau/4$, $\tau/2$, $3\tau/4$).

visible in the temperature field (Fig. 4.56). One complete period for this time-dependent behavior takes a time $1/f$ where f is the characteristic frequency of the temperature oscillations measured at a fixed point.

Figures 4.59–4.61 illustrate how, after the second transition to the azimuthally traveling wave, remarkable differences exist with respect to the prominent features of the standing-wave regime. The behavior is not pulsating, but rotating. The surface-temperature spots do not change their intensity and rotate around the perimeter of the liquid bridge. The vortex cells in the section orthogonal to the axis do not change their sense of rotation and their strength is constant in time (the cells never vanish, see Fig. 4.60). The time-dependent behavior of the velocity field is simply characterized by a full rotation of the entire flow pattern in the azimuthal direction. For the temperature field (Fig. 4.59) the amplitude of the axes e_1 and e_2 is fixed, whereas the inner elliptic region rotates with a center of rotation corresponding to the axis of symmetry. In this case one entire rotation needs a time $2/f$.

The standing modes of convection for $A_H = 0.25$ ($m = 3$) and for $A_H = 0.2$ ($m = 4$) are depicted in Figs. 4.62 and 4.63 and Figs. 4.64 and 4.65, respectively. In this case the behavior is similar to that described for the case $m = 2$, the only difference being the number of spots and vortex cells in the azimuthal direction (Figs. 4.62 and 4.64 show a pulsation of the inner triangle and the quadrangle, respectively: The nodes of

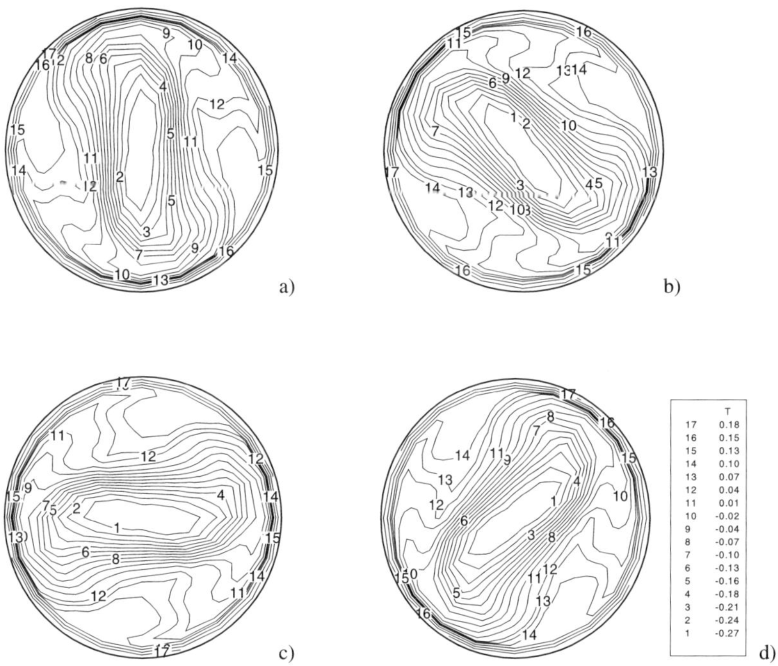

Fig. 4.59 Temperature distribution in the section $z=0.75$ for $A_{\mathrm{H}}=0.4$ ($Pr=30$, $Ma=3.6\times10^4$, zero-g) and traveling-wave regime (the field is shown in figures a–d corresponding to $t=0$, $\tau/4$, $\tau/2$, $3\tau/4$).

these polygons are replaced in time by the presence of sides and vice versa). One complete period for this time-dependent behavior takes a time $1/f$ as in the case $m=2$. The rotating regime for these cases (not shown) is merely characterized by a full rotation of the thermo-fluid-dynamic field as for $m=2$ and the entire rotation of the configuration needs a time m/f.

4.12.3 *Symmetric and asymmetric oscillatory modes of convection*

For a more detailed description of the thermo-fluid-dynamic field, it is necessary to categorize the oscillatory state also with respect to the behavior in the generic meridian plane. In this plane, in fact, complex symmetric and asymmetric models of convection appear according to the azimuthal wave number. In practice, these different oscillatory convective modes are caused by the traveling of, respectively, symmetric or asymmetric structures of periodicity in the azimuthal direction. Thereby, it is possible to simply categorize the symmetric and asymmetric oscillatory modes according to the even and odd wave numbers, respectively, like the case of small Prandtl number liquids.

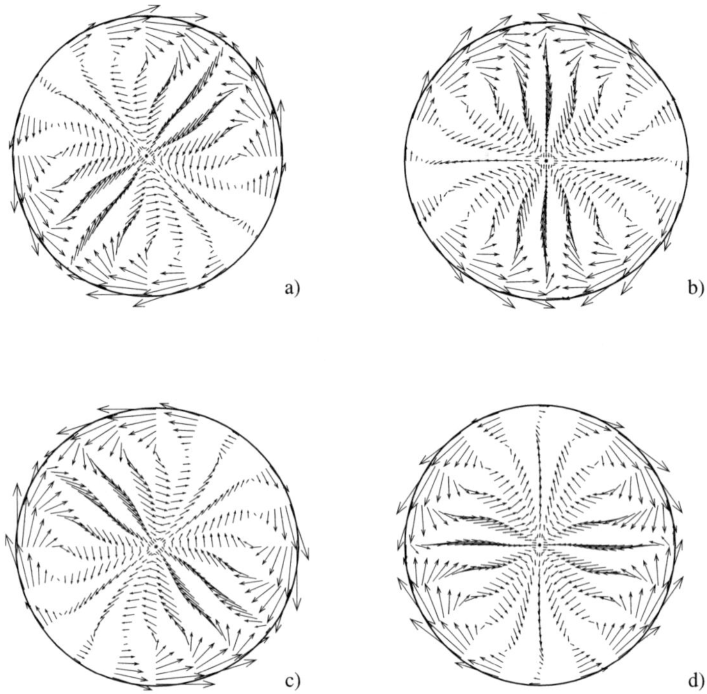

Fig. 4.60 Velocity field in the section $z=0.5$ for $A_{\mathrm{H}}=0.4$ ($Pr=30$, $Ma=3.6\times10^4$, zero-g) and traveling-wave regime (the field is shown in figures a–d corresponding to $t=0$, $\tau/4$, $\tau/2$, $3\tau/4$).

If m is odd the vortex in one half of the zone section appears smaller than the opposite vortex. The time dependence is observed as a periodical interchange of the shape of the vortices in the left and right parts of the zone. After one half of the oscillation period the small vortex and the large vortex change position. The branching streamline of the opposite vortices changes its inclination continuously during an oscillation period when the azimuthal wave number is odd. This behavior is illustrated in Fig. 4.66 ($m=1$), where the streamlines of the projected velocity vectors in the meridian plane $\varphi=0$ are shown for eight subsequent moments in time during one cycle for the case $A_{\mathrm{H}}=0.7$ (the period τ has been divided into 8 parts and the velocity field is shown in figures a–h corresponding to $t=0$, $\tau/8$, $\tau/4$, $3\tau/8$, $\tau/2$, $5\tau/8$, $3\tau/4$, $7\tau/8$). Therein, the reader will clearly recognize that the characteristic sign of a mode $m=1$ is a reciprocal variation of the diameters of the convection vortices on the left side and on the right side; i.e. when the left vortex puffs up, the right roll contracts during a half-period and vice versa (see Savino et al., 1999 and Monti et al., 2000b for similar behavior detected experimentally). For even critical wave numbers, the time dependence is instead observed as a synchronous pulsation of two symmetrical vortices (see also Monti et al., 2000c).

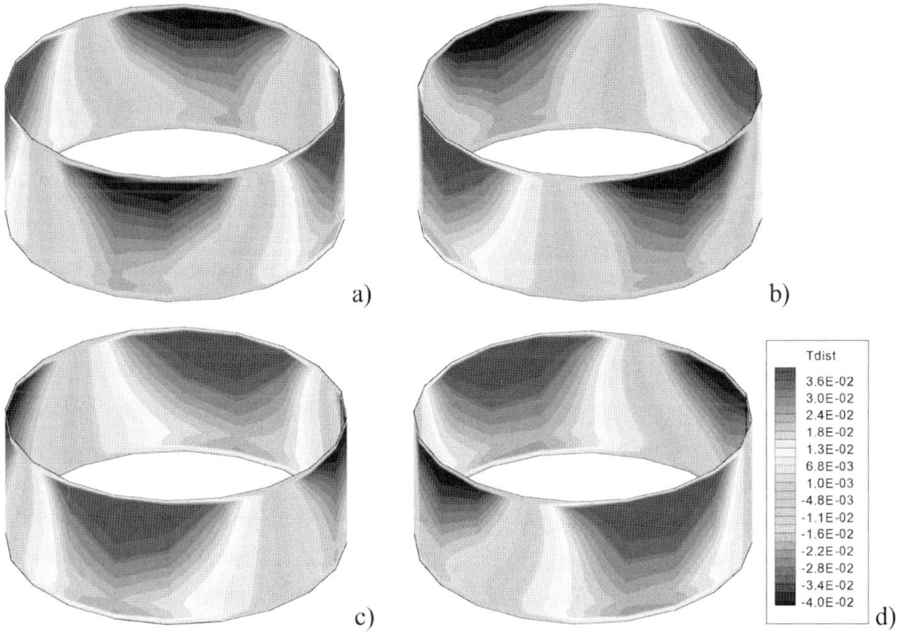

Fig. 4.61 Temperature disturbance on the bridge surface for $A_H=0.4$ ($Pr=30$, $Ma=3.6\times10^4$, zero-g) and traveling-wave regime (the field is shown in figures a–d corresponding to $t=0$, $\tau/4$, $\tau/2$, $3\tau/4$).

Symmetric and asymmetric behaviors of the velocity field in a meridian plane, are shown in Fig. 4.67 for two moments in time ($t=0$, and $t=\tau/2$).

For $A_H=0.7$ ($m=1$) one of the two vortex cells in the section prevails over the other one and is extended along the whole axial plane of the bridge; the other cell contracts and is confined in the hot corner, moreover, since in the boundary layer near the cold disk steep axial gradients of temperature are present, here a small Marangoni cell is still present driven by the surface-temperature distribution (see Fig. 4.67a). For $A_H=0.5$ ($m=2$) the velocity field in the section of Fig. 4.67b is symmetric and the convective cells travel axially up and down. The situations for $A_H=0.25$ ($m=3$) and $A_H=0.2$ ($m=4$) are illustrated in Fig. 4.67c and d, respectively. The liquid bridge in this case is very short and the two driving cells are confined near the free surface, with other counter-rotating vortex cells induced by continuity in the interior of the bridge. Similar results were obtained using the 2D computations by Rybicki and Floryan (1987). They found, for short bridges, the emergence of several layers of vortices, with the strength of each layer decreasing approximately exponentially with the distance from the liquid/air interface.

A characteristic feature of the asymmetric modes is a periodically "staggering ring" of vortex core of the toroidal flow pattern on a meridian section and also a periodically tilting branching streamline that divides the two vortices in the generic

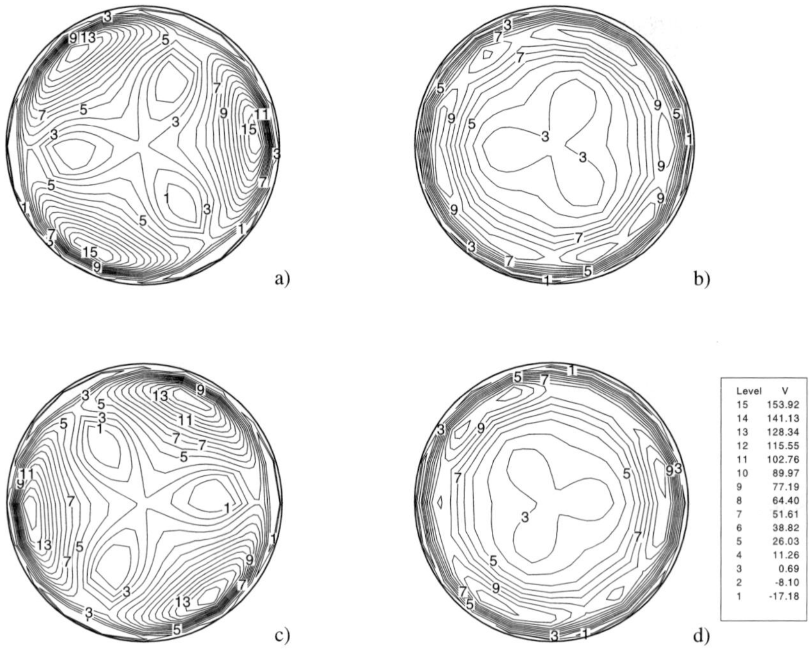

Fig. 4.62 Radial-velocity distribution in the section $z=0.75$ for $A_{\mathrm{H}}=0.25$ ($Pr=30$, $Ma=3.8\times10^4$, zero-g) and standing-wave regime (the field is shown in figures a–d corresponding to $t=0$, $\tau/4$, $\tau/2$, $3\tau/4$).

meridian plane. For the symmetric modes two vortices on the meridian plane are moving up and down simultaneously, and accordingly the tilting of the branching streamline disappears.

For $A_{\mathrm{H}}=0.5$ only one toroidal vortex exists (Fig. 4.67a). When A_{H} is reduced, as explained earlier, small additional vortices located around the axis of rotation and attached to the side walls emerge. These vortices are clearly visible in the bridges of aspect ratio 0.25 and 0.2. For $A_{\mathrm{H}}=0.25$ (Fig. 4.67c) the critical wave number is odd ($m=3$) and the additional counter-rotating vortex cells in the interior of the bridge are attached one to the hot disk and the other to the cold disk; for $A_{\mathrm{H}}=0.2$ (Fig. 4.67d) they are symmetric since the wave number is even ($m=4$).

4.12.4 *System dynamical evolution*

The computed spatiotemporal structures of the three-dimensional oscillatory Marangoni convection previously depicted in detail agree with available experimental observations and, in principle, with the predictions of linear stability analyses, according to which, for large Pr, the basic Marangoni flow loses its stability due to a pair of counterpropagating azimuthal waves, with a phase depending not only on φ and t but also on the radial and axial coordinates r and z (Kuhlmann and Rath, 1993a).

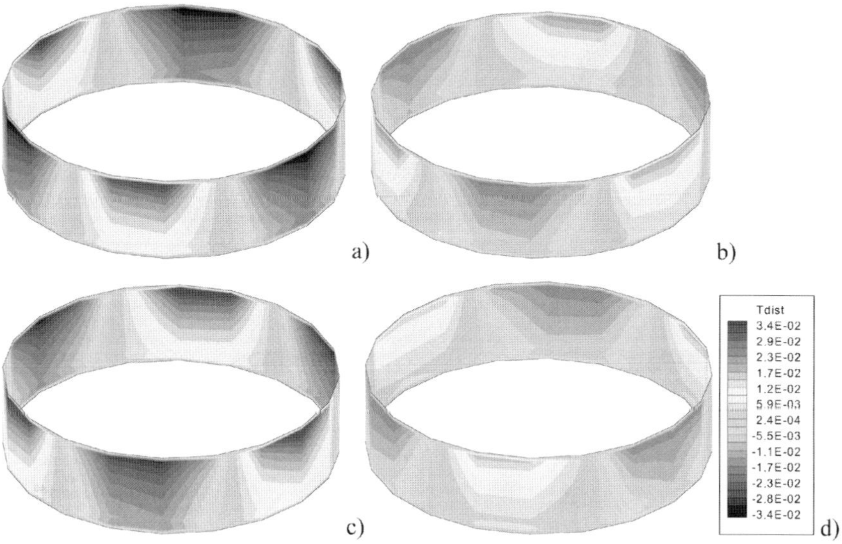

Fig. 4.63 Temperature disturbances on the bridge surface for $A_H = 0.25$ ($Pr = 30$, $Ma = 3.8 \times 10^4$, zero-g) and standing-wave regime (the field is shown in figures a–d corresponding to $t = 0$, $\tau/4$, $\tau/2$, $3\tau/4$).

The wavefront f of such hydrothermal waves is inclined with respect to z. Since the constant-phase surfaces can be imagined as vertical planes that have been twisted around the vertical axis, the twist being given by a phase $G(r, z)$, each wave having an amplitude $B(r, z)$ can be represented according to Kuhlmann and Rath (1993b) as $F_{\pm} = B(r, z)\exp\{i[\pm m\varphi - \omega t + G(r, z)]\}$. A superposition of two counterpropagating waves with the same amplitude results in

$$F = B(r, z)\cos(m\varphi)\cos(\omega t - G(r, z)) \tag{4.24a}$$

Since in this case the oscillatory term does not depend on φ, on the basis of Eq. (4.24a), this situation should be characterized by maximum and minimum disturbances fixed in space with the minimum being continually replaced by the maximum and vice versa (as soon as $\cos(\omega t - G(r, z))$ changes its sign). Since these extrema in the disturbance distribution correspond to hotter and colder zones in the bridge, the three-dimensional temperature disturbance should simply consist of a number m of couples of spots (hot and cold) pulsating at the same azimuthal positions along the interface.

When the amplitude of the two hydrothermal waves is not the same the superposition gives:

$$F \cong B(r, z)\, \mathrm{a}(m\varphi)\cos[\mathrm{b}(m\varphi, G) - \omega t] \tag{4.24b}$$

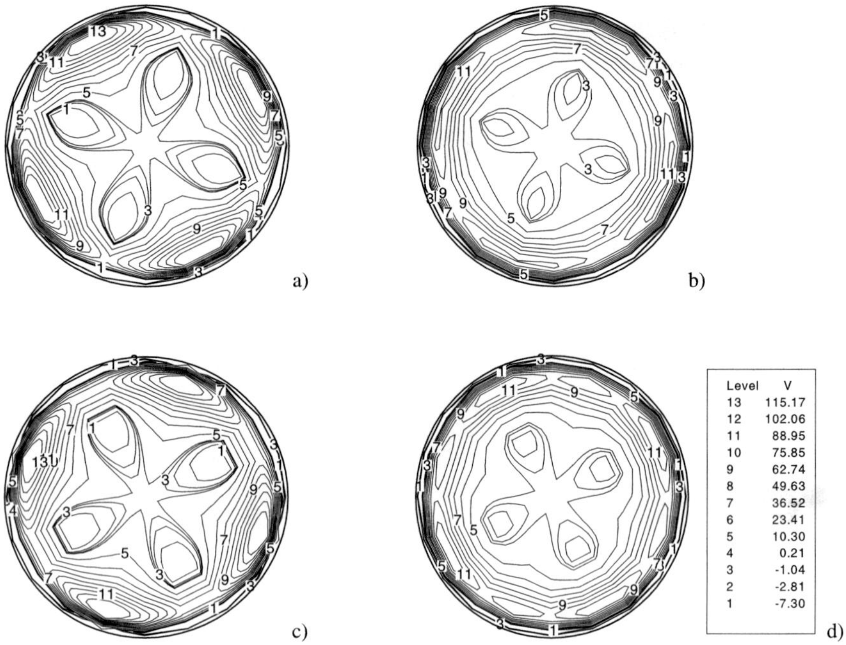

Fig. 4.64 Radial velocity distribution in the section $z=0.75$ for $A_{\mathrm{H}}=0.2$ ($Pr=30$, $Ma=3.9\times10^4$, zero-g) and standing-wave regime (the field is shown in figures a–d corresponding to $t=0$, $\tau/4$, $\tau/2$, $3\tau/4$).

(the functions a and b are reported in Kuhlmann and Rath, 1993b) so that the oscillatory term must depend on φ; in this case, on the basis of Eq. (4.24b) the minimum and maximum disturbances should travel in the azimuthal direction and the phase of the oscillations should depend continuously on φ.

Such a discussion provides theoretical validation for the behaviors captured by the numerical simulations and shown in Figs. 4.56–4.65. It is worthwhile to stress, however, that the 3D numerical simulations in turn take the current comprehension of the problem beyond the point of view provided by the linear stability analyses. There is, in fact, indirect evidence of the fact that at the beginning two counter-propagating waves with equal amplitude are formed, while in a second phase one of these waves prevails over the other. This result is very important if one considers (see also Section 4.12.1) that the standing-wave or the traveling-wave modes were experimentally observed through disjoint glimpses without discerning the relationship between them.

All these arguments also lead to the natural introduction of a simple "numerical" criterion for detecting the regime (pulsating or rotating). In the light of Eqs. (4.24) such a distinction, in fact, could be directly based on the phase shift between some "numerical" probes located on the free surface at the same axial position with a certain angular shift (Monti et al., 1997). If the oscillations are measured in points

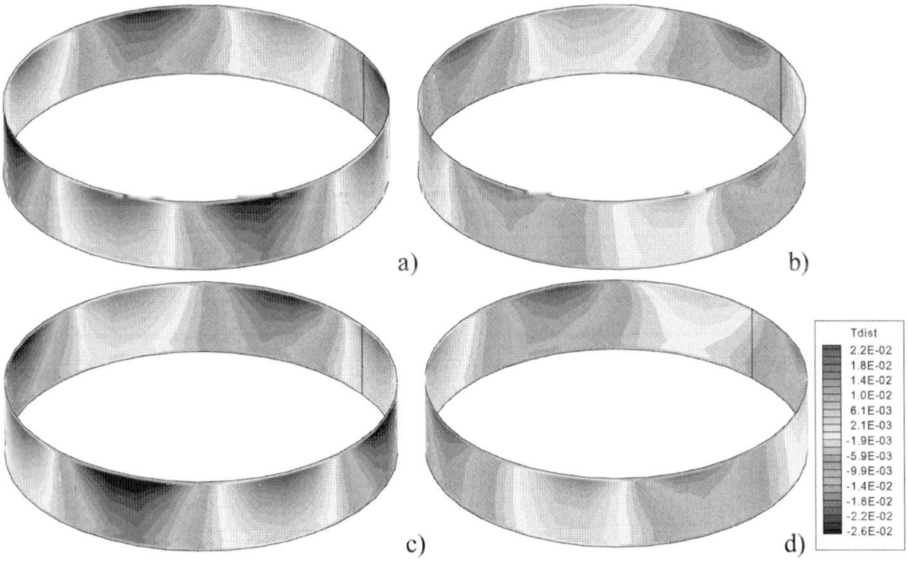

Fig. 4.65 Temperature disturbances on the bridge surface for $A_H = 0.2$ ($Ma = 3.9 \times 10^4$, zero-g) and standing-wave regime (the field is shown in figures a–d corresponding to $t = 0$, $\tau/4$, $\tau/2$, $3\tau/4$).

with the same axial and radial coordinates but at different azimuthal positions, for the standing wave, only two values of the azimuthal phase shift are allowed ($\Delta\varphi = 0$ or $\Delta\varphi = \pi$); on the contrary, for the traveling wave the possible values of the phases are not discrete. Monitoring this phase shift therefore it is possible to detect the regime and, in particular the transition from one to the other.

As an example, Fig. 4.68 shows the signals provided by "numerical" thermocouples located on the surface of the bridge having the same axial coordinates ($r = 1/2A_H$, $z = 0.75$) but different azimuthal positions (with a shift of 90°) for a liquid bridge with $A_H = 1$ and $Ma \cong 1.5Ma_c$. For this case the azimuthal wave number is $m = 1$.

During the early transient stage dominated by the pulsating behavior there is no phase shift between T_1 and T_4 and no phase shift between T_2 and T_3 but T_1 and T_4 measure values with a phase shift of π with respect to T_2 and T_3.

As previously explained, the temperature disturbance can be represented as spots on the liquid-bridge surface, pulsating while maintaining fixed azimuthal positions. Since for $m = 1$ the azimuthal extension of each spot ($360/2m$) is 180°, and the thermocouples considered have an azimuthal shift of 90°, two thermocouples will be placed on a spot and the others on the second spot. This implies $\Delta\varphi = 0$ (if the points are placed on the same spot), $\Delta\varphi = \pi$ (if the points belong to different spots). This simple argument explains why the signals in Fig. 4.68 exhibit opposite phase initially. This figure also shows that the amplitudes of the temperature oscillations T_1, T_2, T_3

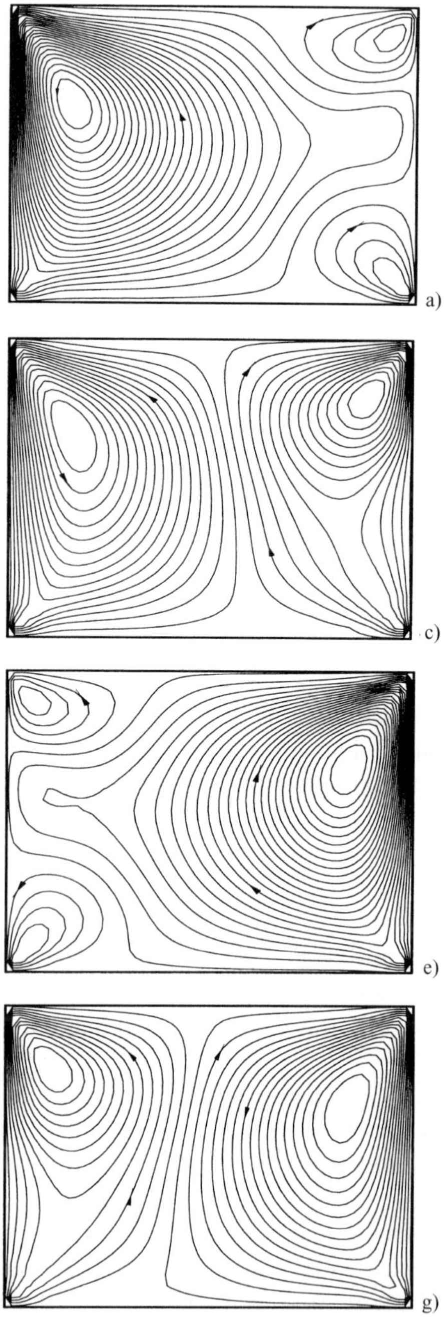

Fig. 4.66 Streamlines of the projected velocity vectors on the meridian plane $\varphi=0$ for $A_H=0.7$, $Pr=30$, $Ma=3.35\times10^4$, zero-g (the field is shown in figures a–h corresponding to $t=0$, $\tau/8$, $\tau/4$, $3\tau/8$, $\tau/2$, $5\tau/8$, $3\tau/4$, $7\tau/8$).

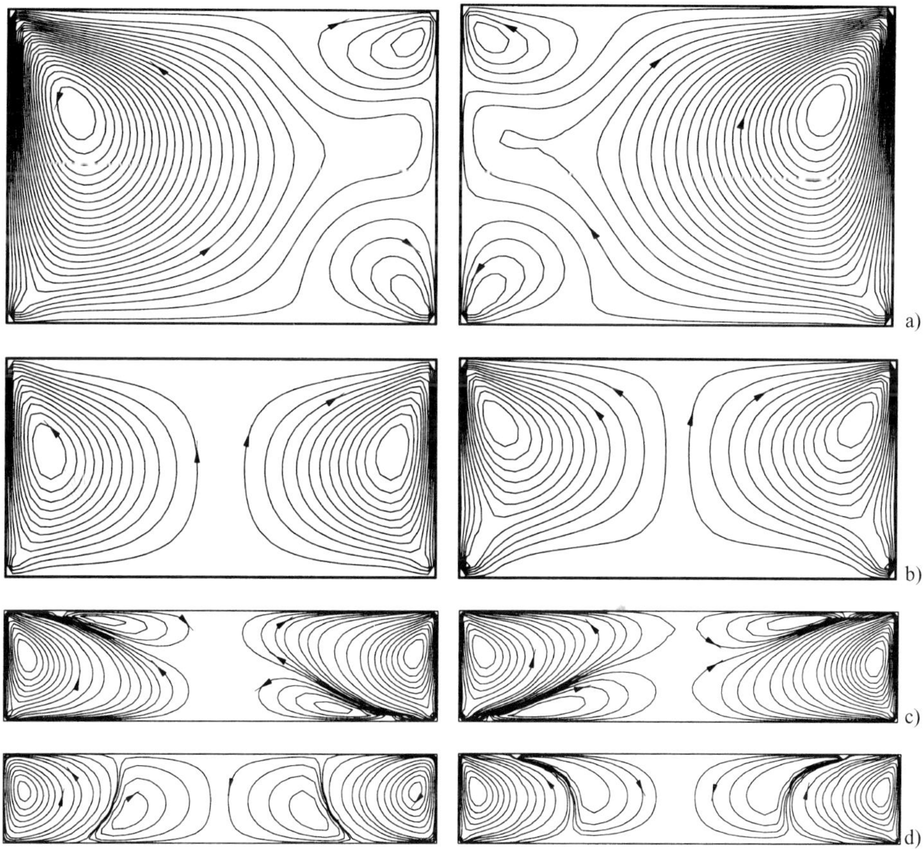

Fig. 4.67 $Pr = 30$, 0g, streamlines of the projected velocity vectors in the meridian plane (the field is shown on the left and on the right for $t = 0$ and $t = \tau/2$): (a) $A_H = 0.7$, $Ma = 3.35 \times 10^4$, (b) $A_H = 0.5$, $Ma = 3.5 \times 10^4$, (c) $A_H = 0.25$, $Ma = 3.8 \times 10^4$, (d) $A_H = 0.2$, $Ma = 3.9 \times 10^4$.

and T_4 are different, i.e. the "numerical" probes with an angular shift of 90°, do not measure the same maximum nor the same minimum.

Also, this behavior is susceptible of an interesting interpretation in terms of the relative position of surface-temperature spots and thermocouples. In fact, since both the positions of the spots and of the probes are fixed and the temperature distribution along each spot is not uniform (there is a peak in the "core" of the spot, see, e.g., Figs. 4.58 and 4.61), each thermocouple will measure a maximum (minimum) value of the temperature depending on its local azimuthal position along the spot.

When the pulsating regime is taken over by the rotating one, the amplitude of the temperature oscillations does not depend on the local azimuthal position of the thermocouples and is the same for all the points having the same radial and axial

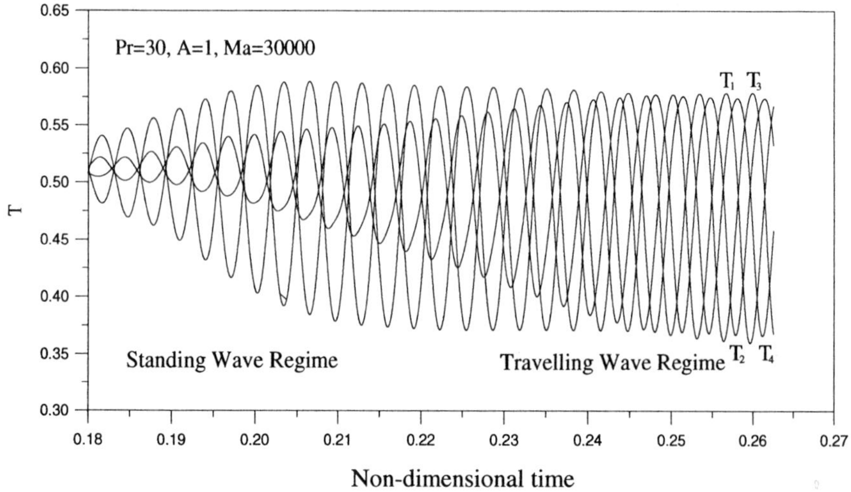

Fig. 4.68 Signals provided by four numerical "probes" for the case $Pr = 30$, $A_H = 1$, $Ma = 3 \times 10^4$.

position (Fig. 4.68). In this stage, the surface spots are not fixed, but rotate in the azimuthal direction, hence each of the four numerical thermocouples will measure the same maximum (minimum) temperature value when the hot (cold) spot passes on it. This is also the reason why, in the rotating regime, the time-temperature profiles exhibit a phase displacement continuously depending on the azimuthal coordinate. For the case chosen as an example, the critical wave number in particular is $m = 1$, the oscillations show a phase displacement of $\pi/2$ between two numerical thermocouples located at an angular distance of 90° and a phase displacement of π between two numerical thermocouples located at an angular distance of 180°.

Figure 4.69c and d are interesting as they show that the behaviors elucidated above in terms of phase shift and amplitude of the signals correspond to well-defined "attractors" in the T_i–T_j plane. These plots are often used as idealized versions of the state of the considered system and its possible dynamical evolution. For instance, of particular interest is the interplay between T_2 and T_1. During the first regime the system evolves along an ellipsoidal spiral (the initial point corresponds to the initial axisymmetric and nonoscillatory condition of the system) whereas in the subsequent rotating regime the attractor is given by a quasicircle. This can be used as a distinguishing mark for detecting the regime. Between these two cases an intermediate (pulsorotating) regime occurs, with a phase shift increasing with time from zero (or π) up to $\pi/2$ (Fig. 4.70). Also this case is interesting since it could explain some unexpected behaviors obtained by Velten et al. (1991). Using the three thermocouples having the same axial and radial coordinates but different azimuthal positions, they measured in many situations phase shifts changing continuously in

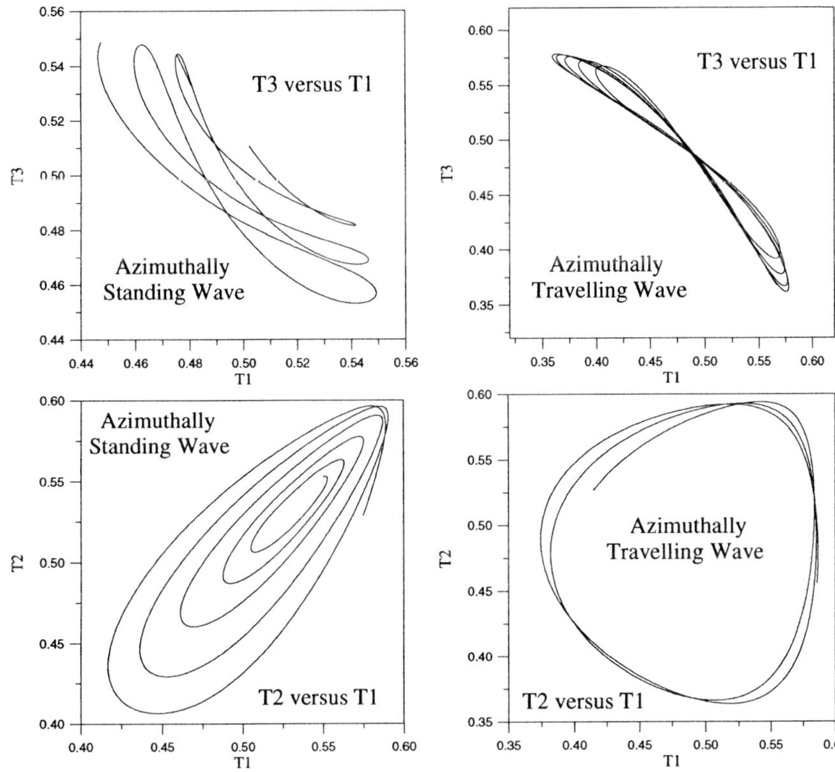

Fig. 4.69 Oscillatory behaviors in the phase space.

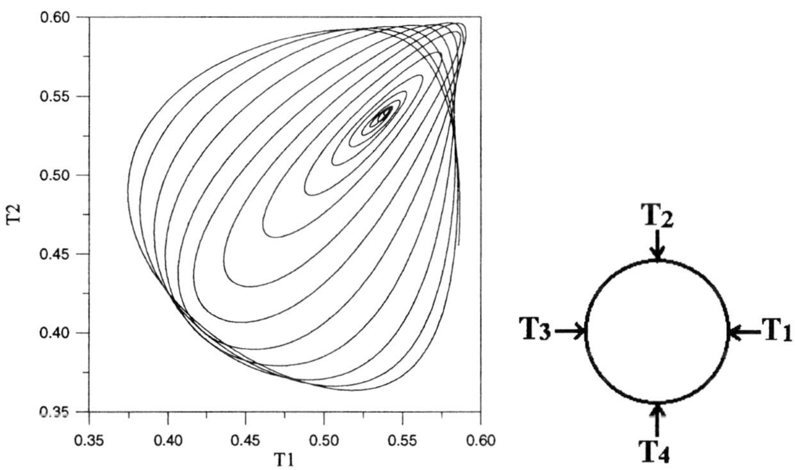

Fig. 4.70 Transition from the standing to the traveling-wave regime ($Pr = 30$, $A_H = 1$, $Ma = 3 \times 10^4$, signals provided by two thermocouples with an angular shift of $90/m°$).

time that, in the light of the above arguments, could be simply interpreted as a transition between the pulsating and the rotating modes of convection.

The criterion (and related spatial interpretation in terms of temperature spots and signals) discussed in the foregoing for $m=1$ focusing on T_1 and T_2 (at 90°), can be generalized to the case $m \neq 1$ in a simple way by choosing thermocouples with an azimuthal shift of $90/m°$.

The study of such dynamical evolutions and of the subsequent possible transitions to even more complex spatiotemporal behaviors ($Ma \gg Ma_c$) falls into the category of problems dealing with fundamental research in fluids illustrated in Section 1.4.1. As also discussed in Chapter 2, the liquid bridge with transparent liquids has often been used, in fact, as a mere pattern-forming dynamical system and as a paradigm for studying the progression from order to chaos (e.g., Ueno et al., 2003; Shevtsova et al., 2003).

4.12.5 *The hydrothermal mechanism*

Besides the aspects mentioned above, numerical simulations are also very useful since they somewhat allow a "reconstruction" of the physical mechanism with which the aforementioned counteracting hydrothermal waves predicted by Smith and Davis (1983) and then by Kuhlmann and Rath (1993a) are initiated and lead to the subsequent evolution with pulsating (and then rotating) temperature spots and azimuthal convective structures.

Before starting to deal with such a description, however, it is worth mentioning some arguments that were pointed out since 1980 by Chun and coworkers.

Chun and Wuest (1979) speculated that the growth or the damping of a small temperature disturbance must depend on the ratio of the rapidity of the heat transfers by conduction and by convection. For the extreme case of very high conductive heat transfer, a temperature disturbance should die, diffusing very rapidly in all directions. In the other extreme case of very high convective heat transfer, the disturbances should move with the flow, "frozen" in the flow motion.

In practice, the first case corresponds to small Prandtl number liquids for which the high thermal diffusivity tends to spread any possible temperature disturbance over a large extension in a very short time. Among other things this may explain why the instability mechanism for $Pr \ll 1$ is not based on the amplification of temperature disturbances but is hydrodynamical in nature (see Section 4.1).

The second case, corresponds to $Pr \gg 1$ for which the thermal diffusivity is relatively small if compared with the kinematic viscosity. It is very interesting within the context of this section since it makes possible the introduction of a simple mechanism explaining the numerically simulated behaviors previously shown.

According to Chun (1980) a temperature disturbance on the surface leads to a disturbance of the temperature gradient and thus to a disturbance of the surface-tension gradient that induces a distortion of the velocity field. The velocity distortion, in turn, is responsible for a distortion of the temperature distribution and this coupling mechanism between the surface-tension gradient and the heat transfer may generate growth or damping of the initial temperature disturbance.

Since, as previously explained, the amplification or damping of temperature disturbances on the free surface must depend on the relative importance between convection and diffusion (i.e. on the order of magnitude of the Marangoni number), the disturbances can grow and instability sets in if a critical temperature difference is exceeded (i.e. if the Marangoni number is large enough).

The concept of a hydrothermal wave is in line with this theory. Hereafter the discussion is supported by some practical (somewhat educational) examples with hot or cold disturbances on the free surface:

If a random hot disturbance is created on the free surface at a rather high position (not too distant from the hot disk) it generates a surface azimuthal flow resulting in two counterpropagating azimuthal flows moving far away from the disturbance (Fig. 4.71a). For continuity a radial flow is generated and, owing to the presence of cold bulk fluid, (the so-called "cold finger" already described in Section 2.4) cold fluid under the surface is carried at the interface, cooling the hot spot (Fig. 4.71b). The phase difference between the temperature and the velocity disturbances is responsible for further cooling of the free surface, resulting in a cold spot (Fig. 4.71c).

Such a mechanism is clearly visible if one compares frame by frame, Figs. 4.56–4.58. The hot disturbance causes a pair of counter-rotating vortex cells in the

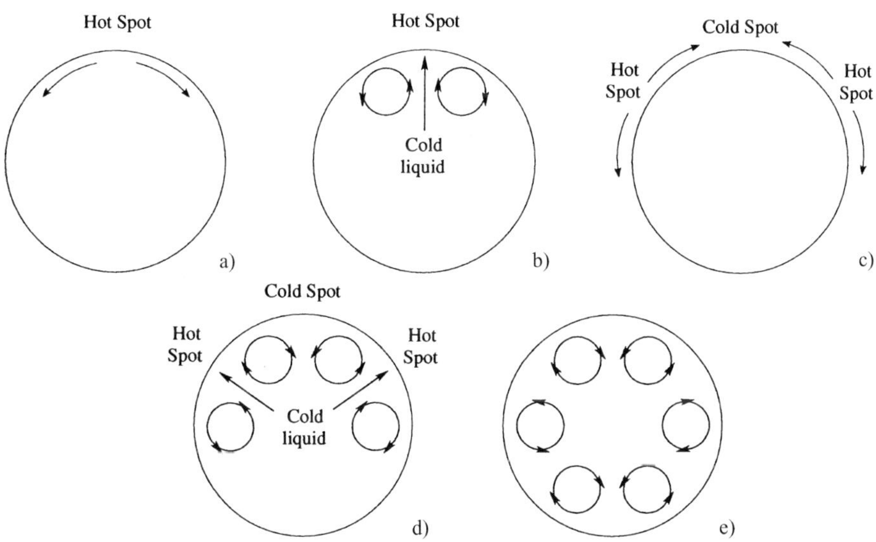

Fig. 4.71 Origin of counterpropagating hydrothermal waves.

cross-section. Due to the phase difference between surface-azimuthal and radial-velocity components when the hot spot disappears the radial flow is different from zero and moves the cold liquid to the surface; therefore a cold spot is generated.

If the initial disturbance starting the mechanism is not hot but cold, its main effect involves a disturbance in the axial component of the surface velocity. The surface velocity directed from the hot disk to the cold one increases locally, leading to an overheating of the initial cold disturbances so that it is replaced by a hot spot (Zeng et al., 2004). Owing to coupling of these mechanisms hot and cold disturbances can replace each other periodically leading to a pulsating behavior.

Since, for the case of high Prandtl number, surface-temperature disturbances are somehow "frozen" (as previously explained), they can be convected on the free surface in the azimuthal direction without being destroyed by the effect of diffusion.

Figure 4.71c shows, for instance, how, when cold fluid is carried at the interface due to continuity effects, the hot disturbance, initially located in that position, is transported by Marangoni flow in clockwise and counterclockwise directions towards new positions.

By these effects, the pulsating mechanisms initiated by a random single disturbance at a certain azimuthal position, can propagate along both the azimuthal directions on the surface (Fig. 4.71d) until it affects the entire circumferential extension (Fig. 4.71e). Among other things such a discussion also illustrates how the counterpropagating azimuthal hydrothermal waves are initiated. Precise mathematical arguments supporting the discussion above have been given in the previous section.

The occurrence of the traveling wave, simply corresponds to a subsequent azimuthal symmetry breaking in the mechanism described above.

The numerical results show, in fact, that during the standing-wave regime the whole body of the liquid bridge does not rotate (i.e. the azimuthally disturbance does not carry on its motion). Standing waves do not have a detectable azimuthal mean flow (i.e. the mean value of the oscillating azimuthal velocity is zero).

On the contrary, in the traveling-wave regime, a small azimuthal mean flow is observed (see Fig. 4.72 where at a certain distance from the onset, when the instability mechanism becomes dominated by a traveling wave, the velocity profile loses its symmetry with respect to the zero value).

This is in agreement with the findings of Schwabe et al. (1996) who under supercritical conditions, found an additional clearly visible oscillation in the azimuthal direction. It was found that the azimuthal oscillation of tracer particles captured in the vortex center was not symmetrical and characterized by a mean value of the azimuthal velocity (i.e. net flow in the azimuthal direction). They observed groups of tracer particles that were trapped in the center of the Marangoni vortex roll, moving azimuthally back and forth asymetrically, such that there was a net movement of the particles in the direction of propagation of the traveling wave.

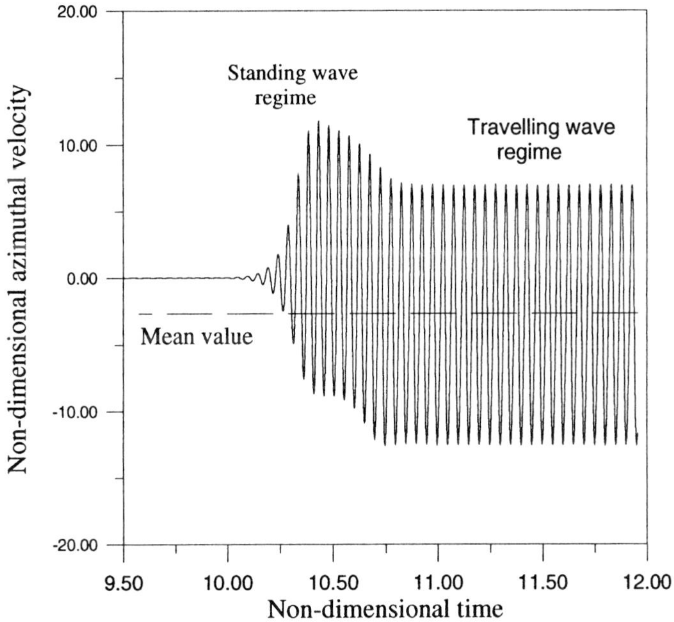

Fig. 4.72 Surface azimuthal velocity in a fixed point for $A_H = 0.5$, $Pr = 4$, $Ma = 4800$.

4.13. The full-zone: state-of-the-art

Till date, in spite of the extensive body of interesting theory and worthy results available for the stability properties of the pure Marangoni flow (microgravity conditions) and of the mixed buoyant–Marangoni convection (on the ground conditions) in the case of simplified liquid bridges, the investigation of these topics in the case of volumes of full-zone extent heated through the melt/gas interface can be considered still an open task.

For the sake of simplicity, the half-zone model has been preferred in recent years. On the contrary, very few studies related to the numerical analysis of these aspects in the case of a complete floating zone configuration have appeared in the literature and many of these analyses were based on the simplifying assumption of axisymmetric flow.

The analytical solutions of the steady and the axisymmetric flow in the case of liquid metals and the small-amplitude Marangoni-driven convection in a floating zone configuration with radiation and in the absence of gravity were obtained by Chen and Roux (1991). They expanded all fields in a power series in terms of capillary number and solved the first-order approximation, a Stokes-type problem, in terms of Papkovich–Fadle functions.

Rybicki and Floryan (1987) investigated the properties of the steady axisymmetric Marangoni convection. In particular, they analyzed numerically the effects of the aspect ratio A_F showing that a decrease in the aspect ratio (short bridges) leads to the emergence of several layers of vortices, with the strength of each layer decreasing approximately exponentially with the distance from the surface. Zhang and Alexander (1992) evaluated the influence of buoyancy forces on the Marangoni convection in the case of axisymmetric flow, melt/gas interface deformed due to gravity and planar supporting rods. Lan and Kou (1990) studied the axisymmetric Marangoni flow taking into account the simulation of the melt/solid interfaces behavior. Chen et al. (1994) studied the interaction between the axisymmetric Marangoni convection and the melt/solid shape for different values of the input power supplied to the heating source. Chen and Chu (1995) carried out a similar study for the case of mixed buoyant–Marangoni convection. Chen and Wu (1996) extended this analysis taking into account both the shape of the melt/solid interfaces and the melt/gas interface. Chenier et al. (1999) investigated the transitions and stability properties of the steady mixed buoyant–Marangoni flow adapting a direct numerical method to accurately compute Hopf (transition to time-dependent regime) and saddle-node bifurcation points, as well as the leading eigenvalues and their corresponding eigenvectors. They focused on the stability of axisymmetric flow with respect to axisymmetric disturbances. The case $Pr=0.01$ and aspect ratio $A_F=1$ was analyzed. Chen and Roux (1992) used linear stability analysis to investigate the Hopf bifurcation of Marangoni flow in low Prandtl number floating zones ($Pr=0.023$) under microgravity conditions. They found a strong stabilization of Marangoni convection with decreasing zone length in the case of axisymmetric disturbances ($m=0$) and predicted the most dangerous azimuthal wave number in the case of nonaxisymmetric disturbances for a zone with unitary aspect ratio. Chen and Chin (1995) used a mixed finite-difference spectral method to solve the linear disturbance equations for the linear stability analysis of Marangoni convection in a floating zone of molybdenum ($Pr=0.243$) in the case of axisymmetric disturbances. No indication was given about the steady/oscillatory nature of the bifurcation since only the real part of the eigenvalues was analyzed in detail. Kasperski et al. (2000) studied Marangoni flow up to the transition to unsteadiness (Hopf bifurcation). The governing equations were solved numerically by a pseudospectral scheme under the constraint of axisymmetric flow. Due to the pseudospectral nature of the numerical method, particular care was taken to make it well-posed from a mathematical point of view. Different flow-structure evolutions were found depending on the value of the Prandtl number ($Pr=0.01$, $Pr=100$).

These analyses were based on the simplifying assumption of permitting only axisymmetric disturbances. Therefore, due to the experimental evidence of the existence of 3D states, further investigations on the effect of nonaxisymmetric disturbances are needed.

Only recently have some three-dimensional (3D) numerical results become available.

Chen and Saghir (1994) and Saghir and Maffei (1999) investigated the three-dimensional convection in the molten zone of bismuth-germanate with temperature differences both in vertical and in azimuthal directions. These computations were carried out to simulate the nonuniform heating profiles that often arise due to the difficulty in fabricating ring heaters with perfect axisymmetric heat flux and structure, and/or due to small misalignments of the ampoule axis with the axis of the ring heater. The azimuthal flow and asymmetrical temperature distribution in the molten zone were examined for various asymmetrical heating conditions under microgravity and normal-gravity conditions.

Other researchers investigated the onset of 3D flow that can be induced by the instability of the basic axisymmetric Marangoni convection. Baumgartl et al. (1990) used a finite difference method in the case of a liquid zone having an aspect ratio $A_F = 1.2$ and Prandtl number $Pr = 0.02$. Levenstam et al. (1996) carried out experiments in the case of a floating-zone-like configuration ($Pr = 0.023$) and simulated the experiments by a finite-element method. Cröll et al. (1998a), during the Spacehab-4 mission (STS-77) performed two experiments on the crucible-free growth of GaSb ($Pr = 0.04$, $A_F = 0.8$–0.9). The experiments were performed in the so-called commercial float zone furnace (CFZF), an ellipsoid–paraboloid mirror furnace with a ring focus of 20 mm diameter. The 3D numerical simulations of the experiments were carried out using the finite-element method (FIDAP commercial software). Lan and Chian (2001) investigated the case of a floating zone having an aspect ratio $A_F = 1.0$ and $Pr = 0.01$. Wanschura et al. (1997a) carried out a linear stability analysis for the case $Pr = 0.02$ and $A_F = 1.0$.

In agreement with the results available in the case of half-zone configuration, the above investigators found that for liquid metals, the first bifurcation is stationary (i.e. the supercritical state is three-dimensional and steady) and that the regime becomes oscillatory only when the Marangoni number is further increased (second oscillatory bifurcation).

Since these authors were interested mainly in the numerical simulations of their experiments and thus in the critical Marangoni number (Ma_{c2}) for the onset of oscillatory flow (known to be the source of undesirable striations in the crystallized materials), the transition to steady nonaxisymmetric flow (Ma_{c1}) was not determined with any accuracy. Only Lan and Chian (2001) determined the first (steady bifurcation) and the second (oscillatory bifurcation) critical Marangoni number. The first critical Marangoni number was also determined by Wanschura et al. (1997a).

However, there seems to be a great lack of information dealing with the 3D structure and the fundamental properties of the Marangoni flow instability in floating zones heated around the equatorial plane. The available 3D results are fragmentary and incomplete. This is true in particular for the first (steady) bifurcation to the 3D flow.

The next sections extend the models, the numerical techniques and the insights into the physics illustrated for the half-zone in the first part of this chapter to the case of the full-zone configuration. Working engineering models and generalized correlations are developed, while providing a rigorous framework for a deeper understanding and effective treatment of relevant phenomena encountered in the FZ process. The difference between the half-zone and the full-zone and the related possibility to use one or the other configuration to capture the FZ fundamental hydrodynamics are elucidated. The attention is focused on microgravity fluid-dynamic aspects since high-quality material processing is expected to be enabled in the space environment. Moreover, since most of the available experimental data come from on the ground experimentation, some effort is spent to elucidate the behavior of the liquid column of full-zone extent also under normal-gravity conditions.

The material considered for the analysis is silicon ($Pr=0.01$) like the cases previously discussed for the half-zone.

4.14. The full-zone: modeling and definitions

4.14.1 *Basic assumptions*

The geometry of the problem is shown in Fig. 4.73. The zone of a full extent is confined between two coaxial cylindrical rods of diameter D placed $2L$ apart, heated at the midheight by an external ring heater. The upper and the lower solid/melt interfaces are assumed to be flat and they are kept at the melting point temperature T_m. The maximum temperature on the free surface is denoted by T_{max}. The temperature difference acting on the free surface is a result of the Marangoni flow itself and follows from the calculations.

The temperature difference achieved for the 2D solution used as the initial condition for the 3D computations is denoted by ΔT ($T_{max} - T_m$) and is used for the definition of the Marangoni number (note that the temperature is kept dimensional in the nondimensional form of the equations). The liquid is assumed to be homogeneous and Newtonian, with constant density and constant coefficients; viscous dissipation is negligible.

The boundary conditions on the rigid disks are the no-slip condition and the temperature conditions. The boundary conditions on the free surface ($r=c(z)$) are the kinematic conditions of a stream surface (zero normal velocity), the well-known Marangoni conditions (shear stress balance, where the Marangoni number Ma is defined as $Ma=Ma_r\,\Delta T$ with $Ma_r=\sigma_T L/\mu\alpha$) and the condition for the heat flux (see Eq. (4.11)).

4.14.2 *The ring heater*

Many models have been proposed in the literature to simulate surface heating ($-1<z<1$) due to a ring heater.

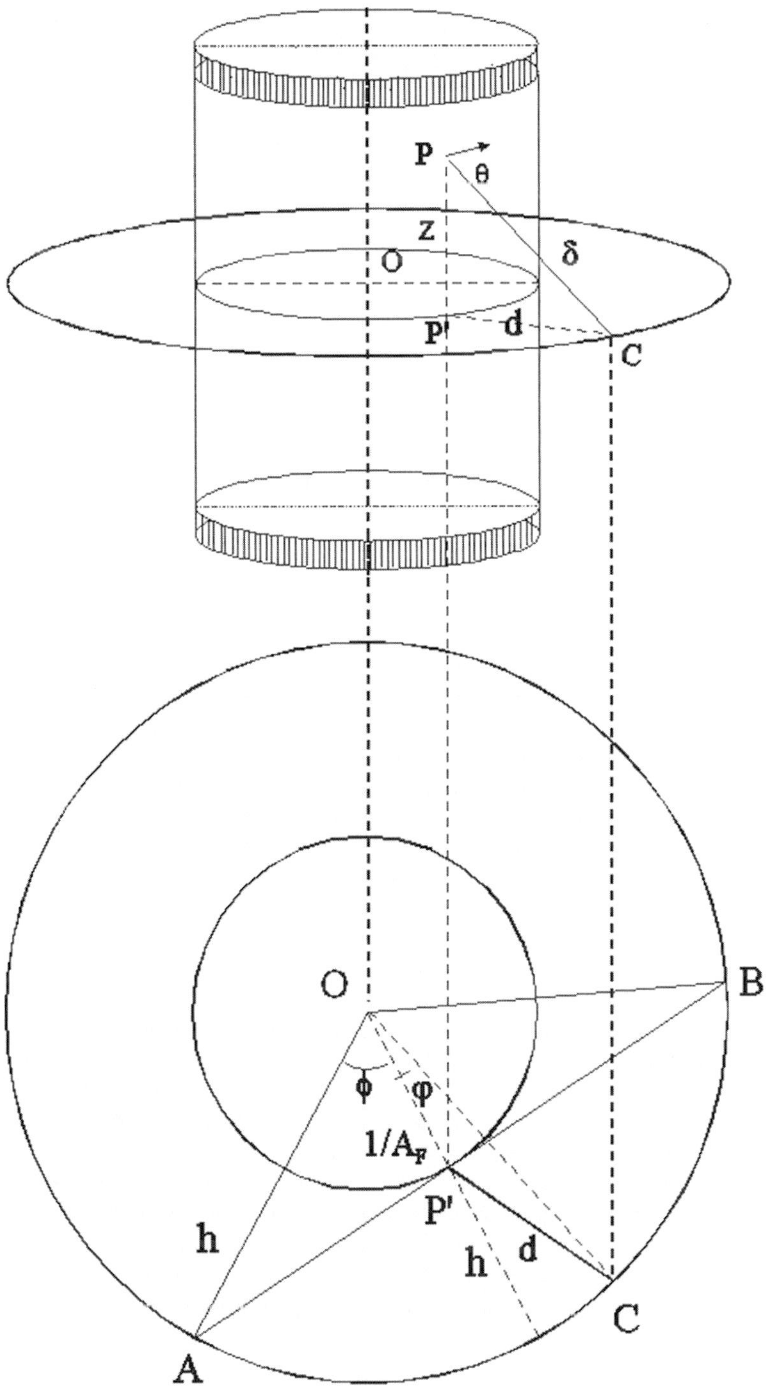

Fig. 4.73 The full-zone and the annular ring heater.

Rybicki and Floryan (1987) pointed out that if the liquid zone has a ring heater about its center, the surface-temperature distribution can be qualitatively described as $T(z) = T_m + 0.5(\cos \pi z + 1)(T_{max} - T_m)$. Wanschura et al. (1997a) used a fixed Gaussian temperature profile. Chen and Saghir (1994) and Saghir and Maffei (1999) assumed this profile to be parabolic.

Baumgartl et al. (1990) pointed out that a fixed temperature distribution in the free surface given as the boundary condition would lead to unrealistic results of the numerical modeling. For this reason they did not use a fixed temperature distribution, but assigned a fixed heat-flux profile across the free surface. In particular, a heat flux with a parabolic flux profile $J(z) \propto ((T_m - T_{max})z^2 + T_{max})$ was used.

In general, the methods introduced by the different investigators to model the heating source (ring heater, circular arrangement of halogen lamps, ellipsoid–paraboloid mirror furnace, laser with axisymmetric ring shape, etc.) fall into two categories: specifying the ambient temperature distribution $T_a(z)$ or finding the power requirement for producing a desired ΔT on the free surface. For the first case the ambient temperature distribution is assumed to be known.

Zhang and Alexander (1992) provided surface heating through a radiative heat transfer at the free surface. The authors took the ambient temperature to be parabolic: $J(z) \propto Bi(T - T_a(z))$ with $T_a(z) = (T_m - T_{max})z^2 + T_{max}$ (Bi is the Biot number). Chen and Roux (1992) assumed $T_a(z) = T_m + \Delta T \exp(-cz^2)$ along the free surface ($\Delta T = T_{max} - T_m$, T_{max} refers to the maximum temperature of the heater, c is a constant value) and the heat flux at the free surface was computed as $J(z) \propto Bi(T - T_a(z))$ with $Bi = (4\varepsilon\sigma T_m^3 + \eta)L/\lambda$ taking into account the radiative and convective heat transfer at the free melt/gas interface (λ is the thermal conductivity, ε the emissivity, σ the Stefan-Boltzmann constant and η the convective heat-transfer coefficient).

According to Chen and Chu (1995), Chen and Wu (1996), however, in real systems the ambient temperature distribution is unknown a priori, whereas the power supplied by the external source is a known parameter. This led some investigators to model the surface heating, assuming a certain law of variation (i.e. parabolic, exponential) and fixing the total input requirement (ΔQ).

Chen and Chu (1995), Chen and Wu (1996), Chen and Chin (1995) and Levenstam et al. (1996) assumed a Gaussian heater profile to model the power input from the heating system, i.e. $J(z) \propto \Delta Q \exp(-cz^2)$. Croll et al. (1998) modeled the radiation heating of the zone by assuming a parabolic heat flux at the free surface, $J(z) \propto \Delta Q(1 - z^2)$ similar to that assumed by Baumgartl et al. (1990) and Zhang and Alexander (1992). In the investigations of Chenier et al. (1999) and Kasperski et al. (2000) the polynomial form $J(z) \propto \Delta Q(1 - z^2)^2$ was introduced to model the heat-flux distribution.

Following the ideas and concepts pointed out in Lappa (2003a, 2004a, b), in the present book a new distribution is considered for the heat flux, corresponding to the radiative flux generated by a ring heater having negligible thickness and positioned around the equatorial plane of the full-zone at a distance $h + R/L$ from the axis of the liquid zone.

Each point C disposed on the ring heater between the points A and B, see Fig. 4.73, gives a radiative contribution to the generic point P of the free surface; this contribution in nondimensional form reads:

$$j(z) = \left(\frac{1}{L\lambda\Delta T}\right)\frac{1}{4\pi}\frac{Q}{\delta^2}\cos(\vartheta) \tag{4.25}$$

where Q is the power supplied to the ring heater, λ the thermal conductivity of the liquid, δ the distance between the points C (on the ring heater) and P (on the free surface) and θ is the angle between $\underline{n}$ (unit vector orthogonal to the free surface in P) and the direction PC.

Thus the total flux in P can be computed as

$$J(z) = \frac{1}{4\pi}\left(\frac{Q}{L\lambda\Delta T}\right)\int_{-\tilde{\phi}}^{\tilde{\phi}}\frac{\cos(\vartheta)}{\delta^2}\,\mathrm{d}\varphi \tag{4.26}$$

where:

$$\delta^2 = \frac{1}{A_\mathrm{F}^2}\left[(1 + hA_\mathrm{F})^2 + A_\mathrm{F}^2 c^2(z) - 2A_\mathrm{F}c(z)(1 + hA_\mathrm{F})\cos\varphi + z^2\right] \tag{4.27a}$$

$$\cos(\vartheta) = \frac{1}{\sqrt{(1 + c'^2)}}\frac{[(1 + hA_\mathrm{F})\cos\varphi - A_\mathrm{F}c(z)] + c'z/\delta}{\sqrt{\left[(1 + hA_\mathrm{F})^2 + A_\mathrm{F}^2 c^2(z) - 2A_\mathrm{F}c(z)(1 + hA_\mathrm{F})\cos\varphi + z^2\right]}} \tag{4.27b}$$

$$\tilde{\phi} = a\cos\frac{1}{(1 + hA_\mathrm{F})} \tag{4.27c}$$

where $h = 0.5$ and ε is the emissivity of the free surface.

Figure 4.74 shows the surface heat-flux profile computed according to Eq. (4.26) and other analytical laws plotted for the same value of the maximum heat flux at the equatorial plane in the case of a cylindrical interface.

The heat transport is, therefore, maximum in the surface region and tries to change the driving temperature gradients in the surface by itself (i.e. via the Marangoni stress acting on the free interface).

Figure 4.75, in particular, show the heat-flux profile and the corresponding surface-temperature distribution for different values of the volume parameter and for a fixed input power supplied to the heating device. The maximum of the curves is an increasing function of the parameter S (for a fixed radius of the ring, in fact, the distance between the heater and the free surface is reduced if the volume of the liquid is increased).

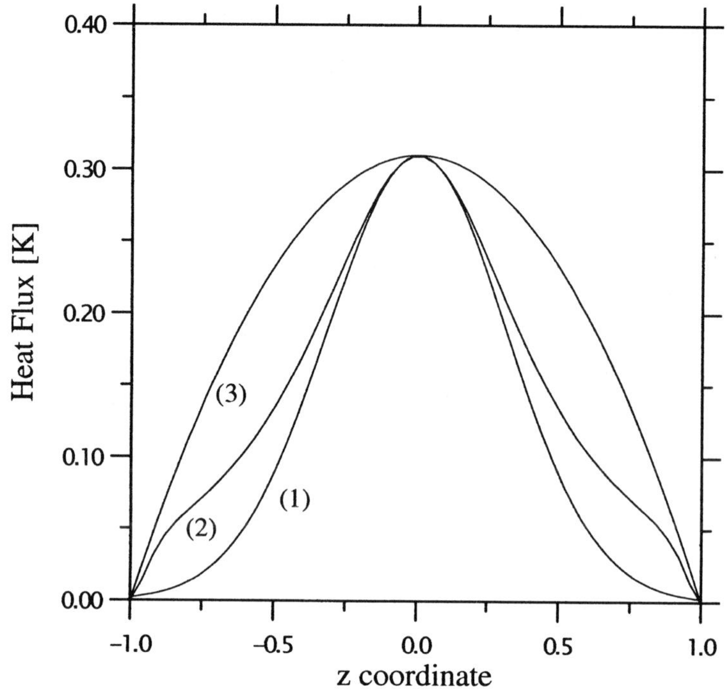

Fig. 4.74 Heat-flux profile $J(z)$: (1) $J \propto \exp(-cz^2)$, (2) flux generated by the ring heater according to Eq. (4.26) for $A_F = 1.0$, cylindrical interface and $\Phi = \varepsilon Q = 1$ w, (3) $J \propto (1 - z^2)$.

4.14.3 *Definition of the critical Marangoni number*

In the analyses of Chen and Roux (1992), Chenier et al. (1999) and Kasperski et al. (2000), the stability boundaries were computed directly as a function of the input power supplied to the heating source (i.e. their Marangoni number is based directly on ΔQ). However, it is worthwhile to point out how the Marangoni number can be based directly on the temperature difference ($T_{max} - T_m$) achieved for the 2D steady solution (basic state) used as the initial condition for the 3D computations. This definition is closer to the experimental criteria and allows one to compare the results with the critical Marangoni numbers obtained previously in the case of half-zone computations.

It has been emphasized in the previous sections (see, in particular, Section 4.3), however, that in the case of a full-zone model the temperature gradient acting on the free surface is not known a priori. Its value, in fact, is affected by the strength of the Marangoni convection. Due to Marangoni convection, the surface-temperature profile departs from the corresponding diffusive one (i.e. obtained in the case $Ma = 0$). This is clearly shown in Fig. 4.76. The ΔT acting on the free surface is

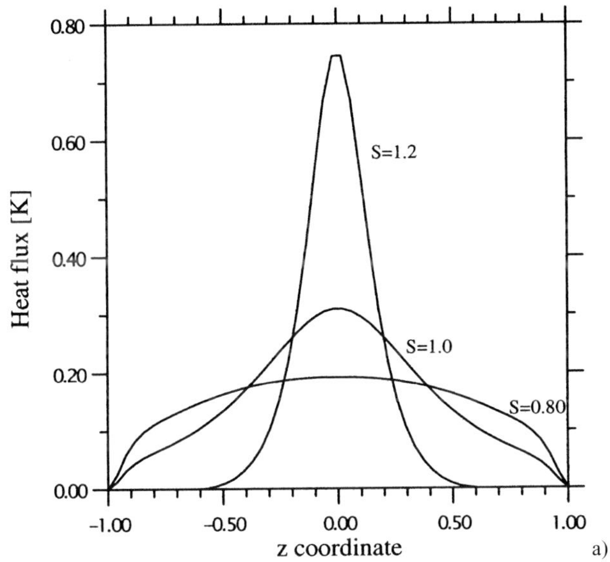

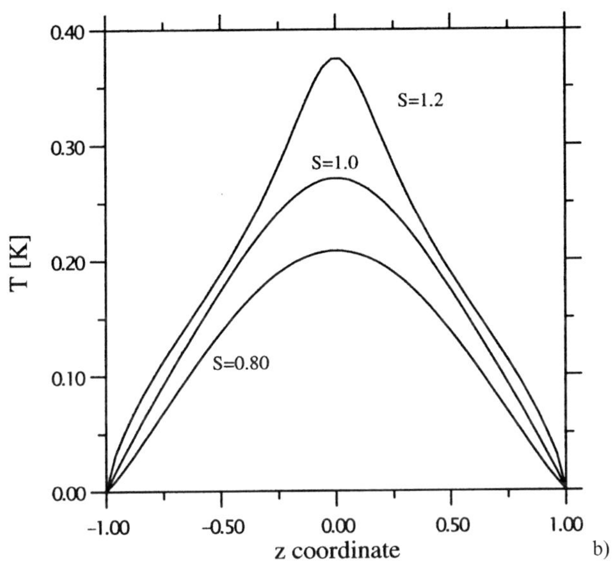

Fig. 4.75 (a) Heat-flux profile versus S ($A_F = 1.0$) for $\Phi = \varepsilon Q = 1$ w, (b) surface-temperature distribution versus S ($A_F = 1.0$) for $\Phi = \varepsilon Q = 1$ w.

smaller than that obtained for $Ma = 0$. For this reason, the ΔT-based Marangoni number cannot be fixed a priori, but has to be computed as a function of the power supplied to the heating device.

As explained in Section 4.5, the critical Marangoni number Ma_c in turn is defined as a threshold value at which the growth rate of the disturbances is zero. Therefore,

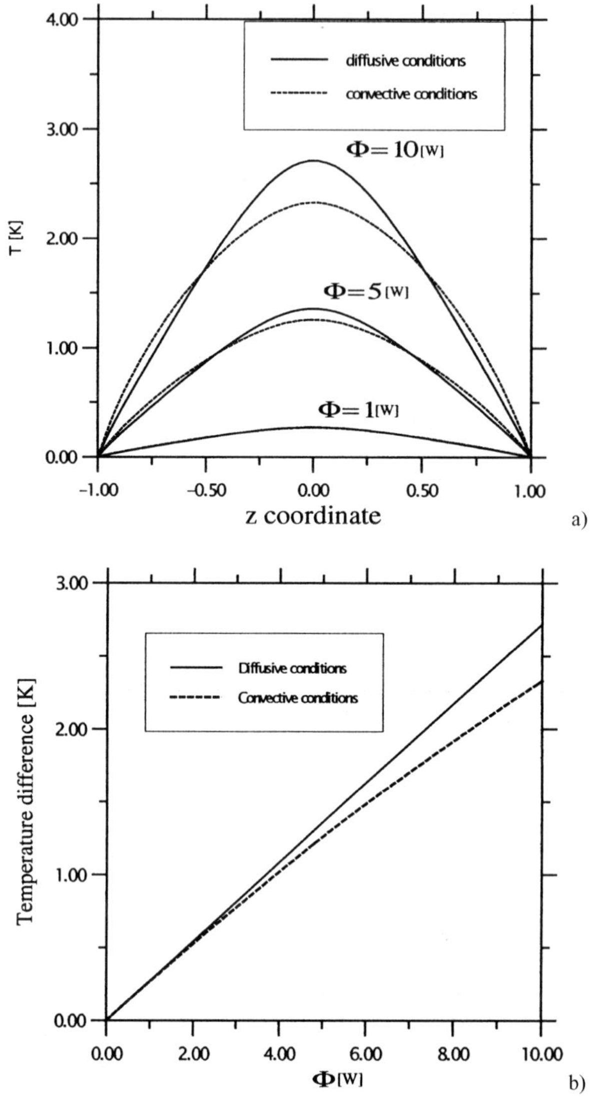

Fig. 4.76 (a) Temperature distribution on the free surface ($Pr = 0.01$, cylindrical interface, $A_F = 1.0$, $\Phi = \varepsilon Q$), (b) temperature difference versus $\Phi = \varepsilon Q$.

in order to determine Ma_c, the power supplied to the system must be changed several times in a stepwise manner. The results of the different simulations give the growth-rate constant as a function of the Marangoni number and according to linear stability criteria (see, e.g., Chen and Roux, 1992), the critical Marangoni number can be finally determined as the Marangoni number at which the growth rate becomes zero.

4.15. Cylindrical interface and microgravity conditions

Under zero-g conditions the cylindrical configuration is only a special case since often the melt/gas interface tends to be concave or convex. This configuration, however, similarly to the case of the half-zone, can be regarded as an archetype model for the study of the stability of the Marangoni flow and a first step towards the full comprehension of the different effects influencing the phenomena under investigation.

4.15.1 *Effect of the aspect ratio and description of the spatial organization*

Table 4.4 and Fig. 4.77 show that the azimuthal wave number of the critical mode (m) is a function of the aspect ratio of the full-zone. In particular, the critical number is $m=1$ for $1.3 \leq A_F \leq 2.0$, whereas higher values appear for lower aspect ratios: $m=2$ for $0.7 \leq A_F \leq 1.2$, $m=3$ for $A_F=0.6$, $m=4$ for $0.4 \leq A_F \leq 0.5$, $m=6$ for $A_F=0.3$ and $m=9$ for $A_F=0.2$ (very short liquid column). These results suggest an empirical correlation between the geometrical aspect ratio and the critical azimuthal wave number of the instability similar to that for the case of the half-zone configuration (see Section 4.8 and Eq. (4.17)). This correlation is

$$mA_F \cong 2 \tag{4.28}$$

The flow structure is related to the value of m and hence depends on the value of the aspect ratio. As m increases, the flow organization becomes more complex. When A_F is decreased, flows with different patterns of symmetry and increasing degree of

Table 4.4 Critical azimuthal wave number and critical Marangoni number versus the aspect ratio ($Pr=0.01$, full-zone, $S=1$, microgravity conditions, Lappa 2003a).

A_F	L (cm)	m	Ma_{c1}
0.2	0.1	9	22.157
0.3	0.15	6	15.340
0.4	0.2	4	12.280
0.5	0.25	4	9.760
0.6	0.3	3	13.597
0.7	0.35	2	10.90
0.8	0.4	2	12.028
0.9	0.45	2	11.326
1.0	0.5	2	12.413
1.1	0.55	2	14.910
1.2	0.6	2	15.086
1.3	0.65	1	11.350
1.5	0.75	1	10.154
1.8	0.9	1	10.364
2.0	1.0	1	9.923

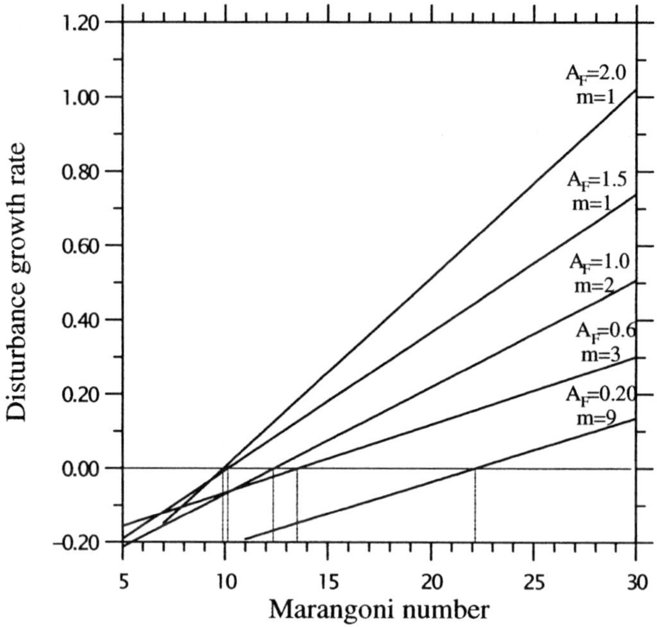

Fig. 4.77 Growth rates as a function of the Marangoni number and determination of the first critical Marangoni number for different values of the aspect ratio (cylindrical interface, microgravity conditions).

complexity arise. The breakdown in the symmetry of solutions is shown for growing values of m in Figs. 4.78–4.81.

For instance, frames (a) in Figs. 4.78–4.80 show for $m=1$, $m=2$ and $m=3$, respectively, the axial velocity in cross-sections orthogonal to the liquid column axis. Frames (b) show the azimuthal velocity in the same sections. Like the half-zone case, in the azimuthal direction there are $2m$ convective cells and the isocontour lines of the axial velocity in the generic section z describe, in this plane, well-defined curves whose shape is related to the value of the azimuthal wave number (Fig. 4.82). For $m=1$ a region whose position is eccentric to the geometrical axis of symmetry of the liquid zone appears; for $m=2$ the inner region is approximately elliptic and is not eccentric; for $m=3$ the inner region has the form of a triangle, whereas for $m=4$ it is a quadrangle.

Table 4.4 shows that, while the critical wave number changes with the aspect ratio, the critical Marangoni number does not change much and tends to increase with decreasing aspect ratio (see also Fig. 4.77 where the growth rates as a function of the Marangoni number and the determination of the first critical Marangoni number are shown for different values of the aspect ratio).

Like the case of the half-zone, also for the full-zone, a more detailed analysis requires different descriptions for even and odd modes of convection.

If m is odd in the generic meridian plane, the mirror symmetry with respect to the axis ($r = 0$) of the liquid zone is broken; on the contrary, if m is even the flow in each meridian plane retains this type of symmetry (the structure of the flow, however, changes according to the angular position, i.e. according to φ). These behaviors

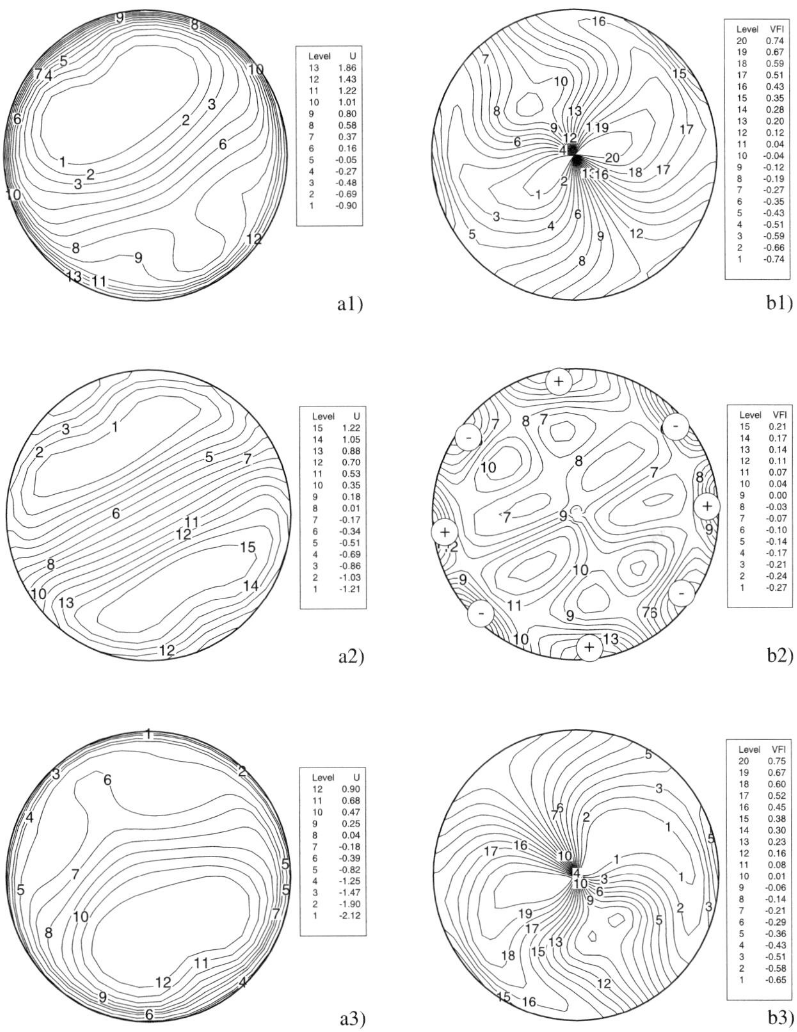

Fig. 4.78 Structure of 3D Marangoni flow at the 3D stationary state with $m = 1$, $A_F = 1.3$, $Ma \cong 30$, zero-g, $\Phi = 2.54$ w: (a1), (a2), (a3) axial velocity in the cross sections $z = 0.5$, $z = 0$, $z = -0.5$; (b1), (b2), (b3) azimuthal velocity in the cross sections $z = 0.5$, $z = 0$, $z = -0.5$; (c1) Surface azimuthal velocity distribution (numerical results); (c2) plot of the analytical expansion of the surface azimuthal velocity: $[\tilde{F}_1(z)\sin(m\varphi + G_o) - \tilde{F}_2(z)\sin(Nm\varphi + NG_o)]$, $\tilde{F}_1(z) = z(1 - z^{20})^2$, $\tilde{F}_2(z) = 0.3(1 - z^2)^2$, $m = 1$, $N = 4$, $G_o = 1$; Velocity field in the meridian plane at (d1) $\varphi \cong 3\pi/4$, (d2) $\varphi \cong 5\pi/4$.

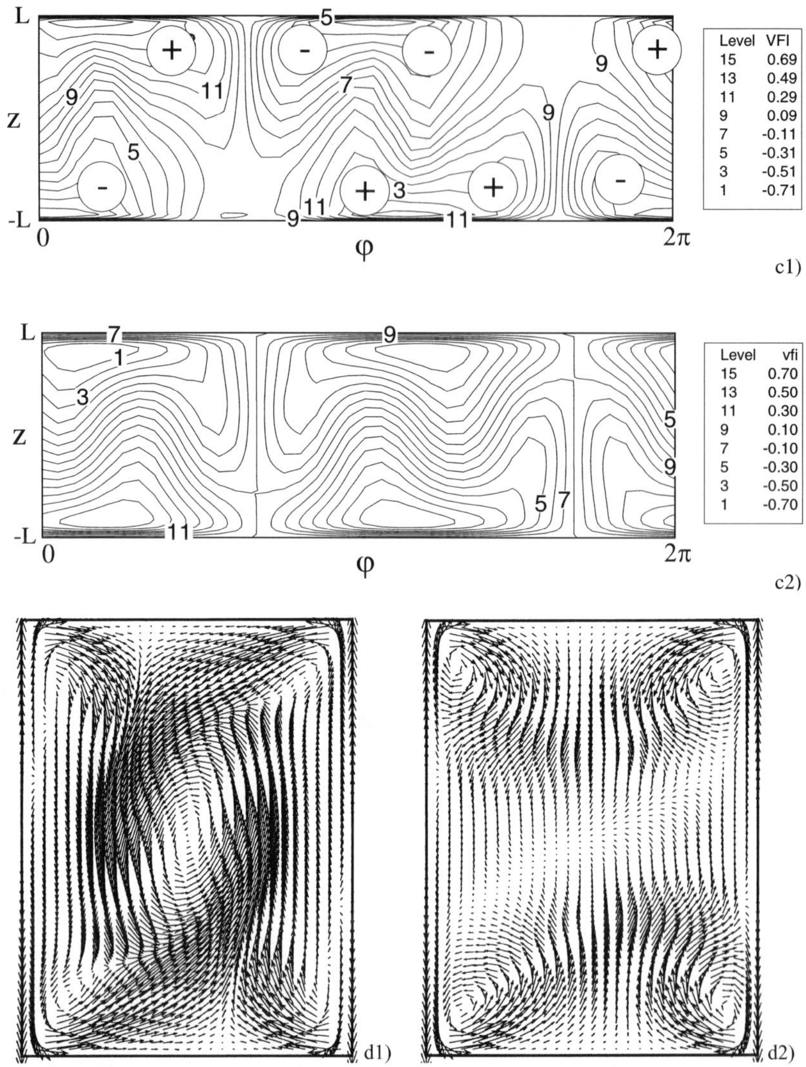

Fig. 4.78 Continued.

resemble those already depicted in the case of the half-zone (Section 4.7). However, for the full-zone also the flow symmetry with respect to the midplane is an additional and relevant part of the problem. It indeed introduces a new degree of freedom for the flow structure.

For instance, in the case $A_F = 1.3$ and $m = 1$ the flow is no longer separated in cells being positioned in the lower and the upper part of the meridian plane. The left vortex cell in the lower part and the right vortex cell in the upper part merge giving rise to a vortex extended along the whole axial plane of the full-zone; the other cells contract up to be confined in the left and right corners of the upper and lower half, respectively (Fig. 4.78-d1). The behavior is similar for $A_F = 0.6$ ($m = 3$); in this case (Fig. 4.80, frames d1–d2 and e1–e2) however, since the liquid zone is short and since the Marangoni convection, being a surface phenomenon, is confined to

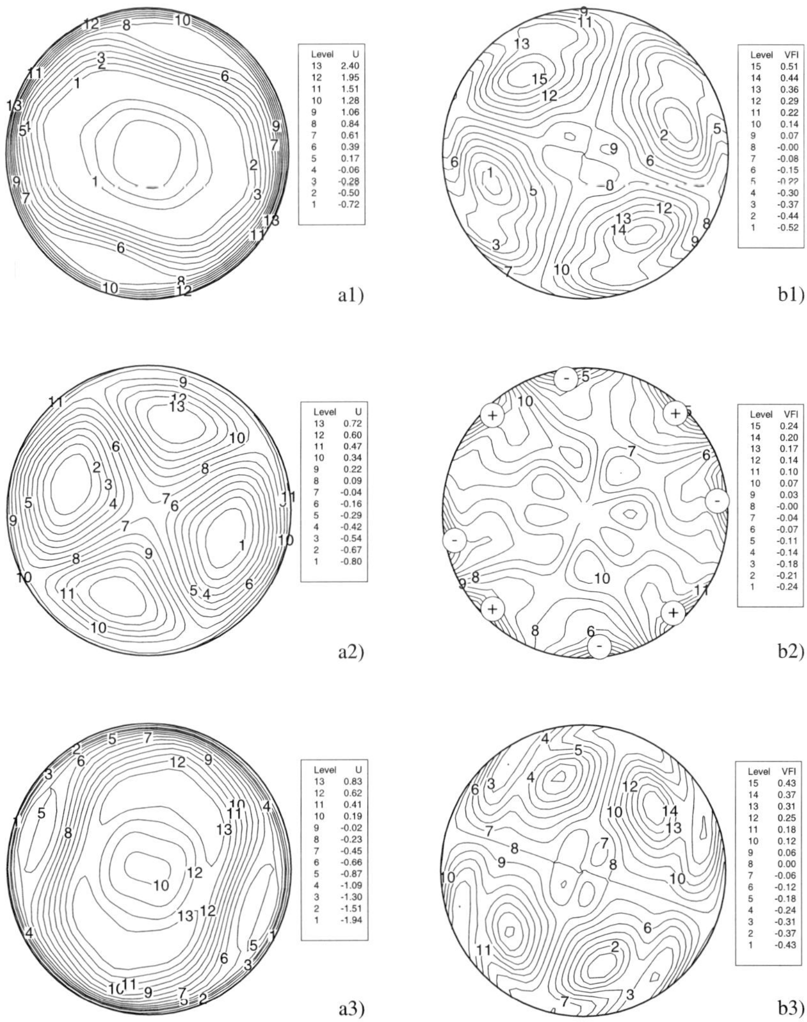

Fig. 4.79 Structure of 3D Marangoni flow at the 3D stationary state with $m = 2$, $A_F = 0.8$, $Ma \cong 30$, zero-g, $\Phi = 4.57$ w: (a1), (a2), (a3) axial velocity in the cross sections $z = 0.5$, $z = 0$, $z = -0.5$; (b1), (b2), (b3) azimuthal velocity in the cross sections $z = 0.5$, $z = 0$, $z = -0.5$c, (c1) Surface azimuthal velocity distribution (numerical results); (c2) plot of the analytical expansion of the surface azimuthal velocity: $[\tilde{F}_1(z)\sin(m\varphi + G_o) - \tilde{F}_2(z)\times\ \sin(Nm\varphi + NG_o)]$, $\tilde{F}_1(z) = z(1 - z^{20})^2$, $\tilde{F}_2(z) = (1 - z^2)^2$, $m = 2$, $N = 2$, $G_o = 1$; Velocity field in the meridian plane at (d1) $\varphi \cong \pi/2$, (d2) $\varphi \cong \pi$; (e) Structure of the toroidal convection rolls for $m = 2$.

regions near the free surface, a coalescence of the opposite convection cells (leading to a single vortex pervasive throughout the meridian plane) does not occur. For $A_F = 0.8$ ($m = 2$) and $A_F = 0.3$ ($m = 6$) the flow is symmetric with respect to the axis (in this case, in fact, m is even) but again asymmetric with respect to the equatorial plane. Figures 4.79-d1 and 4.79-d2, ($m = 2$) show that the vortex cells in the lower (or upper) half puff up protruding in the upper (lower) half of the floating zone.

In the light of these results, for both even and odd critical wave numbers, the mirror symmetry with respect to the midplane is broken.

4.15.2 *Comparison with the half-zone*

The finding that for both odd and even modes, the system loses the flow mirror symmetry with respect to the equatorial plane is of paramount importance if one

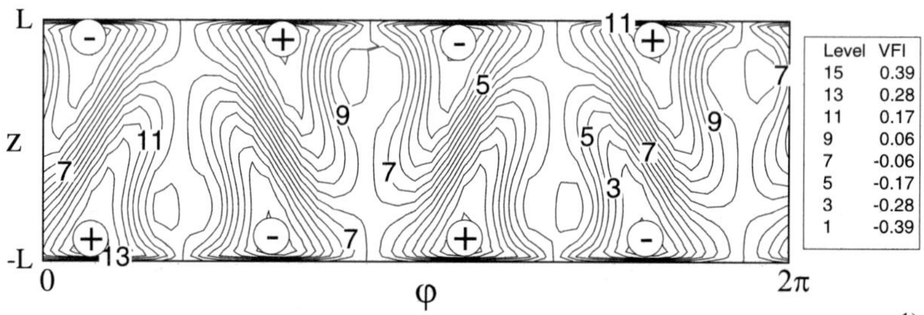

c1)

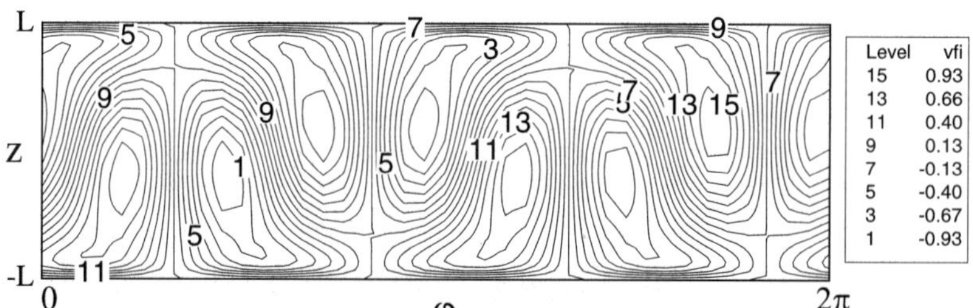

c2)

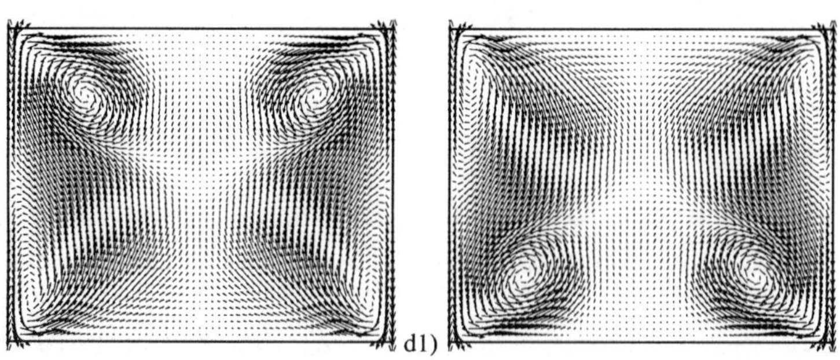

d1) d2)

Fig. 4.79 Continued.

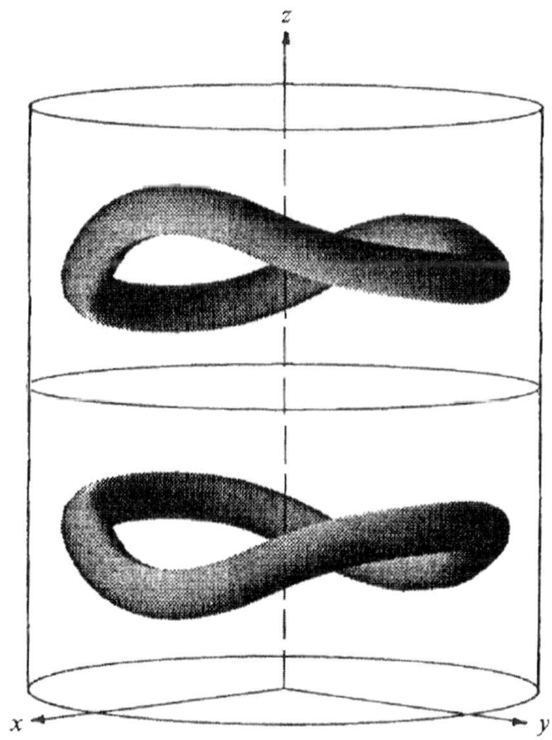

Fig. 4.79 Continued.

considers that, in the light of this result, the half-zone should not be used to obtain quantitative data about the FZ technique.

For instance, this is why the Ma_{c1} values computed for the half-zone model by several investigators (Rupp et al., 1989; Levenstam and Amberg, 1995; Wanschura et al., 1995; Chen et al., 1999; Imaishi et al., 1999, 2000; Leypoldt et al., 2000; Yasushiro et al., 2000) are twice the values calculated for the case of full-zone model (see Fig. 4.83).

The major difference between the two models comes from the instability threshold and the reason for this difference is that the independence and the separation of the flows in the upper and the lower half of a full-zone, which is stated by using a half-zone model, is not true. For this configuration, in fact, the convection roll is bounded by a wall from one side and it is free to interact with the opposite convection roll from the other side as shown in the previous sections.

The Ma_c value for a half-zone is likely to be higher than that for a full-zone because of the additional no-slip condition at the heated disk and the reduced degrees of freedom compared to the full-zone configuration.

Note that, contrary to the above findings, Wanschura et al. (1997a) found that the scenario of development of the instability for small Prandtl numbers in full-zone is similar to that for the half-zone model. This apparent inconsistency can be explained according to the different boundary conditions introduced in that analysis. Wanschura et al. (1997a) used a fixed temperature profile, whereas in the present book a more realistic condition has been adopted: a fixed surface heat-flux (the surface-temperature profile is not fixed a priori, but determined as part of the computations and as a function of the power supplied to the heating device).

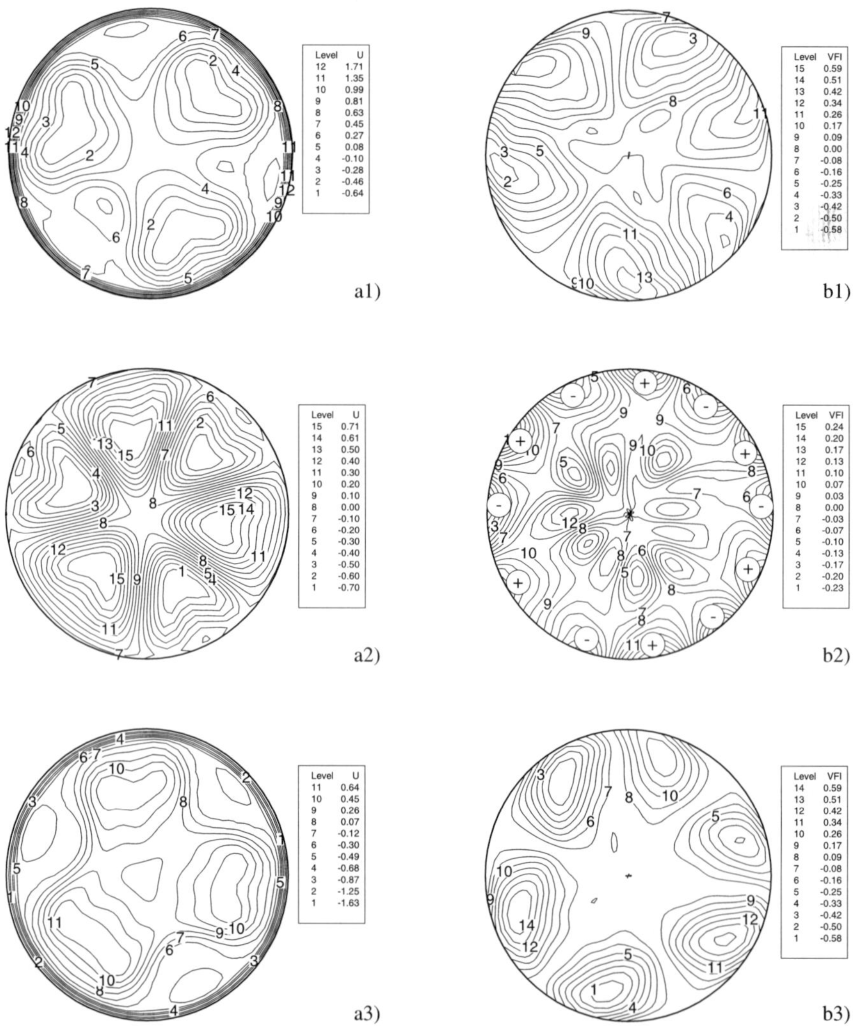

Fig. 4.80 Structure of 3D Marangoni flow at the 3D stationary state with $m=3$, $A_F=0.6$, $Ma \cong 30$, zero-g, $\Phi=6.27$ [w]: (a1), (a2), (a3) axial velocity in the cross sections $z=0.5$, $z=0$, $z=-0.5$; (b1), (b2), (b3) azimuthal velocity in the cross sections $z=0.5$, $z=0$, $z=-0.5$; (c), surface azimuthal velocity distribution; velocity field in the meridian plane at (d1) $\varphi \cong \pi/4$, (d2) $\varphi \cong 7\pi/12$; (e1) axial-velocity distribution, (e2) radial-velocity distribution.

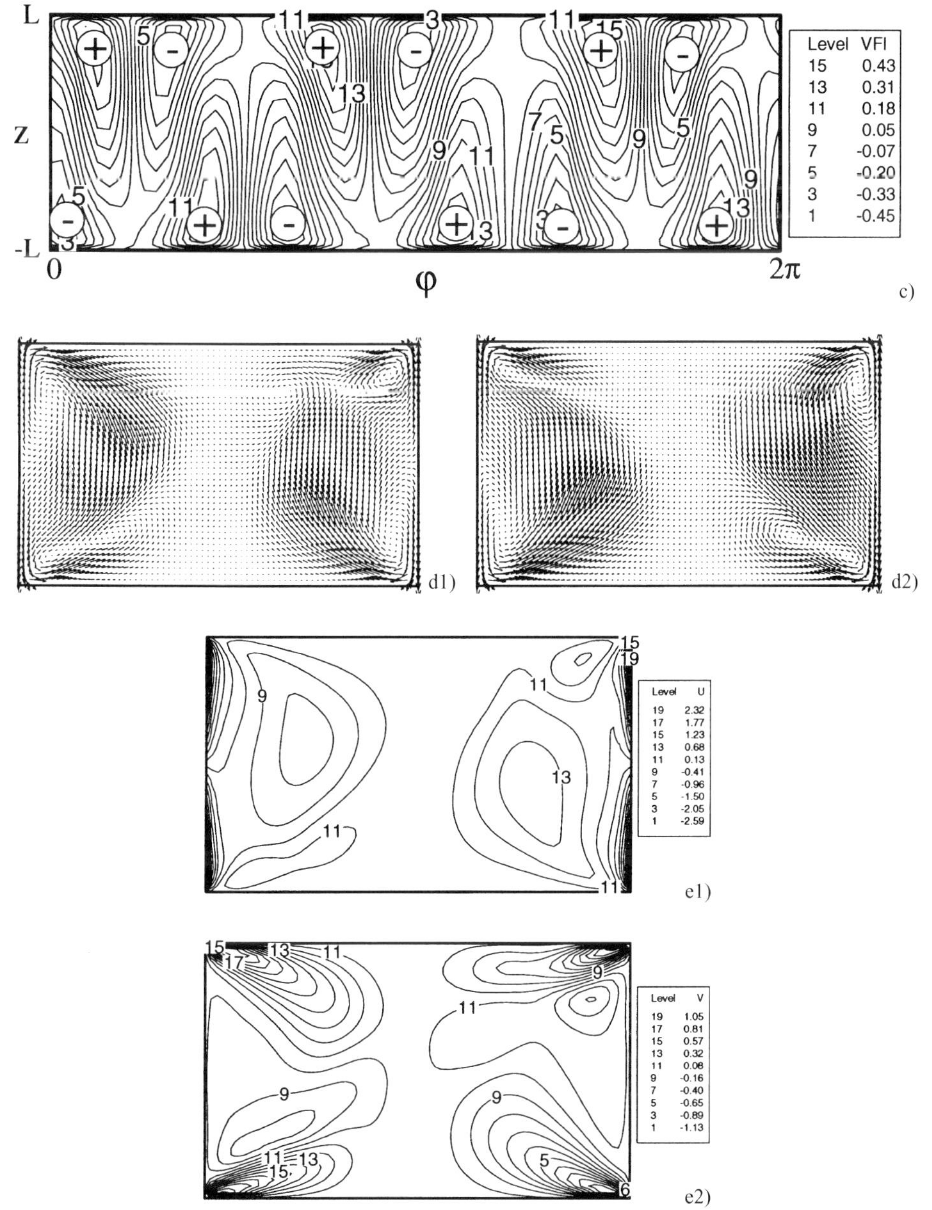

Fig. 4.80 Continued.

Generally the surface-temperature profile in the 3D state is neither axisymmetric nor mirror symmetric with respect to the equatorial plane. If the surface-temperature distribution is fixed, the full liquid zones behave (approximately) as half-zones since interaction between the upper and lower parts is somehow prevented. Fixing the temperature profile leads, in fact, to a surface temperature that is mirror symmetric with respect to the equatorial plane regardless of the flow conditions. Since the surface Marangoni flow is forced to follow the steepness of the imposed temperature profile, this may artificially force the system to maintain symmetrical behavior preventing transition to the three-dimensional state and mutual interaction of the toroidal convection rolls. This, of course, leads to larger values of the critical

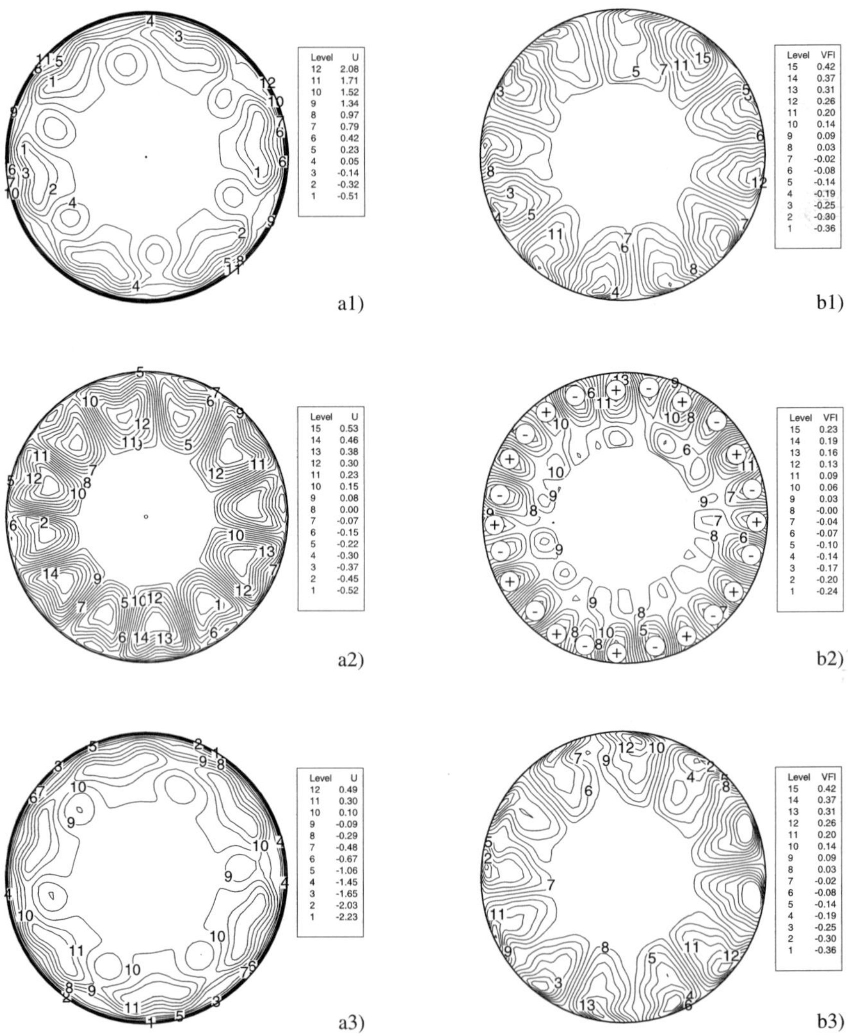

Fig. 4.81 Structure of 3D Marangoni flow at the 3D stationary state with $m = 6$, $A_F = 0.3$, $Ma \cong 30$, zero-g, $\Phi = 12.3$ [w]: (a1), (a2), (a3) axial velocity in the cross sections $z = 0.5$, $z = 0$, $z = -0.5$; (b1), (b2), (b3) azimuthal velocity in the cross sections $z = 0.5$, $z = 0$, $z = -0.5$; (c) Surface azimuthal-velocity distribution (d) velocity field in the generic meridian plane.

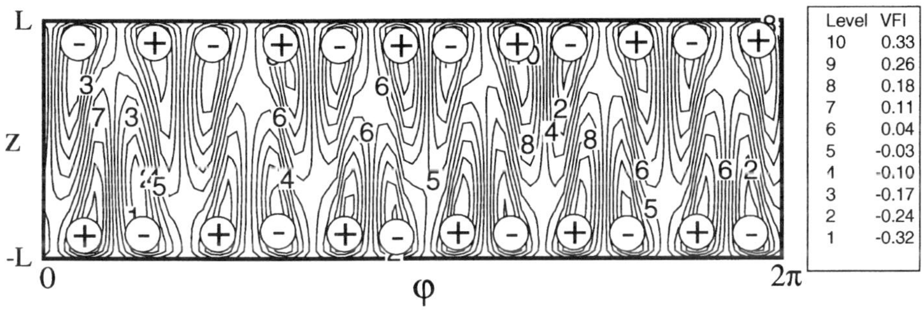

c)

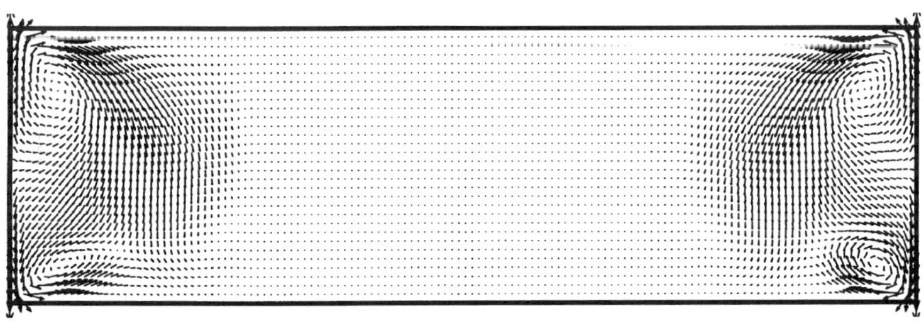

d)

Fig. 4.81 Continued.

Marangoni number (due to the reduced degrees of freedom). If the surface temperature is constraint-free the behavior must be different.

According to Fig. 4.84 for a fixed value of the Prandtl number ($Pr = 0.01$) the azimuthal structure of the flow in the zone of full extension and in the corresponding half-zone does not differ much (with the exception of $A_H = 0.25$, $A_H = 0.35$, and $0.65 \leq A_H \leq 0.85$). The physical arguments elucidated in Section 4.10 (the generalized theory for the azimuthal wave number), in fact, should hold in the case of the cylindrical full-zone model.

Additional insights into the reasons leading to the different behaviors emphasized herein in terms of critical Marangoni number and in terms of different structure of the azimuthal flow within certain ranges of the aspect ratio, are given in the next section, where precise mathematical arguments are provided to support the discussion.

4.15.3 *The symmetry of the disturbances*

By analogy with the half-zone (Section 4.7) after the bifurcation, the flow field can be interpreted as the superposition of steady sinusoidal azimuthal disturbances onto an axisymmetric reference state. For each of the two toroidal convection

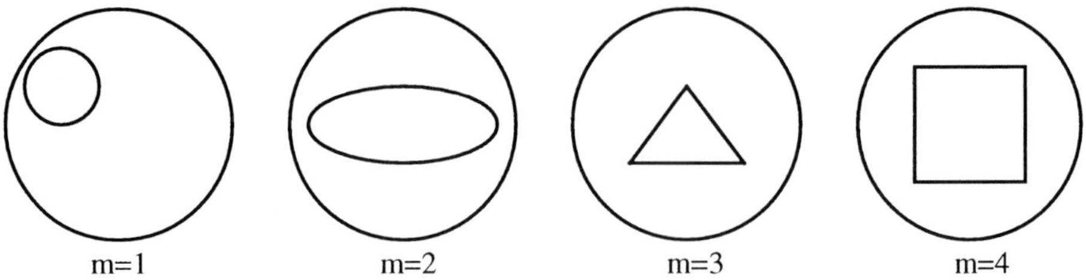

Fig. 4.82 Typical polygonal modal structures of axial velocity in the cross section.

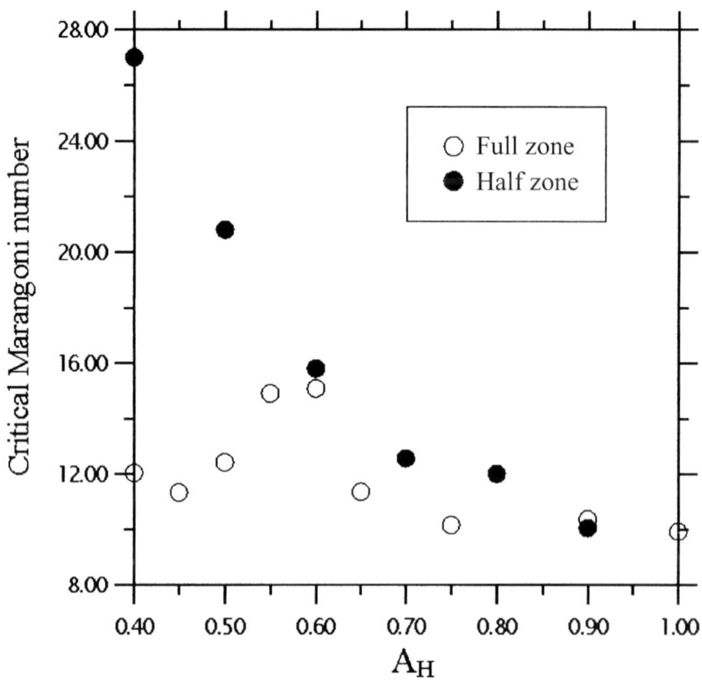

Fig. 4.83 Comparison between the critical Marangoni number of the full-zone configuration (present results) and available results (Imaishi et al., 1999, 2000; Chen et al., 1999) dealing with the half-zone configuration ($Pr = 0.01$).

rolls in the upper and lower part of the floating zone, this superposition can be expressed as:

$$F_i(r, z, \varphi) = F_o(r, z) + f_i(r, z, \varphi) = F_o(r, z) + \tilde{f}_i(r, z) \sin(m\varphi + G_i) \tag{4.29}$$

$i =$ up ($0 < z < 1$), low ($-1 < z < 0$), where $F_i(r, z, \varphi)$ is the generic flow field variable, the subscript (o) refers to the axisymmetric state, m is the azimuthal wave number that represents the number of spatial periods in the azimuthal direction,

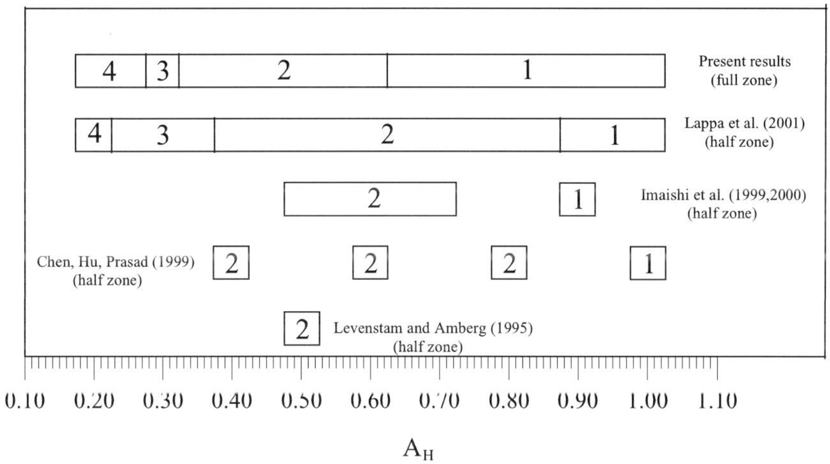

Fig. 4.84 Comparison between the azimuthal wave number of the full-zone configuration (discussed in the present book) and available results dealing with the half-zone configuration ($Pr = 0.01$).

G_i is a constant phase shift related to the azimuthal position of the disturbance extrema for the considered convection roll ($i =$ low, convection roll in the lower part of the full-zone; $i =$ up, convection roll in the upper part of the full-zone). The numerical results provided by direct solution of the 3D Navier–Stokes equations and plotted in Figs. 4.78–4.81 show that $\tilde{f}_{up}(r,z) = \tilde{f}_{low}(r,-z)$ for all the cases investigated under the constraint of a cylindrical interface ($S = 1$, zero-g conditions).

For the basic state, the radial velocity and the temperature are symmetric with respect to the midplane of the full-zone, whereas the axial velocity is antisymmetric. For the disturbance modes, the same symmetry or the opposite case (all fields are antisymmetric except for the symmetric axial velocity) is possible in theory (Wanschura et al., 1997a). For the first case $G_{up} = G_{low}$, for the second, $G_{up} = G_{low} + \pi/m$. Numerical simulation shows that the azimuthal disturbances always exhibit antisymmetric behavior with respect to the equatorial plane, i.e. the latter case occurs (see, e.g., Fig. 4.79-c1).

According to these behaviors, Eq. (4.29) can be rewritten as

$$F(r,z,\varphi) = F_o(r,z) + f(r,z,\varphi) = F_o(r,z) + \tilde{F}_1(r,z)\sin(m\varphi + G_o) \tag{4.30}$$

where G_o is a constant phase shift related to the azimuthal position of the disturbance with respect to $\varphi = 0$ and $\tilde{F}_1(r,z)$ satisfies the conditions $\tilde{F}_1(r,z) -$ $-\tilde{F}_1(r,-z)$ and $\tilde{F}_1\,(r, z = \pm 1) = 0$. The behavior described above in terms of Eq. (4.30) is, however, only a simplified model, in fact, a strong nonlinear interaction occurs between the toroidal convection rolls located in the upper part and lower part, as discussed in the foregoing sections. In particular, Figs. 4.78-c1, 4.79-c1, 4.80-c and 4.81-c show that $f(r,z,\varphi)\neq 0$ for $z\to 0$. Since $\tilde{F}_1(r,z)\to 0$ for $z\to 0$, a second-order function has to be present in the series expansion of $f(r,z,\varphi)$. The surface azimuthal-velocity distribution, in fact, can be expressed approximately as:

$$V\varphi(z,\varphi)=f(z,\varphi)\propto[\tilde{F}_1(z)\sin(m\varphi+G_o)-\tilde{F}_2(z)\sin(Nm\varphi+NG_o)] \quad (4.31)$$

where N is a new parameter, the interaction factor, and $\tilde{F}_2(z)$ is a second-order contribution due to the interaction between the disturbances acting in the lower and upper halves: $\tilde{F}_2(z)$ satisfies the conditions $\tilde{F}_2(z)=\tilde{F}_2(-z)$ and $\tilde{F}_2(z)\to 0$ if $|z|>\xi$ with $0<\xi<1$, where ξ is the distance from the midplane at which the interaction between the upper and lower surface azimuthal flows can be considered negligible.

The numerical results shown in Fig. 4.78-c1 for the case $A_F=1.3$ and $m=1$ are well fitted by Eq. (4.31) with $\tilde{F}_1(z)=z(1-z^{2n1})^2$ with $n_1=10$, $\tilde{F}_2(z)=A_2(1-z^{2n2})^2$ with $n_2=1$, $N=4$ and $A_2=0.3$ (Fig. 4.78-c2 shows the plot of the analytical expansion (4.31) with $\tilde{F}_1(z)$ and $\tilde{F}_2(z)$ given above, to be compared with Fig. 4.78-c1). The value $N=4$ leads to an "apparent" quadrupling of the azimuthal wave number of the azimuthal velocity distribution in the equatorial plane (i.e. the number of surface maxima of the azimuthal velocity in the midplane is quadrupled with respect to cross-sections far from this plane, see, e.g., Fig. 4.78, frames b1, b2, b3).

For $A_F<1.3$, Eq. (4.31) and the previous expressions given for $\tilde{F}_1(z)$ and $\tilde{F}_2(z)$ still hold with $N=2$ instead of $N=4$ and $A_2=1$ (compare, e.g., Fig. 4.79-c1 with Fig. 4.79-c2 where the analytical expansion is plotted). This value leads to an "apparent" doubling of the azimuthal wave number of the azimuthal-velocity distribution in the equatorial plane (see, e.g., Fig. 4.79, and Fig. 4.80 frames b1, b2, b3).

Note that, due to the antisymmetric behavior of the disturbances, the flow pattern at $z=0.5$ appears to be rotated of π/m with respect to the flow pattern at $z=-0.5$ (see, e.g., Figs. 4.78, 4.79, 4.80 frames a1, a2, a3).

Accordingly (see, e.g., Fig. 4.79-e) the part of the toroidal vortex that is displaced towards the disk is also displaced away from the center of the zone and the part that is displaced towards the center is at the same time displaced towards the equatorial plane (the displacement down (up) and up (down) for the upper (lower) convection roll is coupled to displacement in and out, respectively).

A possible strategy for clarifying why disturbances antisymmetric with respect to the equatorial plane are selected and amplified whereas z-symmetric disturbances never occur, consists of repeating some computations (for instance, for $A_F=0.5$, $A_F=1.0$, $A_F=1.5$) under the constraint that the flow is symmetric with respect to the equatorial plane.

For such a case the governing equations (4.4) must be solved for $0\leq z\leq 1$ with the symmetry conditions ($0\leq r\leq 1/A_F$; $0\leq\varphi\leq 2\pi$):

$$u(z=0,r,\varphi)=0 \quad (4.32a)$$

$$\frac{\partial v}{\partial z}(z = 0, r, \varphi) = 0 \tag{4.32b}$$

$$\frac{\partial V_\varphi}{\partial z}(z = 0, r, \varphi) = 0 \tag{4.32c}$$

$$\frac{\partial T}{\partial z}(z = 0, r, \varphi) = 0 \tag{4.32d}$$

Figure 4.85 shows that the growth-rate constants obtained in this case always lie below the corresponding ones obtained without any constraint for the flow. Hereafter the critical Marangoni number obtained under the constraint of disturbance symmetric with respect to the midplane will be referred to as $Ma_c^{(s)}$. The computations show that for $A_F = 0.5$, $Ma_{c1} = 9.76$ and $Ma_{c1}^{(s)} = 18.255$, for $A_F = 1.0$, $Ma_{c1} = 12.413$ and $Ma_{c1}^{(s)} = 13.536$, for $A_F = 1.5$, $Ma_{c1} = 10.154$ and $Ma_{c1}^{(s)} = 13.53$.

Therefore the antisymmetric disturbances are selected and amplified since they are the most dangerous.

These results also show that for the case $A_F = 1.5$ the critical azimuthal wave number is $m = 1$ without constraints and $m = 2$ for the full-zone z-symmetric case. Thus they provide a rigorous theoretical background and justification to the proposed theory in Section 4.15.2 for which the interaction between the upper and the lower rolls can play a critical role in determining Ma_{c1} as well as in the selection of the azimuthal wave number ($A_F = 1.5$ corresponds to $A_H = 0.75$ that falls into the aforementioned range $0.65 \leq A_H \leq 0.85$ in Fig. 4.84).

4.16. Concave and convex volumes in microgravity

4.16.1 *Basic state and instability threshold*

The case of a noncylindrical volume needs preliminary discussion about the axisymmetric state. In this case, in fact, different phenomena occur in terms of symmetry with respect to the midplane according to whether $S < 1$ or $S > 1$.

As for the cylindrical shape, also for $S \neq 1$, in the axisymmetric state (prior to the transition) the maximum of the surface tension occurs at the contact points (disks), and the minimum occurs at the midsection of the floating zone. The liquid is therefore driven along the surface away from the midsection. Because of the geometrical constraints, the liquid has to turn to flow along the side walls, then along the axis, and then it has to flow back toward the surface along the midsection of the liquid zone.

In the case of concave configurations ($S \leq 1$) the resulting two toroidal vortices have symmetric form but rotate in the opposite directions, i.e. the basic state exhibits radial velocity and temperature symmetric with respect to the midplane of the

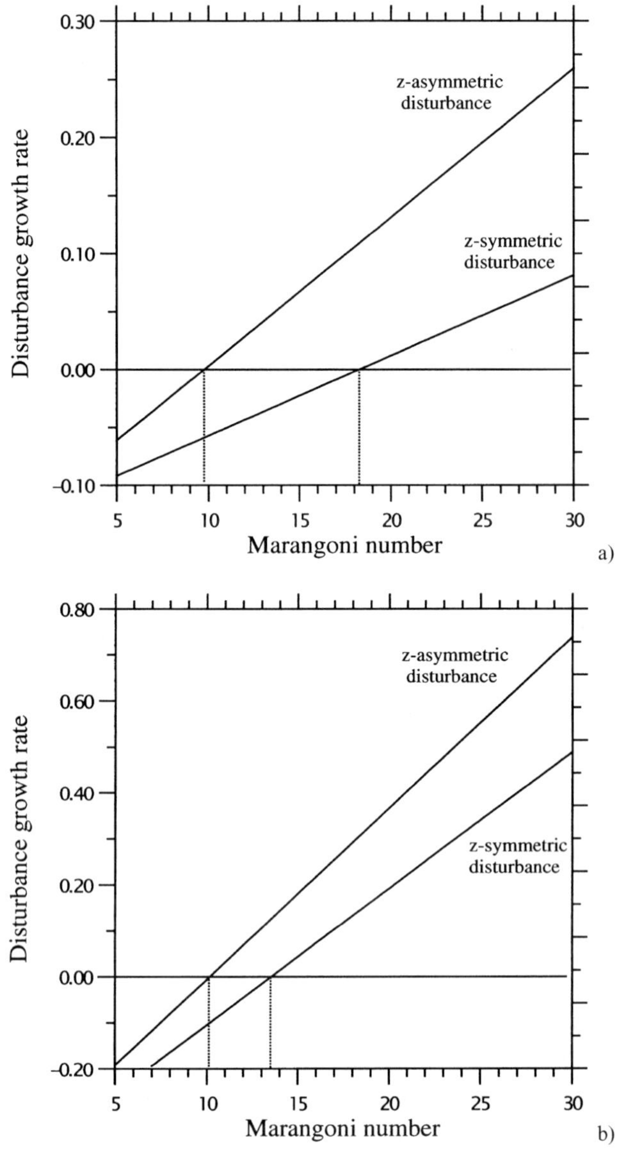

Fig. 4.85 Growth rates as function of the Marangoni number and determination of the first critical Marangoni number (mirror-symmetric and antisymmetric disturbances): (a) $A_F = 0.5$, (b) $A_F = 1.5$.

full-zone, whereas the axial velocity is antisymmetric (as also previously highlighted for the basic state of the cylindrical column).

For convex configurations the behavior is different. The numerical experimentation shows that symmetry breaking with respect to the equatorial plane may occur

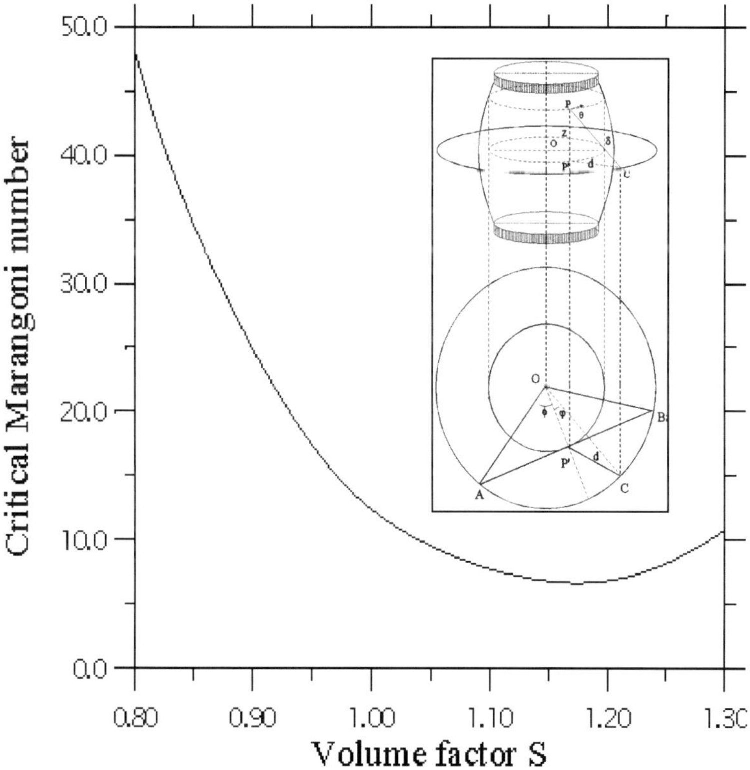

Fig. 4.86 Critical Marangoni number versus S ($A_F = 1.0$) under microgravity.

prior to the onset of azimuthal disturbances, i.e. the basic state does not exhibit mirror symmetry with respect to $z = 0$. Therefore, for this case 3D disturbances superimpose on an axisymmetric field that is not mirror-symmetric with respect to the equatorial plane per se.

Despite this difference, for both convex and concave floating zones (Figs. 4.87–4.90) the toroidal roll in the upper part tends to exhibit the maxima of the azimuthal disturbance around the positions where the other one (lower half) is characterized by the minima and vice versa (see frame (c) in Figs. 4.87–4.90). This means that $G_{up} = G_{low} + \pi/m$ as for the case of a cylindrical interface, i.e. the azimuthal disturbances always exhibit a nonsymmetric-like behavior with respect to the equatorial plane regardless of the interface shape ($S < 1$ or $S > 1$).

With regard to these aspects it should be pointed out that the number of possible solutions appearing at the bifurcation can be directly related to the number of broken symmetries, because the image of a solution by a broken symmetry is also a solution. For the half-zone, as also discussed in Section 4.5, the group of symmetries contains all the rotations around the z axis and the reflections with respect to the meridian planes containing this axis. For the full-zone under microgravity, however, the reflection with respect to the horizontal midplane introduces an additional class of possible solutions (for a given solution, also the upside-down state is a solution).

Figure 4.91 shows the growth rates as a function of the Marangoni number and the determination of the first critical Marangoni number for different values of the volume. The critical Marangoni number (see Table 4.5) is a decreasing function

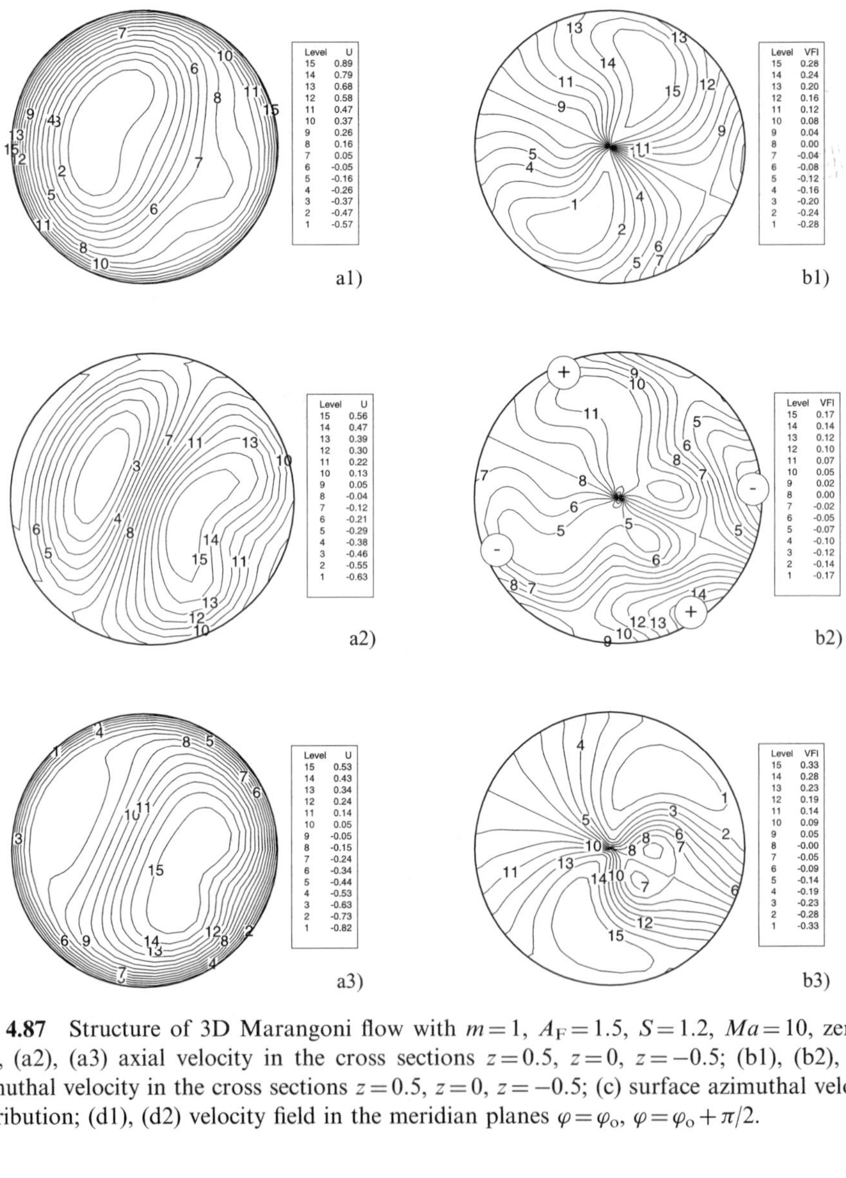

Fig. 4.87 Structure of 3D Marangoni flow with $m=1$, $A_F=1.5$, $S=1.2$, $Ma=10$, zero-g: (a1), (a2), (a3) axial velocity in the cross sections $z=0.5$, $z=0$, $z=-0.5$; (b1), (b2), (b3) azimuthal velocity in the cross sections $z=0.5$, $z=0$, $z=-0.5$; (c) surface azimuthal velocity distribution; (d1), (d2) velocity field in the meridian planes $\varphi=\varphi_o$, $\varphi=\varphi_o+\pi/2$.

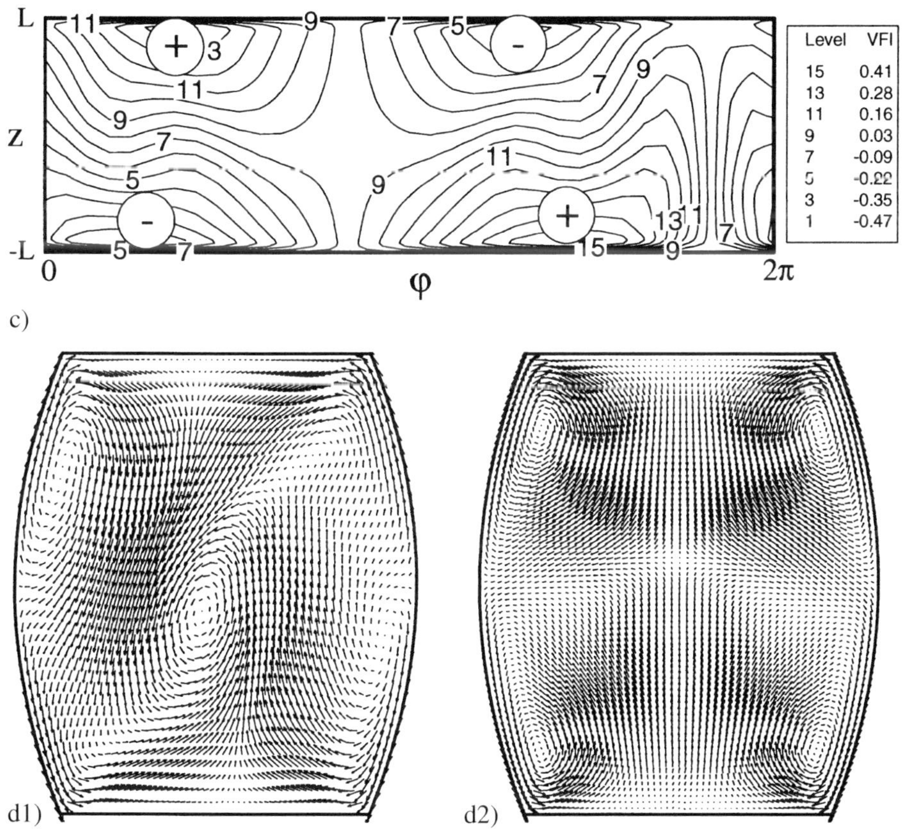

Fig. 4.87 Continued.

of the volume for $S \leq 1.2$. For the case $A_F = 1.0$, it reaches a maximum for $S = 0.8$ ($Ma_{c1} = 48.08$) and a minimum for $S = 1.2$ ($Ma_{c1} = 6.85$) then it increases again for a further increase of S with respect to $S = 1.2$ (Fig. 4.86). A similar behavior occurs for the case $A_F = 1.5$: the critical Marangoni number has a minimum for $S = 1.2$. Therefore the effect of the volume is very important. In the range investigated the per cent variation of the critical Marangoni number may be of the order of 1000% (Ma_{c1} for $S = 0.8$ is seven times Ma_{c1} for $S = 1.2$).

4.16.2 *Structure of the 3D flow*

Table 4.5 shows that the azimuthal wave number m is function of both the aspect ratio and of the volume factor, i.e.:

$$m = m(A_F, S) \tag{4.33}$$

this finding is not unexpected if one considers the previous studies dealing with the half-zone.

The azimuthal wave number of the critical mode (m) is $m=1$ for $A_F=1.5$ and $m=2$ for $A_F=1.0$ in the case of cylindrical shapes ($S=1$). For $A_F=1.0$, m is increased if the volume is decreased with respect to the condition $S=1.0$: $m=2$ for $1.0 \le S \le 1.3$, $m=3$ for $S=0.85$, and $m=4$ for $S=0.80$. For $A_F=1.5$, m is increased if the volume is increased: $m=1$ for $1.0 \le S \le 1.2$ and $m=2$ for $S=1.4$.

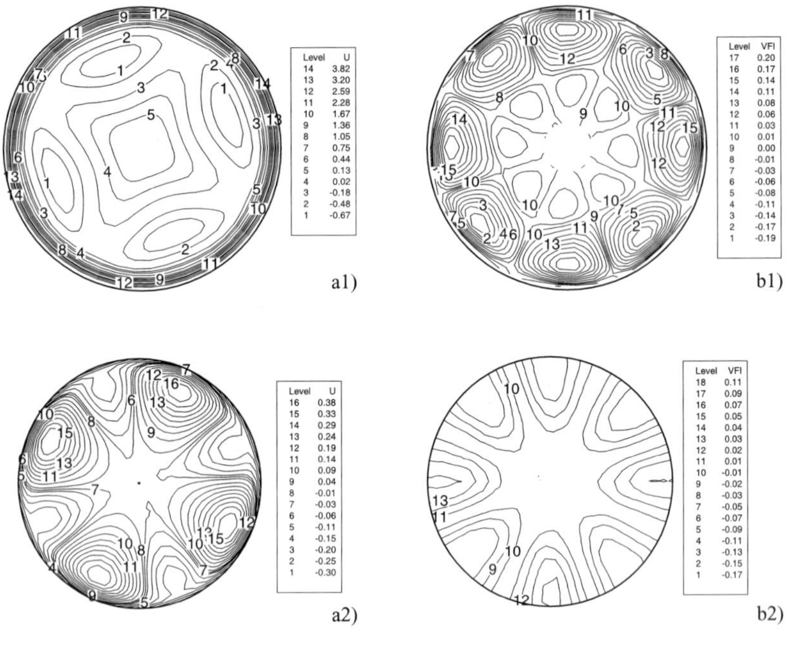

Fig. 4.88 Structure of 3D Marangoni flow at the 3D stationary state with $m=4$, $A_F=1.0$, $S=0.80$, $Ma=100$, zero-g: (a1), (a2), (a3) axial velocity in the cross sections $z=0.5$, $z=0$ and $z=-0.5$; (b1), (b2), (b3) azimuthal velocity in the cross sections $z=0.5$, $z=0$ and $z=-0.5$; (c) surface azimuthal velocity distribution; (d) velocity field in the generic meridian plane.

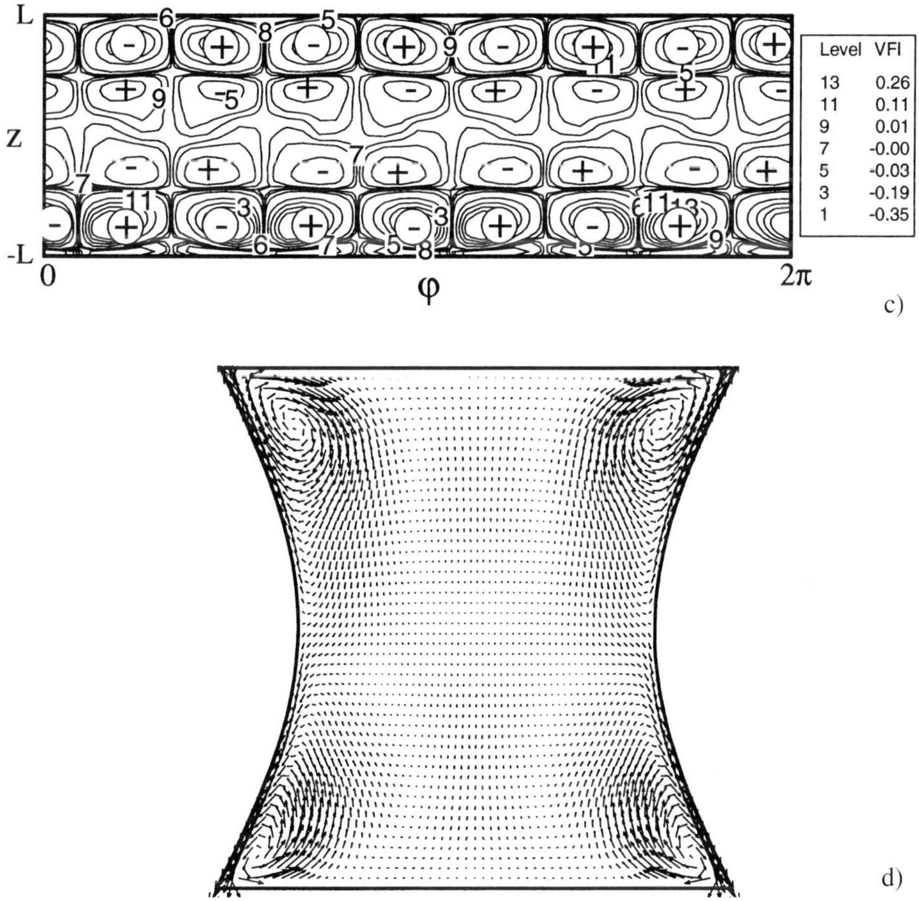

Fig. 4.88 Continued.

The azimuthal wave number changes in the case of deviation from a volume equal to the cylindrical one for both cases $S < 1$ and $S > 1$.

The major outcome of these full-zone-based studies, however, deals with the fact that for $S > 1$, a doubling mechanism of the apparent wave number occurs close to the equatorial plane (frames b1, b2, b3 in Figs. 4.87 and 4.90) that is very similar to that elucidated in Section 4.15.3 for the cylindrical ($S = 1$) case.

This behavior is evident, in particular, for $r > 0.5/A_F$ (i.e. close to the free surface). It is due to the strong interaction that occurs between the upper and the lower rolls. The toroidal vortices are very close or coalesce. For this reason the equatorial cross-section is affected by the azimuthal flow induced by both. Since the angular positions of the disturbance maxima (minima) for the upper roll do not coincide with those of the lower roll (nonsymmetric azimuthal disturbances with respect to the midplane), the liquid region in the middle is characterized by $m+m$ maxima (minima).

On the contrary, the flow exhibits nearly complete separation of the toroidal vortices in the case of a slender configuration ($S<1$) and no doubling and similar mechanisms are observed. This phenomenon behaves as a distinguishing mark between the convex and the concave floating zones.

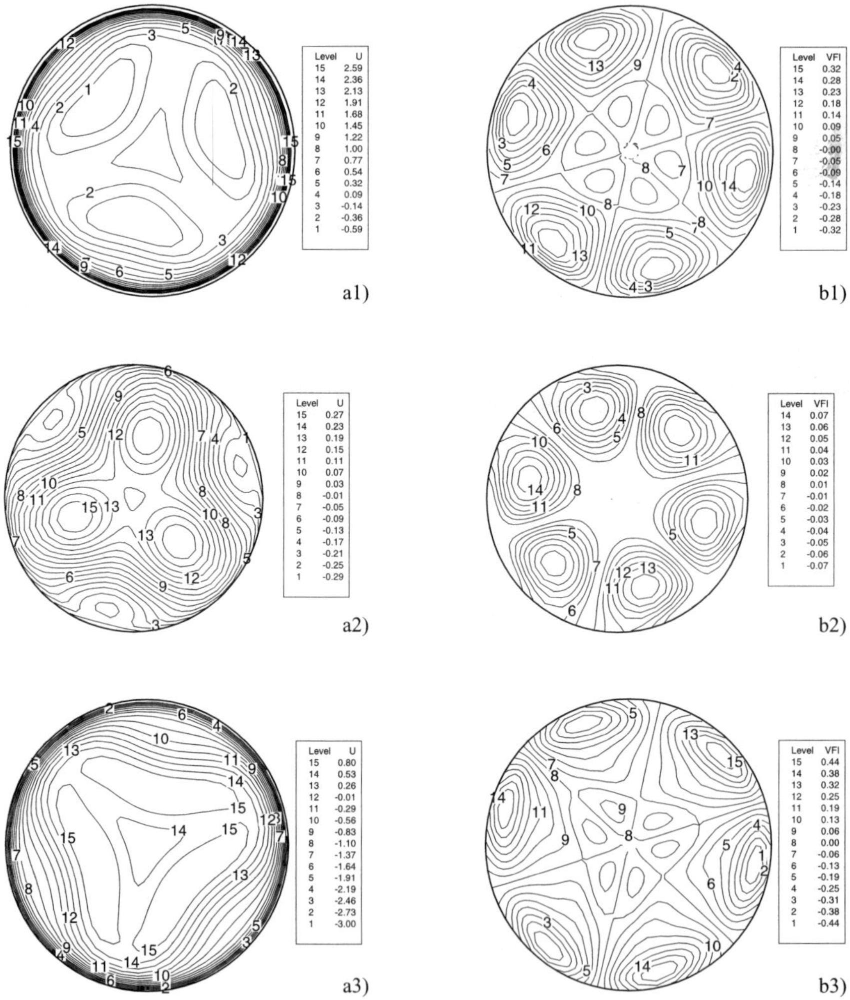

Fig. 4.89 Structure of 3D Marangoni flow with $m=3$, $A_F=1.0$, $S=0.85$, $Ma=50$, zero-g: (a1), (a2), (a3) axial velocity in the cross sections $z=0.5$, $z=0$ and $z=-0.5$; (b1), (b2), (b3) azimuthal velocity in the cross sections $z=0.5$, $z=0$ and $z=-0.5$; (c) surface azimuthal velocity distribution; (d1), (d2) velocity field in the meridian planes $\varphi=\varphi_o$, $\varphi=\varphi_o+\pi/6$.

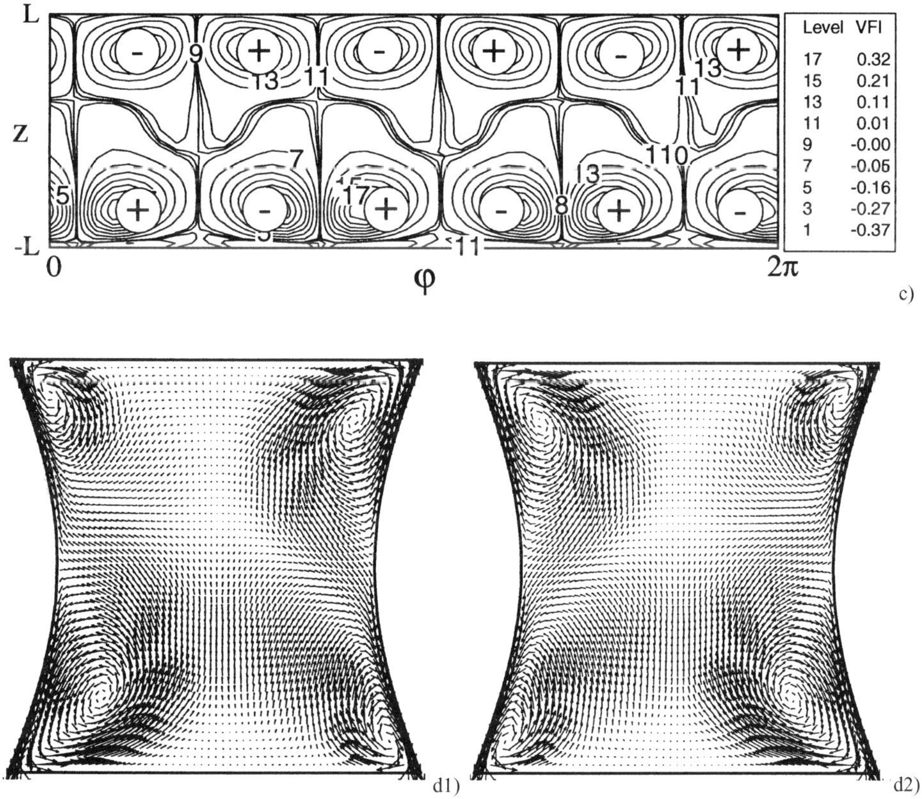

Fig. 4.89 Continued.

4.16.3 *Comparison with the half-zone*

Other results dealing with the noncylindrical zones in the case of the half-configuration available in the literature are of little help in discerning the mechanisms and the underlying physics of the real floating zone process since they deal with a simplified configuration and/or with a different type of transition.

Hirata et al. (1997c) yielded very interesting experimental results about the effect of the volume of the liquid zone under microgravity conditions using a drop shaft tower and transparent liquids (silicone oils). These results, however, cannot be used to draw conclusions about the behavior of liquid metals since in the case of high Prandtl number liquids the intrinsic nature of the instability is hydrothermal rather than hydrodynamic.

Lappa et al. (2001a) found in the case of liquid metals (half-zone, $Pr = 0.01$) under microgravity that for a fixed aspect ratio the critical azimuthal wave number can be shifted to higher values by increasing S (convex shape) or to lower values by decreasing S (concave shape). This behavior is quite different with respect to that of the full-zone where the azimuthal wave number increases in the case of deviation from a volume equal to the cylindrical one for both cases $S < 1$ and $S > 1$. This is an additional proof of the fact that the instability of Marangoni flow in half-zones is not relevant for the full-zone configuration.

Chen et al. (1999) studied the influence of the liquid-bridge (half-zone) volume on the instability of the Marangoni flow in the case of a half-zone configuration

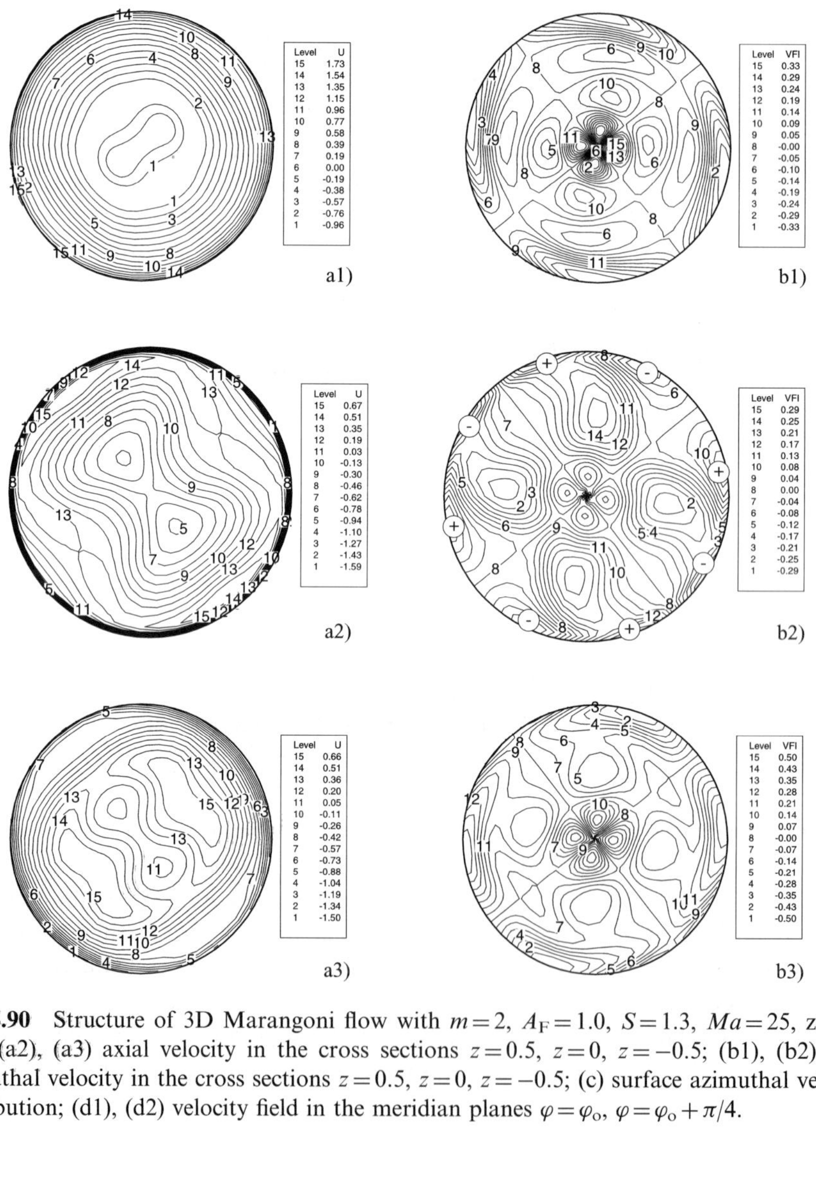

Fig. 4.90 Structure of 3D Marangoni flow with $m=2$, $A_F=1.0$, $S=1.3$, $Ma=25$, zero-g: (a1), (a2), (a3) axial velocity in the cross sections $z=0.5$, $z=0$, $z=-0.5$; (b1), (b2), (b3) azimuthal velocity in the cross sections $z=0.5$, $z=0$, $z=-0.5$; (c) surface azimuthal velocity distribution; (d1), (d2) velocity field in the meridian planes $\varphi=\varphi_o$, $\varphi=\varphi_o+\pi/4$.

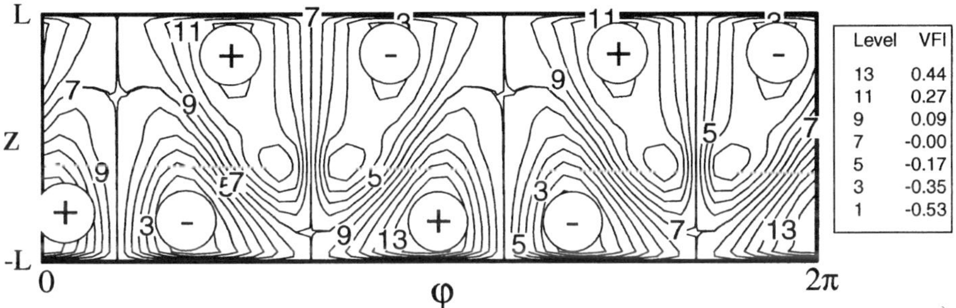

c)

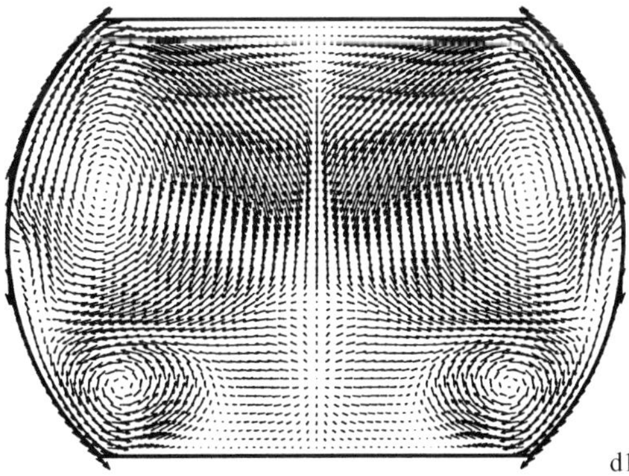

d1)

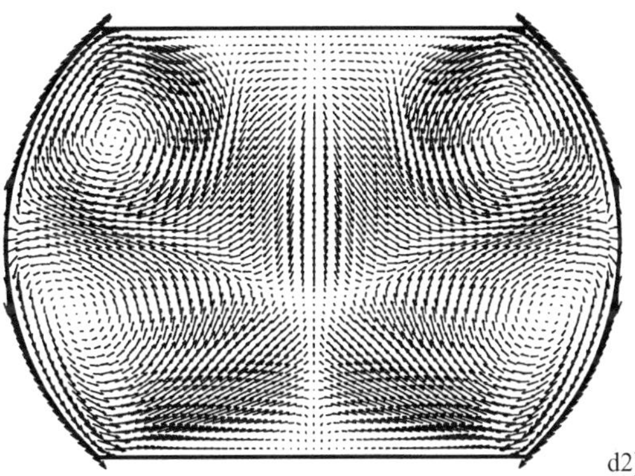

d2)

Fig. 4.90 Continued.

by the technique of linear stability analysis. For the cases $Pr = 0.01$, $0.4 \le A_H \le 1.0$ and $0.5 \le S \le 1.2$ they found that the marginal instability curves decrease rapidly if the volume of the slender liquid bridge ($S < 1$) is increased and that the critical Marangoni number changes smoothly (with a small increase) during the increase of the volume of the fat liquid bridge ($S > 1$). This behavior of the critical Marangoni number is in qualitative agreement with the case of the full-zone (Fig. 4.86). However, further comparison is not possible due to the

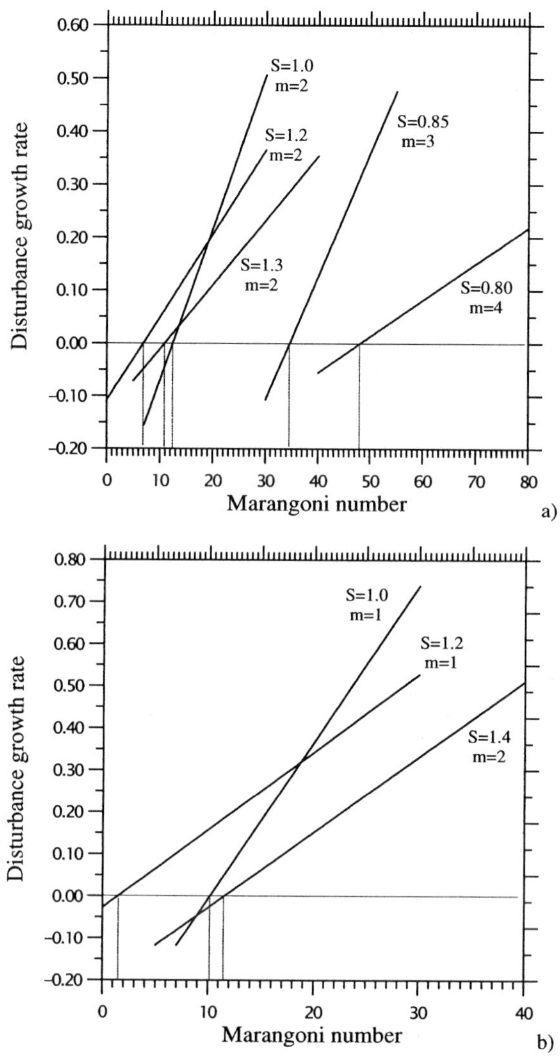

Fig. 4.91 Growth rates as function of the Marangoni number and determination of the first critical Marangoni number for different values of the volume (microgravity conditions: (a) $A_F = 1$, (b) $A_F = 1.5$).

very different fluid dynamics of the two configurations when their noncylindrical volumes are considered. For the full-zone, the free surface is mirror-symmetric with respect to the equatorial plane, but convex or concave with the maximum or minimum diameter occurring at the midplane. Of course, this situation cannot be modeled by noncylindrical half-zones supported by disks having the same diameter.

4.17. The laterally heated column on the ground

4.17.1 *Basic state and instability threshold*

In the axisymmetric state (prior to the transition) the two toroidal vortices have similar form but rotate in the opposite directions. In a zero-gravity environment the shape of the gas/melt interface is symmetric with respect to the heating center. In the Earth environment gravity, however, has a strong influence on this shape. The shape becomes convex to the melt at the upper zone and concave at the lower zone. Owing to this feature, of course, for any solution that breaks reflection symmetry in $z=0$, the upside-down state cannot also be a solution of the problem.

It also causes an asymmetry in the structure of the two rolls in the basic state, with the lower roll being more intense for $S<1$ and vice versa for $S\geq 1$ (Fig. 4.92). For $S<1$ they are not mirror-symmetric with respect to the equatorial plane, but each vortex cell is still confined to the respective half of the floating zone (Fig. 4.92a). For $S\geq 1$ the effect of the shape on the flow field results in a larger vortex cell in the upper region that extends into the lower half, confining the other cell to a small region in the lower half (Fig. 4.92b).

Therefore, also in this case an exhaustive discussion of the full-zone behavior is not possible on the basis of previous studies dealing with the half-zone as the complete structure of the convective field is required to account for nonlinearities and other complex effects arising from the mutual interaction of the two aforementioned facing toroidal vortices (see Sections 4.15.3 and 4.16.2 for additional relevant background information).

The instability threshold changes according to the volume of the liquid. Figure 4.93 shows the growth rates as function of the Marangoni number and the determination of the first critical Marangoni number for different values of S. The critical Marangoni number (see Table 4.6) is a decreasing function of the volume in the range $0.8\leq S\leq 1.1$. Within this range, it reaches a maximum for $S=0.8$ ($Ma_{c1}=33.24$) and a minimum for $S=1.1$ ($Ma_{c1}=3.313$) then it increases again for a further increase of S with respect to $S=1.1$. Therefore, the effect of the volume is of paramount importance also on the ground.

The comparison between the computations in the case of microgravity and normal-gravity conditions gives an insight into the effect of gravity in liquid columns of full-zone extent. For instance, in the case $A_F=1$, microgravity conditions and cylindrical shape, the critical Marangoni number is $Ma_{c1}=12.43$. If the gravity deformation of the shape is taken into account the critical Marangoni number ($Ma_{c1}=8.40$) is significantly decreased with respect to the cases where the cylindrical shape is assumed ($\Delta Ma_{c1}=O(40\%)$). Therefore for a volume equal to the cylindrical one the on the ground Marangoni flow is destabilized with respect to microgravity conditions. The same trend occurs in the case of $S<1$; gravity acts to destabilize the Marangoni flow. Vice versa, stabilization occurs in the case $S>1$ (compare Tables 4.5 and 4.6).

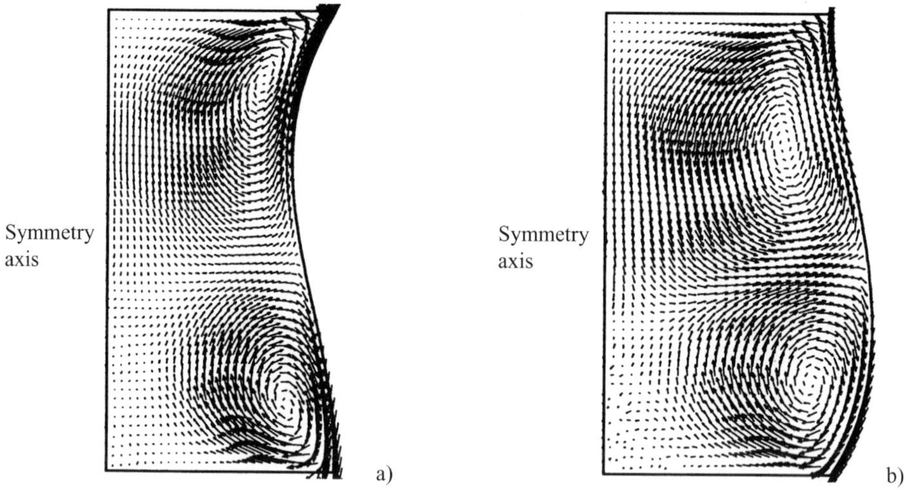

Fig. 4.92 Velocity field in the axisymmetric basic state (on the ground conditions): (a) $A_F = 1$, $S = 0.9$, $Ma = 30$, (b) $A_F = 1$, $S = 1.1$, $Ma = 10$.

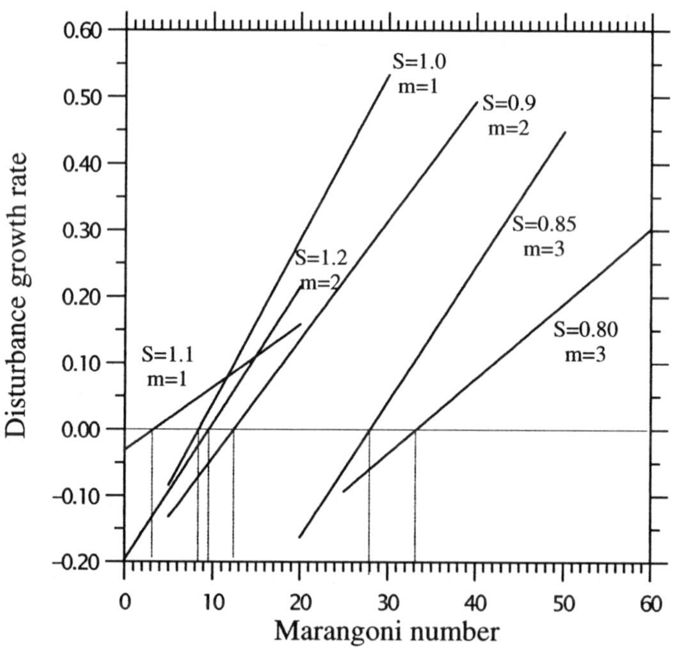

Fig. 4.93 Growth rates as function of the Marangoni number and determination of the first critical Marangoni number for different values of the volume ($A_F = 1$, on the ground conditions).

Table 4.5 Critical azimuthal wave number and critical Marangoni number versus the volume of the liquid zone ($Pr = 0.01$, full-zone, microgravity conditions, Lappa 2004a).

A_F	L cm	$S = V/V_o$	m	Ma_{c1}
1.0	0.5	0.80	1	18.08
1.0	0.5	0.85	3	34.59
1.0	0.5	1	2	12.413
1.0	0.5	1.2	2	6.85
1.0	0.5	1.3	2	10.87
1.5	0.75	1	1	10.15
1.5	0.75	1.2	1	1.454
1.5	0.75	1.4	2	11.53

Table 4.6 Critical azimuthal wave number and critical Marangoni number versus the volume of the liquid zone ($Pr = 0.01$, full-zone, $L = 0.5$ cm, $A_F = 1.0$, on the ground conditions, Lappa 2004b)

$S = V/V_o$	m	Ma_{c1}
0.80	3	33.24
0.85	3	27.99
0.90	2	12.39
1	1	8.40
1.1	1	3.31
1.2	2	9.54

4.17.2 *3D patterns of symmetry*

Figures 4.94–4.97 show some examples.

As expected, the azimuthal wave numbers for this case are different if compared with the situation of microgravity conditions. Different values of m are obtained in well-defined ranges of the volume as for zero-g conditions, but on the ground the ranges corresponding to a fixed value of m are completely different (see again Tables 4.5 and 4.6).

For instance, in the case of microgravity conditions and cylindrical shape ($A_F = 1$), the azimuthal wave number is $m = 2$. If the deformation of the shape is taken into account the most dangerous azimuthal wave number becomes $m = 1$, i.e. for a volume equal to the cylindrical one the on the ground azimuthal wave number is shifted to a lower value. The same trend occurs in the case of $S < 1$.

The results for on the ground conditions suggest an empirical correlation between the volume factor and the critical azimuthal wave number of the instability. This correlation (for $A_F = 1$) is

$$m \cong 1 + \zeta \tag{4.34}$$

where $\zeta \cong 10 \times \text{abs}(S - 1)$ ($\zeta = 10 \times (1 - S)$ for $S < 1$, $\zeta = 10 \times (S - 1.1)$ for $S > 1$).

In addition to the differences with respect to the microgravity case pointed out above, it is interesting to highlight how in Eq. (4.29) $\tilde{f}_{\mathrm{up}}(r,z) \neq \tilde{f}_{\mathrm{low}}(r,-z)$, i.e. the strength of the disturbances at the 3D stationary state is different according to the relative position of the convection roll with respect to the equatorial plane (above or below, see frame (c) in Figs. 4.94–4.97). This is a notable distinguishing feature with respect to the microgravity case. However, $G_{\mathrm{up}} = G_{\mathrm{low}} + \pi/m$ for all the cases, i.e. the azimuthal disturbances always exhibit nonsymmetric behavior with respect to the midplane as for zero-g conditions.

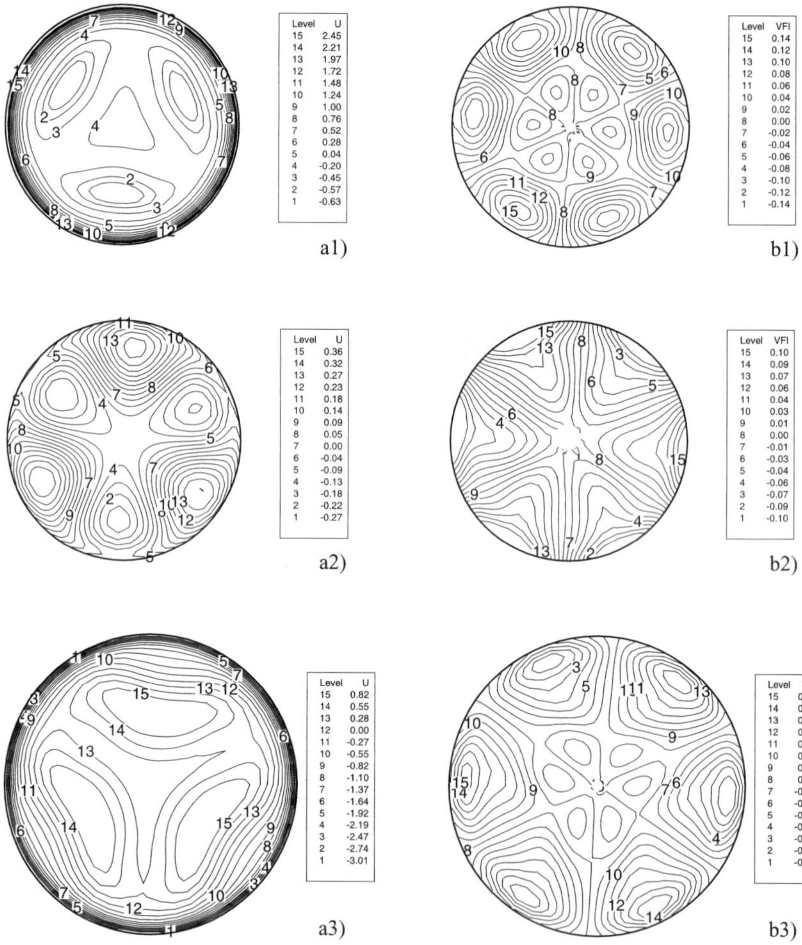

Fig. 4.94 Structure of 3D Marangoni flow with $m=3$, $A_{\mathrm{F}}=1$, $S=0.8$, $Ma=60$ (on the ground conditions): (a1), (a2), (a3) axial velocity in the cross-sections $z=0.5$, $z=0$, $z=-0.5$; (b1), (b2), (b3) azimuthal velocity in the cross sections $z=0.5$, $z=0$, $z=-0.5$; (c) surface azimuthal velocity distribution; (d) velocity field in the generic meridian plane.

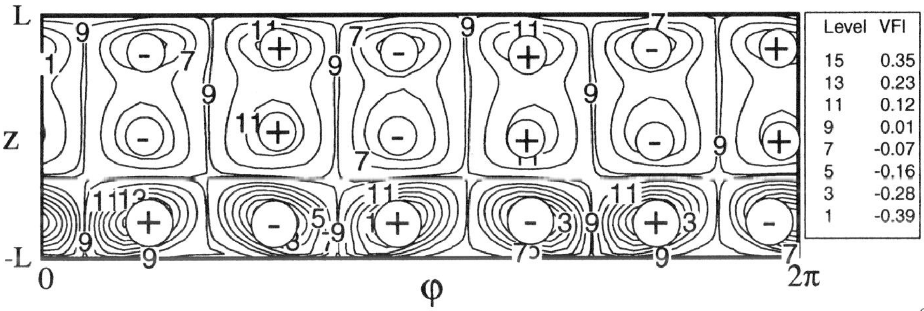

c)

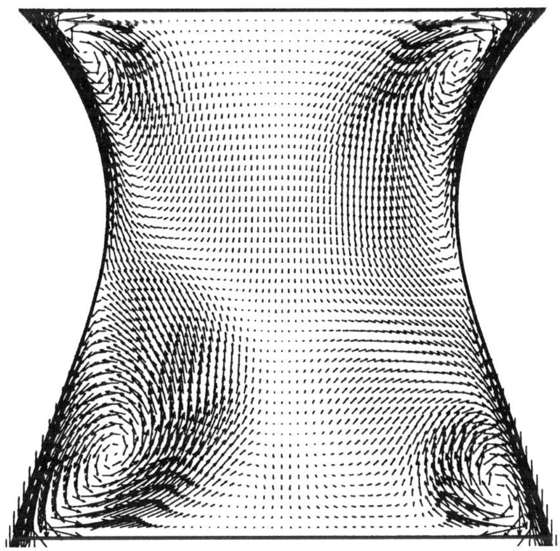

d)

Fig. 4.94 Continued.

4.17.3 *Comparison with the half-zone*

With regard to the half-zone, some interesting results about liquid bridges ($A_H = 1$, $L = 1$ cm) heated from above or from below in the case of low Prandtl number liquids have been discussed in Section 4.9 on the basis of the results of Lappa et al. (2003a).

According to the aforementioned simulations, if the deformation of the shape is taken into account, gravity always acts to stabilize the Marangoni flow regardless of the direction of the gravity vector (parallel or antiparallel to the axis), whereas in the case of a straight surface the heating from below condition leads to destabilization of the flow. The on the ground deformation of the shape also influences the value of the azimuthal organization of the flow field. The azimuthal wave number is shifted to higher values if the heating from above condition is considered.

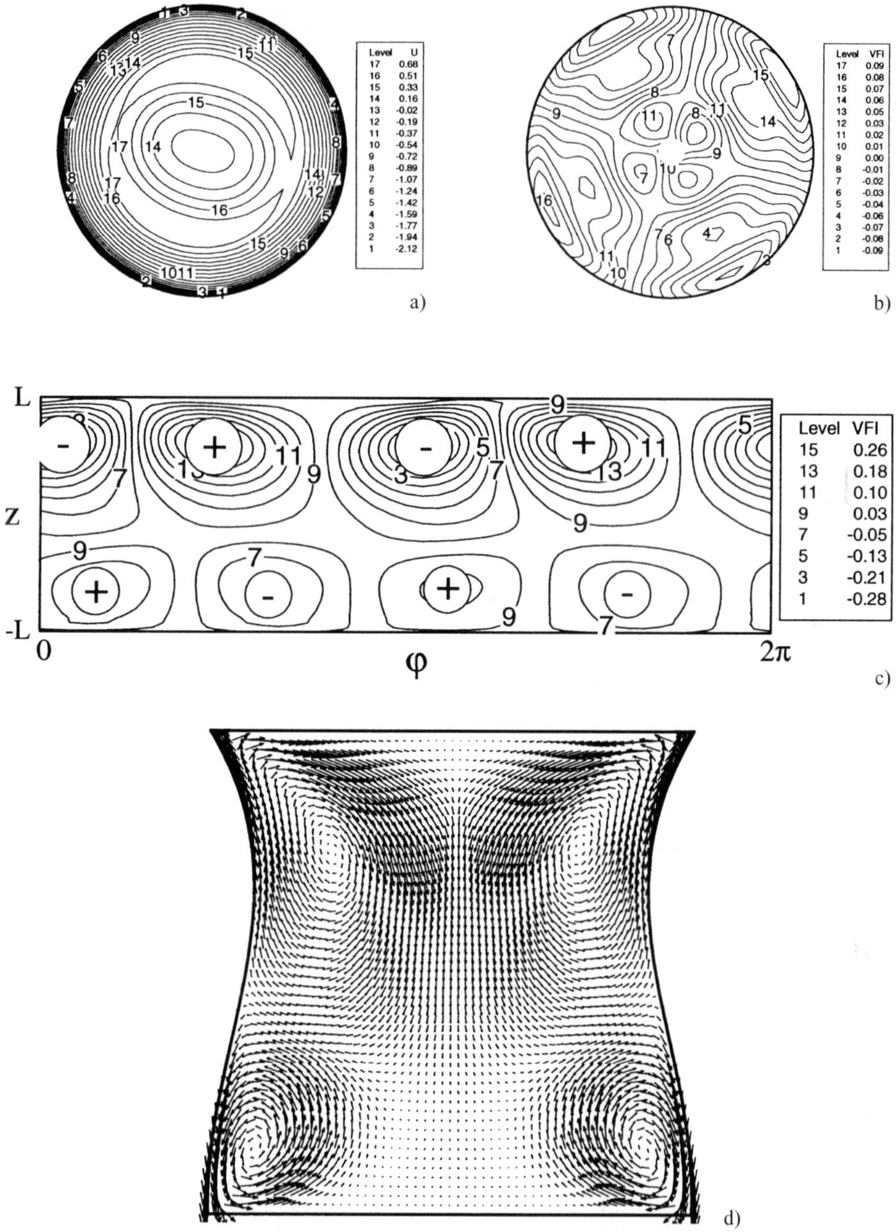

Fig. 4.95 Structure of 3D Marangoni flow at the 3D stationary state with $m=2$, $A_F=1$, $S=0.9$, $Ma=30$ (on the ground conditions): (a) axial velocity in the cross section $z=-0.5$, (b) azimuthal velocity in the cross section $z=-0.5$, (c) surface azimuthal-velocity distribution, (d) velocity field in the generic meridian plane.

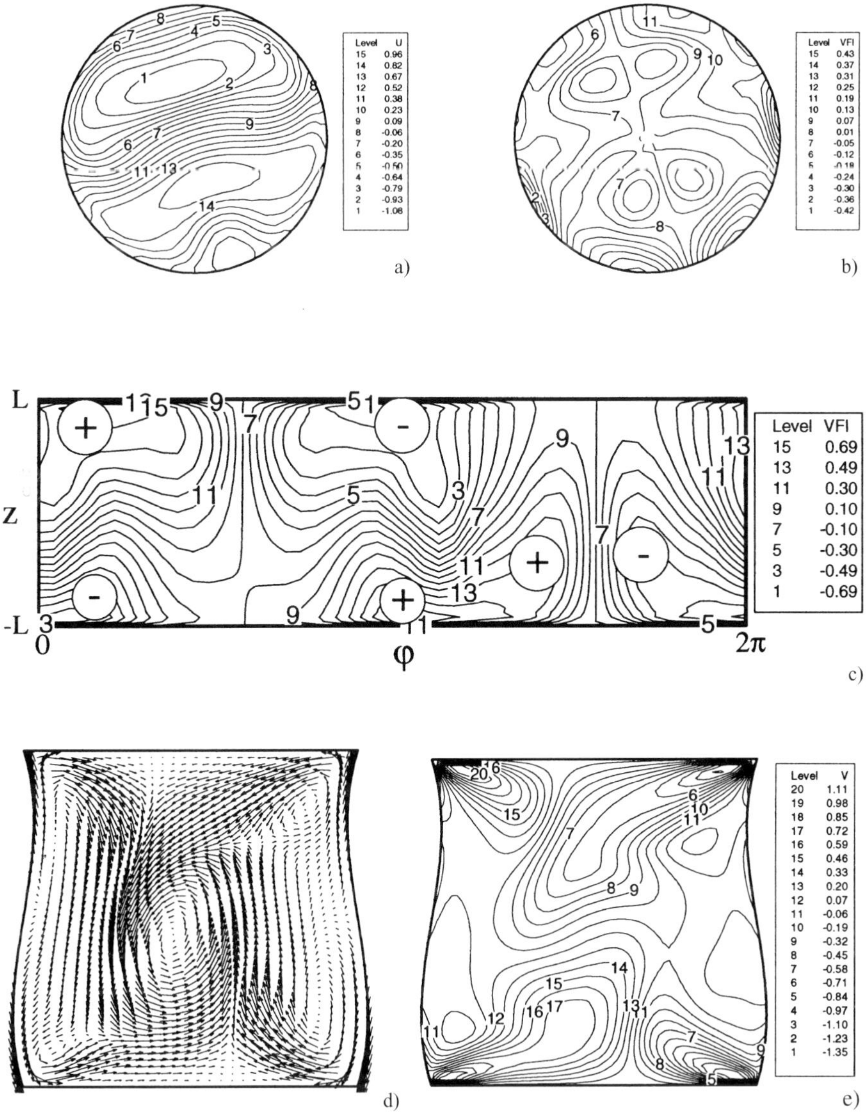

Fig. 4.96 Structure of 3D Marangoni flow at the 3D stationary state with $m=1$, $A_F=1$, $S=1.0$, $Ma=30$ (on the ground conditions): (a) axial velocity in the cross section $z=0$, (b) azimuthal velocity in the cross section $z=0$, (c) surface azimuthal-velocity distribution, (d) velocity field in the generic meridian plane, (e) radial-velocity distribution in the same plane of frame (d).

With regard to the full-zone note that on the ground, the upper half behaves (approximately) as a half-zone heated from below, whereas the lower half behaves (approximately) as a half-zone heated from above. Again, a direct comparison between the available results in the case of half- and full-zones is not possible.

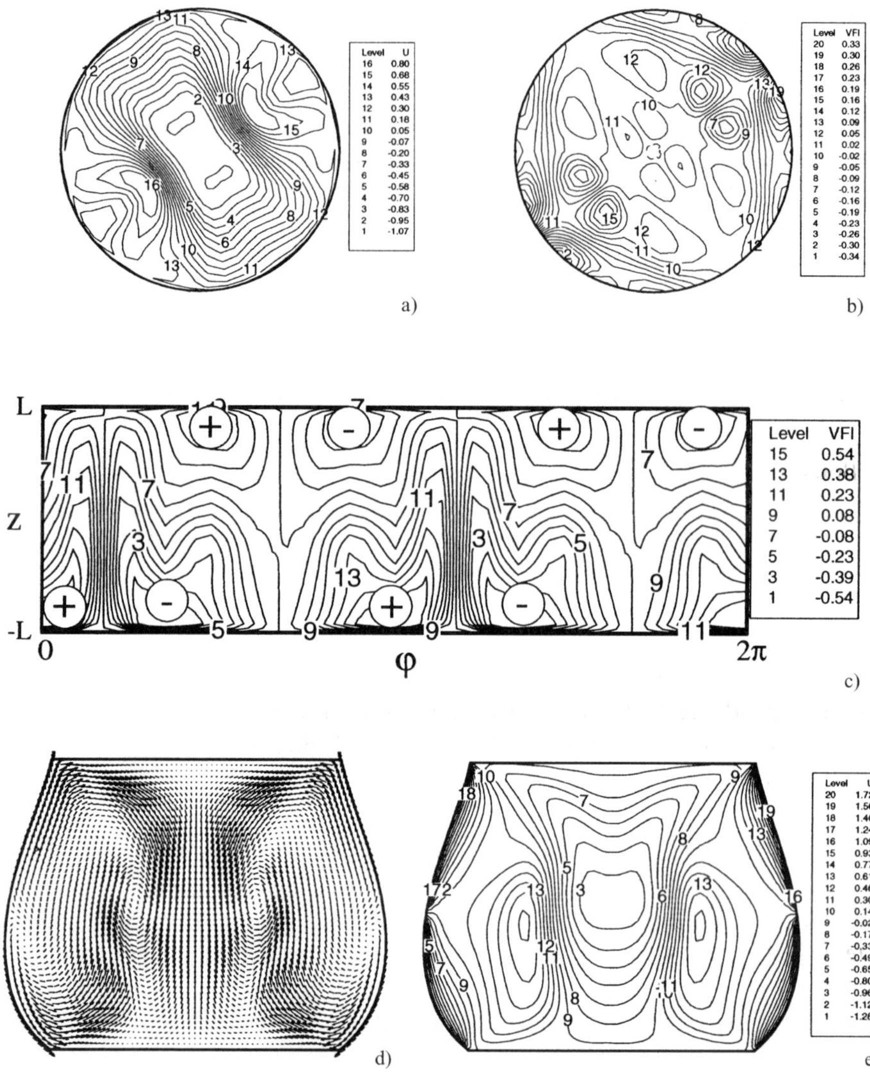

Fig. 4.97 Structure of 3D Marangoni flow at the 3D stationary state with $m=2$, $A_F=1$, $S=1.2$, $Ma=20$ (on the ground conditions): (a) axial velocity in the cross section $z=0$, (b) azimuthal velocity in the cross section $z=0$, (c) surface azimuthal-velocity distribution, (d) velocity field in the generic meridian plane, (e) axial-velocity distribution in the same plane of frame (d).

In the case of the full-zone configuration the surface (amphora-like shape due to the gravity field) is heated at midheight by an external heat source; in the case of the half-zone the amphora-like interface is put in contact with the heat source at the bottom (bridge heated from below) or at the top (bridge heated from above). Moreover, in the case of a full-zone the aforementioned strong interaction between the two toroidal vortices (apparent doubling or quadrupling of the azimuthal wave number around the equatorial plane and/or coalescence of the toroidal convection rolls) occurs, making, for $S \geq 1$, the interplay between the upper half and lower half an essential factor for the correct description of the phenomena under investigation. Because of the complexity of the full-zone with respect to the half model, the important principles and main results underlying half-zone-based investigations are of little help in elucidating the gravity effect in full-zones.

4.17.4 *The Buoyancy effect*

The effect of the buoyancy forces with respect to that induced by the shape can be discerned by artificially neglecting the buoyancy forces ($Ra = 0$).

The numerical simulations repeated after "switching off" the effect of buoyancy (the Boussinesq term in the Navier–Stokes equations) show that for low values of the imposed temperature difference (i.e. near the threshold of the first bifurcation), the influence of the buoyancy forces on the flow field is very weak (the difference between the results lies below 3%).

It is interesting to point out that the same conclusion has been obtained by Nienhüser and Kuhlmann (2002) for the half-zone configuration and a value of the Prandtl number ($Pr = 0.02$) very close to that considered in the present book (in their work the dependence of the critical Reynolds number on the volume fraction, gravity level, and aspect ratio has been explained by analyzing the energy budgets of the neutral modes).

Their result and the present ones, even if obtained with very different techniques, show that both the half-zone and the full-zone of a liquid metal are quite insensitive to the effect of the buoyancy forces near the threshold of the stationary instability.

In summary, for the case of very low values of the Marangoni number (i.e. of the temperature gradient acting along the melt/gas interface), the main effect of the gravitational field on the features of the instability comes through its influence on the shape of the free surface, whereas the role played by the buoyancy effect can be regarded as almost negligible.

Finally, for the sake of completeness it should be pointed out that many studies (Velten et al., 1991; Hu et al., 1995; Hirata et al., 1997a, b; Wanschura et al., 1997b, etc.) have also appeared where the stability of the combined Marangoni buoyant convection was investigated in the case of the simplified half-zone configuration and in the case of transparent ($Pr \gg 1$) liquids. However, they deal with a different type of instability (of the hydrothermal type) that, as previously pointed out, has nothing to do with the FZ technique. Some aspects dealing with the effect of gravity in transparent liquid bridges have been discussed in Section 2.4 in the framework of fundamental fluid-dynamic information provided to the reader. Unlike the first bifurcation of Marangoni flow for small Pr, the buoyancy effect plays a crucial role in determining the structure of the 3D flow and the instability threshold in the case of large Pr.

4.18. Physical explanations

To recapitulate, the critical azimuthal wave number increases if the geometrical aspect ratio of the liquid zone is decreased and it is shifted to higher values by increasing or decreasing the volume with respect to $S=1$ for both microgravity and on the ground conditions. The critical Marangoni number is increased if the volume is reduced with respect to $S=1$. Vice versa, if the volume is increased it is decreased up to $S=S^*$ ($S^*\cong 1.2$ for 0 g, $S^*\cong 1.1$ for 1g) and then it increases again for a further increase of S.

The stabilization and the destabilization mechanisms are not trivial. However, possible explanations could be outlined on the basis of several lines of evidence supported by previous analyses; in particular, in the framework of the ideas and concepts highlighted in Section 4.10, the increase or the decrease of the diameter of the convection rolls due to the static deformation of the shape and the relative importance of bounds and the degrees of freedom of the Marangoni toroidal convection rolls (this relative importance being dependent on the shape of the free surface) can be used to draw general physical principles.

Following the arguments elucidated in Section 4.10, there is a precise relationship between the azimuthal wave number and the aspect ratio of the toroidal vortex. This dependency reads $mA_V\cong\pi/2$ (Eq. (4.18)).

For the case $S<1$ (see, e.g., Fig. 4.92a), the toroidal vortex of the basic state is confined to the corner between the concave free surface and the supporting disk. This decreases the axial extension of the toroidal convection rolls (approximately as $L_V\propto LS$) thus reducing the value of A_V. For this reason, each half of the floating zone behaves (approximately) as a half-zone of low aspect ratio, thus increasing the value of the azimuthal wave number. For $S>1$ (see, e.g., Fig. 4.92b) the increase of the azimuthal wave number is related to the increase of the diameter D_V ($D_V\propto DS$) of the toroidal vortices (in this case, in fact, the shape is convex). This decreases the value of A_V and increases m for the same reasons discussed for the case $S<1$. From a mathematical point of view:

$$S<1,\qquad L_V\propto LS \quad \text{and} \quad D_V\cong D\rightarrow A_V\propto\frac{LS}{D}=\frac{A_F S}{2} \tag{4.35a}$$

$$S>1,\qquad L_V\cong L \quad \text{and} \quad D_V\propto DS\rightarrow A_V\propto\frac{L}{DS}=\frac{A_F}{2S} \tag{4.35b}$$

And since $mA_V\cong\pi/2$, Eqs. (4.18) read:

$$S<1,\qquad m\propto\frac{\pi}{A_F S} \tag{4.36a}$$

$$S>1,\qquad m\propto\frac{\pi S}{A_F} \tag{4.36b}$$

Equations (4.36) provide a qualitative physical explanation for the observed trend of the azimuthal wave number that increases if the aspect ratio is reduced and/or if the parameter S is increased (decreased) for $S > 1$ (for $S < 1$).

Regarding the trend of the critical Marangoni number shown in Tables 4.5 and 4.6, however, the attention has to be focused directly on the degrees of freedom of the toroidal vortices.

As previously discussed, in principle, each convection roll is bounded by a wall from one side and it is free to interact in a nonlinear way with the opposite convection roll from the other side. It can expand, crossing the midplane and protruding in the opposite side of the liquid zone (as shown in Figs. 4.87-d1, 4.90-d1, 4.97-d, for $S > 1$), moreover, coalescence of the toroidal convection rolls can occur; however, this degree of freedom is somewhat reduced in the case $S < 1$.

In this case, in fact, due to the concave shape of the free surface, the convection rolls are confined each to the respective half of the full-zone configuration (this prevents mutual interference and strengthens the constraints for the flow field leading to a larger value of the critical Marangoni number). In the case $S > 1$ each toroidal roll is free to interact with the opposite roll and to become pervasive throughout the system lowering the instability threshold.

This explains why the critical Marangoni number tends to be higher for $S < 1$.

A final aspect needs clarification: the minimum value of the critical Marangoni number occurs at intermediate values of the volume ($S^* \cong 1.2$ for 0g, $S^* \cong 1.1$ for 1g). A possible explanation of this trend can be based on the findings of Nienhüser and Kuhlmann (2002).

They concluded that in the case of liquid metals and noncylindrical half-zones, two different (straining and centrifugal) mechanisms are responsible for the inertial instability of Marangoni flow. The first contribution is due to the strain in the vortex center near the free surface (as previously pointed out also by Levenstam and Amberg (1995) and Leypoldt et al. (2000) for the case of a cylindrical interface). The second contribution is due to the centrifugal effects related to the curvature of the surface.

For small volumes the basic vortices are stretched in the axial direction and displaced towards the walls (as shown in Fig. 4.92a). In this case according to their analysis, the enhanced strain on the vortex due to the geometrical constraints leads to a high contribution of the strain-induced-instability mechanism. On the other hand, due to the position of the basic vortices, the contribution of the centrifugal destabilization mechanism is less important.

On the contrary, when the volume increases, the basic streamlines become more circular (as shown in Fig. 4.92b), indicating less strain, and the efficiency of the mechanism in the vortex core is substantially reduced. For this case, however, the centrifugal mechanism becomes more important (Nienhüser and Kuhlmann, 2002).

From the foregoing, the efficiency of the two instability mechanisms depends on the volume in an opposing manner. At intermediate values of S both the mechanisms are operative leading to an optimum (minimum) critical Marangoni number.

The proposed principles exhibit interesting capabilities to elucidate the effect of geometrical parameters and to identify the cause-and-effect relationships, i.e. to give insights into the physical mechanisms driving the phenomena under investigation. Along these lines, the trends highlighted by the numerical results could also be directly used to postulate strategies for optimization of the floating-zone process. For instance, the occurrence of the aforementioned low value of the critical Marangoni number around $S = S^*$ would suggest to maintain the crystallization system as apart as possible from this condition in order to avoid the undesirable effects of the steady 3D flow. Similarly the condition $S < 1$ would favor crystal growth preventing the instability onset as well as other nonlinear effects associated with the mutual interaction of the toroidal convection rolls. Other possible strategies for stabilizing the flow are briefly discussed in the next section.

4.19. Control of Marangoni convection

As emphasized throughout this chapter, during a crystal growth from the melt, the flow regime in the liquid is the dominating factor for the heat and mass transport during the growth process and finally for the quality and yield of the product (see, e.g., Section 4.11).

For this reason the possibility to somehow control the intensity of these flows as well as their bifurcations has become, a topic of great interest, over the years.

This is particularly true for the case of the FZ technique since the 3D instabilities of the flow discussed in the previous sections can appear even if the temperature gradient along the free surface is very small (in practice, in the case of semiconductor melts and liquid metals, a critical Marangoni number $Ma_c = O(10)$ corresponds for a typical length $L = O(1)$ cm to $\Delta T = O(1)$ K). Along these lines, control of the flow is usually regarded by the investigators as an essential and growing part of the problem and accordingly different approaches with increasing complexity have been proposed during the last decade (many of which are still in an embryonic condition).

4.19.1 *Suppression of hydrothermal waves*

Since, in the case of high Prandtl number liquids, the instability is hydrothermal (Section 4.12) and the related mechanism involves a communication between free-surface temperature perturbations and bulk-liquid temperature (see, in particular, Section 4.12.5), it was speculated (e.g., Benz et al., 1998) that by eliminating the free-surface temperature oscillations caused by hydrothermal waves this coupling could be broken and the oscillations would cease.

In the experiments of Benz et al. (1998) concerning the case of shallow layers, free-surface temperature oscillations of a hydrothermal wave state were sensed at two locations using an infrared camera, allowing the phase speed of each wave to be determined. At a location further downstream (in the sense of the wave-propagation direction), a CO_2 laser used the actual temperature signature of each hydrothermal wave to supply heat to troughs of disturbance temperature. When done successfully, hydrothermal waves disappeared downstream of the periodic-heating location.

Along these lines, some investigators tried to suppress the oscillatory Marangoni flow by an active control of the surface-temperature distribution on the liquid-bridge surface. Experiments by Petrov et al. (1998) demonstrated suppression of the oscillatory convection in liquid bridges through the sensing of surface-temperature variations and the application of surface heating using externally placed elements.

Ideas from nonlinear dynamics were applied to effect control, with the control law constructed from observations of the bridge response to randomly applied perturbations. Attempts to effect control of a helical traveling-wave state using a single sensor/heater pair were unsuccessful: It changed the mode into a standing wave with a node at the sensor location. This was resolved with the addition of a second sensor/heater pair that permits the algorithm to distinguish between clockwise- and counterclockwise-propagating waves.

Within this context it is also worth mentioning the landmark study of Shiomi et al. (2003).

An active feedback control was applied to suppress oscillations in the Marangoni convection in a half-zone of a transparent liquid. The active control was realized by locally modifying the surface temperature using the local temperature measured at different locations fed back through a simple control law. The performance of the control process was quantified by analyzing the local temperature signals, and the flow structure was simultaneously identified by flow visualization. With the optimal placement of sensors and heaters, and proportional control, these researchers were able to raise the critical Marangoni number by more than 40%. The amplitude of the oscillation was suppressed to less than 30% of the initial value for a wide range of Marangoni number, up to 90% of the critical value.

Owing to the intrinsic nature of the control mechanism (Fig. 4.98) and of the instability of the Marangoni flow for $Pr > 1$, however, these procedures become inapplicable in the case of low Prandtl number fluids for which the instability, being hydrodynamic, is not driven by surface azimuthal temperature gradients.

4.19.2 *Magnetic fields*

Since molten semiconductors are excellent electrical conductors, in this case magnetic fields are widely used.

This effect was used for the first time in materials science to suppress temperature fluctuations in the melt during the horizontal Bridgman growth process and is currently used by commercial crystal growers around the world to grow more homogeneous Si, GaAs and InP crystals with various techniques. In many material processing (such as Bridgman (HB or VB) and Czochralski techniques), a magnetic field is commonly used to control the liquid flows: Its action can lead to the braking of the flow (i.e. the reduction of the rate of convective transport) or to the damping of possible oscillatory convective instabilities, the first effect being important with respect to the macroscopic homogeneity, the latter with respect to the formation of striations.

For these reasons, a number of theoretical studies have appeared during recent years concerning the effect of magnetic fields on the "basic" flow motion in several geometrical models of semiconductor growth techniques (the term "basic" is used herein to indicate the various types of steady or unsteady, two-dimensional or three-dimensional buoyant, Marangoni or "mixed" convection without the presence of magnetic fields that have been discussed in detail for various geometrical configurations and heating conditions in Sections 2.2.4 and 2.3.3).

For instance, the effect of a constant magnetic field on the electrically conducting liquid-metal flows in heated cavities has been mainly studied for the lateral heating corresponding to horizontal Bridgman crystal growth configurations. Ben Hadid et al. (1997a) and Ben Hadid and Henry (1997b) investigated numerically the case of parallelepipedic or cylindrical cavities respectively. Different magnetic field orientations were considered, as well as different situations in the parallelepipedic case, namely buoyancy driven convection, in a confined cavity or in a cavity with a stress-free surface at the top boundary, and Marangoni convection in cavities with the upper boundary subjected to surface-tension variation. In the case of shallow layers, the magnetic damping was found to be more effective with a vertical field.

Similar studies concerning the effect of static magnetic fields in laterally heated cylindrical vessels introduced as models of the vertical Bridgman technique have been carried out by Baumgartl et al. (1990) and more recently by Gelfgat et al. (2001) in the case of a axial direction of the fields. In particular in Baumgartl et al. (1990) the use of magnetic fields was proposed as a possible alternative to the expensive microgravity and a critical comparison of the vantages and advantages provided by these two different approaches were discussed in a quite exhaustive way.

The corresponding Rayleigh–Bènard problem for the case of cylinders heated from below has been considered recently by Touihri et al. (1999b) in the case of axial and horizontal magnetic fields.

The influence of vertical magnetic fields on convection arising in a fluid layer heated from below (Marangoni–Bènard problem) was studied by Chandrasekhar (1981).

Some interesting numerical results for similar effects on buoyancy convection and Marangoni flow in Czochralski configurations/models have been reported in Khine and Walker (1995) and Khine and Walker (1994), respectively; the combined effect of the Buoyancy forces, Marangoni effects and static magnetic fields has been analyzed by Jing et al. (2000).

Recently, the possibility to use magnetic fields has been considered also for the FZ technique. It has been shown by Baumgartl et al. (1990) and Prange et al. (1999), in fact, that the application of axial magnetic fields leads to a significant increase of the critical Marangoni number for full- and half-zones, respectively.

The motion of the electrically conducting melt under a magnetic field induces electric currents. Lorentz forces, resulting from the interaction between the electric currents and the magnetic field, affect the flow. The complete theoretical description is rather complex; it is comprehensively treated in the monograph of Chandrasekhar (1981) on magnetohydrodynamics. Different possible degrees of approximation for the magnetohydrodynamic equations have been discussed by Baumgartl and Muller (1992).

In the following the "physics" of the problem and the introduction of the corresponding model equations are considered for the simple and representative case of constant magnetic fields:

A uniform magnetic field *Bo* generates a damping Lorentz force through the electric currents induced by the motion across the magnetic field. Like the case of other body forces (e.g., the buoyancy forces), this force can be added to the momentum equation.

Scaling the magnetic flux density with *Bo* and the electric current density with $\sigma_e V_{ref} Bo$, this equation (see Eq. (4.4b)) including the Lorentz force, can be written in the nondimensional form (for the present case $V_{ref}=\alpha/L$) and in absence of phase transition as:

$$\frac{\partial \underline{V}}{\partial t} = -\underline{\nabla}\pi - \underline{\nabla}\cdot[\underline{V}\,\underline{V}] + Pr\nabla^2\underline{V} + \mathrm{Pr}\cdot Ra_r(T - T_{ref})\underline{i}_g + \mathrm{Pr}\cdot Ha^2\left(\underline{J}\wedge \underline{i}_{Bo}\right) \quad (4.37)$$

where *Ha* is the Hartmann number:

$$Ha = BoL\left(\frac{\sigma_e}{\rho\nu}\right)^{1/2} \quad (4.38)$$

In these relations, σ_e is the electrical conductivity and $\underline{i}_{Bo}$ the unit vector in the direction of $\underline{Bo}$.

In practice, in Eq. (4.37) the magnetic field density vector $\underline{B}$ is set equal to $\underline{Bo}$ (the applied magnetic field) since the induced field $\underline{b}$ is small for $Re_m \ll 1$, where Re_m is the magnetic Reynolds number (μ_o is the magnetic field constant):

$$Re_m = \mu_o\sigma_e V_{ref} L \quad (4.39)$$

It is known (see, e.g., Baumgartl et al., 1990) that Re_m behaves as a measure of the "bending" of field lines of the original applied undisturbed magnetic field $\underline{Bo}$ by the fluid flow with velocity $\underline{V}$: $|\underline{B}| = |\underline{Bo}| + O(Re_m)$; this deviation of the disturbed field $\underline{B}$ from the original applied field $\underline{Bo}$ can be estimated to be very small in the case of semiconductor melts and related processing ($Re_m = O(10^{-3})$). In fact, the magnetic diffusion coefficient $(\mu_o\sigma_e)^{-1}$ is about 4 orders of magnitude greater than the thermal diffusivity α which in turn is 10^2 times higher than the kinematic viscosity ν (values given for liquid silicon). So on a time-scale defined by α or ν which gives a natural time-scale for transport processes in the fluid (if the flow velocity is not too high) the magnetic field $\underline{B}$ retains always its steady-state value.

Accordingly, the nondimensional electric current density $\underline{J}$ is given by Ohm's law for a moving fluid:

$$\underline{J} = \underline{E} + \underline{V} \wedge \underline{i}_{Bo} \tag{4.40}$$

where $\underline{E}$ is the electric field normalized by $V_{\text{ref}} \cdot Bo$. Since, as mentioned before, the unsteady induced field $\underline{b}$ is negligible, in particular, the electric field can be written as the gradient of an electric potential:

$$\underline{E} = -\underline{\nabla}\Phi_{\text{e}} \tag{4.41}$$

The conservation of the electric current density gives:

$$\underline{\nabla} \cdot \underline{J} = 0 \tag{4.42}$$

which combined with Eq. (4.41) gives a Poisson equation for the electric potential:

$$\nabla^2\Phi_{\text{e}} = \underline{\nabla} \cdot \left(\underline{V} \wedge \underline{i}_{Bo}\right) = \underline{i}_{Bo} \cdot \underline{\nabla} \wedge \underline{V} \tag{4.43}$$

Finally, the momentum equation can be rewritten as:

$$\frac{\partial \underline{V}}{\partial t} = -\underline{\nabla}\pi - \underline{\nabla} \cdot [\underline{V}\,\underline{V}] + \text{Pr}\nabla^2\underline{V} + \text{Pr} \cdot Ra_{\text{r}}(T - T_{\text{ref}})\underline{i}_{\text{g}} + \text{Pr} \cdot Ha^2\left(-\underline{\nabla}\Phi_{\text{e}} \wedge \underline{i}_{Bo} + \underline{V} \wedge \underline{i}_{Bo} \wedge \underline{i}_{Bo}\right) \tag{4.44}$$

In agreement with the findings of Baumgartl et al. (1990), Prange et al. (1999) and Saghir and Viviani (2001), Fig. 4.99 show that axial magnetic fields lead to a concentration of the Marangoni flow close to the free surface as well as to the dumping of the flow velocity (see Fig. 4.100).

Owing to this effect, the diameter of the toroidal convection rolls is increased and correspondingly its effective aspect ratio is decreased (see Section 4.10). As a consequence of this behavior, axial magnetic fields tend to favor modes with larger wave numbers (Prange et al., 1999). Table 4.7 and Fig. 4.101 show that Ma_{c1} is doubled for $Ha = 10$.

Generally, it is possible to reduce or suppress Marangoni convection in semiconductor melts, and thus the formation of dopant striations, through the use of either static (Cröll et al., 1998b; Dold et al., 1998) or dynamic (Dold et al., 2001) magnetic fields. The transition from a time-dependent to a laminar flow regime in these cases is usually coupled with a resymmetrization of the flow geometry. Over recent years, the first choice had nearly always been the application of static magnetic fields to suppress time-dependent flows; transverse rotating magnetic fields, however, follow a different approach: it is not intended to damp the fluid motion, but to

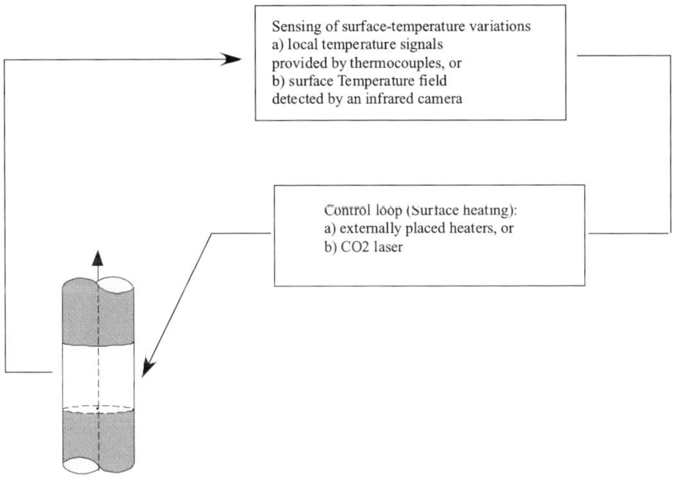

Fig. 4.98 Conceptual sketch of the control loop for the suppression of hydrothermal waves.

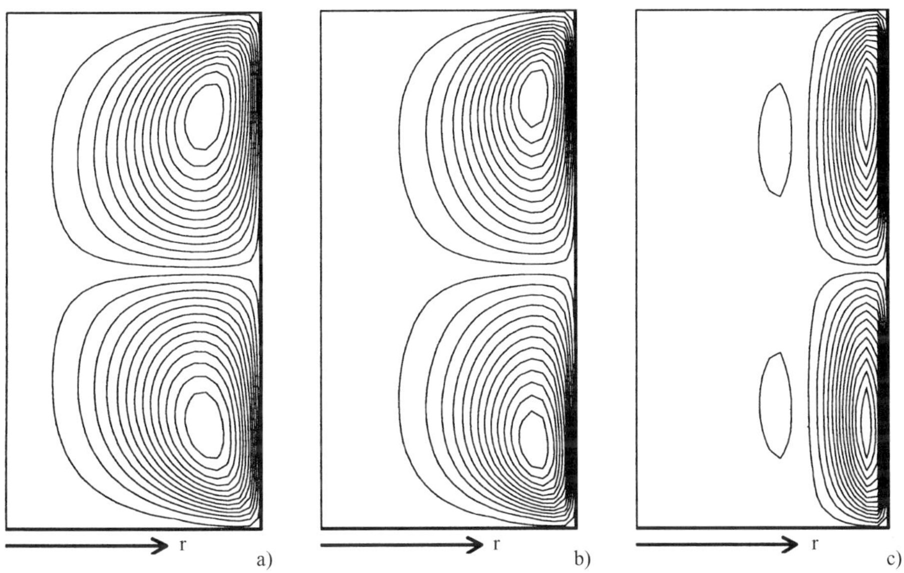

Fig. 4.99 Streamlines of axisymmetric convection in a silicon floating zone ($Pr = 0.01$, $A_F = 1$, $S = 1$, $\Phi = \varepsilon Q = 1$ w): (a) pure Marangoni convection, (b) $Ha = 10$, (c) $Ha = 50$.

dominate the irregular flow structure by overlaying it with fast azimuthal, axisymmetric flows.

A number of other methods exist to control melt flow; for example, by rotating or accelerating the crystal/crucible (see, e.g., the discussion about the Czochralski method in Section 1.5.2). Another method, more recently suggested, is to use axial

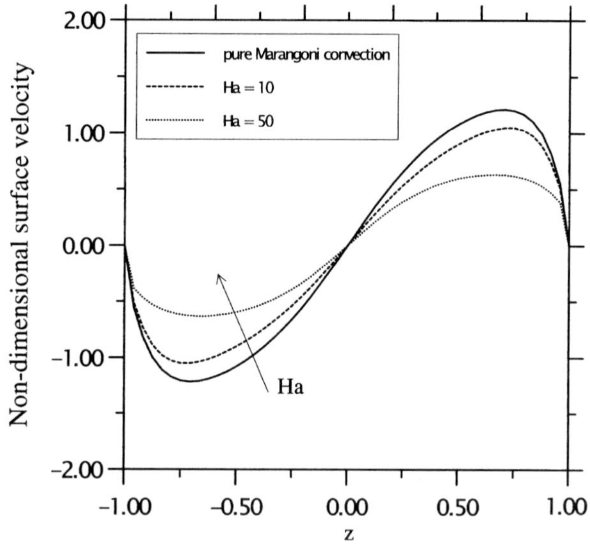

Fig. 4.100 Surface-velocity distribution ($Pr = 0.01$, $A_F = 1$, $S = 1$, $\Phi = \varepsilon Q = 1$ w) versus the Hartmann number (axial static magnetic field).

Table 4.7 Critical azimuthal wave number and critical Marangoni number versus the Hartmann number ($Pr = 0.01$, full-zone, $A_F = 1$, $S = 1$, microgravity conditions).

Ha	m	Ma_{c1}
0	2	12.41
10	2	25.42

vibrations. In some cases Marangoni and thermovibrational mechanisms produce flow motions of opposite direction, so that their competition is expected to be useful for controlling the intensity of the flow. Moreover, vibrational fields can change the Marangoni flow structure and are capable of exerting a strong influence on the flow stability.

These aspects were pointed out for the first time by Birikh et al. (1992). The analysis of the joint action of vibrational and thermal Marangoni convection mechanisms demonstrated the possibilities for providing effective control of flow in an infinite fluid layer (it was found that longitudinal vibrations stabilize the flow by deforming the velocity profiles). The effect on the hydrothermal instability in liquid bridges was demonstrated by Tang et al. (1996) in the case of silicone oils.

The possible utilization of such a strategy for the FZ problem (liquid metals) has been suggested by Gershuni et al. (1994), Lyubimova et al. (1994), and Lyubimova et al. (2003). Like the case of magnetic fields, vibrations allow a contactless control of the melt flow. It is a rather new and as yet less investigated technique that can be

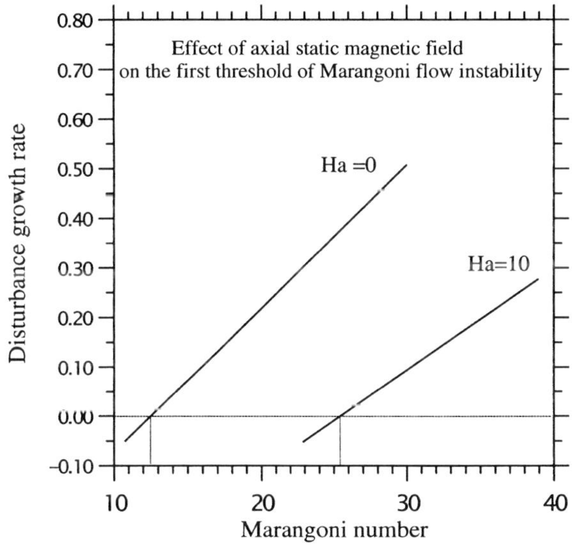

Fig. 4.101 Disturbance growth rate versus the Marangoni number for two different values of the Hartmann number.

used more universally because it is not restricted to the electrically conductive melts as is the case for magnetic fields or to high Prandtl number liquids as is the case for active temperature control.

For this reason, it is considered in the next section as well. Besides the aforementioned technological application, the interaction of the thermal Marangoni flow and the thermovibrational convection investigated in Section 4.20 complements, from a theoretical point of view, the similar topics of a mixed buoyancy/Marangoni convection and a mixed buoyancy/thermovibrational treated in Sections 2.4 and 2.7, respectively. In this case new mechanisms of flow generation arise, and by varying the vibrational parameters (i.e. amplitude or vibration frequency), it is possible to change the flow characteristics (not only to damp the melt flow but also to generate a specific one) in a wide range.

4.20. Mixed Marangoni/thermovibrational convection

As in the case of magnetic fields treated in the previous section, a full-zone of silicon with $A_F = 1$ ($S = 1$, zero-g) heated by an equatorial ring heater with $\Phi = \varepsilon Q = 1$ w is considered as a reference case (this corresponds to $Ma = O(10)$). The entire system is supposed to oscillate in the axial direction with displacement amplitude b and angular frequency ω (the reader will find additional theoretical background in Section 2.6).

The time-averaged formulation (the Gershuni formulation already illustrated in Chapter 2) is used for the treatment of a thermovibrational convection since it considerably simplifies the initial problem in the case of small amplitude and high frequency of the periodic disturbances (see Section 2.6.4) and can be applied with considerable saving of computing time.

Figures 4.102a and 4.103a show the classical Marangoni vortex cells generated in the upper and lower parts of the liquid zone in the case of vibrations being absent ($Ra_V = 0$). On the surface, the flow is directed from the ring heater towards the supporting disks. For the considered conditions convection is axisymmetric since the Marangoni number is $Ma \cong 9 < Ma_{c1}$ (see Table 4.4).

If the case of pure thermovibrational convection is considered (surface-tension effects "switched off"), Figs. 4.102b and 4.103b show that the flow has the form of two vortices with the direction of circulation opposite to that of Marangoni flow. The location of these vortices almost coincides with that of Marangoni vortices.

These features suggest that if the vibrational Rayleigh number is tuned in such a way that the magnitude of the surface velocities is comparable to that related to the Marangoni flow, it should be possible to considerably damp Marangoni convection.

Both types of convection structures, in fact (as shown in Fig. 4.102a and b) are incompatible in the sense that their respective transport mechanisms exclude each other.

Figure 4.104 shows that the vibrational Rayleigh number providing such a condition is $Ra_V \cong 150$. The velocity field resulting from the combined action of surface tension and vibrations is shown in Figs. 4.102c and 4.103c. Figure 4.103c shows that two small thermocapillary rolls are still present; they are confined

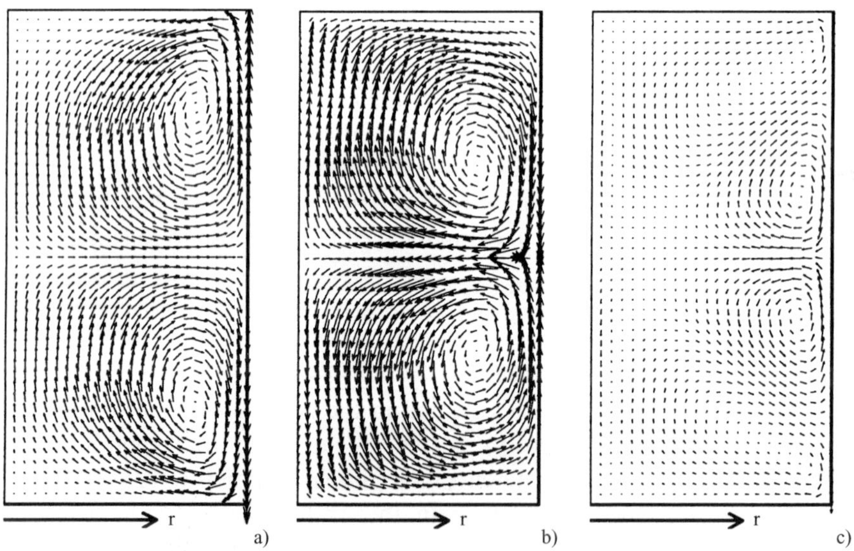

Fig. 4.102 Velocity field in a silicon floating zone ($Pr = 0.01$, $A_F = 1$, $S = 1$, $\Phi = \varepsilon Q = 1$ w): (a) pure Marangoni convection, (b) pure thermovibrational convection $Ra_V = 150$, (c) combined convection.

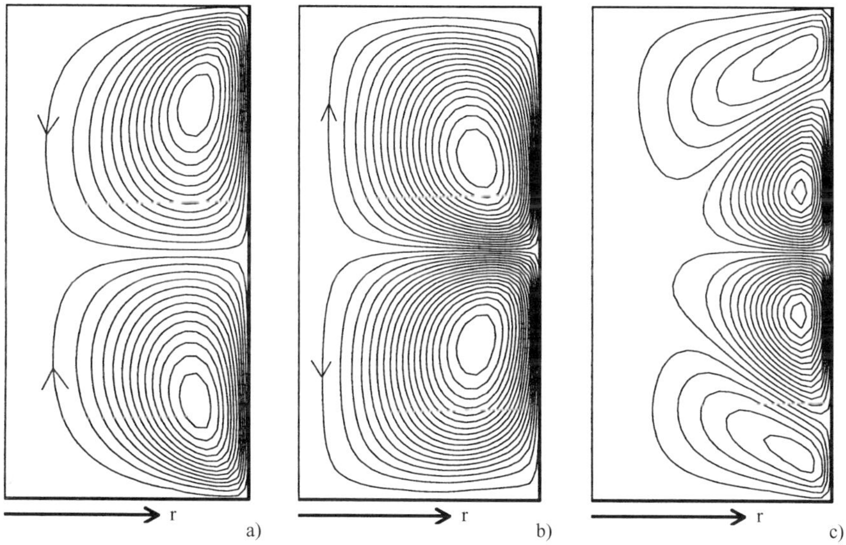

Fig. 4.103 Streamlines of axisymmetric convection in a silicon floating zone ($Pr = 0.01$, $A_F = 1$, $S = 1$, $\Phi = \varepsilon Q = 1$ w): (a) pure Marangoni convection, (b) pure thermovibrational convection $Ra_V \cong 150$, (c) combined convection.

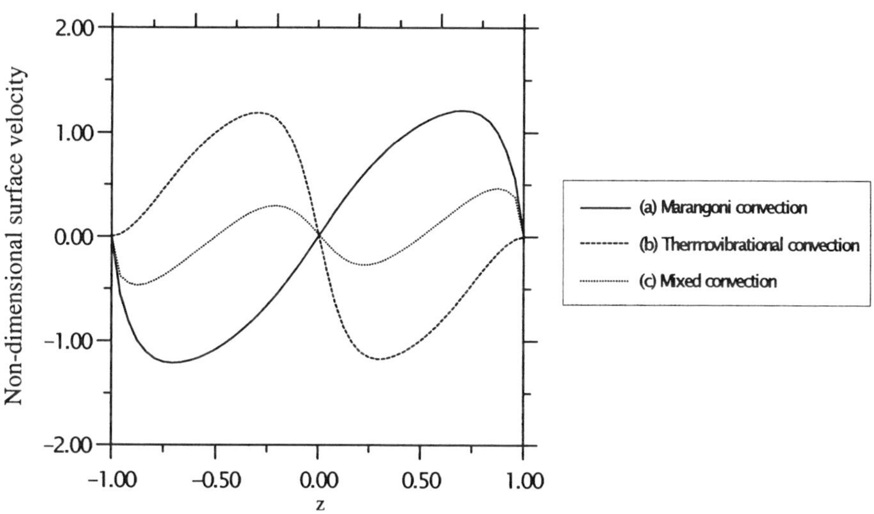

Fig. 4.104 Surface-velocity distribution $\Phi = \varepsilon Q = 1$ w: (a) pure Marangoni convection, (b) pure thermovibrational convection $Ra_V = 150$, (c) combined convection.

close to the disks. Figure 4.102c, however, illustrates that these cells are very weak. Other counter-rotating cells driven by thermovibrational convection are located in a zone close to the ring heater. This simple example makes it possible to conclude that the vibrational mechanism can be effectively used for the suppression of the

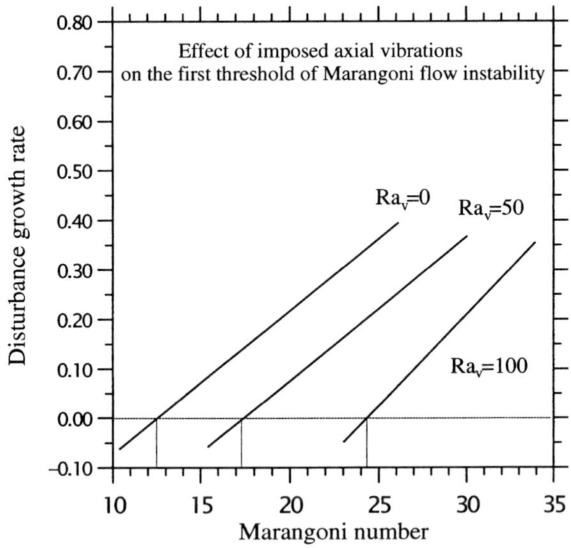

Fig. 4.105 Disturbance growth rate versus the Marangoni number for different values of the vibrational Rayleigh number.

Table 4.8 Critical azimuthal wave number and critical Marangoni number versus the vibrational Rayleigh number ($Pr = 0.01$, full-zone, $A_F = 1$, $S = 1$, microgravity conditions).

Ra_V	m	Ma_{c1}
0	2	12.41
50	2	17.26
100	2	24.17

Marangoni flow. The vibration effects reduce the intensity of the circulation cell, the local temperature deformations and the shear stresses below the free surface.

Like the case of magnetic fields, the application of axial vibrations does not alter the fact that the first instability of the Marangoni flow is axisymmetry-breaking. Table 4.8 shows the significant increase of the critical threshold for the onset of 3D convection as Ra_V is increased (see also Fig. 4.105).

The case of g-jitters perpendicular to the free surface is not considered herein since it is not of interest from a technological point of view. For the sake of completeness, however, it should be pointed out that the application of such g-jitters to the case of high Prandtl number liquid bridges results in a strong stabilization of the standing-wave regime (not shown) with hot and cold surface-temperature disturbances pulsating along the direction of the imposed vibrations.

CHAPTER 5

Macromolecular crystal growth: surface kinetics and morphological studies

5.1. Introduction

In recent years great interest has been directed towards crystals of biological macromolecules and towards the crystallization process of protein substances. In fact, single crystals with good diffraction properties and structural quality are needed to achieve high-resolution data on protein structure, needed for progress in biotechnology and drug industries. Typically, the crystals are obtained by precipitation from supersaturated solutions with a number of techniques. Producing protein crystals of adequate size, however, is often the "bottleneck" for three-dimensional, atomic-resolution structure studies of protein molecules. Superimposed on this is the poor state of our current understanding of macromolecular-crystallization mechanisms and the forces that promote and maintain these mechanisms (see also the introductory information provided in Section 1.6). For this reason further investigation is required.

Because of the large size and complexity of protein molecules and the weakness of the bonding forces between them, it is often believed that the experimental principles and methodologies (and associated mathematical models) underlying high-quality inorganic crystal growth are of little help in protein crystallization.

Macromolecules are extremely complex physical–chemical systems whose properties vary as a function of many environmental influences. Moreover, the various parameters under one's control are not independent of one another and their interrelations may be difficult to discern. It is, therefore, not easy to elaborate rational guidelines relating to physical factors that can increase the probability of success in crystallizing a particular protein.

Along these lines, today the major effort is devoted to a better understanding of the conditions that may influence the growth process. Such knowledge, in fact, may support the introduction of new crystallization techniques and the design of new reactors, leading to crystals with improved properties.

With regard to these aspects it is worthwhile to point out how research is generally carried out at two distinct length scales.

If the investigators are interested in the analysis of macroscopic behaviors, the shape evolution of the crystals is ignored and the crystals are considered only in terms of mass and position. On the ground, typically, investigators let reagents diffuse in a gel. In this environment, reagents react to produce spaced deposits and the gel prevents sedimentation and the onset of convection. In such conditions, much data can be obtained dealing with the "ensemble behavior" of the systems as a function of many parameters (effect of initial concentration of precipitant agents and/or of protein on the number of nuclei that are created when nucleation occurs, their spacing and size, effect of the size of the reactor, etc.). Usually in these analyses the mechanisms of nucleation (initial crystal formation) also form a relevant and important part of the investigation (these aspects will be discussed in detail in the next chapter).

If the local evolution of the crystal surface is the subject of the investigation (distribution of the local growth rate along a crystal face, shape instabilities, onset of surface depressions due to diffusive and/or convective effects, local pluming effects, i.e. all those factors dealing with the local history of the shape) the approach undertaken by researchers is very different. It must, in fact, take into account microscopic details. The present chapter is devoted to the mathematical treatment of such local aspects.

5.2. Surface-attachment kinetics and convective effects

Since it is expected that growth at the microscopic scale is driven by the phenomena occurring in a thin "halo" surrounding the crystal (the so-called depletion zone) where the concentration of the nutrients is lowered due to solute absorption, the investigators tend to focus on "what happens close to the crystal." Several studies have shown that the growth rates are sensitive to a very large number of parameters: e.g., temperature, supersaturation, defect formation, shear forces, etc. In particular, it has been shown that the growth rates can be crystal-surface-orientation dependent, i.e. the different faces of a growing crystal may exhibit different growth according to the crystallographic structure ("faceted" or anisotropic growth; see, e.g., Coriell et al., 1998); correspondingly crystal "habit" change may occur due to changes in the level of supersaturation. All these aspects are usually known as "surface kinetics" of the growth process. The problem is very complex and far from being well understood. An exhaustive review of the earlier experimental studies on fundamental protein crystal growth and morphology has been given in Monaco and Rosenberger (1993) and in Vekilov and Chernov (2002). These reviews show, in particular, that earlier experimental reports have offered conflicting results on the "controlling forces" in protein crystal growth.

With regard to the rate of growth of protein crystals, it is well known that there are two important effects to consider: the transport of molecules to the face of a growing nucleus or crystal, and the frequency with which the molecules orient and attach themselves to the growing surface (i.e. crystal growth rates can be considered in terms of mass transport in the liquid phase and attachment kinetics). However, the results provided by the different investigators do not agree about the relative importance of these two mechanisms. The presence of a distinct boundary layer around the crystals supports the idea of competitive transport and attachment kinetics in limiting the crystal growth rate. However, the interplay between these effects is still unknown, even if the case of mass transport in the liquid phase simply governed by diffusion is considered. Superimposed on this is the poor state of our current understanding of the effect of convective transport in the liquid phase. Some interesting information has been provided by recent experiments in space (see Littke and John, 1986; Day and McPherson, 1992; McPherson, 1993; Bi et al., 1994 and the recent overview of Vargara et al., 2003).

Crystallization experiments carried out in microgravity conditions, in fact, yielded protein crystals that resulted in diffraction data of significantly higher resolution than the best crystals of these proteins grown on the Earth. Since an obvious difference between the space and Earth-based experiments is the magnitude of the buoyancy forces in the solution, the role of solutal convection in protein crystal growth has been proved to be very important.

Protein crystals are typically grown from solutions that, in addition to the solvent and protein, contain a precipitant of well-defined bulk concentration. The interactions between these species in the solution are not well understood. Neither is how convection affects these interactions. In general, convection enhances mass transport. As a consequence, crystal growth rates in most inorganic systems are increased by convection. Some experimental results obtained in the case of protein substances suggest that convection seems to influence interface behavior. Convective effects in protein precipitation processes, however, are very poorly understood. It is known that they impact protein structures in precipitates, particle size and morphology (Broom et al., 1988), but the mechanisms are still unknown. The available experimental results are fragmented, incomplete and often contradictory (e.g., Otàlora et al., 2001 have found a decrease of the growth rates due to convection).

While efficient numerical methods have been developed and made available to the scientific community in the case of growth of inorganic substances (in particular for the case of crystals grown by thermal solidification of melts under pure diffusive regime or under various convective effects, see, for instance, Chapter 4), the modeling of the growth of protein crystals and of the interplay between the growth and the presence of convective fields is still an open task.

5.3. Morphological studies

An interesting effort in the understanding of the interaction between organic crystal growth and convection has been made by Lin et al. (1995). They devised an excellent comprehensive model of convective–diffusive transport in protein crystal growth under standard and zero-gravity conditions. Time-dependent solutions were obtained for the convective velocity field, concentration distributions of the protein and precipitant in the solution, the growth rate and salt distribution in the crystal. The protein concentration at the crystal interface was computed using an attachment-kinetics-based surface condition. The increase of mass (size) of the growing crystal due to protein depletion in the liquid phase, however, was not taken into account.

According to different experimental studies (see, e.g., Pusey et al., 1986), on the other hand, it seems that the increasing size of a growing crystal plays a "critical role" in the phenomena under investigation and, in particular, in determining the relative importance of surface-attachment kinetics and mass transport in the liquid phase under diffusive or convective conditions.

From a numerical point of view, strictly speaking, the growing crystal gives rise to a moving-boundary problem. As illustrated in the earlier chapters, moving-boundary problems are still a challenging task for numerical simulation, have instigated much research, and have led to many different solutions.

A very interesting and pioneering study related to the interplay between crystal growth and convection and to the associated moving boundary problem has been carried out recently by Noh et al. (1998). They investigated the growth behaviors of a precipitate particle in a supersaturated solution under the effect of "a priori" (well-defined) imposed "ambient" flows. The particle was assumed to have a spherical shape initially. The Stokes-flow approximation was assumed to simplify the model. Numerical solutions were obtained through a numerically generated orthogonal curvilinear co-ordinate system, automatically adjusted to fit the boundary shape at any instant. Crystal growth rates were computed through consideration of the isotropic and anisotropic interfacial free energies. The initially spherical particle was found to evolve into a peach-like shape if posed in a uniform streaming flow.

Coriell et al. (1998) investigated the effect of a parallel shear flow and anisotropic interface kinetics on the onset of morphological instability during growth from a supersaturated solution. Their excellent linear stability analysis was carried out taking into account a kinetic coefficient that is a nonlinear function of the supersaturation and crystallographic orientation.

The first method capable of modeling complex multiphase flow separated by a moving interface and able to undertake a fixed-grid solution without resorting to mathematical manipulations and transformations (i.e. moving-grid boundary-fitted methodology), was the Marker and Cell (MAC) of Harlow and Welch (1965). Instead of MAC, however, volume-tracking methods (e.g., volume of fluid, SLIC simple line interface calculation, and PLIC piecewise linear interface calculation), front-tracking and level-set methods have become popular in recent years.

The philosophy underlying, in particular, volume-tracking methods has been elucidated in Chapter 3; for a very comprehensive discussion on the genesis and evolution of these methods the reader may also consider Hirt and Nichols (1981), Osher and Sethian (1988), Brackbill et al. (1992), Unverdi and Tryggvason (1992), Sussman et al. (1994), Rider and Kothe (1998) and references therein,

Gueyffier et al. (1999), Sethian (1999), Osher and Fedkiw (2002), etc. As discussed in Section 3.5, in particular, they have enjoyed widespread use in the simulation of typical problems associated with gas–liquid or liquid–liquid systems where the surface-tension effects play a "critical role" in determining the shape of the fluid/fluid interface and/or its motion (e.g., gas–bubble or drops in liquid matrices, see also Nobari et al., 1996; Ory et al., 2000, etc.).

On the other hand, "enthalpy methods" and similar techniques taking into account the release or absorption of latent heat have been successfully applied to the case of thermal phase-change problems characterized by the presence of moving solid/melt interfaces due to the heating or cooling of the system under investigation (see Chapter 4, and in particular, Section 4.2).

Despite these worthy and ambitious contributions, however, a complete numerical method aimed at the ("moving boundary") simulation of the growth of organic crystals (due to the addition of solute molecules at the solid/liquid interface) by solubility modulation is still not available.

The aims of the present chapter are (1) to introduce a mathematical model to handle the complex phenomena related to the growth or resolution of protein crystals according to the protein concentration and solubility distribution in the feeding solution, taking into account the mass variation (and the morphological evolution) of the crystal due to the incorporation of nutrients and the possibility of (faceted) surface-orientation-dependent growth; (2) shed some light on the complex interplay and relative importance of the surface-attachment kinetics and the mass transport (diffusive and/or convective) in the liquid phase; the present chapter deals, in fact, with a novel numerical technique to investigate the combined effect of crystal growth and gravitational convection and, in particular, the interaction of this motion with the concentration field around the crystal and with the growth rates. No simplifying assumptions (such as Stokes flow, "a priori" imposed ambient flows or "fixed" crystal size) are assumed in the development of the model.

5.4. Differences between organic and inorganic crystal growth

Before embarking on the detailed coverage of the aforementioned aspects, the present section tries to elucidate the major differences between inorganic and organic crystal growth. Such an effort has a two-fold purpose: an attempt to create a link between this chapter and the earlier ones and to introduce the reader who may not be an expert in these topics to the field of macromolecular crystallization.

Along these lines, it should be pointed out that, while an extensive body of theory and experimental results is available for the crystal growth rate analysis of low molecular weight species, there is less experimental data for higher molecular weight material and, in particular, for the crystallization of proteins. Factors such as

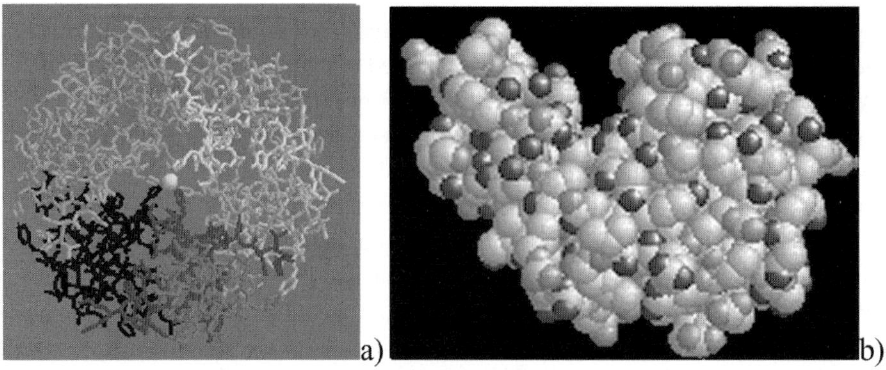

Fig. 5.1 Protein molecules: (a) insulin, (b) lysozyme (Courtesy of CMMR).

limited availability of material and fragility of the crystals, in fact, have all combined to limit investigations into protein crystal growth kinetics.

The perplexing difficulties that arise in the crystallization of macromolecules in comparison with conventional small molecules stem from the greater complexity (see, e.g., Fig. 5.1), lability and dynamic properties of proteins. Proteins are very sensitive to their environment and if exposed to sufficiently severe conditions may denature and/or degrade. They must be constantly maintained in a thoroughly hydrated state at or near physiological pH and temperature. Thus common methods for the crystallization of conventional molecules such as evaporation of solvent, dramatic temperature variation, or addition of strong organic solvents are unsuitable and destructive (McPherson, 1985, 1990; Kuznetsov et al., 1995). They must be supplanted with more gentle and restricted techniques. Because proteins are sensitive and labile macromolecules that readily lose their native structures, the only conditions that can support crystal growth are those that cause little or no perturbation of the molecular properties. The protein crystals must be grown from a solution to which they are tolerant. As anticipated in Section 1.6, this is called the mother liquor. Because complete hydration is essential for the maintenance of structure, protein crystals are always bathed in the mother liquor.

The strategy employed to bring about crystallization is to guide the system very slowly toward a state of reduced solubility by modifying the properties of the solvent.

The classical procedure for inducing proteins to separate from the solution and produce a solid phase is to gradually increase the level of saturation of the precipitant agent.

Protein precipitants fall into four broad categories: (a) salts; (b) organic solvents; (c) long-chain polymers; and (d) low molecular mass polymers and nonvolatile organic compounds (Arakawa and Thimasheff, 1985).

Salts exert their effect by dehydrating proteins through competition for water molecules (their ability to do this being proportional to the square of the valences of the ionic species composing the salt).

For a specific protein, the precipitation point (or solubility minima), however, is usually critically dependent on the pH, temperature, the chemical composition of the precipitant and the properties of both the protein and the solvent. Just as proteins may be driven from solution at constant pH and temperature by the addition or removal of salt, they can similarly be crystallized or precipitated at constant ionic strength by changes in pH or temperature. This is because the electrostatic character of the macromolecule, its surface features, or its conformation may change as a function of pH, temperature and other variables as well.

By these techniques a "suitable" limited degree of supersaturation can be achieved. In very concentrated solutions, in fact, the macromolecules may aggregate as an amorphous precipitate. This result is to be avoided if possible and is indicative that supersaturation has proceeded too extensively or too swiftly.

Many other distinguishing features exist with respect to the crystallization of inorganic substances (see, e.g., Rosenberger, 1986; Feigelson, 1989). In many protein crystallization systems, interfacial kinetics are slow and are considered a major source of difficulties in crystal growth. In many inorganic crystal growth systems, however, the growth rate is limited by the transport of species to the interface so that the fluid dynamics of the nutrient phase is at the root of growth problems. The inorganic crystal growth, in general, is not plagued by problems related to interfacial kinetics.

Superimposed on this is the fact that the rate of growth of macromolecular crystals is determined by the nature of the growing crystal surface. Addition of molecules to a rough surface requires less energy than addition to a smooth surface, where surface nucleation is required for addition.

According to the Periodic Bond Chain Theory (Boistelle and Astier, 1988), in particular three different types of growth faces exist: flat faces, stepped faces and kinked faces. Flat faces require two-dimensional nucleation (the formation of growing sheets of molecules) in order to induce growth, and thus grow the slowest. Stepped faces grow as columns of molecules, which requires only one-dimensional nucleation, and thus have intermediate growth rates. Stepped faces typically occur as a result of a crystallographic screw axis causing spiral growth patterns to occur at the surface of the crystal. Finally, kinked faces are growth sites that do not require nucleation to promote further growth, and therefore grow faster than the other two face types. Thus the type of the growing crystal face (flat, stepped, or kinked) strongly influences the rate at which crystal growth occurs (for mathematical models dealing with these aspects see Section 5.6.2).

In protein crystallization, the problem is complicated even further by the incorporation of solvent (water) and solution (also containing precipitant) into the crystal. Thus, in contrast to anhydrous inorganic crystallization, there is only partial (often negligible) rejection of the solvent at the crystal/solution interface.

5.5. Moving-boundary approach

In the present section, ancient "moving-boundary" numerical methods and, in particular, possible novel strategies are briefly outlined in order to point out analogies, similarities and differences with the sophisticated numerical algorithm proposed here for the case of crystallization of organic substances due to solubility modulation (OCGVOF-Organic Crystal Growth Volume of Fraction Method).

Such discussion, being propaedeutical to the subsequent sections, is, however, mainly addressed to organic crystal growers, who generally do not know much about the philosophy of modeling and the numerical treatment of moving interfaces.

The numerical simulations of these problems require a discretization or nodalization to allow numerical treatment on computers. There are two fundamentally different approaches: Eulerian methods use a frame of reference (discretization grid or mesh, control volumes, etc.) fixed in space, and matter moving through this frame of reference. Lagrangian methods instead use a frame of reference (marker particles) fixed to and moving with the matter.

The first method capable of modeling multiphase flow, separated by a moving interface, was the aforementioned Marker and Cell (MAC) of Harlow and Welch (1965). This was, in fact, a combination of a Eulerian solution of the basic flow field, with Lagrangian marker particles attached to one phase to distinguish it from the other phase. While the staggered mesh layout and other features of MAC have become a model for many other Eulerian codes, the marker particles proved to be computationally too expensive and have been rarely used.

In the specific case of mass crystallization from a supersaturated solution and in order to introduce novel numerical techniques, one must generally accomplish at least two things simultaneously: (a) determine the concentration fields of the organic substance and precipitant in the liquid phase and (b) determine the position of the interface between the solid and liquid phases. According to the technique used to address (a) and (b), in principle, the numerical procedures able to solve these problems can be divided into two groups:

1. Multiple-region solutions utilizing independent equations for each phase and coupling them with appropriate boundary conditions at the crystal/liquid interface. This approach to the problem takes the point of view that the interface separating the bulk phases is a mathematical boundary of zero thickness where interfacial conditions are applied. These interfacial conditions couple to the concentration equations in the bulk and this system of equations and boundary conditions provides a means to address (a) and (b). Difficulties arise when this technique is employed since in this case in the vicinity of the crystallization front (phase change), conditions on mass flux, velocity, and solubility evolution have to be accounted for. This effectively rules out the application of a fixed-grid

numerical solution, as deforming grids or transformed (body-fitted) coordinate systems are required to account for the position of the crystallization front (see, e.g., the approach undertaken by Noh et al., 1998).

2. Single-region (continuum) formulations (or "phase-field" models) that eliminate the need for separate equations in each phase, by establishing conservation equations that are universally valid. From a theoretical point of view the major advantage of the single-region formulations is that they do not require the use of quasisteady approximations, numerical remeshing and coordinate mapping.

In a phase-field model, a phase-field variable ϕ that varies in space and time is introduced to characterize the phase of the material. In place of the "sharp" transition from crystal to liquid phases that would characterize the multiple-region formulations, here the phase field varies smoothly but rapidly through an interfacial region. Additionally, in place of the interfacial jump conditions used in the multiple description a differential equation applied over the entire computational domain governs the evolution of ϕ. The effect is a formulation of the free-boundary problem that does not require the explicit application of interfacial conditions at the unknown location of a phase boundary. For this reason, this strategy is adopted in the present book.

As will be discussed in the following, it exhibits an outstanding theoretical kinship with the Volume of Fluid methods introduced in Chapter 3.

5.6. OCGVOF – Organic Crystal Growth Volume of Fraction Method

5.6.1 *General properties*

The OCGVOF method, which, similar to volume-tracking (see, e.g., the VOF methods in Chapter 3) and enthalpy methods (see Chapter 4), is a single-region formulation, allows a fixed-grid solution to be undertaken and is therefore able to utilize standard solution procedures for the fluid flow and species equations directly, without resorting to mathematical manipulations and transformations (Lappa, 2003b).

This method accounts for the solid mass stored in the generic computational cell by assigning an appropriate value of ϕ to each mesh point ($\phi=1$ crystal, $\phi=0$ feeding solution and $0<\phi<1$ for an interfacial cell). The key element for the OCGVOF method is its technique for adjourning ϕ. Upon changing phase, the ϕ-value of the cell is adjusted to account for mass release or absorption, this adjustment being reflected in the protein-concentration distribution as either a source or sink. The modeling of these phenomena leads to the introduction of a group of differential equations, strictly related, from a mathematical point of view, to the "kinetic conditions" used to model mass transfer at the crystal surface.

5.6.2 *Mathematical formulation of the surface-attachment kinetics*

On a molecular level, crystal growth consists of the successive addition of "growth units" or "building blocks" (nutrients) to a lattice. Growth units can consist of single molecules or possibly of clusters of molecules. In solution, growth units are typically solvated, i.e. are surrounded by highly regularly arranged shells of solvent molecules that interact with the growth unit in a bond-like fashion. Protein crystallization does not fit any of the traditional (inorganic) models without major modification. Realistic models for protein growth kinetics will have to accommodate that protein crystals consist of (see, e.g., Rosenberger, 1986):

(a) protein molecules on regular lattice sites forming a mesh-like network;
(b) "bound water" that lines the large channels formed by the wide-open protein network in bond-like arrangements that are regular enough to be identified by X-ray diffraction;
(c) "bulk water."

These complicating features make it very unlikely that protein growth kinetics will follow any unmodified inorganic growth models. Hence, not surprisingly, the few available results for protein growth-rate dependence on bulk supersaturation deviate in form as well as magnitude considerably from any inorganic model predictions.

As anticipated in Section 5.4, the "attachment rates" in protein crystallization are considerably lower than in most inorganic systems (this slow interface kinetics have been interpreted in terms of the low symmetry and the small binding energies involved in protein crystallization).

Because of these theoretical difficulties, Pusey et al. (1986) expressed the growth rate of a protein substance as an "empirical function" of supersaturation introducing a "kinetic coefficient" (cm s^{-1}) dependent on the physical properties of the protein. This model is adopted in the present book and further improved to take into account anisotropic growth. This situation occurs for crystalline proteins that have high anisotropy (preferred orientations) in either their surface energy or atomic-attachment kinetics.

Surface-attachment kinetics at the crystal surface depend on the local value of solubility and on a coefficient λ (kinetic coefficient) having the dimension of a velocity whose value may be different according to the local orientation of the crystal surface (the aforementioned surface-orientation-dependent growth); using mass balance (see, e.g., Rosenberger, 1986), and assuming a linear dependence of the growth rate by the interface supersaturation (see, e.g., Lin et al., 1995), one obtains:

$$\left(\frac{D}{\rho_{\mathrm{P}} - \rho_{\mathrm{C}} C_i/\rho_{\mathrm{S}}}\right) \frac{\partial C}{\partial n}\bigg|_i = \lambda(\hat{\underline{n}}) \left(\frac{C_i - S}{S} - \delta_{\mathrm{O}}\right) \tag{5.1}$$

where C_i is the concentration of the protein at the crystal surface, D is the related diffusion coefficient, S is the solubility (its value is a function of the local concentration of the precipitant agent and/or of temperature), ρ_P and ρ_C are the protein mass density and the total mass density in the crystal, ρ_S is the total density of the solution, δ_O is the width of the supersaturation zone in which no growth occurs, $\lambda(\underline{\hat{n}})$ is the aforementioned kinetic coefficient and $\underline{\hat{n}}$ is the unit vector perpendicular to the crystal surface pointing into the liquid. The parameter δ_O takes into account the so-called "dead zone', that according to many investigators, in the case of organic substances, is due to the absorption of impurities that lead to strong retardation of the growth kinetics. In the case of protein substances, such "impurities" are not necessarily extrinsic contaminants, being often the result of protein "microheterogeneity" (Monaco and Rosenberger, 1993).

Whenever protein in solute phase and solid crystal coexist in equilibrium (saturation condition):

$$C_i = S \tag{5.2}$$

In a saturated solution, two states exist in equilibrium, the solid phase, and one consisting of free molecules in solution. At saturation, no net increase in the proportion of the solid phase can accrue since it would be counterbalanced by an equivalent dissolution. Thus "crystals do not grow from a saturated solution." The system must be in a nonequilibrium, or supersaturated state to provide the thermodynamic driving force for crystallization (McPherson, 1990).

Solution must by some means be transformed or brought into the supersaturation state whereby its return to equilibrium forces exclusion of solute molecules into the solid state, the crystal. As long as $C_i < S$, more solid material will dissolve, if any. If, on the other hand, $C_i > S$, material will condense on any material *already existing* and augment its size. The "growth regime" may be very complex and nonlinear (see, e.g., Vekilov et al., 1996). Its features depend on several parameters and, in particular, it is a function of the ratio between surface-attachment kinetics (modeled by Eq. (5.1)) and mass transport in the liquid phase (diffusive or convective).

The nucleation of new material is something else and the features of this phenomenon will be discussed in the next chapter where ensemble behaviors are the subject of investigation (for theoretical analysis of this phenomenon see also the excellent paper of McPherson, 1990); the term "precipitation" refers, in fact, to the composite phenomenon of nucleation and subsequent growth. Growth can take place at concentrations lower than those needed for nucleation, as long as $C_i > S$. The solution is said to be *supersaturated* when the solute content is greater than S, and the degree of supersaturation σ is defined by $\sigma = C/S$.

5.6.3 *Governing field equations*

The model is based on the mass-balance equations. Therefore, in the absence of convection, the diffusion of protein is governed by the equation

$$\frac{\partial C}{\partial t} = \nabla^2 C \qquad \text{if } \phi = 0 \tag{5.3a}$$

The diffusion of the precipitant agent is governed by the equation (it is assumed that precipitant does not precipitate in the solid phase):

$$\frac{\partial C_{\text{salt}}}{\partial t} = \frac{D_{\text{salt}}}{D} \nabla^2 C_{\text{salt}} \qquad \text{if } \phi = 0 \tag{5.3b}$$

The nondimensional form of the equations results from scaling the lengths by a reference distance (L), the time by L^2/D; the initial value of protein and precipitant agent (salt) are $C_{(o)}$ and $C_{\text{salt}(o)}$, respectively. Note that concentrations are not posed in nondimensional form (g cm^{-3}).

In the presence of convection, the flow is governed by the continuity, Navier–Stokes and species equations, that in nondimensional conservative form read:

$$\underline{\nabla} \cdot \underline{V} = 0 \tag{5.4}$$

$$\begin{aligned} \frac{\partial \underline{V}}{\partial t} = &-\underline{\nabla} p - \underline{\nabla} \cdot [\underline{V}\,\underline{V}] + Sc\, \nabla^2 \underline{V} + Sc \cdot Ra\left(\frac{C}{C_{(o)}} - 1\right)\underline{i}_g \\ &+ \frac{D_{\text{salt}}}{D} Sc \cdot Ra_{\text{salt}}\left(\frac{C_{\text{salt}}}{C_{\text{salt}(o)}} - 1\right)\underline{i}_g - Sc\,\frac{1}{\eta}\,\underline{V} \end{aligned} \tag{5.5}$$

where ν is the kinematic viscosity, $Sc = \nu/D$, $Ra = g\beta_{\text{prot}} C_{(o)} L^3/\nu D$ and $Ra_{\text{salt}} = g\beta_{\text{salt}} C_{\text{salt}(o)} L^3/\nu D_{\text{salt}}$ (following the concepts in Chapter 2, the Boussinesq approximation is used to model the buoyancy forces, β_{prot} and β_{salt} are the solutal expansion coefficients related to organic substance and salt, respectively);

$$\frac{\partial C}{\partial t} = [-\underline{\nabla} \cdot (\underline{V} C) + \nabla^2 C] \qquad \text{if } \phi = 0 \tag{5.6}$$

$$\frac{\partial C_{\text{salt}}}{\partial t} = [-\underline{\nabla} \cdot (\underline{V} C_{\text{salt}}) + (D_{\text{salt}}/D)\nabla^2 C_{\text{salt}}] \qquad \text{if } \phi = 0 \tag{5.7}$$

where $\underline{V}$ and p are the nondimensional velocity and pressure. The reference pressure is $\rho_S D^2/L^2$. Note that when solubility modulation is induced by temperature control, Eq. (5.7) must be replaced by the energy equation.

Assumptions invoked in the development of equations for this continuum model include: laminar flow, Newtonian behavior of the phases (this implies that solids should be treated as highly viscous fluids), constant phase densities (and Boussinesq approximation).

Moreover, the solid phase (crystal) is assumed to be nondeforming and free of internal stress, while the multiphase region (region where phase change occurs, i.e. region where the "growth units" are added to the pre-existent crystalline structure) is viewed as a porous solid characterized by an isotropic permeability η by analogy with the enthalpy methods (for additional insight into this aspect see, e.g., Section 4.2, Voller and Prakash, 1987; Bennon and Incropera, 1987a, b; Brent et al., 1988).

The species equations (5.6) and (5.7) are not solved for the domain occupied by solid phase since macromolecular solute and salt cannot diffuse through the lattice; nutrient is incorporated (attached) at the crystal surface for $0<\phi<1$ according to Eq. (5.1), i.e. the crystal surface behaves as a sink for the protein available in the liquid phase.

The rate of absorption for the above-mentioned phenomenon is usually measured by the so-called growth rate θ^* (that can also be seen also as a measure of the velocity of the moving solid/liquid interface, see the next section for further details).

An important problem with fixed-grid solution procedures, however, is accommodating the zero-velocity condition, which is required as a liquid region turns to solid. Various methods can be used, in principle, to "switch off" velocities in computational cells that are becoming solid (or "switch on" velocities in the reverse case, see Section 4.2 for further details). A possible approach (porosity approach) requires that computational cells that are undergoing a phase change are modeled as a pseudoporous media, with the pseudopermeability η being a function of ϕ ranging between 0 (fully liquid cell) and 1 (fully solid cell). For the present case of macromolecular organic crystal growth, this assumption can be based directly on the fact that solid formation occurs as a "permeable" crystalline matrix that coexists with the liquid phase. The term $-ScV/\eta$ in Eq. (5.5) in fact, is the Darcy term added to the momentum equation to eliminate convection in the solid phase. The pseudopermeability can be assumed to vary according to the Carman–Kozeny equation (Bennon and Incropera, 1987b; Brent et al., 1988).

$$\eta = \frac{(1-\phi+\varepsilon)^3}{(\phi+\varepsilon)^2}, \qquad \varepsilon = 10^{-5} \tag{5.8}$$

In the pure solid ($\phi=1$) and pure liquid ($\phi=0$), Eq. (5.8) reduces to the appropriate limits, namely $\eta \cong 0$ and $\eta \rightarrow \infty$, respectively. In practice, the effect of η is as follows: in full-liquid elements $1/\eta$ is very small and has no influence; in elements that are changing phase, the value of $1/\eta$ will dominate over the transient, convective and diffusive components of the momentum equation, thereby forcing them to imitate the Carman–Kozeny law; in totally solid elements, the final large value of $1/\eta$ will swamp all the terms in the governing equations and force any velocity predictions effectively to zero.

Since the momentum equation is valid throughout the entire domain, explicit consideration need not be given to boundaries between solid, multiphase and liquid regions.

5.6.4 *Phase-field equation*

On the surface of the crystal ($|\underline{\nabla}\phi| \neq 0$, $0<\phi<1$), the protein concentration must satisfy the kinetic condition that in nondimensional form reads:

$$\left(\frac{1}{\rho_{\mathrm{P}} - \rho_{\mathrm{C}} C_i/\rho_{\mathrm{S}}}\right)\frac{\partial C}{\partial n}\bigg|_i = \tilde{\lambda}(\underline{\hat{n}})(\sigma_i - 1 - \delta_{\mathrm{O}}) \tag{5.9}$$

where $\tilde{\lambda} = \lambda L/D$, σ (the aforementioned degree of supersaturation) is a function of the local precipitant concentration and

$$\underline{\hat{n}} = -\frac{\underline{\nabla}\phi}{|\underline{\nabla}\phi|} = (\alpha, \beta) \tag{5.10}$$

$$\alpha = -\frac{\partial\phi}{\partial x}\bigg/\sqrt{\left(\frac{\partial\phi}{\partial x}\right)^2 + \left(\frac{\partial\phi}{\partial y}\right)^2}, \qquad \beta = -\frac{\partial\phi}{\partial y}\bigg/\sqrt{\left(\frac{\partial\phi}{\partial x}\right)^2 + \left(\frac{\partial\phi}{\partial y}\right)^2} \tag{5.11}$$

since $\partial C/\partial n = \alpha(\partial C/\partial x) + \beta(\partial C/\partial y)$ (hereafter the subscript "i" is omitted), Eq. (5.9) can be rewritten as

$$\alpha\frac{\partial C}{\partial x} + \beta\frac{\partial C}{\partial y} = \tilde{\lambda}(\alpha, \beta)(\rho_{\mathrm{P}} - \rho_{\mathrm{C}} C/\rho_{\mathrm{S}})(C/S - 1 - \delta_{\mathrm{O}}) \tag{5.12}$$

$\tilde{\lambda}(\rho_{\mathrm{P}} - \rho_{\mathrm{C}} C/\rho_{\mathrm{S}})(C/S - 1 - \delta_{\mathrm{O}})$ represents the mass-exchange flux between solid and the liquid phase (i.e. crystal and protein solution). The mass stored in computational cells that are undergoing phase change can be computed according to the equation:

$$\frac{\partial M}{\partial t} = \frac{\lambda(\alpha, \beta)L^4}{D}(\rho_{\mathrm{P}} - \rho_{\mathrm{C}} C/\rho_{\mathrm{S}})(C/S - 1 - \delta_{\mathrm{O}})\,\mathrm{d}s \tag{5.13}$$

where ds is the "reconstructed" portion of the crystal surface (by definition perpendicular to the interface normal vector $\underline{\hat{n}}$) "bounded" by the frontier of the control volume (computational cell) located astride the crystal surface.

The nondimensional volume of the crystal mass stored in a grid cell can be computed as (ρ_P is the protein mass density in the crystal):

$$\mathrm{d}v|_{\text{stored}} = \frac{1}{L^3}\frac{M}{\rho_P} \tag{5.14}$$

Correspondingly:

$$\phi = \frac{\mathrm{d}v|_{\text{stored}}}{\mathrm{d}v} \tag{5.15}$$

where dv is the volume of the computational cell.

Therefore, the phase-field equation reads

$$\begin{aligned} &\partial\phi/\partial t = 0, && \text{if } |\underline{\nabla}\phi| = 0 \\ &\frac{\partial\phi}{\partial t} = \frac{\tilde{\lambda}(\alpha,\beta)(\rho_P - \rho_C C/\rho_S)(C/S - 1 - \delta_O)\,\mathrm{d}s}{\rho_P\,\mathrm{d}v}, && \text{if } |\underline{\nabla}\phi| \neq 0,\ 0 < \phi < 1 \end{aligned} \tag{5.16}$$

with C satisfying Eq. (5.12).

From mass balance the solution for the normal velocity at the interface is:

$$\underline{V}\cdot\underline{\hat{n}} = (\rho_C/\rho_S)\theta^*, \qquad \text{if } |\underline{\nabla}\phi| \neq 0,\ 0 < \phi < 1 \tag{5.17a}$$

where $\rho_S = \rho_{H_2O}(1 - \beta_{\text{prot}}C_{(o)} - \beta_{\text{salt}}C_{\text{salt}(o)})$ and the nondimensional growth rate (see, e.g., Pusey et al., 1986) can be computed as:

$$\theta^* = \left(\frac{1}{\rho_P - \rho_C C/\rho_S}\right)\frac{\partial C}{\partial n} = \tilde{\lambda}(\underline{\hat{n}})(\sigma - 1 - \delta_O) \tag{5.17b}$$

with C satisfying Eq. (5.12).

As anticipated, the parameter θ is a measure of the rate of absorption of the protein at the crystal surface (it is proportional to $\partial C/\partial n$) and can also be seen as a measure of the velocity of the advancing solid/liquid interface. It is called growth rate by organic crystal growers and usually is expressed in dimensional form as [Å s^{-1}]. According to the present method it can be computed on the basis of Eq. (5.17b) after the computation of the protein concentration at the crystal surface according to Eq. (5.12).

Equations (5.12), (5.16), and (5.17) behave as "moving-boundary conditions," their solution being strictly associated with the computational check on the value of ϕ and its gradient. Note that, typically, for organic crystal growth, the normal velocity at the interface is very small with respect to buoyancy-induced velocities even if the gravity acceleration is reduced by several orders of magnitude in space, and for this reason it could be neglected with regard to the solution of the momentum equation (see the order of magnitude analysis in Lappa, 2003b).

5.6.5 *Discretization*

Equations (5.4–5.7) subjected to the initial and boundary conditions may be solved numerically, for instance, by a primitive-variables method. For educational purposes and for the sake of clarity, simple forward differences in time are used to discretize the partial differential equations, in the following (superscript n indicates time step):

$$\underline{V}^{n+1} = \underline{V}^n + \Delta t\left[-\underline{\nabla}\cdot(\underline{V}\,\underline{V}) + Sc\,\nabla^2\underline{V} + Sc\cdot Ra\left(\frac{C}{C_{(o)}} - 1\right)\underline{i}_g + \frac{D_{salt}}{D}Sc\cdot Ra_{salt}\left(\frac{C_{salt}}{C_{salt(o)}} - 1\right)\underline{i}_g\right]^n - \Delta t\frac{1}{\eta}\underline{V}^{n+1} - \Delta t\,\underline{\nabla}p^n \tag{5.18}$$

$$C^{n+1} = C^n + \Delta t[-\underline{\nabla}\cdot[\underline{V}C] + \nabla^2 C]^n \tag{5.19}$$

$$C^{n+1}_{salt} = C^n_{salt} + \Delta t\left[-\underline{\nabla}\cdot[\underline{V}C_{salt}] + \frac{D_{salt}}{D}\nabla^2 C_{salt}\right]^n \tag{5.20}$$

In addition to these aspects, the method proposed herein requires that the orientation of the surface of the crystal is used in determining the face fluxes for the computation of C at the crystal surface (Eq. (5.12)). The interface orientation depends on the direction of the volume fraction gradient of the phase within the cell, and that of the neighbor cell (or cells) sharing the face in question.

In practice, the unit vector $\hat{\underline{n}}$ results from the gradient of a smoothed phase field $\hat{\phi}$, where the transition from one phase to the other takes place continuously over several cells. The smoothed phase field $\hat{\phi}$, in turn, is obtained by convolution of the unsmoothed field ϕ with an interpolation function (see, e.g., Eq. (3.22) in Chapter 3).

Depending on the interface's orientation, concentration gradients must be discretized by forward or backward schemes. For this reason, Eq. (5.12) in discretized form reads:

$$\alpha > 0,\quad \beta > 0:\quad C^{n+1}_{i,j} = \frac{\left[\tilde{\lambda}(1+\delta_O)\left(\rho_P - \rho_C C^{n+1}_{i,j}/\rho_S\right) + \alpha C^{n+1}_{i+1,j}/\Delta x + \beta C^{n+1}_{i,j+1}/\Delta y\right]}{\left[\tilde{\lambda}\left(\rho_P - \rho_C C^{n+1}_{i,j}/\rho_S\right)/S + \alpha/\Delta x + \beta/\Delta y\right]} \tag{5.21a}$$

$$\alpha < 0, \quad \beta > 0: \quad C_{i,j}^{n+1} = \frac{\left[\tilde{\lambda}(1+\delta_{\mathrm{O}})\left(\rho_{\mathrm{P}} - \rho_{\mathrm{C}} C_{i,j}^{n+1}/\rho_{\mathrm{S}}\right) - \alpha C_{i-1,j}^{n+1}/\Delta x + \beta C_{i,j+1}^{n+1}/\Delta y\right]}{\left[\tilde{\lambda}\left(\rho_{\mathrm{P}} - \rho_{\mathrm{C}} C_{i,j}^{n+1}/\rho_{\mathrm{S}}\right)\Big/ S - \alpha/\Delta x + \beta/\Delta y\right]} \tag{5.21b}$$

$$\alpha > 0, \quad \beta < 0: \quad C_{i,j}^{n+1} = \frac{\left[\tilde{\lambda}(1+\delta_{\mathrm{O}})\left(\rho_{\mathrm{P}} - \rho_{\mathrm{C}} C_{i,j}^{n+1}/\rho_{\mathrm{S}}\right) + \alpha C_{i+1,j}^{n+1}/\Delta x - \beta C_{i,j-1}^{n+1}/\Delta y\right]}{\left[\tilde{\lambda}\left(\rho_{\mathrm{P}} - \rho_{\mathrm{C}} C_{i,j}^{n+1}/\rho_{\mathrm{S}}\right)\Big/ S + \alpha/\Delta x - \beta/\Delta y\right]} \tag{5.21c}$$

$$\alpha < 0, \quad \beta < 0: \quad C_{i,j}^{n+1} = \frac{\left[\tilde{\lambda}(1+\delta_{\mathrm{O}})\left(\rho_{\mathrm{P}} - \rho_{\mathrm{C}} C_{i,j}^{n+1}/\rho_{\mathrm{S}}\right) - \alpha C_{i-1,j}^{n+1}/\Delta x - \beta C_{i,j-1}^{n+1}/\Delta y\right]}{\left[\tilde{\lambda}\left(\rho_{\mathrm{P}} - \rho_{\mathrm{C}} C_{i,j}^{n+1}/\rho_{\mathrm{S}}\right)\Big/ S - \alpha/\Delta x - \beta/\Delta y\right]} \tag{5.21d}$$

$C_{i,j}^{n+1}$ is computed from Eq. (5.21) iterating to reach convergence, then the phase variable can be updated using Eq. (5.16):

$$\phi_{i,j}^{n+1} = \phi_{i,j}^{n} + \Delta t \, \frac{\tilde{\lambda}\left(\rho_{\mathrm{P}} - \rho_{\mathrm{C}} C_{i,j}^{n+1}/\rho_{\mathrm{S}}\right)\left(C_{i,j}^{n+1}/S - 1 - \delta_{\mathrm{O}}\right)\Delta s}{\rho_{\mathrm{P}} \Delta x \Delta y} \tag{5.22}$$

According to Eqs. (5.21) and (5.22), if the protein concentration is locally depleted, correspondingly, the solid mass stored in the computational cell grows and the phase variable is increased; on the other hand if mass stored in the cell begins to redissolve, protein is released in the solute phase and the local value of protein concentration is increased. These phenomena are driven by the attachment-kinetic condition, i.e. the existing deposit grows if the protein concentration is larger than S and the mass exchange is proportional to the local value of the orientation-dependent kinetic coefficient; in the opposite situation, i.e. when the protein concentration becomes smaller than S, deposit begins to redissolve. The solubility is a function of the precipitant agent (salt) concentration and for this reason the phenomena are mainly governed by the distribution of this agent (i.e. by Eq. (5.20), or by temperature when solubility modulation is induced by thermal control).

In Eq. (5.22) Δs is the "reconstructed" portion of the solid wall "bounded" by the frontier of the control volume (computational cell) located astride the crystal surface. The determination of Δs requires a well-defined "interface-reconstruction" technique (the shape of the crystal for a fixed time is not known a priori and must be determined as part of the solution). The reader will find quite an exhaustive discussion of these aspects in Chapter 3.

For the present case, a reconstruction is a geometrical approximation of the true solid/liquid interface. In particular, the interface can be approximated by a straight line of appropriate inclination in each cell (PLIC piecewise linear interface approximation, see Section 3.5.3) and this reconstruction technique is recommended herein. Various conditions can be used to determine such a reconstruction, one of which is that the straight lines connect at cell faces. With this condition, the resulting interface representation becomes continuous. However, for strong distortions and topology changes similar to those involved in organic crystal growth, this condition becomes virtually unenforceable. If the PLIC approach with nonconnecting straight lines discussed in the article of Gueyffier et al. (1999) (each line is determined independently of the neighbor lines and their ends need not necessarily connect at the cell faces) is used, this guarantees maximum robustness and simplicity, and allows extension to three dimensions, while sacrificing little in accuracy.

In practice, two conditions must be used for the straight line in cell (i,j), which always yield an unambiguous solution: (a) it is perpendicular to the interface-normal-unit vector $\underline{\hat{n}}$; (b) it delimits a solid area "matching" the given $\phi^n_{i,j}$ for the cell; i.e. the slope of the line is given by the interface normal (gradient of the volume fractions), and the intercept follows from invoking volume conservation.

The reconstructed interface then can be used to compute the fluxes (i.e. for the present case the mass flux of growth units that are incorporated by the crystal) necessary to integrate the volume-evolution equations.

Within this section it is also important to remember how the system of Eqs. (5.4) and (5.5) for incompressible flow presents a particular situation in which one of the unknowns, namely the pressure, does not appear under a time-dependent form due to the nonevolutionary character of Eq. (5.4). This actually creates a difficult situation for the numerical schemes and special techniques have to be adapted in order to treat the continuity equation (see the SMAC method in Section 2.8). According to this method, the computation of the velocity field at each time step can be split into two substeps.

In the first, an approximate nonsolenoidal velocity field $\underline{V}^*$ that corresponds to the correct vorticity of the field is computed at time $(n+1)$ neglecting the pressure-gradient term in the momentum equation (5.18). In the second substep, the pressure field is computed by introducing the divergence-free velocity condition:

$$\nabla^2 p^n = \frac{1}{\Delta t} \underline{\nabla} \cdot \underline{V}^* \tag{5.23}$$

This equation can be solved with a SOR (Successive Over Relaxation) iterative method. For further details on the numerical method, see Section 2.8. On the solid walls the $\partial P/\partial n=0$ condition is imposed; at the crystal surface, in particular, this condition must be discretized by forward or backward schemes according to the interface's orientation:

$$0 < \phi < 1, \quad \alpha > 0, \quad \beta > 0 \qquad p_{i,j} = \frac{(\alpha p_{i+1,j}/\Delta x + \beta p_{i,j+1}/\Delta y)}{(\alpha/\Delta x + \beta/\Delta y)} \tag{5.24a}$$

$$0 < \phi < 1, \quad \alpha < 0, \quad \beta > 0 \qquad p_{i,j} = \frac{(-\alpha p_{i-1,j}/\Delta x + \beta p_{i,j+1}/\Delta y)}{(-\alpha/\Delta x + \beta/\Delta y)} \tag{5.24b}$$

$$0 < \phi < 1, \quad \alpha > 0, \quad \beta < 0 \qquad p_{i,j} = \frac{(\alpha p_{i+1,j}/\Delta x - \beta p_{i,j-1}/\Delta y)}{(\alpha/\Delta x - \beta/\Delta y)} \tag{5.24c}$$

$$0 < \phi < 1, \quad \alpha < 0, \quad \beta < 0 \qquad p_{i,j} = \frac{(-\alpha p_{i-1,j}/\Delta x - \beta p_{i,j-1}/\Delta y)}{(-\alpha/\Delta x - \beta/\Delta y)} \tag{5.24d}$$

Finally, the correct solenoidal velocity field can be updated using the computed pressure field to account for continuity:

$$\underline{V}^{n+1} = \underline{V}^{*} - \Delta t \,\underline{\nabla}\, p^{n} \tag{5.25}$$

The protein and salt distributions at time $(n+1)$ can be obtained from Eqs. (5.19) and (5.20) after the velocity calculation.

5.7. Comparison with other methods

The volume-tracking and enthalpy methods, which have become so popular in recent years, are single-region formulations. In the present section, the OCGVOF method is briefly compared with these methods with which it shares some features. Moreover, the applicability of the method is discussed.

Volume of fluid methods do not use marker particles. The liquid volume-fraction field ϕ, is used. It indicates for each computational cell its liquid fraction (e.g., $\phi = 1$ liquid, $\phi = 0$, gas and $0 < \phi < 1$ for an interfacial cell). The key element for a volume of fluid method is its technique for advecting ϕ: standard finite-difference methods suffer from numerical diffusion, which would soon smooth ϕ excessively. The ϕ-field is the only phase information stored in volume of fluid methods. Approximate interface locations are found from a so-called interface reconstruction. As discussed in Section 3.5, this is needed for "advecting" ϕ (a transport equation is solved at each time step to determine the relative position of the different phases, i.e. the ϕ distribution, see, e.g., Eq. (3.14) in Chapter 3), for determining the local properties of the fluid (e.g., viscosity) and for better graphical representation.

In a phase-field model for the thermal solidification of a liquid ("enthalpy method"), as in the case of volume of fluid methods, a phase-field variable ϕ that varies in space and time is introduced to characterize the phase of the material. Unlike in the case of volume of fluid methods, however, in this case the phase-field variable ϕ is not "advected" by the motion of the fluid, but computed as a function of the temperature distribution (see, e.g., Eq. (4.2) in Chapter 4). An energy partial differential equation applied throughout the computational domain and taking into account the release or absorption of latent heat associated with the phase change, governs the evolution of ϕ (see, e.g., Eq. (4.3)).

In the volume of fluid methods, ϕ is "advected" by solving an appropriate continuity transport (i.e. volume-tracking) equation (see, e.g., Section 3.5); in the enthalpy methods ϕ is computed as a function of the temperature, obtained by solving an appropriate energy equation taking into account the release or absorption of latent heat (Section 4.2), in the OCGVOF method, ϕ is computed using attachment kinetics at the surface of the crystal, i.e. by taking into account the phenomenon of addition/incorporation of solute molecules (building blocks or growth units) to the crystal. These conditions lead to the introduction of a group of differential equations for the protein concentration at the crystal surface and the evolution of the solid-mass displacement.

The OCGVOF technique shares with the aforementioned classical methods the interface-orientation analysis, which in these methods is needed to compute the surface-tension force perpendicular to the interface separating the fluid phases (expressed as a corresponding volume force, which can be included in the momentum equation), whereas in the OCGVOF method it is needed for the orientation-dependent kinetic coefficients and the discretization of the protein concentration gradients perpendicular to the surface (fluxes of nutrients). Moreover, it shares with the enthalpy methods the technique used to accommodate the zero-velocity condition in liquid regions turning to solid phase (the so-called "mushy" regions). This discussion clarifies the analogies and differences among these methods.

It is also interesting to point out how the ideas and concepts at the root of the OCGVOF technique have led to the introduction of a similar method for the simulation of the growth of organic living tissues in bioreactors (see Chapter 7). In this case, however, the mathematical model is quite different since the living-construct exhibits high sensitivity to the effect of the fluid-dynamic shear stress at the surface (shear-stress-dependent kinetics). In the case of macromolecular crystals the growth units are merely added to the solid (the molecules orient and attach themselves to the growing surface) without changing their initial composition. In the case of biological tissues, of course, the mechanism is more complex. The growth units are incorporated and converted into the main tissue components. Consequently, tissue enlargement occurs due to "internal cell division" and "production of extracellular matrix" supported by the aforementioned incorporation of nutrients in the biological cells. The surface shear stress acts by modifying the manner in which absorbed nutrients are used to produce the new cells and the extracellular matrix (it induces changes in the tissue metabolism and function eliciting a physiologic response from the biological cells, see Chapter 7 for further details on the modeling of these aspects and the related volume-tracking method).

Note that, in principle, since constant phase densities are assumed, the OCGVOF method provides reliable results only if the decrease of solution volume due to the advancing interface is negligible. In the case of inorganic crystal growth, since solvent inclusions in the solid mass are very small, this would occur if the decrease of solution volume was only a few per cent of its initial volume (i.e. if the crystal undergoes only small dimensional changes with respect to the overall dimensions of the cell containing the feeding solution). Fortunately, this problem is not significant for the case of protein substances. Protein crystals, in fact (as pointed out in Section 5.6.2), have widely open structures and incorporate up to 90% by volume of solvent within the network of protein molecules. Macromolecular crystals are mainly composed of solvent. The protein occupies the remaining volume so that the entire crystal is in many ways an ordered gel with extensive interstitial spaces through which solvent and other "small" molecules may freely diffuse (see, e.g., Rosenberger, 1986; McPherson, 1990).

5.8. The OCGLSET – Organic Crystal Growth Level-Set Method

It is worthwhile to point out how, in addition to the worthy contributions cited in the previous section, level-set methods have recently enjoyed a widespread use and are currently becoming the most popular techniques for the simulation of moving-boundary problems (see, e.g., the landmark papers of Osher and Sethian, 1988; Sethian, 1999; Sussman et al., 1994; Osher and Fedkiw, 2002). The great success of level-set methods can be attributed to their versatility and simplicity as well as to significant advantages offered in taking into account some aspects (Sethian, 1999; Osher and Fedkiw, 2002). For these reasons, the methodology and the formalism underlying modern level-set methods has been recently extended also to the aforementioned techniques dealing with the solidification of melts (see, e.g., the excellent works of Chen et al., 1997b; Kim et al., 2000).

Along these lines, the objective of the present section is to reformulate the OCGVOF kinetic model in the framework of a level-set technique (in order to benefit from the advantages offered by this technique, see also Lappa, 2004d).

5.8.1 *General properties*

This method, first introduced by Osher and Sethian (1988), is conceptually similar to a phase-field model in that the solid/liquid interface Γ is represented as the zero contour of a level-set function, $\varphi(\underline{r}, t)$, which has its own equation of motion. The movement of the interface is taken care of implicitly through an advection equation for $\varphi(\underline{r}, t)$. Unlike the phase-field model, however, there is no arbitrary interface width introduced in the level-set method; the sharp-interface equations can be solved directly and, as a result, no interface-reconstruction techniques (e.g., PLIC) are required (Kim et al., 2000).

The goal is to compute and analyze the subsequent motion of Γ under a velocity field q (Osher and Fedkiw, 2002). This velocity can depend on the position of the interface of the organic crystal (Ω) and the external physics (for the case under investigation it depends on the surface-incorporation kinetics, i.e. on the mechanisms of attachment of solute molecules to the growing solid surface, i.e. Eq. (5.1)). The interface is captured for later time as the zero level-set of the function $\varphi(\underline{r}, t)$, i.e. $\Gamma(t) = \underline{r}|_{\varphi(\underline{r},t)=0}$.

In practice, the level-set function must be defined as the signed normal distance from the solid/liquid interface such that φ is positive in the liquid phase, negative in the crystal, and zero at the solid/feeding-solution interface:

$$\begin{aligned} &\varphi(\underline{r}, t) > 0 && \text{for } \underline{r} \notin \Omega \\ &\varphi(\underline{r}, t) = 0 && \text{for } \underline{r} \in \partial\Omega = \Gamma(t) \\ &\varphi(\underline{r}, t) < 0 && \text{for } \underline{r} \in \Omega \end{aligned} \tag{5.26}$$

Thus, the interface is to be captured for all later time, by merely locating the set $\Gamma(t)$ for which φ vanishes. Its motion is analyzed by convecting the φ values (levels) with the velocity field $\underline{q}$:

$$\frac{\partial \varphi}{\partial t} + \underline{q} \cdot \underline{\nabla}\varphi = 0 \tag{5.27}$$

where $\underline{q}$ is the desired velocity on the interface, and is arbitrary elsewhere.

Actually, only the normal component of $\underline{q}$ is needed: $q_n = \underline{q} \cdot (\underline{\nabla}\varphi / |\underline{\nabla}\varphi|)$ so Eq. (5.27) becomes

$$\frac{\partial \varphi}{\partial t} + F|\underline{\nabla}\varphi| = 0 \tag{5.28}$$

Integrating Eq. (5.28) for one time step results in moving the contours of φ along the directions normal to the interface according to the velocity field F, which varies in space. In turn, F must be constructed to be an extension of the interface velocity, q_n, such that $F = q_n$ for points on the interface and the lines of constant F are normal to the interface (see Eq. (5.30)). Thus, advecting φ according to Eq. (5.28) moves the front with the correct velocity.

After solving Eq. (5.28) for one time step, the level-set function will no longer be equal to the distance away from the interface. It is necessary to *reinitialize* φ to be a signed distance function. This step can be accomplished by solving

$$\begin{aligned} &\frac{\partial \psi}{\partial \tau} + \mathrm{sqn}(\varphi)[|\underline{\nabla}\psi| - 1] = 0 \\ &\psi(\underline{r}, 0) = \varphi(\underline{r}, t) \end{aligned} \tag{5.29}$$

Following Osher and Fedkiw (2002), in order to define the distance φ in a band of width ε around $\Gamma(t)$, Eq. (5.29) should be solved only for $\tau = O(\varepsilon)$.

The basic level-set method concerns a function $\varphi(\underline{r}, t)$ that is defined throughout space. Clearly this is useless if one only cares about information near the zero level-set. The local level-set method defines φ only near the zero level-set. In practice, Eq. (5.28) is solved in a neighborhood of $\Gamma(t)$ of width $m\Delta r$, where m is typically 5 or 6. Points outside this neighborhood need not be updated by this motion. The function φ *is* reinitialized to be signed distance to $\Gamma(t)$, only near the boundary, smoothly extending the velocity field q_n off the front $\Gamma(t)$ (see Eq. (5.30)) and solving Eq. (5.28) only locally near the interface $\Gamma(t)$, thus lowering the complexity of this calculation by an order of magnitude (Osher and Fedkiw, 2002). This makes the cost of level-set methods competitive with other techniques.

With regard to the smooth extension of the quantity q_n on $\Gamma(t)$ to a neighborhood of $\Gamma(t)$, this step is accomplished by solving:

$$\begin{aligned} &\frac{\partial F}{\partial t} + \mathrm{sqn}(\varphi)\left[\frac{\underline{\nabla}\varphi}{|\underline{\nabla}\varphi|}\cdot\underline{\nabla}F\right] = 0 \\ &F(\underline{r}, 0) = q_n(\underline{r}, t) \end{aligned} \tag{5.30}$$

Again, this equation is solved only for $\tau = O(\varepsilon)$ in order to extend q_n to be constant in the direction normal to the interface in a region of width ε.

For the case under investigation q_n is given by Eq. (5.17a), i.e.

$$q_n = \frac{\rho_C}{\rho_S}\theta^* \tag{5.31}$$

It is worthwhile to stress how the function q_n is not given a priori but has to be computed at any instant as part of the problem (it changes dynamically during the growth process according to the residual protein available in the liquid phase, and according to the steepness of the concentration gradient at the solid/liquid interface that, in turn, is changed by the motion of the fluid surrounding the macromolecular seed). For further details on the computation of q_n see the next section.

5.8.2 *Solution procedure*

The solution procedure is summarized in the flow-scheme below; it proceeds in five major stages: (i) advancing the interface (Eq. (5.28)), (ii) reinitializing the level-set function to be a signed distance function (Eq. (5.29)), (iii) solving for the new concentration and velocity fields in the liquid (Eqs. 5.4–5.7), (iv) adjourning of the local values of C at the crystal/feeding-solution interface (Eq. (5.12), this stage accounts for solute depletion), (v) computing the surface growth-rate distribution (the growth velocity is not directly imposed but it results from internal conditions related to solute transport).

5.8.3 *Discussion and comparison with the OCGVOF method*

To conclude this section, it is worthwhile to compare the present OCGLSET with the corresponding phase-field method introduced in Section 5.6 in order to show analogies and differences, advantages and limitations.

As previously elucidated, the OCGVOF method is a volume of fraction method. It accounts for the solid mass stored in the generic computational cell by assigning an appropriate value of a phase-field variable (ϕ) to each mesh point ($\phi = 1$ crystal, $\phi = 0$ feeding solution and $0 < \phi < 1$ for an interfacial cell). The key element for the OCGVOF method is its technique for adjourning ϕ. Upon changing phase, the ϕ-value of the cell is adjusted to account for mass release or absorption, this adjustment being reflected in the protein-concentration distribution as either a source or sink. As previously explained, the modeling of these phenomena leads to the introduction of two differential equations, strictly related, from a mathematical point of view, to the "kinetic conditions" used to model mass transfer at the crystal surface. The first equation is Eq. (5.12), the second one is a phase-field equation where the mass-exchange flux $\tilde{\lambda}(\rho_P - \rho_C C/\rho_S)(C/S - 1 - \delta_O)$ between crystal and protein solution is used to update the value of ϕ in the computational cell located astride the crystal surface (Eq. (5.16)).

According to these equations, if the protein in solution is locally depleted (released), correspondingly, the solid mass stored in the computational cell grows (redissolves) and the phase variable is increased (decreased). These phenomena are driven by the attachment kinetic condition and by $(\sigma - 1 - \delta_O)$.

The OCGLSET technique leaves aside mass-exchange phenomena, i.e. it does not take into account the solid mass stored in each computational cell and the evolution equations for this quantity. Rather, the surface-attachment kinetics are directly used to compute the normal velocity at the crystal/solution interface (i.e. the growth rate). Then this velocity is used to advect φ and calculations of this function are performed only in a narrow region around the interface.

If the protein concentration is locally depleted, correspondingly, the concentration gradient in Eq. (5.9) and the associated growth rate in Eq. (5.17b) are positive and the interface Γ is transported outward; on the other hand if mass stored in the cell begins to redissolve, protein is released in the solute phase and the concentration gradient in Eq. (5.9) and the associated growth rate in Eq. (5.17b) are negative (the interface Γ is transported inward). These phenomena are driven by Eqs. (5.12) and (5.28).

Although the volume of fraction model can lead to very interesting results, there are still some limitations in this approach. The proper use of these models requires, in fact, that an asymptotic analysis is performed in order to obtain a mapping between the parameters of the phase-field equations and the sharp-interface equations, i.e. a very accurate interface-reconstruction technique is required (for further details on this aspect see the excellent paper of Kim et al., 2000, and the discussion dealing with the PLIC technique in Section 5.6). Moreover, computationally, the grid spacing must be small enough to resolve the interfacial region. The multiphase region (region where phase change occurs, i.e. region where the growth units are added to the pre-existent crystalline structure) is, in fact, defined by the condition $0 < \phi < 1$ and is therefore associated with a somewhat arbitrary thickness (the width depends on the resolution of the computational mesh).

These reconstruction techniques are very laborious and not easy to implement. On the contrary, the level-set computational approach has the capability to track the motion of the interface without resorting to mathematical manipulations and complex reconstruction techniques. Moreover, there is no arbitrary thickness associated with the region where liquid turns to solid since the interface is captured as the zero level-set of the function (it is sharp). Shape changes, corner and cusp development, and accurate determination of geometric properties are naturally obtained in this setting. Moreover, topological merging and breaking are well defined and easily performed.

Preliminary results provided by the present level-set method have been compared by the author with simulations of crystal growth performed using the corresponding VOF model. The results exhibit a satisfactory agreement. In conclusion, the level-set method should be considered as an interesting alternative to the use of volume of fraction models for the simulation of problems dealing with the crystallization of organic crystals and the related slow attachment kinetics. The level-set method offers significant advantages in terms of versatility, robustness and simplicity and exhibits wide capabilities to provide all the parameters of interest for organic crystal growers.

5.9. Prototype applications and physical aspects

Hen egg-white lysozyme is considered herein as a model protein, being a well-characterized molecule (see Table 5.1 for the properties). The precipitant agent is NaCl. The tetragonal form of hen egg-white lysozyme typically grows in a habit formed by a combination of prismatic (110) and pyramidal (101–011) faces (see, e.g., Monaco and Rosenberger, 1993).

The subsequent sections deal with some prototype simulations. In the spirit of the book, systems of growing complexity are considered as the analysis progresses.

Particular care is devoted to the physical context of the examples provided, i.e. to the real conditions encountered in the experiments and to the aspects of the problem that are of particular interest for organic crystal growers.

It is also worthwhile to stress how the proposed simulations, exhibiting significant capabilities to predict and elucidate experimental observations, lead to identifying

Table 5.1 Properties of lysozyme and NaCl and operating conditions.

D_{lys} ($cm^2 s^{-1}$)	10^{-6}
D_{NaCl} ($cm^2 s^{-1}$)	10^{-5}
ν ($cm^2 s^{-1}$)	8.63×10^{-3}
ρ_c ($g cm^{-3}$)	1.2
β_{lys} ($g^{-1} cm^3$)	−0.3
β_{NaCl} ($g^{-1} cm^3$)	−0.6
ρ_P ($mg\, ml^{-1}$)	820
λ ($Å\, s^{-1}$)	$\cong 10$
δ_O (−)	2
T (°C)	18
pH	4.5

cause-and-effect relationships, i.e. they are propaedeutical to the discerning of heretofore unknown physical mechanisms. Thus, the prototype examples can also be regarded as an additional step in the evolutionary progress towards the full understanding of the physics of protein crystallization and related processes.

In order to show the capabilities of the proposed numerical methods, their robustness and versatility and how they can deal with very different situations, two "typical" cases are considered. In the first case, analysis of growth-habit change is carried out in the case of microcrystals having a size of the order of 50 μm (study of the microscopic facet morphologies) for typical experimental conditions and in the absence of convection (diffusive transport). In the second case, the growth of one or many lysozyme macrocrystals (size 1 mm) is investigated under the effect of buoyancy convection induced by typical values of residual gravity occurring in space.

For the first time the (still unknown in the literature and contradictory) interplay of surface-attachment kinetics and mass transport in the liquid phase is estimated through direct computation of the "quantitative" parameter γ:

$$\gamma = \frac{(C_B - C_i)}{(C_B - S)} \tag{5.32}$$

If the incorporation rate is much slower than the transport rate in the concentration boundary layer surrounding the crystal, then the concentration close to the crystal face is maximal, i.e. it is equal to the bulk protein concentration. If, to the contrary, the incorporation rate is much faster than the transport rate, the protein concentration at the crystal face is minimal, i.e. it is equal to the saturation concentration. From a mathematical point of view: if attachment kinetics is the rate-limiting step in the growth process, $C_i = C_B$ (C_B bulk protein concentration) and the local supersaturation will be approximately the bulk supersaturation; on the other hand, if the growth rate is limited by transport (convective or diffusive) of solute to the growth interface, $C_i = S$. Therefore, the nondimensional coefficient γ can be regarded as a measure of the relative importance of surface kinetics versus transport as the rate-limiting step in the growth of crystals. In fact, for growth controlled by surface kinetics $C_i \rightarrow C_B$ and $\gamma \rightarrow 0$; for a transport-controlled process, $C_i \cong S$ and $\gamma \rightarrow 1$.

Transport and surface kinetic processes can be considered competitive for $\gamma \cong 0.3$–0.5.

5.10. Growth-habit simulation and microscopic facet-morphology study

In order to obtain crystals of high compositional uniformity it is important that the growth rate is kept constant over the whole advancing interface, i.e. the crystal shape remains unchanged during growth. The conditions required for morphological (i.e. shape) stability of interfaces are well understood for the diffusive "transport-limited" growth of nonfaceted crystals (in this case, particle growth is mainly controlled by mass transport to the interface and the growth behavior depends on the local mass flux at the interface, Rosenberger, 1986). There the growth kinetics are rather isotropic and interfacial processes are relatively unimportant for the stability. However, most protein crystals, when grown not too rapidly, are faceted and the present section deals with the shape evolution of precipitate particles in this case ("faceted growth").

This implies that the growth kinetics are anisotropic, i.e. that the kinetics of incorporation at the crystal surface are strictly linked to the different crystallographic directions. Then morphological stability is governed by the interaction between the nonuniform interfacial processes and mass transfer to the interface.

The experimental results of Monaco and Rosenberger (1993) are considered for comparison.

Monaco and Rosenberger (1993) found experimentally that the habits of lysozyme crystallites and larger crystals (up to 500 μm) were bound by (101 or 011) and (110) faces (see Fig. 5.2).

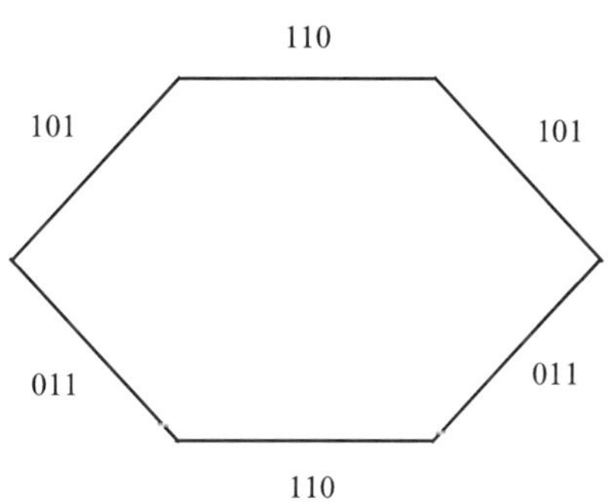

Fig. 5.2 Tetragonal form of hen egg-white typically growing in a habit formed by a combination of prismatic (110) and pyramidal (101–011) faces.

This overall habit formed regardless of the supersaturation prevailing during nucleation and growth. However, the aspect ratios of the (110) faces were found to be dependent on the growth conditions. At low σs, the distance between opposite (101–011) pyramids far exceeded the distance between opposite (110) faces. At high σs, the reverse occurred. They found differences in the average growth rate (θ) of adjacent faces according to the space-orientation of the faces and according to the value of the supersaturation. In particular, the macroscopic growth rates were found to be almost equal for $\sigma_{(o)} < 8$ with the (110) growth rate located below the (101) one (e.g., $\theta_{101} = \theta_{011} = 1.3 \times \theta_{110}$ for $\sigma_{(o)} = 6$), whereas the opposite situation occurred for $\sigma_{(o)} > 8$ with θ_{110} located far above θ_{101} ($\theta_{110} = 1.86 \times \theta_{101}$ for $\sigma_{(o)} = 11$).

According to Monaco and Rosenberger (1993), the observed phenomena for $\sigma_{(o)} \leq 11$ were controlled mainly by surface-attachment kinetics (interface-kinetically-limited phenomena); therefore it is reasonable to assume that the different behavior of the differently oriented faces were strictly associated with different values of the kinetic coefficient. In the case of pure kinetically driven situations, the kinetics of incorporation at the crystal surface are strictly linked to the bond configuration of the crystallographic structure, whereas other effects are negligible. This is a simplified case of the more general situation that can be handled by the mathematical model presented in Section 5.5.

The numerical results presented in this section have been obtained for two different situations. In the first case ($\sigma_{(o)} = 6$), the kinetic coefficients (in the literature this value is affected by a large uncertainty, many investigators found values around 10^{-7}cm s^{-1}) $\lambda_{101} = \lambda_{011} \cong 1.5 \times 10^{-7}$ cm s^{-1} and $\lambda_{110} = 0.75 \times \lambda_{101}$ are assumed for the different faces of the crystal (this corresponds to Monaco and Rosenberger's $\sigma_{(o)} < 8$ condition). For the second case ($\sigma_{(o)} = 11$), the orientation-dependent crystal growth is simulated by assuming $\lambda_{101} = \lambda_{011} \cong 10^{-7}$ cm s^{-1} and $\lambda_{110} = 1.86 \times \lambda_{101}$ (this information being extrapolated from the experimental behavior of the kinetically limited average growth rates).

In the experiments of Monaco and Rosenberger (1993) crystal growth was obtained from a supersaturated solution with $C_{(o)} = 5 \times 10^{-2}$ g cm^{-3} and solubility modulation was controlled by temperature. For the present simulations solubility modulation is induced by setting an appropriate salt concentration whose value is supposed to be constant in the bulk liquid phase. The dependence of S of lysozyme on C_{NaCl} has been determined experimentally by other investigators (see, e.g., Otàlora and Garcìa Ruiz, 1997).

Correspondingly the values of C_{NaCl} needed to obtain $\sigma_{(o)} = 6$ and $\sigma_{(o)} = 11$ and used for the present numerical simulations, are $C_{NaCl(o)} = 2.65 \times 10^{-2}$ g cm^{-3} and $C_{NaCl(o)} = 3.3 \times 10^{-2}$ g cm^{-3}, respectively. For both cases, a seed crystal having a size of 50 μm is placed in the center of a computational domain having a width of 10^3 μm and the concentration of protein is supposed to be constant at the initial time. The frontier of the computational domain is supposed to be impermeable to protein and salt. A 200 × 200 grid is used to achieve good resolution.

The results, grouped in Figs. 5.3 ($\sigma_{(o)}=6$) and 5.4 ($\sigma_{(o)}=11$) show the different growth habit of the crystals according to the value of the supersaturation and according to the corresponding surface-kinetic-coefficient distribution.

The computed face growth rates versus time are shown in Figs. 5.3(b) and 5.4(b). The computed time-averaged growth rates ($\sigma_{(o)}=6$, $\theta_{101}\cong 18\,\text{Å}\,\text{s}^{-1}$, $\sigma_{(o)}=11$, $\theta_{110}\simeq 59\,\text{Å}\,\text{s}^{-1}$) are in good agreement with the experimental ones ($\sigma_{(o)}=6$, $\theta_{101}\cong 21\,\text{Å}\,\text{s}^{-1}$, $\sigma_{(o)}=11$, $\theta_{110}\cong 65\,\text{Å}\,\text{s}^{-1}$).

Finally Figs. 5.3(d) and 5.4(d) show the time evolution of the parameter γ. Since $\gamma<0.2$, these plots confirm the assumption of Monaco and Rosenberger (1993) (i.e. the phenomena under investigation are mainly driven by attachment kinetics)

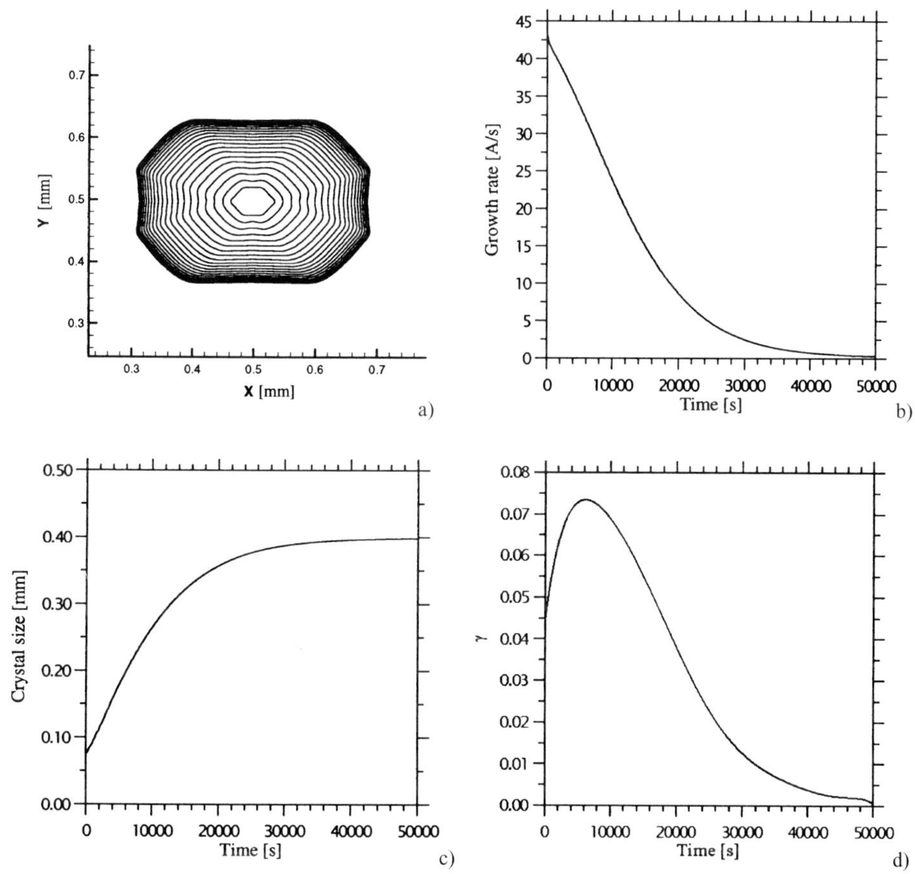

Fig. 5.3 Growth-habit simulation and microscopic facet morphology study of a faceted seed of lysozyme: $C_{(o)}=5\times10^{-2}\,\text{g}\,\text{cm}^{-3}$, $C_{\text{NaCl}(o)}=2.65\times10^{-2}$ ($\sigma_{(o)}=6$), $\lambda_{101}=\lambda_{011}\cong 1.5\times 10^{-7}\,\text{cm}\,\text{s}^{-1}$, $\lambda_{110}=0.75\times\lambda_{101}$: (a) snapshots of the crystal shape versus time ($\Delta t=1220\,\text{s}$); (b) growth rate of the 101 face versus time; (c) crystal size along x versus time; (d) γ versus time.

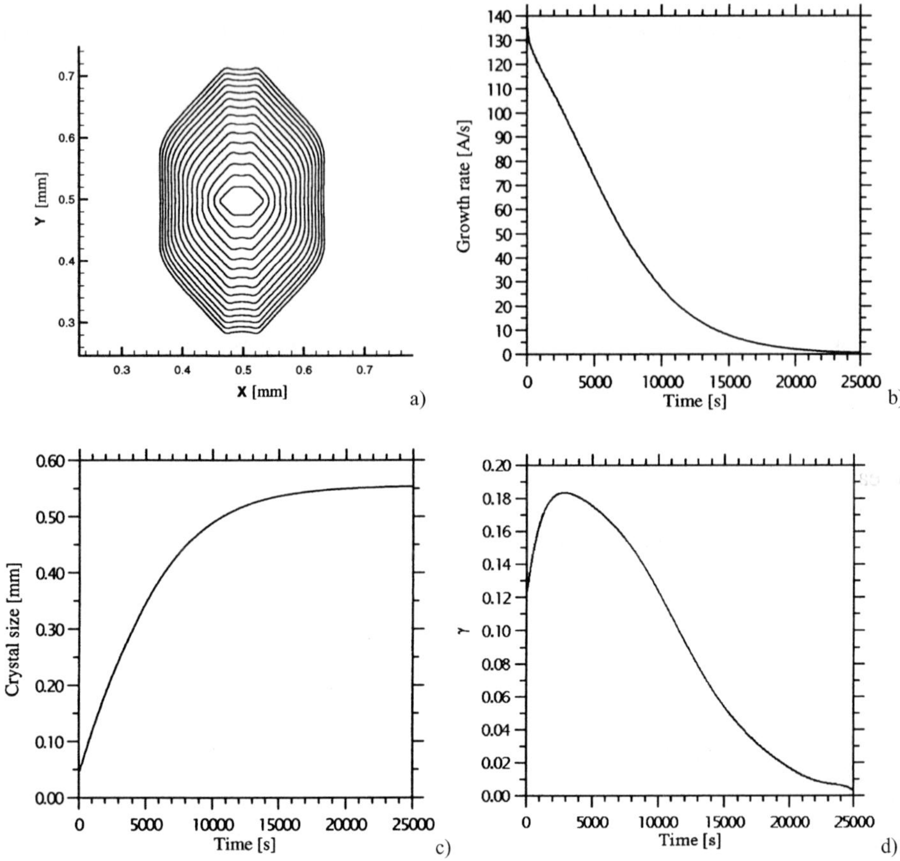

Fig. 5.4 Growth-habit simulation and microscopic facet-morphology study of a faceted seed of lysozyme: $C_{(o)} = 5 \times 10^{-2}\,\mathrm{g\,cm^{-3}}$, $C_{NaCl(o)} = 3.3 \times 10^{-2}$ $(\sigma_{(o)} = 11)$, $\lambda_{101} = \lambda_{011} \cong 10^{-7}\,\mathrm{cm\,s^{-1}}$, $\lambda_{110} = 1.86 \times \lambda_{101}$: (a) snapshots of the crystal shape versus time ($\Delta t = 510\,\mathrm{s}$); (b) growth rate of the 110 face versus time; (c) crystal size along y versus time; (d) γ versus time.

and at the same time provide further validation and justification for the numerical results and the agreement with the experimental ones (in fact, in principle, the relative ratio of the kinetic coefficients for the different faces can be extrapolated from experimental macroscopic averaged growth rates only if the phenomena are kinetically limited). Both experimental observation and numerical modeling exhibit comparable pieces of evidence of the phenomena under investigation. Quantitative comparison between experimental observation and numerical simulation provides a validation of the present mathematical model and gives insight into the crystallization process, providing, for instance, evaluation of the kinetic coefficients; the uncertainty related to the value to assign to the kinetic coefficients (in the literature located around 10^{-7} cm s^{-1}), in fact, can be overcome (through OCGVOF or OCGLSET) changing these values to match exactly the macroscopic experimental growth rates.

5.11. Seeds for morphological studies and space experiments

When organic crystal growers are interested in shape/morphology analysis, kinetics studies, etc., the procedure usually employed consists in growing a very limited number of "seeds" (see, for instance, the studies of Garcia Ruiz and coworkers, Lee and Chernov, 2002, etc.).

In practice, multiple nucleation events are obtained in gel ($N = O(100)$) usually by a counterdiffusion technique; then the nuclei are grown in this environment up to a desired size (this growth technique is known to produce reinforced crystals that can be easily handled and can even be glued to rigid substrates, Otàlora et al., 2001, 2002). At this stage some crystals with large size (not many) are placed in a new (small) reactor expressly conceived and equipped for surface analysis (in-situ, nonintrusive monitoring techniques). At this stage they are called seeds. Particular care is devoted to avoid further nucleation phenomena (i.e. the formation of new nuclei) that would spoil interferometry data.

Often, gel is not used for this second stage of growth since it is well known that it can have detrimental effects on the final quality of the specimens (see, e.g., the discussions in Littke and John, 1986; Bi et al., 1994). Usually these experiments are carried out taking advantage of microgravity conditions (in order to benefit from the quiescent or quasiquiescent conditions that are established in this environment); microgravity has the advantage over gels of being free of macromolecular interactions that could modify in unexpected ways the surface mechanisms; but often convection arises due to residual gravity disturbances.

As previously pointed out, a concentration depletion zone of considerable width develops about the growing crystal (the growth process, in fact, depletes the liquid surrounding the growing crystal of protein). Such solutal boundary layers are responsible for the aforementioned onset of convection under the effect of gravity acceleration. The related concentration gradient implies, in fact, a density gradient.

The less-dense layer adjacent to the growth interface tends to rise, causing a convective flow to occur about the growing crystal, which considerably alters the boundary-layer conditions. Convection, in fact, strongly changes the (diffusive) concentration distribution about the crystal. Depending on its strength and direction with respect to the crystal surface, convection has been observed to modify growth rates and morphological stability with respect to diffusion-controlled (bulk) transport.

These phenomena also occur during experiments in space. In fact, the residual gravity estimated on board the Space Shuttle and the International Space Station (ISS) is not zero, but typically of the order of 10^{-5} of Earth's gravity (see the rich educational environment provided by Chapters 1 and 2). Therefore these effects have to be evaluated also under microgravity conditions, in order to properly prepare and interpret the results of the aforementioned crystal growth experiments in space.

Lysozyme is selected again as the model protein for the prototype applications discussed in the next sections. This molecule has become over the years, in fact, a very important paradigm model protein for fundamental research (Durbin and Feher, 1986). Usually it is selected as the protein to crystallize because its solubility curve is perfectly known and because of the aforementioned current experience of organic crystal growers in producing gel reinforced crystals to be used as seeds for surface studies.

For the prototype simulations the frontier of the reactor is supposed to be impermeable to protein and salt (therefore the salt concentration is constant). For the sake of simplicity the initial shape of the seed crystal is supposed to be quadrate and the kinetic coefficient is supposed to be the same for the different sides of the crystal (however, the numerical method can handle complex shapes and surface-orientation-dependent growth, as shown in Section 5.10).

Growth is obtained from a supersaturated solution with $\sigma_{(o)} \cong 5$ or 6. With regard to this choice of initial conditions it should be pointed out that usually the protein concentration in the nutrient liquid is selected by organic crystal growers in order to favor growth and prevent new crystal formation (nucleation). For this reason the range of protein concentration and of supersaturation investigated is very limited. If protein concentration is too low, the seeds do not grow due to the "dead zone" (δ_O in Eq. (5.1)); if it is too high, new nuclei are formed and this phenomenon behaves as a disturbance for the growth and study of the seeds. For the case under investigation $\sigma_{(o)} \cong 5$ or 6 corresponds to a suitable and realistic condition.

Moreover, though the acceleration environment on the ISS exhibits a rich variation in magnitude and orientation over the course of a mission with a broad range of frequency content (see Chapter 2 and/or Monti et al., 1998b, 2001; Savino and Lappa, 2003a) a residual (steady) $g = 10^{-4} g_o$ (100 μg) oriented in a single direction is considered herein. This is a conservative assumption. The presence of g-jitters will be considered in the next chapter.

5.12. Convective transport under microgravity and shape instabilities

In the present section, the growth of a single seed macrocrystal of lysozyme under microgravity conditions is simulated. In order to discern the effect of convection with respect to other effects, the same simulation is carried out also under a pure-diffusive regime (i.e. in the absence of convection).

The crystal (1 mm × 1 mm) is supposed to be fixed (e.g., by glue) to the mean point of the left wall of the reactor (10 mm high and 30 mm wide test cell) used for the

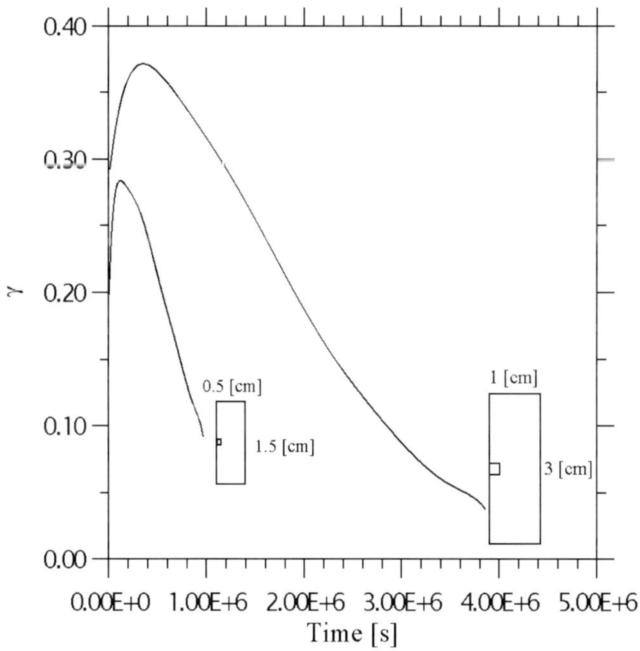

Fig. 5.5 Effect of the size of the reactor on the nondimensional parameter γ (diffusive conditions).

growth process (see Fig. 5.5). Growth is obtained from a supersaturated solution with $C_{(o)} = 5 \times 10^{-2}\,\text{g cm}^{-3}$ and $C_{NaCl(o)} = 2.5 \times 10^{-2}\,\text{g cm}^{-3}$ ($\sigma_{(o)} \cong 5$).

5.12.1 *Growth under diffusive conditions*

Figure 5.5 shows the time evolution of the parameter γ (defined in Section 5.9). Since in particular $\gamma \cong 0.3$–0.40, surface-attachment kinetics and mass transport in the liquid phase can be considered competitive in limiting the growth rate for the case under investigation.

Regarding this aspect (which is crucial for the problem under investigation), the same simulation has been repeated halving the dimensions of the system in order to shed some light on the effect of the size of the reactor on the nondimensional parameter γ. Figure 5.5 shows that large systems tend to exhibit transport-dependent (limited) growth rates (γ tends to be high), whereas small systems tend to be governed by surface-attachment kinetics (for these systems γ is small; this behavior is also confirmed by the results described in Section 5.10, see, e.g., Figs. 5.3(d) and 5.4(d)).

This trend can be explained on the basis of the fact that, if the dimensions of the system are increased, a long time is required to transport available protein (in the solute phase) from far regions (i.e. the frontier of the reactor) towards the crystal surface. This increases the relative importance of mass transport with respect to surface-attachment kinetics in limiting the growth rate.

As a crystal grows from the solution, it depletes the concentration of the growth units that are incorporated into the crystal lattice producing a concentration-depleted zone around it. In this region, the solute concentration changes continuously from the concentration at the crystal face to the concentration in the bulk of the solution. The concentration profile in the concentration-depleted zone varies with time as the crystal grows and is controlled by the balance between the flow of growth units towards the crystal face and the rate of incorporation of these growth units into the crystal lattice. The kinetics of incorporation at the crystal surface are linked to the bond configuration of the crystallographic structure as discussed in Sections 5.2 and 5.6.2 (for the present case of isotropic growth, the kinetics of incorporation are simply linked to the value of the kinetic coefficient and to the supersaturation level), while the flow towards the crystal face is highly dependent on the mass-transport properties in the bulk solution (diffusive or convective), which turn out to be crucial for the overall crystal-growth process.

When convection is absent, Fig. 5.6(b) shows that the face growth rate drops with increasing face size (correspondingly, in Fig. 5.6(a), the size of the crystal tends to an asymptotic constant value). This behavior can be explained according to two effects: (a) the finite volume of the test cell (the amount of solute protein available for the growth of the solid phase is finite and decreases as time passes); (b) as the size of the face increases, more solute should be added to maintain a given face growth rate.

Among other things, Fig. 5.6(a) also shows a deviation from diffusion-controlled growth (which would show a square-root dependence of the crystal size on time). This illustrates the fact that even when diffusive transport is ensured, no guarantee exists that the crystal will grow in the diffusion-controlled regime: in the light of the aforementioned arguments the growth regime for the case under investigation is a function of the ratio between diffusion transport and surface attachment kinetics.

Figures 5.6(c) and 5.6(d) show that after a short initial transient, γ decreases. This is due to diffusive transport that decreases in intensity the concentration gradient between the crystal and regions far from its surface (in "finite size" systems the relative importance of mass transport in the liquid phase with respect to surface-attachment kinetics decreases as time passes due to the decrease of available protein and related gradients).

These results refer to macroscopic parameters; from a "local" point of view, the simulations show that corners and edges of the crystal are more readily supplied with solute than the center of faces (this leads to a macroscopic depression around the center of the faces, see Figs. 5.7(a) and 5.8). As previously discussed, this is due to the diffusive pattern of the protein concentration field around the crystal. Incorporation

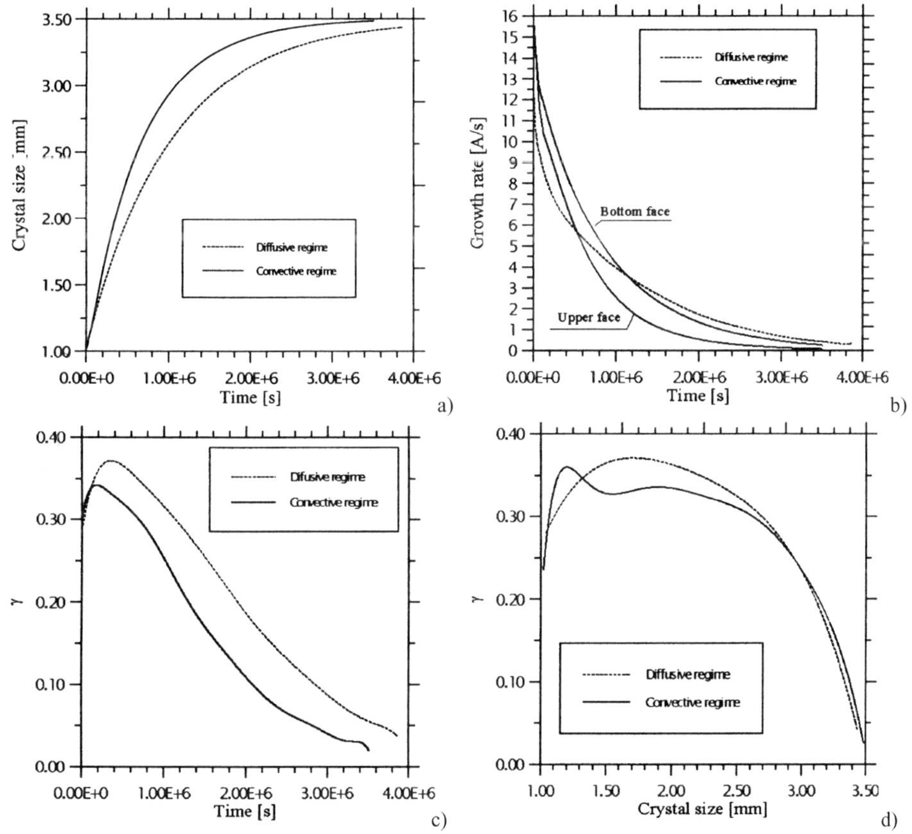

Fig. 5.6 Isothermal protein crystal-growth system under diffusive and microgravity (convective) conditions (crystal of initial dimensions 1 mm × 1 mm fixed to the left wall of a 3 cm high and 1 cm wide growth cell; $C_{(o)} = 5 \times 10^{-2}\,\mathrm{g\,cm^{-3}}$ and $C_{\mathrm{NaCl(o)}} = 2.5 \times 10^{-2}\,\mathrm{g\,cm^{-3}}$): (a) average size (along y) versus time; (b) growth rate versus time; (c) γ versus time; (d) γ versus crystal size.

of the solute into the crystal causes a local depletion in concentration and a solutal concentration gradient to form between the bulk solution and the growth interface. The "steepness" of the gradient determines the rate of solute transport to the growth interface, the steepness being maximum around the corners. Superimposed on this is the fact that a protuberance on the interface sees a higher supersaturation and grows faster than a depression, which sees a lower supersaturation. Note that the existence of "depressions" around the center of the faces of growing crystals (shown by the present numerical results) has often been observed experimentally (Monaco and Rosenberger, 1993). Figure 5.7(a) shows that the difference in gradient steepness between face corner and face center (and consequently the "depth" of the depressions) is proportional to the size of the crystal.

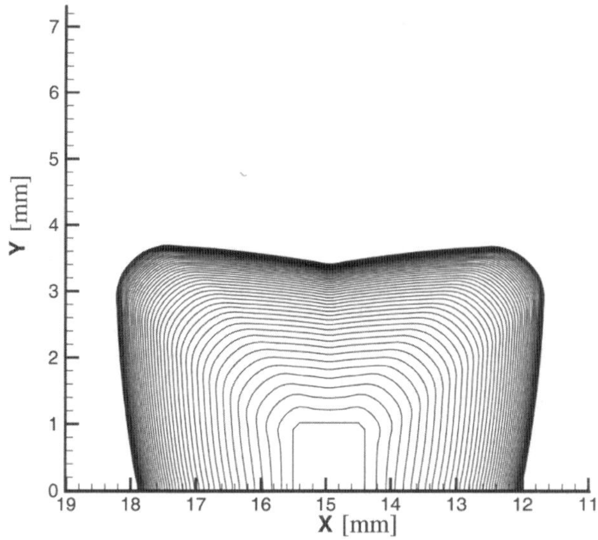

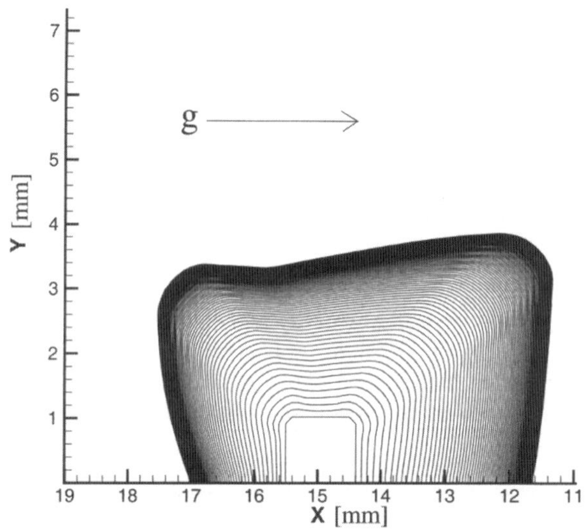

Fig. 5.7 Growth-habit simulation of a seed of lysozyme ($C_{(o)} = 5 \times 10^{-2}\,\mathrm{g\,cm^{-3}}$, $C_{NaCl(o)} = 2.5 \times 10^{-2}$): (a) diffusive regime, snapshots of the crystal shape versus time ($\Delta t = 7.7 \times 10^{4}$ s), (b) microgravity convective conditions, snapshots of the crystal shape versus time ($\Delta t = 3.5 \times 10^{4}$ s).

5.12.2 *Residual gravity-orientation-dependent growth*

If the presence of convection is taken into account a flow field is driven by the density gradient around the growing crystal (as already emphasized in Section 5.11). Since isothermal conditions are assumed throughout the growth process, only

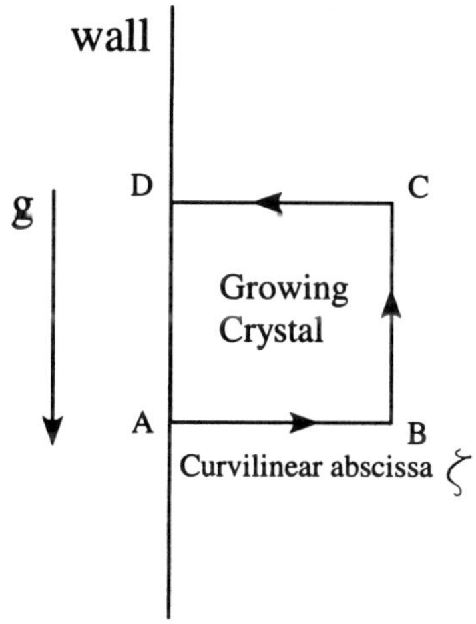

a)

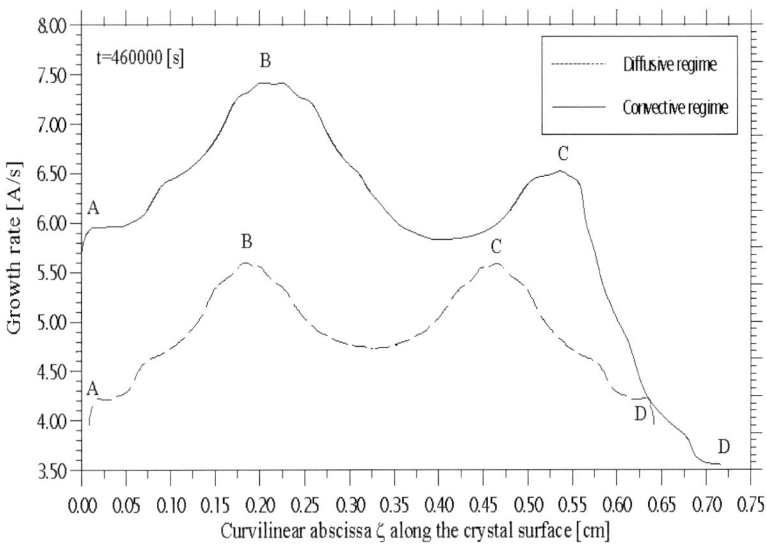

b)

Fig. 5.8 (a) Curvilinear abscissa along the crystal surface, (b) surface growth-rate distribution under diffusive and microgravity (convective) conditions.

compositional nonuniformities, which evolve in the solution due to the crystallization process, act as a source for buoyancy-driven convection. This flow field modifies the protein distribution around the crystal leading to a nonsymmetrical concentration pattern (Fig. 5.9). One recognizes the well-known convective flow pattern for solution growth with a rising "plume" above the crystal. As mentioned before, when growth sets in, the solute concentration around the crystal seed

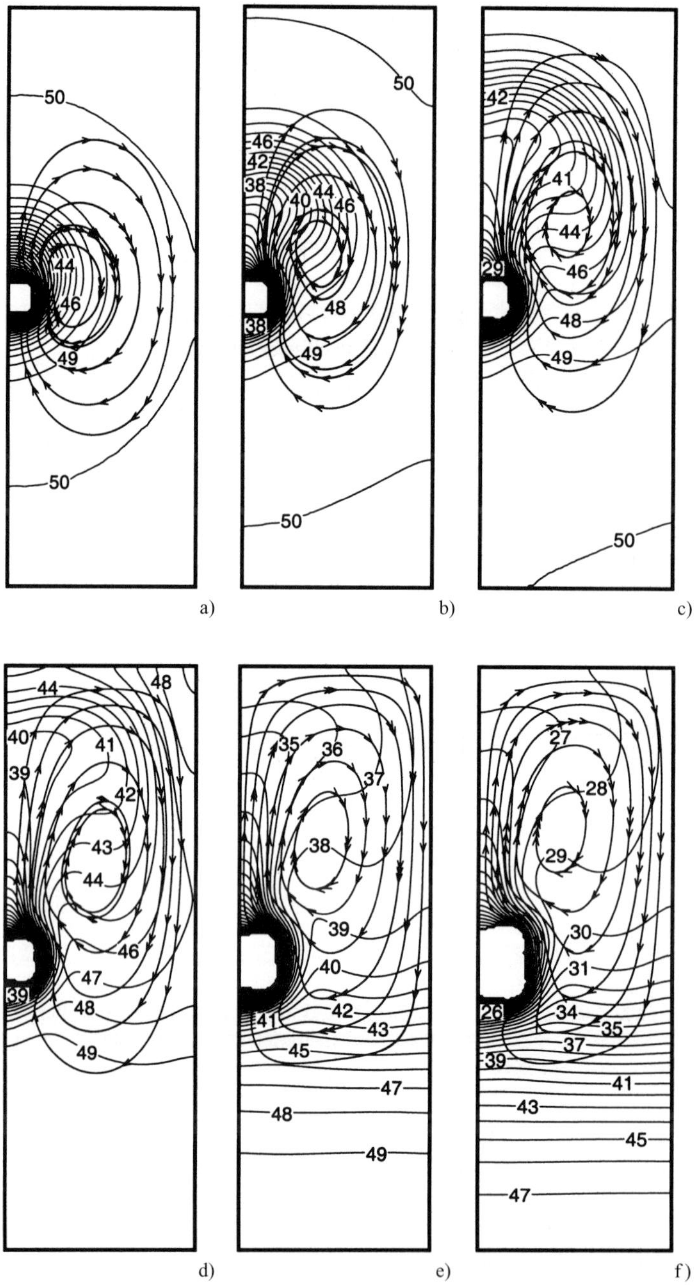

Fig. 5.9 Snapshots of growing crystal, concentration distribution and velocity field under microgravity conditions ($g = 10^{-4} g_o$): (a) $t = 3.5 \times 10^4$ s, (b) $t = 7 \times 10^4$ s, (c) $t = 10.5 \times 10^4$ s, (d) $t = 14 \times 10^4$ s, (e) $t = 2.8 \times 10^5$ s, (f) $t = 4.9 \times 10^5$ s, (g) $t = 8.75 \times 10^5$ s, (h) $t = 1.26 \times 10^6$ s, (i) $t = 1.64 \times 10^6$ s, (l) $t = 2.0 \times 10^6$ s, (m) $t = 2.42 \times 10^6$ s, (n) $t = 2.8 \times 10^6$ s, (o) $t = 3.19 \times 10^6$ s, (p) $t = 3.5 \times 10^6$ s.

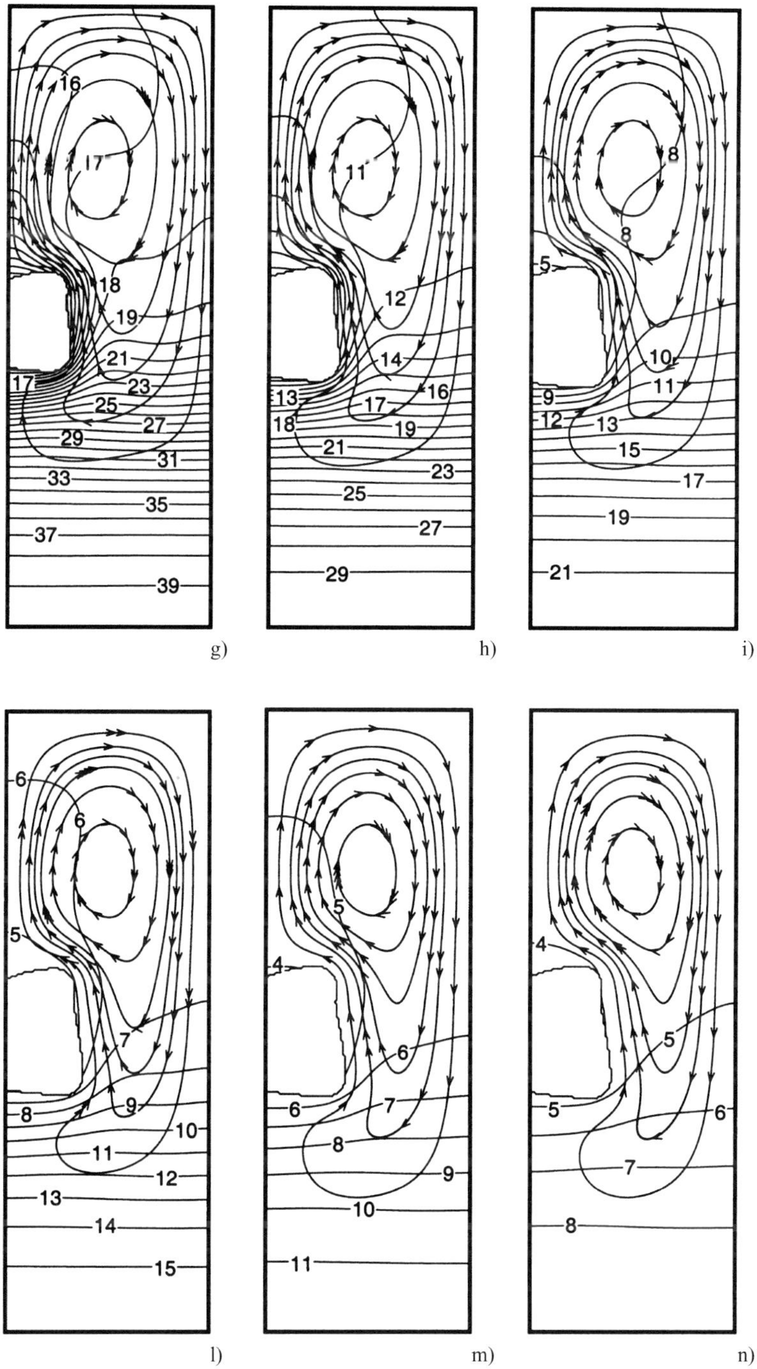

Fig. 5.9 Continued.

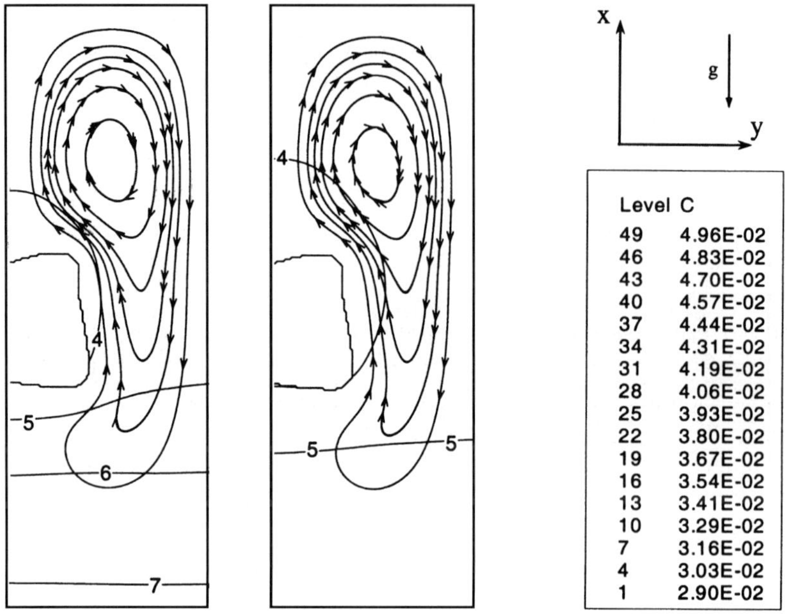

o) p)

Fig. 5.9 Continued.

decreases. With this depletion of the heavier solute, the solution around the crystal becomes lighter and, thus, rises.

Figure 5.9 show that the convective cell, however, does not extend throughout the bulk of the protein chamber and is confined to the zone between the right face of the growing crystal and the right wall of the test cell. As time passes, the vortex roll shrinks while migrating towards the right wall of the protein chamber. The contraction of the cell and its confinement to a zone close to the right wall are strictly related to the behavior of the crystal that puffs out due to growth.

Figure 5.6(b) shows that, during the first part of the growth process, the growth rate under the effect of convection is considerably higher than under diffusive conditions. The growth rates decrease with increasing growth time and crystal size. This is due to the effects already pointed out for the case of absence of convection.

The growth rate under the effect of convection is located above the corresponding one obtained under diffusive regime since convection enhances mass transport in the liquid phase thus increasing the rate of incorporation of the solute into the crystal.

In addition to these effects one must keep in mind that there is an effect related to the fact that the magnitude of the flow is directly proportional to the size of the face upon which it is being generated. Thus, as crystals grow, solutal-gradient-driven flows can become a major factor for the growth rates.

As time passes, however, an inversion occurs in the behavior described above (Fig. 5.6(b)). The growth rate under the diffusive regime becomes higher than the convective ones. This behavior is strictly related to the "finite" amount of protein available for the crystal growth (i.e. the "finite" reactor size). At the beginning, the convective growth rate is located above the diffusive one since solute protein is transported towards the crystal surface more rapidly with respect to a pure diffusive situation (convective motion causes fluxes in the bulk of the solution that exceed the mere diffusive ones). Due to this effect, however, the effect of the "finite" cell size is enhanced.

In the case of reactors having an "infinite" size, the convective growth rates would always be located above the diffusive ones during the crystal-growth process. However, if the initial amount of protein available for the growth is finite, this does not happen. This explains why after an initial transient time the diffusive growth rate becomes higher than the convective one and at the same time could explain some contradictory results in the literature. Otàlora et al. (2001) found a decrease of the growth rates under the effect of convection; this result is in contrast with the generalized idea that convection enhances mass transport in the liquid phase increasing the growth rate. According to the present results, this counterintuitive behavior could be explained by the finite size of their reactor.

Figures 5.6(c) and 5.6(d) clearly show (as expected) that the parameter γ for the case of convection is located below the corresponding one obtained in the case of diffusive conditions (convection reduces the time required to transport available protein from far regions towards the crystal surface and this decreases the relative importance of mass transport with respect to surface-attachment kinetics in limiting the growth rate).

Figure 5.6(b) shows, moreover, that the "face" (average) growth rate exhibits a different value according to the orientation of the face with respect to the direction of the residual g.

This means that an important parameter of the problem is the relative direction of the residual gravity vector relative to the different facets, which induces varying conditions for each facet. The isotropic solutal field around a crystal is deformed by gravity. This leads to flow parallel to the vertical facets and to pluming on the upper horizontal one. Different facets, therefore, grow with different growth rates.

This clearly demonstrates that orientation-dependent growth may occur even if the dependence of the kinetic coefficients on the orientation of the faces is neglected. In this case, it is due to the nonsymmetrical concentration pattern around the crystal distorted by the effect of convection according to the direction of the residual g. The growth rate is strongly affected by the mass transport in the liquid phase and as a natural consequence, g-orientation-dependent growth occurs.

Accordingly, Fig. 5.7(b) shows that the increase of volume of the crystal is more pronounced for the lower side than for the upper side. Figure 5.8 shows in detail how the convection effect results in higher local growth rates near the surface where the flow is incoming and lower local growth rates near the surface where the flow is outgoing. This behavior can be explained in particular according to two different effects.

Due to the convective structure of the flow pattern, in fact, liquid is transported towards the bottom side and in the opposite direction around the upper side. According to this behavior, liquid regions where the amount of protein available in the liquid phase is still large are transported towards the bottom side of the crystal. This increases the growth rate of the lower face (a large amount of protein in the solute phase is available for crystal growth). On the other hand, due to the outgoing flow, the depletion zone close to the upper face is distorted and elongated towards the top of the test cell ("plume"). This leads to a decrease of the concentration gradient (the depletion layer becomes thicker with respect to the bottom side of the crystal) and hence to a decrease of the mass-exchange flux between solid and liquid phases. This explains the occurrence of a higher value of the growth rate on the bottom face.

The analysis of the surface growth-rate distribution in Fig. 5.8 shows, as for the case of diffusive regime, that the growth rate is nonuniform across the crystal faces. However, in the presence of convection the growth-rate distribution is not symmetric and, in particular, the maximum of the growth rate occurs at the corner B. In practice this occurs since, as previously elucidated, the "steepness" of the concentration gradient determines the rate of incorporation of the solute into the crystal. The larger asymmetry in the case of the presence of convection can be explained according to the fact that around the corners the convective flows are very strong and tend to increase the steepness of the concentration gradient there.

Qualitatively, the predicted higher growth rates at crystal corners are consistent with experimental observations on lysozyme (Rosenberger, 1986).

It is also worthwhile to stress how according to the results described and discussed above and in the previous pages, one can speculate that the faster (high supersaturation) a crystal grows the more readily it can lose morphological stability (formation of macroscopic depressions). Moreover, the bigger a crystal, in the presence of convection, the lower may be the critical growth rate for morphological instability.

5.12.3 *Relative importance of mass transport and surface kinetics*

Many investigators found that growth was limited by surface kinetics alone. The present results clearly show that claims that "the role" of mass transport in the liquid phase across the depletion layer can be eliminated are fluid-dynamically unsupportable. Crystal-growth rates have to be considered in terms of both mass transport and attachment kinetics.

This is also the major reason why in this category of problems the growth velocity cannot be directly imposed but must be computed at any instant from internal conditions related to the interplay between solute transport and kinetic barrier (Eq. (5.12)). The simulations, however, confirm that for protein crystals that grow relatively slowly and for typical dimensions of the reactors used for growth (of the order of few millimeters or centimeters), mass transport in the liquid phase, dependent primarily on physical forces and movements in the solution phase, is almost certainly the less important of the two (for all the cases investigated, in fact, $\gamma < 0.4$).

There is an outstanding difference with respect to the case of inorganic crystal growth (see, e.g., the growth of semiconductor crystals by the floating zone technique discussed in Section 4.11) where, in practice (owing to the large thermal diffusivity of liquid metals) the velocity of the advancing solid/melt interface is determined by the ramping rate used to cool one side of the system (or by the velocity of translation of the specimen through the heating device) and thus can be fixed a priori in the growth process. This is why in the solidification of melts the analysis is generally limited to the fluid dynamics of the process.

In the case of macromolecular crystals the effects of fluid dynamics do not behave as a dominant factor, but they are not negligible either ($\gamma \cong 0.35$).

These discussions represent a quite exhaustive and systematic attempt to help organic crystal growers to discern the complex interrelations among the various parameters of interest (which are not independent of one another). A further aspect, however, that has not been considered above would need clarification: The understanding of the effect of the mutual interference among growing seeds. It is expected, in fact, that such an interplay can affect the local growth rates as well as the mechanisms driving growth. Under a pure-diffusive regime, the depletion zones associated with the different growing seeds may intersect and overlap leading to competition for growth; under convective conditions the behavior may be very complex (see the next sections).

5.13. Two interacting crystals

In the present section, the growth of two seeds of lysozyme is simulated. The crystals (1 mm × 1 mm initial size) are supposed to be fixed (e.g., by glue) to the points $x = H/3$ and $x = 2H/3$ of the left wall of the reactor (10 mm high and 30 mm wide test cell) used for the growth process (see Fig. 5.10). Growth is obtained from a supersaturated solution with $C_{(o)} = 6 \times 10^{-2}\ \mathrm{g\,cm^{-3}}$ and $C_{NaCl(o)} = 2.5 \times 10^{-2}\ \mathrm{g\,cm^{-3}}$ ($\sigma_{(o)} \cong 6$).

5.13.1 *Mutual interference under diffusive-transport regime*

Under diffusive conditions (absence of convection), after an initial transient behavior, the depletion zones for the two growing seeds intersect and overlap (Fig. 5.11). At this stage the growth dynamics of the two crystals cannot be considered independent and mutual interference occurs. Both the crystals, in fact, absorb protein from the intermediate region (Fig. 5.11b and c). This leads to competition for growth and to

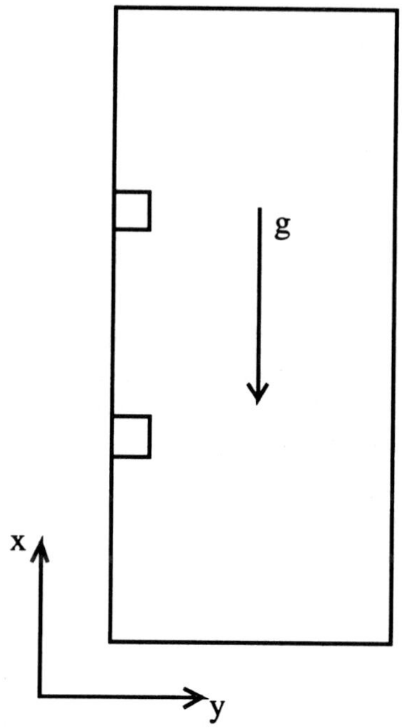

Fig. 5.10 Sketch of the growth reactor and the relative position of protein seeds (isothermal protein crystal-growth system: crystals of initial dimensions 1 mm × 1 mm fixed to the left wall of a 3 cm high and 1 cm wide growth cell at $x = H/3$ and $x = 2H/3$, respectively; $C_{(o)} = 6 \times 10^{-2}\,\mathrm{g\,cm^{-3}}$ and $C_{\mathrm{NaCl(o)}} = 2.5 \times 10^{-2}\,\mathrm{g\,cm^{-3}}$).

a lower value of the protein concentration there. A mirror-symmetric behavior occurs, however, with respect to $x = H/2$ due to the intrinsic symmetry of the system and the absence of convection.

According to the "double depletion" and consequent limited availability of protein in the intermediate zone, the crystals exhibit lower growth rates for the sides facing this region (Fig. 5.12).

As time passes, the face growth rates drop with increasing face size (correspondingly, in Fig. 5.13 the size of the crystal tends to an asymptotic constant value).

Since $\gamma \cong 0.3$–0.5 (Fig. 5.14), surface-attachment kinetics and mass transport in the liquid phase can be considered competitive in limiting the growth rate for both the growing crystals. After a short initial transient, however, γ decreases. This is due to diffusive transport that decreases in intensity the concentration gradient between the crystals and regions far from their surface; as already pointed out in Section 5.12 for the case of a single seed, in "finite size" systems, in fact, the relative importance of mass transport in the liquid phase with respect to surface-attachment kinetics decreases as time passes due to the decrease of available protein and related gradients. This effect, of course, is enhanced if the simultaneous presence of different growing seeds is considered.

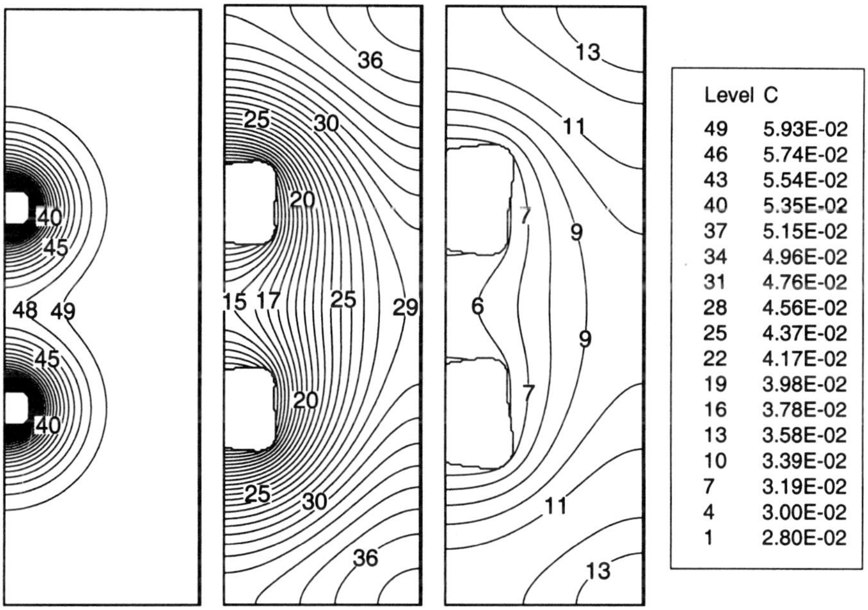

Fig. 5.11 Snapshots of growing crystals and concentration distribution under diffusive-transport regime: (a) $t = 3.6 \times 10^4$ s, (b) $t = 5.4 \times 10^5$ s, (c) $t = 1.33 \times 10^6$ s.

Therefore, in the light of these findings, the growth regime depends on at least three factors: surface-attachment kinetics, mass transport in the liquid phase and competition for growth if more than one crystal share the same reactor.

5.13.2 *Crystal interaction in the presence of convection*

If the presence of convection ($g = 10^{-4} g_o$) is taken into account, like the case in Section 5.12, convection arises around the growing crystals. Note that two different cases are considered herein: residual-g parallel to the x axis, (vertical configuration) and parallel to the y axis (horizontal configuration). Both situations give insight into the effect of steady residual accelerations on the phenomena under investigation.

First the attention is focused on the vertical configuration, then the horizontal condition is discussed and compared with the vertical one.

Figure 5.15 show the results for the first case (vertical configuration). The flow field modifies the protein distribution around the seeds leading to a nonsymmetrical concentration pattern. The mirror symmetry with respect to the midplane $x = H/2$ is broken. For each crystal, one recognizes the well-known convective flow pattern for solution growth with a rising "plume" above the seed.

The structure of the convective field is very complex. As time passes, the vortex shrinks, the contraction being strictly related to the behavior of the crystals that puff out due to growth.

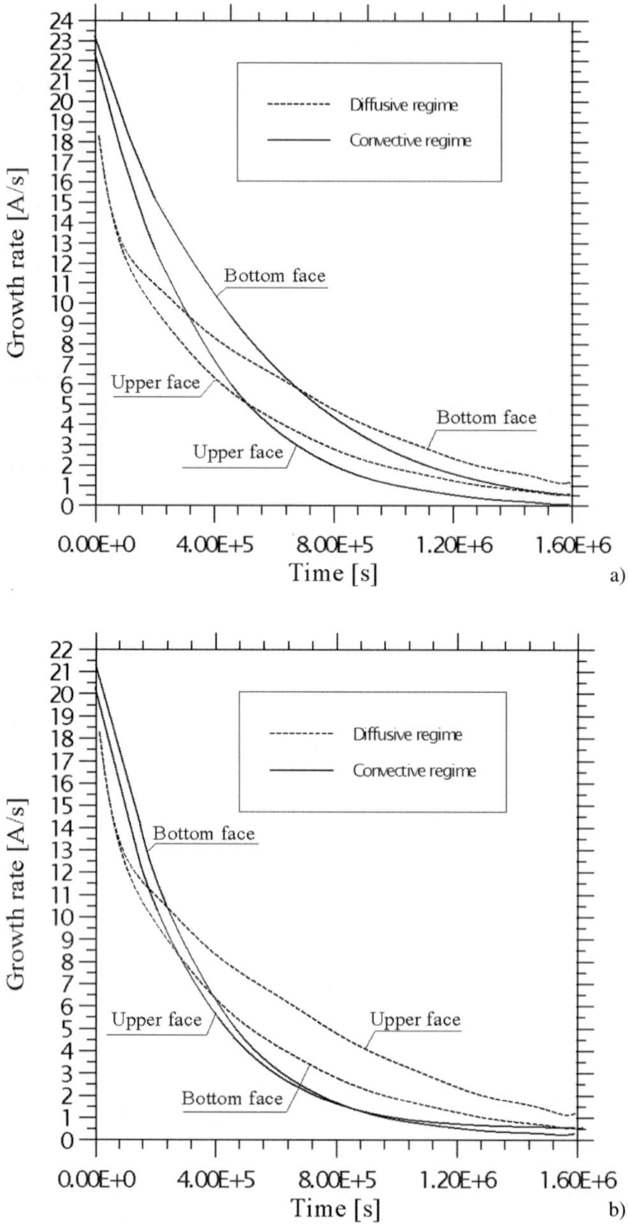

Fig. 5.12 Face growth rates versus time: (a) lower crystal; (b) upper crystal.

Note that the rising plume created by the lower seed laps against the bottom side of the upper seed (Fig. 5.15a and b). This phenomenon turns out to be crucial for the overall crystal-growth process. In fact, the plume induces a uniformly impoverished region close to the surface of the upper crystal. For this reason the protein

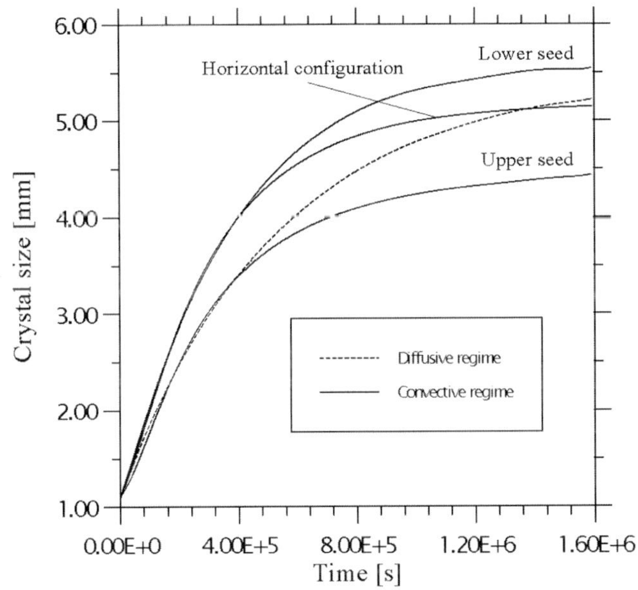

Fig. 5.13 Average size (along x) of the growing seeds versus time (under diffusive and convective transport regimes).

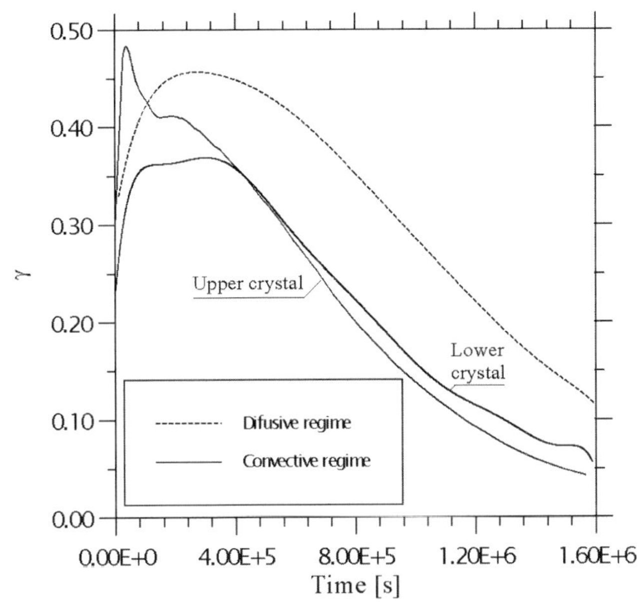

Fig. 5.14 Plot of γ versus time (nondimensional parameter taking into account the relative importance of surface-attachment kinetics and mass transport in the liquid phase).

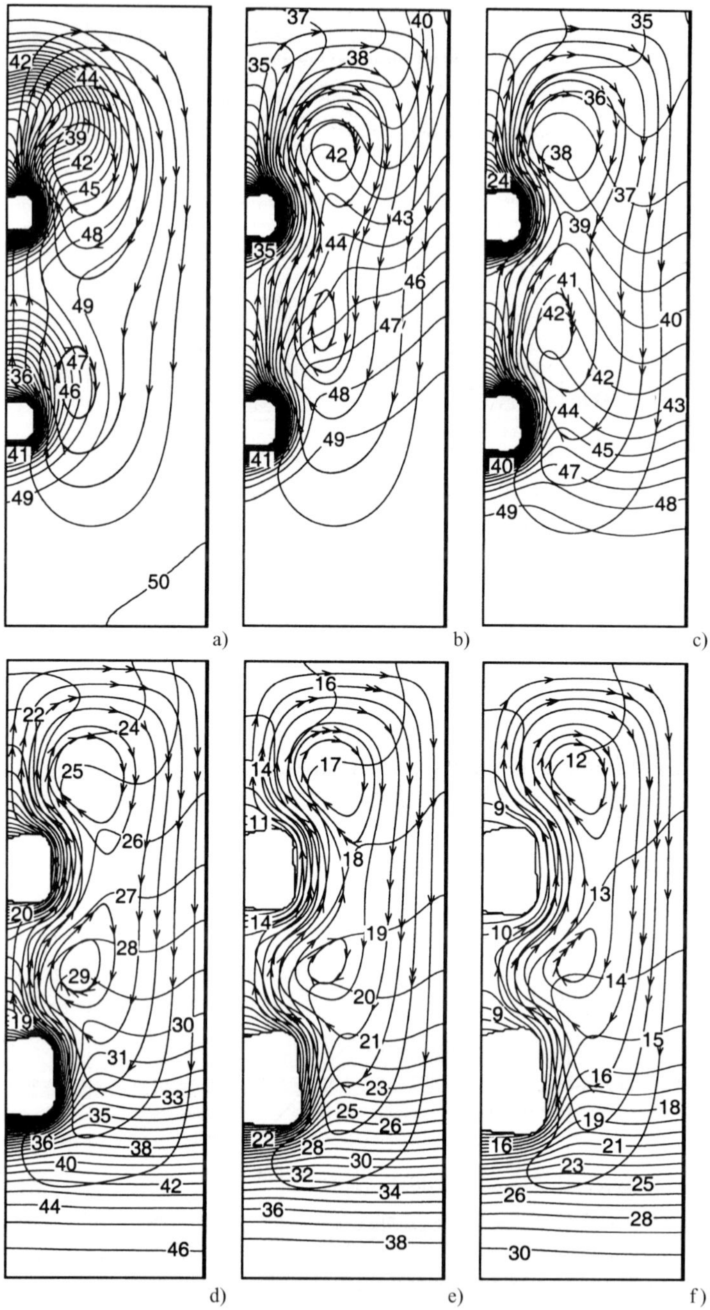

Fig. 5.15 Snapshots of growing crystals, concentration distribution and velocity field under microgravity conditions ($g = 10^{-4} g_o$): (a) $t = 4.8 \times 10^4$ s, (b) $t = 9.6 \times 10^4$ s, (c) $t = 1.44 \times 10^5$ s, (d) $t = 3.04 \times 10^5$ s, (e) $t = 4.64 \times 10^5$ s, (f) $t = 6.24 \times 10^5$ s, (g) $t = 1.0 \times 10^6$ s, (h) $t = 1.5 \times 10^6$ s (level $1 \rightarrow 2.8 \times 10^{-2}$ g cm^{-3}, level $50 \rightarrow 6 \times 10^{-2}$ g cm^{-3}, $\Delta C = 6.4 \times 10^{-4}$ g cm^{-3}).

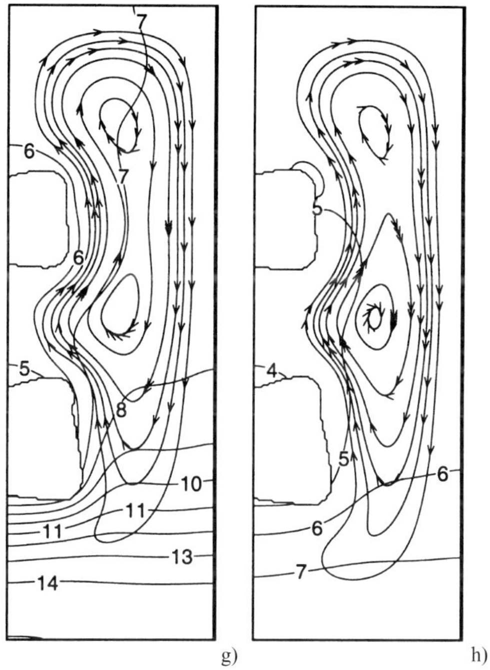

Fig. 5.15 Continued.

concentration gradients causing its growth are weakened and correspondingly the growth rate is reduced.

In order to understand the effect of convection on the system, however, it is necessary to distinguish the detailed description of the lower crystal behavior from the upper one.

Figure 5.15 show that the increase of volume of the lower crystal is more pronounced for the bottom side than for the top side. Figure 5.12a shows in detail how the convection effect results, in fact, in a higher local growth rate near the surface where the flow is incoming and in a lower local growth rate near the surface where the flow is outgoing (see Section 5.12 for details about the influence of such a flow on the face growth rates).

For the upper crystal the increase of size and the face growth rates do not show significant difference for the top and bottom sides (Figs. 5.15 and 5.12b). This occurs since in this case the incoming flow (rising plume of the lower seed) carries already impoverished liquid. The crystal is bounded by its own plume on one side and by the plume of the lower crystal on the other side. The available protein in the liquid phase is small regardless of whether the flow is incoming or outgoing. This, on the other hand, explains why the size of the lower crystal tends to be larger than that of the upper one (Figs. 5.13 and 5.15) and why face depressions do not occur for the latter. The upper seed, in fact, is embedded in a largely depleted liquid zone where the protein concentration is almost constant and the gradients are weakened; in the light of the arguments proposed in Section 5.12, this effect prevents the onset of morphological instabilities.

Since transport of protein in the liquid phase is dominated by convection, protein diffusion in the streamwise direction can be considered negligible. Thus there is no propagation of information in the upstream direction. Hence the lower crystal is not influenced directly by the upper crystal. Influence by the upper seed merely occurs via the return flow and via the total decrease of protein in the liquid phase due to the simultaneous growth of two seeds in the same reactor.

Figure 5.12 show that, during the first part of the growth process, the face growth rates of both the crystals under the effect of convection are considerably higher than under diffusive conditions. The growth rates decrease with increasing growth time and crystal size (the decrease being more rapid in the diffusion-dominated case). This is due to the effects already pointed out for the case of a single seed (Section 5.12).

Figure 5.14 shows that the time evolution of γ under convective effects exhibits a different behavior for each seed according to its own different growth history. The role of incorporation kinetics in limiting the growth rate tends to be more important for the lower seed during the first part of the growth process and for the upper seed later.

This clearly demonstrates that, in the case of simultaneous growth of different seeds, the relative importance of mass transport in the liquid phase and attachment kinetics does not behave as a fixed parameter since it changes according to the considered seed.

Figure 5.16 show the results obtained in the case of the horizontal configuration. As for the vertical condition, at the beginning one recognizes the well-known convective flow pattern with a rising plume above the seeds. Because of the geometrical constraints the rising liquid has to turn to flow along the side walls.

The resulting two main vortices have symmetric form but rotate in the opposite direction, i.e. the flow is symmetric by reflection about the midplane $x = H/2$.

After 4.6×10^5 s (see Fig. 5.16c), the solutal plumes in the left and right halves of the reactor merge giving rise to a single plume located at $x = H/2$. After a further increase in time, the simulations show that, in addition to morphological instabilities of the crystals (due to nonuniform transport of nutrients in the feeding solution) a true fluid-dynamic instability of the convective field appears. This instability leads to nonsymmetric patterns with respect to the midplane. It is steady in nature (stationary bifurcation) since the dynamic process of symmetry breaking is not associated with any oscillatory behavior. Prior to the onset of the bifurcation, the two vortices are the mirror image of the other. After the transition, the flow does not exhibit mirror structure with respect to $x = H/2$ though the geometrical configuration and the applied external forces are symmetric (Fig. 5.16d and e). This behavior shows the high sensitivity of the growth process (and associated fluid mechanics) to small

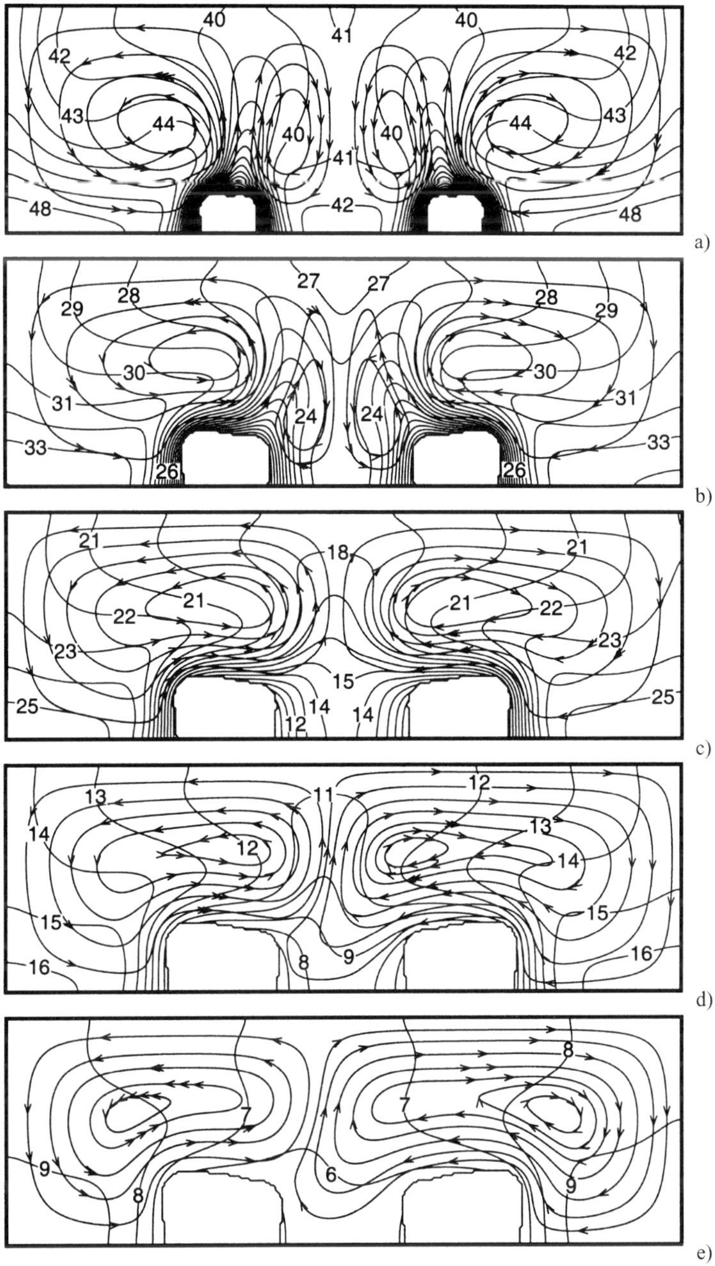

Fig. 5.16 Snapshots of growing crystals, concentration distribution and velocity field under microgravity conditions ($g = 10^{-4} g_o$, horizontal configuration): (a) $t = 1.28 \times 10^5$ s, (b) $t = 3.2 \times 10^5$ s, (c) $t = 4.64 \times 10^5$ s, (d) $t = 6.88 \times 10^5$ s, (e) $t = 9.92 \times 10^5$ s (level $1 \rightarrow 2.8 \times 10^{-2}$ g cm^{-3}, level $50 \rightarrow 6 \times 10^{-2}$ g cm^{-3}, $\Delta C = 6.4 \times 10^{-4}$ g cm^{-3}).

asymmetries of the boundary conditions and/or to the onset of fluid-dynamic instabilities (readers who want to know more about these instabilities and the concepts of transition and symmetry breaking, can find additional useful information in Section 1.4.1).

Figure 5.13 shows that the size of the crystals in the case of the horizontal configuration lies (as expected) between the size of the upper and lower crystals yielded for the case of vertical configuration.

In the light of these results some general trends can be pointed out about the conditions determined by the interplay of seeds simultaneously growing in the same protein reactor for both cases of horizontal and vertical distribution of the crystals.

The mutual interaction of crystals plays a critical role in determining the structure of the convective field. Convective effects, in turn, impact growth rates, macroscopic structures of precipitates, particle size and morphology as well as the mechanisms driving growth (γ). The face growth rates, in particular, depend on the complex multicellular structure of the convective field and on associated "pluming phenomena."

Under residual gravity and for the case of the vertical configuration, the crystals interact mainly via the influence of the solutal plume of the lower crystal on the upper crystal. The rising fluid coming from the lower crystal wraps around the upper crystal and has a major role in determining the concentration distribution on its surface. The protein concentration gradients are weakened and for this reason the upper seed exhibits low values of the growth rate and the onset of morphological instabilities is prevented for this seed.

In the case of the horizontal configuration the crystals interact mainly via the competition for growth in the intermediate region. Moreover, for this case the convective field can undergo a transition to a nonsymmetric flow pattern. A typical flow configuration obtained prior to the onset of the instability consists of two separate mirror vortices in the left and right halves of the reactor with opposite senses of rotation. The bifurcation causes an asymmetry in the structure of the two rolls.

From a global point of view, for both cases (vertical and horizontal configurations) convection enhances mass transport in the liquid phase thus increasing the rate of incorporation (growth rate) of the solute into the crystal, but at the same time, it can lead to a more rapid decrease of the finite amount of protein available in the solute phase and thus to a decrease of the macroscopic growth rate (Fig. 5.12). Different crystallization conditions (in terms of γ) may occur in the same reactor during the crystallization process due to mutual interference of the growing seeds, complex convective effects and the finite size of the reactor (see also Lappa, 2003c).

5.14. *N* interacting crystals

In this section the growth of several macromolecular seeds uniformly distributed on the bottom of the protein reactor (i.e. a discrete layer of *N* crystals) is investigated

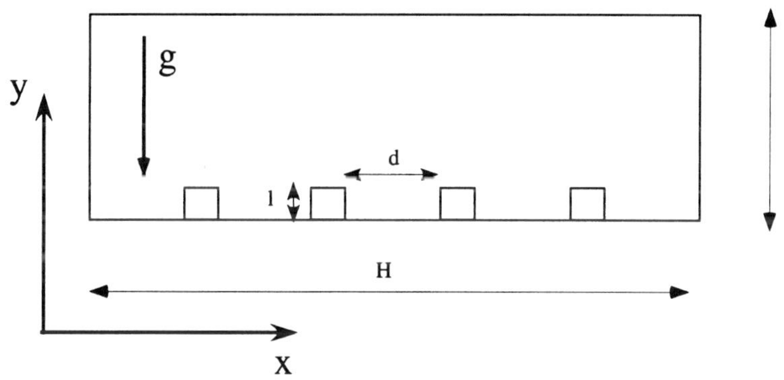

Fig. 5.17 Sketch of the growth reactor with the protein seeds and of the relative direction of the residual g.

for different values of N. Relevant nondimensional geometrical parameters are the aspect ratio $A = H/L$ (see Fig. 5.17) of the container and the parameter $\kappa = l/d$ (ratio of the crystals size and related distance).

Following typical experimental procedures the seeds are assumed to be fixed (e.g., by glue) to the bottom of the reactor.

The section in particular focuses on the case of seeds located one away from the other at a distance of the order of the crystal size ($\kappa = O(1)$) changing their number and considering two different values of the aspect ratio of the reactor ($A = 3$ and $A = 6$).

The condition $\kappa = O(1)$ is investigated since (owing to the small size of the reactors used for morphological analyses) this is the typical situation occurring in this kind of study. The number of seeds is changed in order to study the sensitivity of the overall system to this parameter. The dimension of the reactor along x is doubled ($A = 3 \rightarrow A = 6$ retaining the same extension along y) in order to investigate the possible existence of regularities associated with the transport process in a system with a large number of seeds, i.e. to discern if the domain can be split into a central area where the pattern is almost periodic and sides strongly affected by edge effects.

Growth is obtained from a supersaturated solution with $\sigma_{(o)} \cong 6$.

5.14.1 *Structure of the convective field*

First, attention is focused on the case of a fixed aspect ratio ($A = 3$) and κ not constant, then the case with κ fixed and two different values of the aspect ratio is taken into account.

For the case $A = 3$ ($g = 10^{-4} g_o$), a flow field is driven by the density gradient around the growing crystals (see Figs. 5.18 and 5.19).

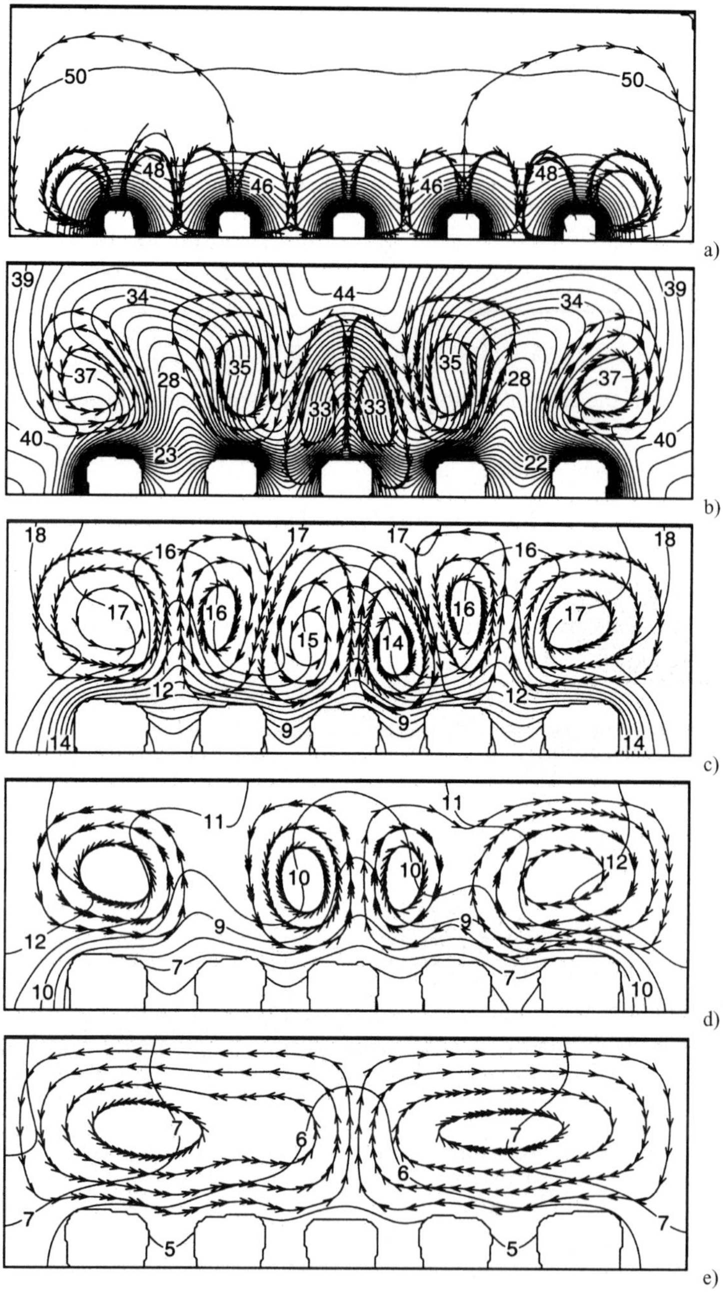

Fig. 5.18 Snapshots of growing crystals ($A=3$, $N=5$), concentration distribution and velocity field under microgravity conditions ($g=10^{-4}\,g_o$): (a) $t=1.38\times10^4$ s, (b) $t=1.38\times10^5$ s, (c) $t=3.59\times10^5$ s, (d) $t=4.97\times10^5$ s, (e) $t=7.18\times10^5$ s (level $1\rightarrow2.8\times10^{-2}\,\mathrm{g\,cm^{-3}}$, level $50\rightarrow6\times10^{-2}\,\mathrm{g\,cm^{-3}}$, $\Delta C=6.4\times10^{-4}\,\mathrm{g\,cm^{-3}}$).

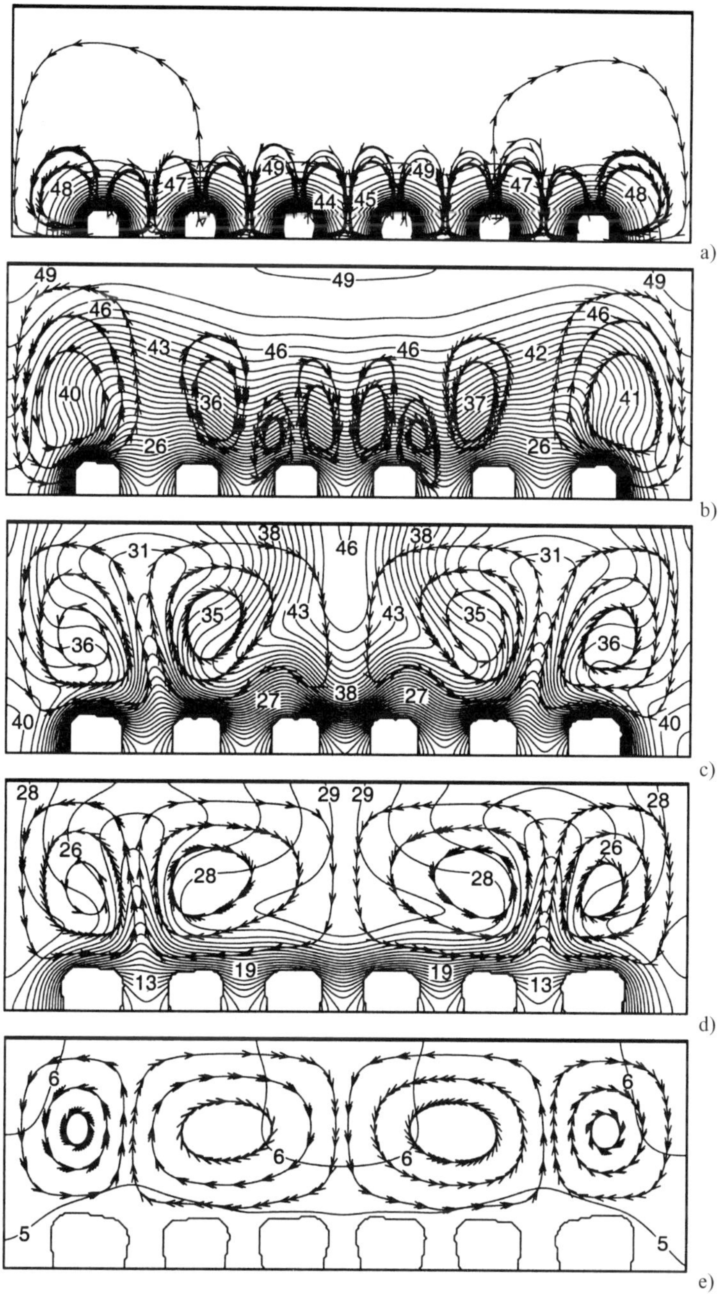

Fig. 5.19 Snapshots of growing crystals ($A = 3$, $N = 6$), concentration distribution and velocity field under microgravity conditions ($g = 10^{-4} g_o$): (a) $t = 1.0 \times 10^4$ s, (b) $t = 8.3 \times 10^4$ s, (c) $t = 1.35 \times 10^5$ s, (d) $t = 2.075 \times 10^5$ s, (e) $t = 8.3 \times 10^5$ s (level $1 \rightarrow 2.8 \times 10^{-2}$ g cm^{-3}, level $50 \rightarrow 6 \times 10^{-2}$ g cm^{-3}, $\Delta C = 6.4 \times 10^{-4}$ g cm^{-3}) .

At the beginning (Figs. 5.18a and 5.19a) each crystal is characterized by its own rising solutal plume. In this early phase of the growth process, i.e. when depletion zones are separated, the growth dynamics of the different seeds can be considered independent. Two convective cells are generated around each crystal. The formation of halos due to the lowered solute concentration around growing seeds, in fact, has the effect of producing density gradients in these areas. These in turn (under the effects of gravity) result in the onset of convection.

After this initial transient behavior, however, the depletion zones for two growing consecutive seeds intersect and overlap (Figs. 5.18b and 5.19b). At this stage mutual interference occurs. Crystals, in fact, absorb protein from common regions (this aspect has already been highlighted in Section 5.13 for the case of two crystals).

For $N>2$, however, the delicate evolutionary equilibrium among crystal growth (absorption process), solution-depletion phenomena and competition among several seeds is coupled to significant and intriguing adjustments in the roll pattern inside the reactor.

In fact, a very complex multicellular structure is created above the interacting crystals and this convective field undergoes interesting subsequent transitions to different regimes as time passes. It is also worthwhile to stress how the major stages of this spatiotemporal evolution change according to the value of N ($N=5$, $\kappa=0.42$ in Fig. 5.18 and $N=6$, $\kappa=0.34$ in Fig. 5.19).

For the case $N=5$, at the beginning the wave number of the multicellular field is $m=10$ (Fig. 5.18a) and five rising convective jets occur above the seeds (hereafter the crystals will be referred to as crystal "i" with $i=1\rightarrow N$, respectively, starting from the left side of the reactor).

After 1.38×10^5 s (see Fig. 5.18b), the solutal plumes originated from crystals 1 and 2 merge giving rise to a single plume located at $x=H/4$. The same behavior holds for seeds 4 and 5 since the system exhibits mirror symmetry with respect to the midsection $x=H/2$. For this reason the depleted lighter fluid is transported upwards by three solutal rising jets only. They are located approximately at $x=H/4$, $x=H/2$ and $x=3H/4$, respectively (instead of the five solutal plumes that characterize the system behavior in the early phase located at $x=iH/6$, $i=1\rightarrow 5$). Correspondingly, the initial number of rolls is reduced from 10 to 6. Two convective cells still exist above the central seed ($i=3$), whereas only two vortices characterize the right and left halves of the reactor with the core located approximately above the seeds 1, 2 and 4, 5, respectively. As time passes new regimes evolve that exhibit different number of rolls; the wave number decreases: $m=4$ ($t\cong5\times10^5$ s, Fig. 5.18d) and finally $m=2$ ($t>5\times10^5$ s, Fig. 5.18e) with only one vortex for each half of the reactor. The resulting two final convective cells have symmetric form but rotate in the opposite direction, i.e. the flow is (approximately, owing to a symmetry-breaking phenomenon similar to that described in Section 5.13) symmetric by reflection about the midplane $x=H/2$.

For the case $N=6$ the roll-pattern formation and evolution are quite different. At the beginning the wave number of the multicellular convective field is $m=12$ (Fig. 5.19a) and six rising plumes appear above the seeds ($i=1\rightarrow 6$). For $t=8.3\times 10^4$ s (see Fig. 5.19b), a large depletion zone surrounding all the crystals is formed and the wave number is decreased to $m=8$; for $t=1.35\times 10^5$ s (Fig. 5.19c) the depleted lighter fluid is transported upwards by two solutal rising jets only and the $m=8$ mode is taken over by a new regime with $m=4$. The surviving solutal plumes are located approximately at $x=H/4$, and $x=3H/4$, respectively, whereas the onset of pluming phenomena at $x=H/2$ is prevented. This is an intriguing and unexpected difference with respect to the case $N=5$. For $N=5$ around the midsection ($x=H/2$) the fluid is carried upwards. For $N=6$ the opposite situation occurs. For $t>1.35\times 10^5$ s (Fig. 5.19d and e) the structure with four vortex cells seems to be quite stable without further transitions.

In the light of the above results, the prominent features of the complex scenario related to the growth exhibit a high sensitivity on the number of crystals simultaneously growing in the protein reactor if this number is not sufficiently large (this case corresponds to typical experiments carried out by organic crystal growers interested in studying morphological and kinetic details). It, in fact, plays a crucial role in determining the wave number of the convective field (i.e. the number of vortex cells), the number and positions of rising solutal jets and the related time evolution.

It is worthwhile to underline how the differences between the cases $N=5$ and $N=6$ could be used to explain, in principle, the scarce reproducibility that seems to characterize this kind of experiment (see also Ramachandran et al., 1995).

On the other hand, the computations carried out increasing the extension of the protein reactor along x provide additional information about the effect of geometrical parameters, in particular with regard to the possible existence of a core zone not affected by the finite size of the reactor.

Along these lines, Fig. 5.20a shows that for the case $A=6$, $N=10$ at the beginning each crystal is characterized by its own rising solutal plume with the exception of the crystals located at the extremities. After a transient time four solutal jets are created above the seeds (Fig. 5.20c) and the wave number is $m=8$. This situation seems to be quite stable (see Fig. 5.20d).

As expected, with the exception of the two crystals at the edges, an interesting spatial periodic behavior arises along x in this case. The seeds seem to "work" in groups of two. Rising solutal jets are created above the couples (2, 3) (4, 5) (6, 7) (8, 9), descending jets occur above the couples (3, 4) (5, 6) (7, 8). Therefore a certain degree of periodicity in space can be highlighted with a central zone affected by a somewhat repetitive behavior.

This trend is confirmed by the simulations carried out for $A=6$ and $N=12$ (not shown).

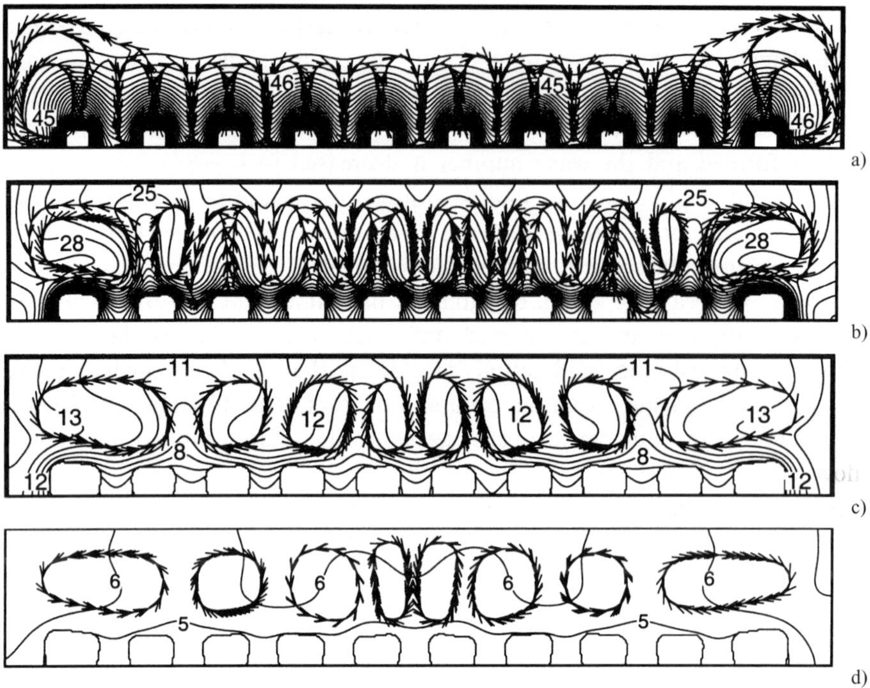

Fig. 5.20 Snapshots of growing crystals ($A=6$, $N=10$), concentration distribution and velocity field under microgravity conditions ($g=10^{-4}g_{\mathrm{o}}$): (a) $t=4.6\times10^{4}$ s, (b) $t=2.3\times10^{5}$ s, (c) $t=4.83\times10^{5}$ s, (d) $t=8.05\times10^{5}$ s (level $1\rightarrow2.8\times10^{-2}\,\mathrm{g\,cm^{-3}}$, level $50\rightarrow6\times10^{-2}\,\mathrm{g\,cm^{-3}}$, $\Delta C=6.4\times10^{-4}\,\mathrm{g\,cm^{-3}}$).

The direction of the jets is always modulated in space with a periodic alternation of rising and descending jets. This behavior holds for an intermediate region not affected by edge effects.

Accordingly, some general rules can be introduced. If the aspect ratio is sufficiently large and $\kappa=\mathrm{O}(1)$, in fact, the wave number can be roughly computed as $m=(N-2)$ and the number of rising solutal jets as $(N-2)/2$.

The cases with a very large number of seeds ($\mathrm{O}(10^2)$) and/or very high aspect ratio are not investigated herein since they are not relevant. As previously discussed, in fact, very large number of crystals are grown by organic crystal growers using gellified configurations where convective effects cannot occur. Moreover, for such a case, usually shape details are not a relevant part of the analysis since researchers are interested in analyzing other aspects of the growth (number of nuclei created and their spacing, effect of different levels of initial supersaturation, velocity of propagation of the nucleation phenomena, etc.). Nevertheless, it can be speculated that when the number of seeds becomes very large in a system with $\kappa=\mathrm{O}(1)$, the wave number $m\rightarrow N$ (since the relative importance of the edge effects tends to become very small).

It is noteworthy how, from a fluid-dynamic point of view, the problem under investigation exhibits a quite interesting theoretical kinship with the Rayleigh–Bènard problem in which convection is induced in a horizontal fluid layer by uniform heating from below (see Section 2.2.4 for additional details about this category of phenomena).

The temporal evolution of two-dimensional Rayleigh–Bènard convection in a shallow fluid layer of aspect ratio $A = H/L$ (i.e. ratio of length and height) confined laterally by rigid sidewalls has been deeply analyzed by many investigators (see, e.g., the excellent work of Daniels, 1984). Several studies have shown that for Rayleigh numbers in certain ranges there exists a class of steady-state solutions that correspond physically to the possibility of an adjustment in the number of rolls in the container as the local value of the Rayleigh number is varied.

Nonlinear effects, geometrical constraints are all essential ingredients of this type of flow (see the discussion in Section 2.2.4).

For the Rayleigh–Bénard problem the wave number, in fact, behaves as a function of two parameters: The Rayleigh number and the aspect ratio (it is well known that the presence of perfectly insulated sidewalls has a significant influence in determining the flow pattern that develops when the Rayleigh number exceeds its critical value wave number, i.e. the finite extension leads to quite a different behavior with respect to the case of infinite layers).

Despite the macroscopic similarities, however, the dynamics and the mechanisms underlying the behaviors of the present system are quite different.

In the case of uniformly heated layers the driving force for the onset of convection is the lighter fluid on the bottom of the container, this force being replaced by the presence of a discrete distribution of sinks of solute for the case of a discrete layer of growing macromolecular crystals (the incorporation of protein is the source of lighter fluid at the crystal surface).

For the growth of many aligned macromolecular seeds there are at least four independent parameters influencing the aforementioned evolution of the multi-cellular structure, i.e. the aspect ratio of the protein reactor, the time evolution of the phenomenon of protein absorption (this would correspond to a quasisteady variation of the Rayleigh number for the case of the layer of fluid heated from below) and finally (as previously highlighted) the number of interacting protein crystals and the growth process itself (which leads to a confinement of the roll pattern to the top of the reactor due to the increasing size of the underlying crystals).

For this case the "heating" is not uniform. The depletion zones do not exhibit uniform behavior. For the case of many interacting crystals, adjustment in the number of rolls occurs as a consequence of the coalescence mechanism of the depletion zones that merge as time passes (the samples grow and the related halos become larger and overlap) and according to the space occupied by the crystals that increases as time passes (this behavior is responsible for the confinement of the overall convective structure to the upper part of the reactor).

Finally, it should be pointed out that an initial nonuniform distribution of the seeds could deeply affect the initial interaction mode. In that case, in fact, initial interaction is expected to occur between the seeds with small distance.

5.14.2 *Growth-rate distribution and morphological instabilities*

The discussion above is quite a comprehensive analysis of the fluid dynamics associated with the growth of the seeds and may provide interesting information and data for fundamental fluid-dynamic research and for investigation of nonlinear behaviors. The present section deals with the growth process itself. Before starting to deal with the local detailed analysis of each crystal according to its position, however, some common features can be pointed out.

Figures 5.18 and 5.19 show that the "depth" of the face protuberances is proportional to the size of the crystals (competition for growth in the intermediate regions and corner effects are responsible for the tooth-shape of the crystals). Moreover, the increase of volume is more pronounced for the crystals at the extremities of the reactor ($i=1$ and $i=5$ for $N=5$, $i=1$ and $i=6$ for $N=6$, $i=1$ and $i=10$ for $N=10$) than for the central crystals ($i=3$ for $N=5$, $i=3$ and $i=4$ for $N=6$). This trend is confirmed by Fig. 5.21 where the size of the seeds is plotted as a function of time. This plot also highlights that the increase of volume is less pronounced in the case $N=6$ with respect to $N=5$. This apparent lack of solid mass,

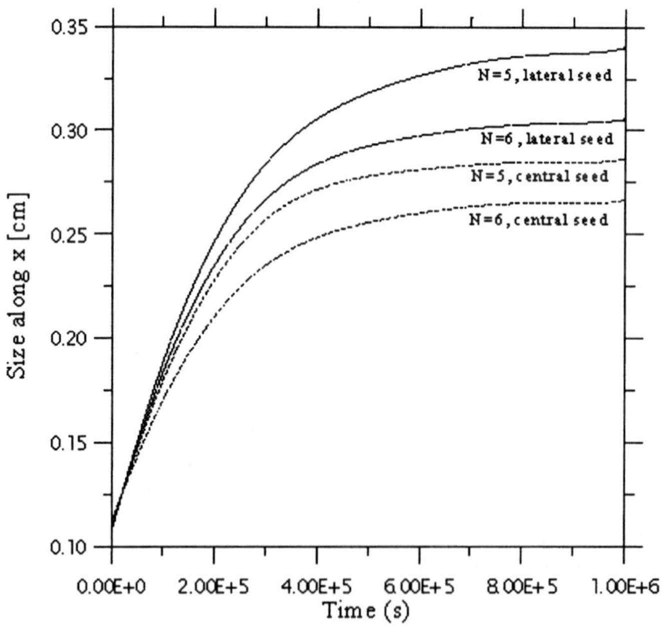

Fig. 5.21 Average size (along x) of the growing seeds versus time ($A=3$).

of course, follows from the presence of a higher number of crystals simultaneously absorbing protein from the liquid phase in the case $N=6$ (i.e. for a fixed initial amount of protein available in the liquid phase, the final size of the crystals decreases as their number increases).

Possible explanations of the above trends can be outlined on the basis of the local growth-rate distributions plotted in Figs. 5.22 and 5.23.

In agreement with the earlier results these figures show that corners and edges of the crystals are more readily supplied with solute than the center of sides (this is responsible for the morphological instability and the presence of a macroscopic depression around the center of the faces shown in Figs. 5.18 and 5.19).

As previously discussed, incorporation of the solute into the crystal causes a local depletion in concentration and a solutal concentration gradient to form between the bulk solution and the growth interface. As expected, the growth rate is always lower at the center than at the corners where the convective flows (and the associated shear stresses) are the strongest, and hence, the interfacial concentration gradient the steepest.

Figures 5.22 and 5.23 also allow us to point out that the difference in gradient steepness (i.e. local growth rate) between the face corner and face center increases as time passes; this explains why the "depth" of the face depressions is proportional to the volume of the samples. Moreover, the convection effects tend to increase the local growth rates for the crystals at the extremities. This trend finally provides a justification for the behaviors shown in Fig. 5.21.

Despite the macroscopic analogies and similarities pointed out above, however, each crystal exhibits different morphological evolution and different growth history according to its position within the reactor and according to the value of N.

For instance, in the case $N=5$ the final ridge (slope of the upper side) of the crystals with $i\neq 3$ is inclined with respect to the direction of the residual g (on the contrary for $i=3$ the ridge is almost flat and horizontal, Fig. 5.18e). For $i=1$ and $i=5$ the ridge is inclined inwardly, vice versa for $i=2$ and $i=4$ it is inclined outwardly. For the case $N=6$, the ridge is inclined inwardly for $i=1$ and $i=6$, and outwardly for $i=2,3,4,5$ (Fig. 5.19e). Edge effects are also very evident in Fig. 5.20d.

In practice, the characters of the final samples change depending on the local fluid-dynamic conditions under which they are operated. In addition to the effect associated with the corners more readily supplied with solute, there is an effect due to the local convective pattern around the crystal. The cross-comparison frame by frame of Figs. 5.18 and 5.22 for the case $N=5$ (5.19 and 5.23 for the case $N=6$) in particular elucidates that the convection effect results in higher local growth rates near the surface where the flow is directed downwards and lower local growth rates near the surface where the flow is carried upwards (if the number of seeds is not sufficiently large these fluid-dynamic conditions in turn change in time and exhibit a different spatial evolution according to the number of interacting crystals simultaneously growing in the protein reactor, see also Lappa, 2004e).

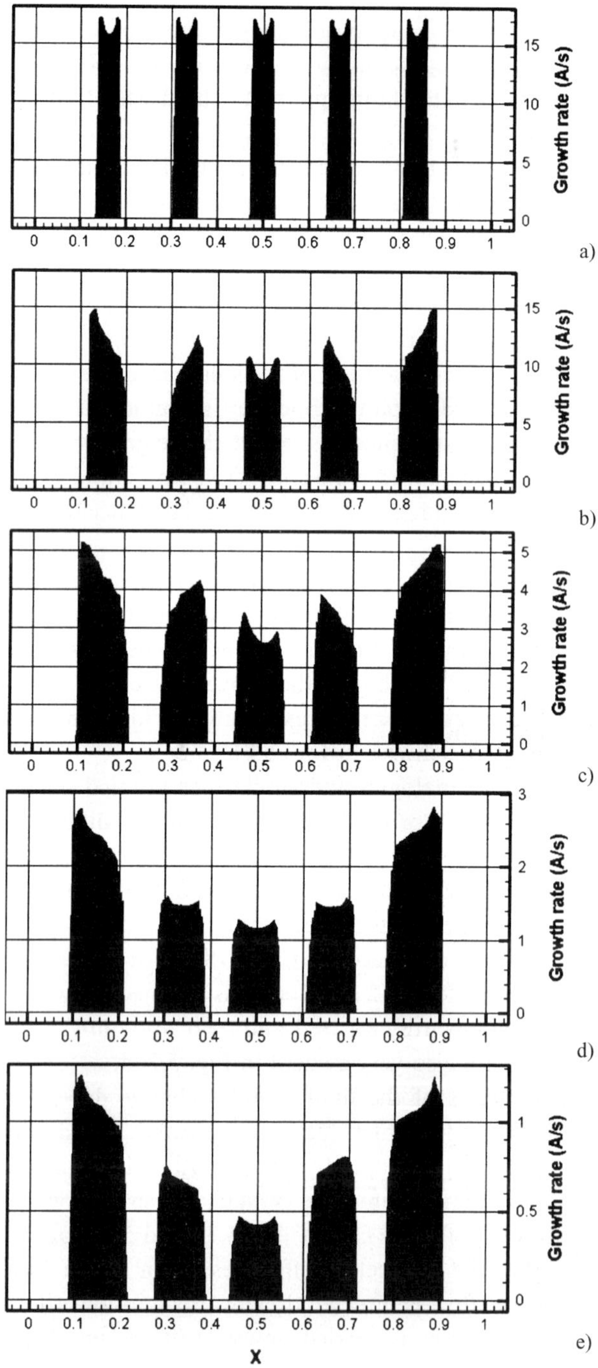

Fig. 5.22 Snapshots of growth-rate distribution, $A=3$, $N=5$: (a) $t=1.38\times10^4$ s, (b) $t=1.38\times10^5$ s, (c) $t=3.59\times10^5$ s, (d) $t=4.97\times10^5$ s, (e) $t=7.18\times10^5$ s.

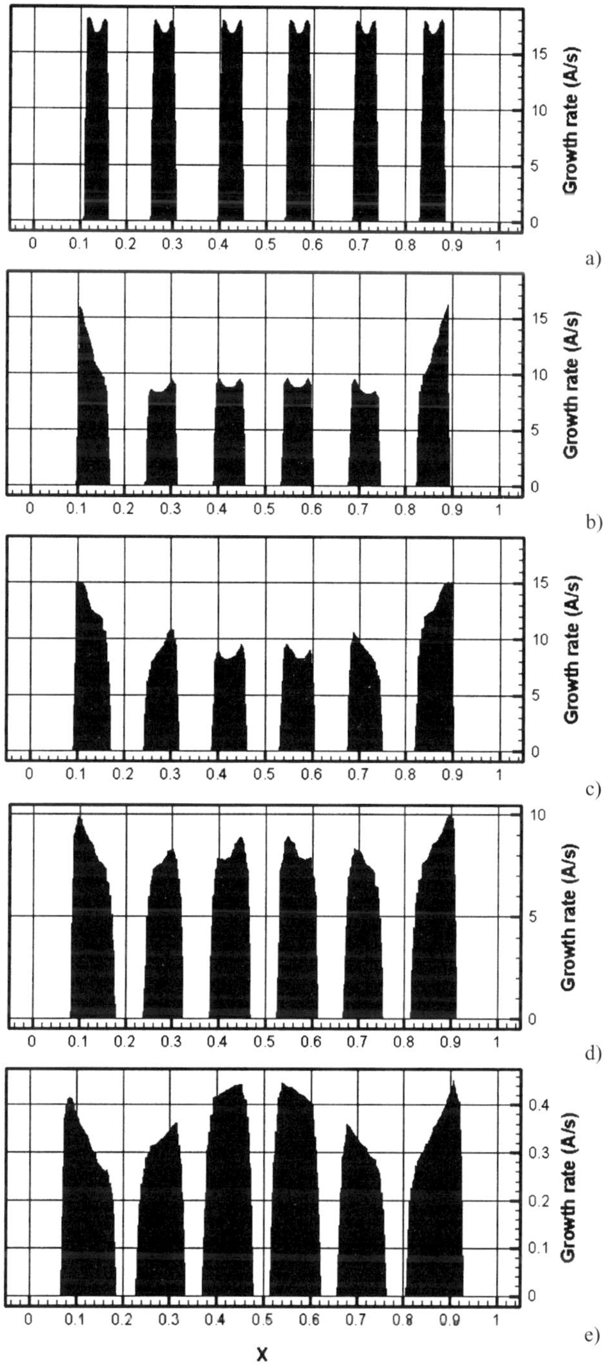

Fig. 5.23 Snapshots of growth-rate distribution, $A = 3$, $N = 6$: (a) $t = 1.0 \times 10^4$ s, (b) $t = 8.3 \times 10^4$ s, (c) $t = 1.35 \times 10^5$ s, (d) $t = 2.075 \times 10^5$ s, (e) $t = 8.3 \times 10^5$ s.

The simulations show that the morphological evolution of the seeds is strictly coupled to the sequence of transitions that characterize the fluid motion. The growth of the crystals, in turn, influences the stages of evolution of the velocity field being responsible for the confinement of the overall convective structure to the upper part of the reactor.

It is very important to point out how the correspondence elucidated above between surface growth rates and roll-pattern evolution is indirect evidence of the fact that for the case under investigation, mass transport in the bulk and surface-attachment kinetics are competitive as rate-limiting steps for growth (in the case of phenomena strictly governed by the kinetic barrier, in fact, the effect of convection on the overall growth process is almost negligible).

5.15. Conclusions, possible improvements and extension to the case $N \gg 1$

The studies presented in the previous pages in the form of prototype applications and numerical examples represent a first (and quite exhaustive) attempt to characterize with precise mathematical arguments the relative importance of surface-attachment kinetics and mass transport in the liquid phase (that is still poorly understood in the literature and is contradictory) and to discern the cause-and-effect relationships underlying the complexity of growth carried out in the presence of interacting crystals. Determining the related growth laws is central for discerning how fluid-dynamic and other environmental conditions affect macromolecular crystallization.

The present chapter, however, does not provide a complete coverage for the subject of macromolecular crystallization. There are, in fact, other important aspects that should be elucidated. Most of them, in particular, deal with the fact that the behavior at the macroscopic scale length in systems with a very large number of crystals (hundreds or thousands) is also a relevant and important part of the problem. This is the subject of the next chapter where the counterdiffusion technique and related numerical models are discussed focusing on ensemble (macroscopic) aspects. Some models dealing with the case $N \gg 1$ are discussed therein.

CHAPTER 6

Macromolecular crystal growth at macroscopic length scales

6.1. Introduction

Macromolecular crystallization is a matter of searching, as systematically as possible, the ranges of the individual parameters that impact upon crystal formation, finding a set or multiple sets of these factors that yield some kind of crystals, and then optimizing the variable sets to obtain the best possible crystals for X-ray analysis (see Chapter 1 for additional theoretical background).

Along these lines, ensemble behaviors must also be regarded as an important aspect of the problem.

In many circumstances, in fact, the optimization of the crystallization process cannot be addressed in an exhaustive way if the investigation is limited to "what happens close to the surface of the crystals," but requires analysis at least at two distinct length scales: on the microscopic length scale of the crystal interface the process is governed by the surface-incorporation kinetics and these aspects and their implications have been exhaustively elucidated in Chapter 5; on the macroscopic scale, the process is controlled by many other factors, e.g., the transport of nutrients available in the liquid phase, competition among different interacting crystals (some initial effort made dealing with this aspect has been provided in the last sections of Chapter 5), nucleation and ensuing growth, sedimentation and convection, etc., and related interplay among the different phenomena.

Fundamental methods for identifying favorable protein crystallization conditions still remain poorly developed (as pointed out by Rosenbaum and Zukoski, 1996; Giege and Ducruix, 1999; McPherson, 1999) and this is probably due to the absence of a macroscopic point of view in the analyses of the related production process.

Owing to the disappointing lack of mathematical models and numerical methods able to provide such a comprehensive description, the present chapter focuses, in particular, on these unresolved issues. As anticipated, such a global analysis may provide additional support to the aforementioned optimization of growth techniques, i.e. it may help to discern interrelations between various macroscopic parameters of interest (for instance, the size of the crystals, their distribution and number, the spatiotemporal behavior of the crystallization process, etc.), and the underlying mechanisms.

It is worthwhile to stress how the computational models introduced within this context, of course, must also take into account the surface incorporation kinetics (which must be regarded as a fundamental and undeniable aspect of the problem, Lappa, 2002b). However, as will be illustrated in the following sections, the strategy used herein relies on the application of an *integral* form of the kinetic balance law for each protein crystal rather than on differential models. The shape of the crystals, in fact, is ignored.

In practice, this integral approach comes through major modification and extension of the kinetic differential model discussed in Chapter 5 (able to elucidate the distribution of the local growth rate along the sides of the crystal as well as to predict shape instabilities, the onset of surface depressions due to diffusive and/or convective effects, etc., i.e. all those factors dealing with the local history of the shape).

The philosophy relying on integral models is particularly suitable for conditions in which the size of the crystals is negligible with respect to the size of the reactor, i.e. if they are small and undergo only small dimensional changes with respect to the overall dimensions of the cell containing the feeding solution and/or in situations where the surface details simply are not a relevant part of the analysis. This is the case, in particular, of counterdiffusion techniques. In such techniques, in fact, usually hundreds of crystals are produced in a single process (among other things, these techniques have enjoyed a widespread use in the literature over recent years) and the investigators are interested in determining their size and distribution rather than focusing on the shape and morphological evolution.

To summarize, the aim of the present chapter is to introduce new mathematical models and tools devoted to a somewhat macroscopic approach to the problem, focusing on phenomena not considered in Chapter 5 (i.e. nucleation, sedimentation, g-jitter effects, and the case of a very large number of crystals simultaneously growing in the reactor). In the spirit of the book, critical information and insights into the physics are provided that may lead to rational guidelines that can increase the probability of success in crystallizing protein substances.

A significant effort is also made throughout the chapter to show how the prototype numerical examples fit the corresponding experimental processes. The consistency of numerical predictions with experimental data suggests, in fact, that rate-controlling steps are taken into account, that the integral approach used to handle the surface kinetics does not distort actual behavior and finally provides validation for the theoretical models and techniques.

In particular, comparison between experimental observations and numerical simulations in the presence of convection and sedimentation and without them is considered to give insight into the potential influence that crystal motion within the surrounding liquid may have on the ground as well as in space (on the International Space Station).

6.2. The use of gel as a substitute for microgravity and true microgravity conditions

On the ground, typically, investigators let reagents diffuse in some (preferably inert) environment. In this environment reagents react to produce spaced deposits.

Under normal-gravity conditions, sedimentation and convection, that would alter patterns of crystals, in practice are prevented by the use of gel (see, e.g., Robert and Lefaucheux, 1988; Robert et al., 1992). Agarose gel turns out to be a particularly suitable diffusion medium, because of its mechanical flexibility and chemical inertness. Due to its mechanical properties, crystals produced due to protein nucleation are locked on the matrix of agarose gel. If the gel matrix is not present, since crystals are denser than the feeding solution, sedimentation occurs and the crystals settle in the growth chamber.

With respect to these aspects, gel might represent a method to simulate microgravity on the ground; however, it is not a real substitute for microgravity, because it might influence the crystallization process. Besides positive effects, gels can also have a negative influence leading sometimes to detrimental effects in the quality of protein crystals.

On the other hand, the opportunity, in recent years, of accessing a low-gravity environment to grow crystals of higher quality is still an open question. Buoyancy-driven convection and sedimentation, the most important causes of imperfections and crystal defects are, in fact, reduced under microgravity conditions, but not suppressed (see Sections 5.11–5.14).

In addition, even if the residual gravity is reduced by several orders of magnitude in space, the presence of time-dependent accelerations (the so-called g-jitter disturbances: periodic accelerations of different amplitudes and frequencies) can induce convective motions.

A large number of space experiments involving protein crystal growth have been carried out in recent years and many have produced conflicting results. Some of these experiments produced crystals that were larger and better ordered than the best ever grown on Earth. For this reason, the use of crystallization techniques under microgravity has been proven to be, in principle, a good candidate for growing crystals in a convection-free scenario. In spite of the perceived advantages of growing macromolecular crystals in space, however, there have also been a number of flight experiments that yielded crystals that exhibited no improvement in internal order, even for growth systems that had previously demonstrated significant improvements (Ramachandran et al., 1995).

Such apparent scarce reproducibility could be explained by the high sensitivity that these systems exhibit under the effect of steady residual accelerations to the number of crystals simultaneously growing in a same reactor (as elucidated in Section 5.14) and/or on the basis of the aforementioned g-jitters that (contrary to the case of steady g components) are specifically mission-dependent disturbances.

Oscillatory g-jitter includes all the periodic time-dependent accelerations that can be approximated by sinusoidal functions. The most common sources of these disturbances are structural vibrations (e.g., at the fundamental natural frequencies), equipment operations and crew activity (e.g., repetitive exercises that induce cyclic displacement of the position of the growth reactor). The reader will find additional and relevant theoretical background about these aspects in Chapter 2.

For the reasons highlighted above, the understanding of the role of microgravity for protein crystal growth and its eventual benefit for obtaining high-quality crystals need further investigation.

Particularly interesting is the interplay between nucleation and growth, on the one hand, and transport phenomena in the fluid phase induced by oscillatory accelerations, on the other.

While it is well known (see, e.g., Qi and Wakayama, 2000) that steady gravity levels may induce sedimentation of the freely suspended crystals and distortion of the concentration field (see the prototype applications in Chapter 5), the role played by periodic accelerations in the dynamic process of macromolecular crystallization is still poorly understood.

Recently, the influence of these g-jitters on the dynamics of protein crystal growth in microgravity has been observed directly. Otàlora et al. (2001) reported novel results from microgravity-conducted experiments. These experiments were specially designed to monitor (for the first time) the development of the depletion zone around protein crystals and for observing separate crystal and fluid motion induced by g-jitter. The depletion zone was observed to be distorted by crystal movements.

6.3. Macroscopic analysis and integral formulation of the kinetic conditions

As anticipated, diversification of the model is needed according to the desired scale length. The mathematical conditions accounting for the surface kinetics, in fact, behave as *differential* equations in the case of microscopic approach (as illustrated in Chapter 5) and as *integral* balance laws if the analysis is carried out at the macroscopic scale (present chapter).

If the shape of the crystal and its morphological time evolution are not an essential part of the problem, the shape can be ignored and each solid particle can be considered in terms of its mass and position within the reactor. Accordingly, to account for the kinetic condition, Eq. (5.1) must be rewritten as an integral condition to be applied along the frontier $\partial\Omega$ of the computational cell containing the solid particle.

In practice, the model assumes that the concentration at the crystal surface (C_{cs}) is equal to the value of the concentration computed along the frontier of the computational cell containing the crystal. The concentration C_{cs} is not obliged to be uniform along the frontier; it may have a different value depending on the side where it is computed; this means that the integral form of Eq. (5.1) is designed in order to take into account the relative direction of the different faces, which induces varying conditions for each face. In particular, for each face C_{cs} is expressed as average value (C_i^{average}) with respect to the concentration $\tilde{C}$ of the cell accommodating the crystal and the neighboring cell sharing the face in question C_i^{neighbor} (Lappa et al., 2003b).

Integrating over the frontier of the computational cell Eq. (5.1) reads:

$$\int_{\partial\Omega} D\,\underline{\nabla}C\cdot\underline{\hat{n}}\,\mathrm{d}A = \int_{\partial\Omega} \lambda\left(\rho_{\mathrm{P}} - \rho_{\mathrm{C}}\frac{C_{\mathrm{cs}}}{\rho_{\mathrm{S}}}\right)\left(\frac{C_{\mathrm{cs}}}{S} - 1\right)\mathrm{d}A \tag{6.1}$$

now $\underline{\hat{n}}$ is the unit vector perpendicular to the computational cell frontier pointing into the generic neighboring cell. In discretized form:

$$\sum_i D\left.\frac{C_i^{\text{neighbor}} - \tilde{C}}{\Delta n}\right|_i A_i = \sum_i \lambda\left(\rho_{\mathrm{P}} - \rho_{\mathrm{C}}\frac{C_i^{\text{neighbor}} + \tilde{C}}{2\rho_{\mathrm{S}}}\right)\left(\frac{C_i^{\text{neighbor}} + \tilde{C}}{2S} - 1\right)A_i \tag{6.2}$$

The concentration $\tilde{C}$ satisfying the integral form of Eq. (5.1) is obtained from Eq. (6.1), then the mass stored in the computational cell can be updated according to the equation:

$$\frac{\partial M}{\partial t} = \int_{\partial\Omega} \lambda\left(\rho_{\mathrm{P}} - \rho_{\mathrm{C}}\frac{C_{\mathrm{cs}}}{\rho_{\mathrm{S}}}\right)\left(\frac{C_{\mathrm{cs}}}{S} - 1\right)\mathrm{d}A$$

However, a more elegant approach consists in introducing the nondimensional phase-field variable ϕ, already defined in Chapter 5 according to the relationship:

$$\mathrm{d}v|_{\text{stored}} = \frac{M}{\rho_{\mathrm{P}}} \quad\rightarrow\quad \phi = \frac{\mathrm{d}v|_{\text{stored}}}{\mathrm{d}v} \tag{6.3}$$

The evolution equation for the phase variable reads:

$$\frac{\partial\phi}{\partial t} = \frac{1}{\Delta v}\int_{\partial\Omega} \lambda\left(1 - \frac{\rho_{\mathrm{C}}}{\rho_{\mathrm{P}}}\frac{C_{\mathrm{cs}}}{\rho_{\mathrm{S}}}\right)\left(\frac{C_{\mathrm{cs}}}{S} - 1\right)\mathrm{d}A \tag{6.4}$$

where ρ_{P} is the protein mass density in the crystal and Δv is the volume of the computational cell. In discretized form Eq. (6.4) reads (n superscript denotes time step):

$$\phi^{n+1} = \phi^n + \Delta t\cdot\frac{1}{\Delta v}\sum_i \lambda\left(1 - \frac{\rho_{\mathrm{C}}}{\rho_{\mathrm{P}}}\frac{C_i^{\text{neighbor}} + \tilde{C}}{2\rho_{\mathrm{S}}}\right)\left(\frac{C_i^{\text{neighbor}} + \tilde{C}}{2S} - 1\right)A_i \tag{6.5}$$

with $\tilde{C}$ satisfying Eq. (6.2).

According to Eqs. (6.2) and (6.5) the growth velocity is not directly imposed but it results from internal conditions related to solute transport and incorporation kinetics. If the protein (solute) concentration is locally depleted, correspondingly, the solid mass stored in the computational cell grows and the phase variable is increased; on the other hand if mass stored in the cell begins to redissolve, protein is released in the solute phase and the local value of protein concentration is increased.

The ϕ variable is the only information associated with each crystal: $\phi \neq 0$ characterizes computational cells accommodating a crystal (i.e. its position), the value ($0 < \phi < 1$) is proportional to the stored mass.

Note that, should the volume of the crystal stored in a grid cell (M_{grain}/ρ_P) become larger than the volume of the computational cell (i.e. $\phi^{n+1} > 1$), the exceeding mass ($\phi^{n+1} - 1$) has to be placed in one of the surrounding computational cells according to the protein concentration distribution (say in the surrounding cell characterized by the lower value of the protein concentration).

The shape of the crystals is ignored while sacrificing little in accuracy for the macroscopic description of the spatiotemporal behaviors. Evidence of the reliability of the present assumptions is provided by the good agreement between experimental and numerical results as will be shown in the sections devoted to numerical examples and applications.

6.4. Nucleation models

The numerical techniques introduced in this chapter and devoted to the case of crystals modeled only in terms of mass and position, allow a simple and efficient treatment of the nucleation phenomena.

It is important to stress how such phenomena are usually a relevant and significant part of the analysis when it focuses on ensemble behaviors (as also anticipated in Chapter 5). From a mathematical point of view it is known that when there is no pre-existing deposit, the concentration of protein has to be greater than the solubility S to create one spontaneously, say

$$\sigma \geq \eta \tag{6.6}$$

where σ is the supersaturation degree (introduced in Section 5.6) and η is the so-called supersaturation limit (it is obvious that $\eta > 1$); once nuclei are created, precipitation can continue if $1 < \sigma \leq \eta$ and vice versa material can come back to the solute condition if $0 < \sigma < 1$. Further details on the well-known physics of these phenomena can be found in the excellent theoretical analysis of McPherson (1990).

Despite the simplicity of these mathematical conditions governing, in principle, nucleation and subsequent growth, it is important to stress how the problem is very complex and far from being well understood if one looks at the kinetics of the nucleation phenomena. Recently, much progress has been made in the understanding of the protein crystal-growth mechanism. By comparison, insight into protein nucleation processes is still very limited.

Many detailed studies of the nucleation process that appeared in the literature described crystallization in terms of engineering mass-crystallization models, rather than crystal-nucleation theories. Other authors spent some effort to interpret their results in terms of the classical nucleation theories and found appreciable differences. Other authors argued that nuclei possess fractal structure and that such structure could significantly alter the nucleation mechanism and kinetics with respect to classical theories.

In the light of these arguments there is still a disappointing lack of a theory of nucleation able to provide a definitive explanation of the nucleation-process kinetics as well as a mathematical model to be applied to predict a priori the experimental behaviors.

In conclusion, the state-of-the-art does not allow one to handle the phenomena of nucleation kinetics from a computational point of view, in particular at the macroscopic scale.

Fortunately, it is well known (Galkin and Vekilov, 1999) that at the supersaturations required for nucleation, crystals typically grow extremely rapidly. Such information is employed herein as a theoretical basis and justification for neglecting the kinetics of the nucleation process, while taking into account the kinetics of the ensuing growth process (known to be very slow, see Chapter 5).

The present model assumes that if the supersaturation limit is exceeded, the amount of supersaturated protein in the liquid phase is posed in solid without any time delay. This assumption, although very crude, however, is supported by the very good agreement between numerical and experimental results shown in the next sections.

In particular, the present model takes advantage of the crystallization criteria introduced by Henisch and Garcìa Ruiz (1986a,b) and Henisch (1991) (hereafter referred to as the algebraic model) for which once the solid particles are formed it is assumed that they are in equilibrium with protein and salt in the liquid phase:

$$\begin{gathered}\text{If} \quad \phi^n = 0 \quad \text{and} \quad \sigma^n = (C/S)^n \geq \eta \rightarrow M_{\text{grain(o)}} = (C - S)^n \Delta\Omega, \\ \phi^{n+1} = \frac{1}{\Delta\Omega}\frac{M_{\text{grain(o)}}}{\rho_{\text{P}}} = \frac{S^n}{\rho_{\text{P}}}(\sigma^n - 1), \quad C^{n+1} = S^n\end{gathered} \tag{6.7}$$

this means that when conditions for nucleation are attained (absence of pre-existing solid mass, i.e. $\phi = 0$ and $\sigma \geq \eta$), then the solute content of the solution is decremented from the original C value to the solubility value S (equilibrium) and the corresponding difference is converted into solid mass ($M_{\text{grain(o)}}$).

6.5. Moving crystals – The OCSVOF (Organic Crystal Sedimentation Volume of Fraction) method

If the problem under investigation deals with the motion of different phases (absence of gel), as previously discussed in Section 5.5, there are two fundamentally different approaches that can be used in principle: Eulerian methods use a frame of reference (discretization grid or mesh, control volumes, etc.) fixed in space, with matter moving through this frame of reference. Lagrangian methods instead use a frame of reference (marker particles) fixed to and moving with the matter.

Although the Eulerian description is at the root of the OCGVOF method introduced in Chapter 5 and devoted to the morphological evolution of the crystals, it is not suitable for problems dealing with the motion of multiple crystals treated in the present chapter.

In the OCGVOF method, the Eulerian description is supported by a special interface reconstruction technique (the PLIC approach treated in Section 3.5.3) that is used, in particular, for the evaluation of the flux needed for the corresponding solid-mass-evolution equation. This guarantees accuracy and reliability of the method.

If the interface is ignored, as in the present case, Eulerian methods would inevitably suffer from numerical diffusion, which would soon smooth ϕ excessively. Hybrid techniques, however, can be used to overcome these problems.

Contrary to the case of the OCGVOF method, for the present case the time evolution of the phase variable ϕ is computed according to a Lagrangian description (particle tracking) even if this approach is computationally very expensive and cumbersome. The phase variable ϕ is "transported" according to a Lagrangian equation solved for each of the solid crystals (marker-particles technique: in practice the positions and the velocities of the moving crystals are stored at any computational step in special arrays). Its value is adjourned according to the integral formulation of the surface-attachment kinetics introduced in Section 6.3. Regarding this aspect, note that interpolations (for instance, simple first-order polynomial functions) are required to link the Lagrangian position of the solid particle with the fixed (Eulerian) points of the computational mesh where the protein concentration is computed.

Within the context of the present section it is also interesting to note how the proposed method for the sedimentation/motion of the crystals does not exhibit any theoretical kinship with the corresponding method discussed in Chapter 3 for the sedimentation of the minority phase in liquid metal alloys and similar problems. The VOF method in Section 3.5, in fact, is basically a Eulerian method that does not require any special care for the tracking of the phase motion. These dynamics are captured in a natural manner without the necessity of Lagrangian markers. On the contrary, these markers (the aforementioned special arrays used to store positions and velocities of the crystals at any time) are the bases of the OCSVOF method that therefore behaves as a hybrid technique combining Eulerian and Lagrangian approaches in a single numerical treatment with special interpolation for linking crystal trajectories with the underlying fixed Eulerian grid.

The concentration and velocity fields are still updated, however, according to a Eulerian description.

6.6. A mathematical model for sedimentation

This section deals with the modeling of the sedimentation process under the assumption that solutal convection cannot become established in the protein chamber. The model for the combined sedimentation–convection will be discussed in the next section (such an approach in progressive steps is used for educational purposes as well as for the sake of clarity).

Crystal sedimentation occurs due to a difference between the gravitational force and the viscous resistance. The viscous resistance acting on a solid-spherical particle (the crystals are assumed to be spheres) moving with a velocity V_{grain} is given by Stokes law as $\underline{F}_{\text{D}} = 6\pi R\mu \underline{V}_{\text{grain}}$ where μ is the coefficient of dynamic viscosity of the solution, and V_{grain} the velocity of the solid particle of radius R. Nonslip conditions are assumed at the particle surface (for the sake of simplicity herein the crystals are assumed to be spheres; the treatment for the case of different shape is analogous; the viscous resistance for a number of other shapes is reported in the book of Clift et al., 1978). The gravitational force is given by

$$\underline{F}_{\text{g}} = \frac{4}{3}\pi R^3(\rho_{\text{C}} - \rho_{\text{S}})\underline{g}$$

The equation of motion of sedimenting crystals reads:

$$\frac{4}{3}\pi\rho_{\text{C}}\frac{\text{d}\left(R(t)^3\underline{V}_{\text{grain}}\right)}{\text{d}t} = \frac{4}{3}\pi R^3(\rho_{\text{C}} - \rho_{\text{S}})\underline{g} - 6\pi R\mu\underline{V}_{\text{grain}} \tag{6.8}$$

where $\rho_{\text{S}} = \rho_{\text{H}_2\text{O}}(1 - \beta_{\text{lys}}C - \beta_{\text{NaCl}}C_{\text{NaCl}})$ is the density of the feeding solution and β_{lys} and β_{NaCl} are the solutal expansion coefficients related to lysozyme and salt, respectively. Note that, if a g-jitter acceleration is present in addition to a steady g, the term $(4/3)\pi R(t)^3(\rho_{\text{C}} - \rho_{\text{S}})b\omega^2\sin(\omega t)\underline{n}$ (see Section 2.6 for the meaning of b, ω, and $\underline{n}$) should be added to the second member of Eq. (6.8); it, however, does not give any average contribution to the motion of the crystal (a solid particle under the effect of a high-frequency sinusoidal acceleration simply moves back and forth around its position but its average location does not change) and for this reason it can be dropped; for the general case Eq. (6.8) reads:

$$\frac{\text{d}\underline{V}_{\text{grain}}}{\text{d}t} = \underline{g}\left(1 - \frac{\rho_{\text{S}}}{\rho_{\text{C}}}\right) - \frac{9}{2}\nu\frac{\rho_{\text{S}}}{\rho_{\text{C}}}\frac{\underline{V}_{\text{grain}}}{R(t)^2} - 3\frac{\underline{V}_{\text{grain}}}{R(t)}\frac{\text{d}R}{\text{d}t} \tag{6.9}$$

where,

$$R(t) = \sqrt[3]{\frac{3}{4\pi}\frac{M_{\text{grain}}(t)}{\rho_P}} = \sqrt[3]{\frac{3}{4\pi}\phi(t)\Delta\Omega},$$

$M_{\text{grain}}(t)$ is the mass of protein associated with each grain. In Eqs. (6.8) and (6.9) the dependence of the radius of the solid particle on time has been highlighted in order to point out that the size of a sedimenting particle may change during sedimentation due to further absorption (or release) of mass (M_{grain} is a function of time) as explained in the previous section. Note that the viscous force is computed in the model according to Stokes formula that is valid only when the velocity does not change in time (i.e. under steady conditions for the motion of the solid particle); Since the time required by the generic solid particle to reach the bottom of the protein chamber is negligible when compared with the characteristic time of the phenomena under investigation (very slow), it is reasonable to suppose that the approximations introduced in Eq. (6.9) do not alter significantly the phenomena under investigation.

Equation (6.9) must be solved for each crystal assuming $\underline{V}_{\text{grain}} = 0$ and $x = x_{\text{nucl}}$, $y = y_{\text{nucl}}$ as initial conditions, where $(x, y)_{\text{nucl}}$ characterize the position at which nucleation occurs.

Since protein crystals are denser than the feeding solution, on Earth they settle in the computational cells on the bottom of the growth chamber.

6.7. Sedimentation–convection model

The sedimentation model discussed in the previous section is now improved and further refined by taking into account the effect of convection arising in the protein chamber due to buoyancy forces. The flow is governed by the continuity, Navier–Stokes and species equations, that in dimensional conservative form read:

$$\underline{\nabla} \cdot \underline{V} = 0 \tag{6.10}$$

$$\begin{aligned}\frac{\partial \underline{V}}{\partial t} = &-\frac{1}{\rho}\underline{\nabla} p - \underline{\nabla} \cdot [\underline{V}\,\underline{V}] + \nu \nabla^2 \underline{V} + g\beta_{\text{lys}}\left(C - C_{(\text{o})}\right)\underline{i}_g \\ &+ g\beta_{\text{NaCl}}\left(C_{\text{NaCl}} - C_{\text{NaCl(o)}}\right)\underline{i}_g\end{aligned} \tag{6.11}$$

$$\frac{\partial C}{\partial t} = -\underline{\nabla} \cdot [\underline{V}C] + D\nabla^2 C \tag{6.12}$$

$$\frac{\partial C_{\text{NaCl}}}{\partial t} = -\underline{\nabla} \cdot [\underline{V}C_{\text{NaCl}}] + D_{\text{NaCl}}\nabla^2 C_{\text{NaCl}} \tag{6.13}$$

The nondimensional Rayleigh numbers are defined as $Ra_{\text{lys}} = g\beta_{\text{lys}}L^3 C_{(\text{o})}/(\nu D)$ and $Ra_{\text{NaCl}} = g\beta_{\text{NaCl}}L^3 C_{\text{NaCl(o)}}/(\nu D_{\text{NaCl}})$. Nonslip conditions must be imposed on the walls.

The protein and salt distributions at time $(n+1)$ can be obtained from Eqs. (6.12) and (6.13) after the velocity calculation. The nucleation and kinetic incorporation phenomena can be taken into account according to the technique described in Sections 6.3 and 6.4. The equation of motion of sedimenting crystals has to be modified (with respect to Section 6.6) in order to take into account the motion of fluid due to the buoyancy (and/or vibrational) forces:

$$\frac{\mathrm{d}\underline{V}_{\text{grain}}}{\mathrm{d}t} = \underline{g}\left(1 - \frac{\rho_S}{\rho_C}\right) - \frac{9}{2}\nu\frac{\rho_S}{\rho_C}\frac{\underline{V}_{\text{grain}} - \underline{V}}{R(t)^2} - 3\frac{\underline{V}_{\text{grain}}}{R(t)}\frac{\mathrm{d}R}{\mathrm{d}t} \tag{6.14}$$

which in discrete form reads:

$$\underline{V}^{n+1}_{\text{grain}} = \underline{V}^{n}_{\text{grain}} + \Delta t\underline{g}\left(1 - \frac{\rho_S}{\rho_C}\right) - \Delta t\frac{9}{2}\nu\frac{\rho_S}{\rho_C}\frac{\underline{V}^{n+1}_{\text{grain}} - \underline{V}^{n+1}}{(R^{n+1})^2} - 3\frac{\underline{V}^{n+1}_{\text{grain}}}{R^{n+1}}\left(R^{n+1} - R^{n}\right) \tag{6.15}$$

i.e. it is assumed that, due to the small size of the sedimenting crystals, the velocity $\underline{V}$ of the fluid is not affected by the motion of these particles and that, vice versa, the moving particles may be accelerated or decelerated by the effect of the external velocity field $\underline{V}$. In the following section a possible organization of a numerical code for practical application of the proposed models is described.

Program organization The numerical solution can proceed in principle in five major stages:

1. Solution of the Navier–Stokes equations ($\underline{V}^{n+1}$ as a function of $\underline{V}^n$, C^n and C^n_{NaCl}).
2. Solution of the species equations (6.12) and (6.13) for protein and salt, respectively (C^{n+1} and C^{n+1}_{NaCl} are computed as function of the corresponding distributions and of the velocity field at time n).
3. Calculation of the solubility distribution S^{n+1} (as a function of the salt distribution C^{n+1}_{NaCl}) and of the supersaturation $\sigma^{n+1} = C^{n+1}/S^{n+1}$.
4. Numerical simulation of the nucleation and further growth or resolution phenomena according to the supersaturation distribution (algebraic and integral models, respectively) and consequent adjournment of the local values of protein concentration, phase field ϕ and radius R at time $n+1$ (see Eqs. (6.2), (6.5), and (6.7)). When new seeds are formed, the size of the arrays storing the information (position, velocity and ϕ) related to moving solid particles must be increased in order to take into account the new particles.
5. Solution of Eq. (6.14) for each solid particle; then $\underline{V}^{n+1}_{\text{grain}}$ is used to compute the new position at time $n+1$: $\underline{r}^{n+1} = \underline{r}^n + \Delta t(\underline{V}^{n+1}_{\text{grain}} + \underline{V}^{n}_{\text{grain}})/2$.

Note that in the case of gellified phenomena the phase variable ϕ can be handled as a Eulerian variable (it is computed in the same points of the computational grid where the other quantities are evaluated) whereas in the case of sedimentation, as already discussed ϕ behaves as a Lagrangian variable associated with marker particles and special arrays used to store its position and velocity (Lappa and Castagnolo, 2003).

6.8. Examples and insights into the physics: the counterdiffusion technique

A counterdiffusion technique is considered herein as a reference case (crystallization occurs by progressive diffusion of salt in the protein reactor). This choice is motivated by the fact that this technique has enjoyed a widespread use in recent years and can be particularly suitable for producing macromolecular crystals, because: (a) the counteracting phenomena that take place during the process self-limit the maximum value of supersaturation that is actually reached during the crystallization, thus establishing better conditions for obtaining relatively large single crystals; (b) wide ranges of initial conditions (supersaturation) can be considered; (c) for fixed initial conditions, a range of supersaturation values is established in different locations along the protein solution during the process, allowing different crystallization conditions to be scanned in the same experiment. All these features are of interest for the growth of macromolecular crystals.

If such a technique is used, "periodic precipitation" occurs. Periodic precipitation is a generic term for material deposition processes that occur intermittently in terms of time or space or (generally) both. The ordered spatial distribution of these deposits and the related formation mechanisms exhibit interesting analogies with the category of nonlinear phenomena known as Liesegang patterns (see Section 6.9). Like the case of the fluid-dynamic systems described in the previous chapters (exhibiting a wealth of different spatiotemporal patterns, transitions and bifurcations), these aspects (with the due differences) represent a significant example of pattern-forming dynamical systems (Lappa et al., 2002; Piccolo et al., 2002).

System dynamics depend on the interplay between transport in the fluid phase and formation and growth of solid phases. These processes also represent a special case of a fashionable topic: oscillatory reactions, with practical implications in crystal growth and material preparation, and a theoretical kinship with the complex problems that come under the heading "order out of chaos."

Both the conditions of gellified and nongellified protein solution are considered herein in order to discern by comparison the effect of convection and, in particular, of sedimentation.

Usually, this technique involves two chambers containing two different solutions (protein solution and precipitant agent solution) that are put in contact directly or via a membrane.

In order to prevent convective mixing and to obtain a stable interface between the two solutions on the ground, gel is added to the lower solution. For this reason the configuration under investigation simply consists of a protein solution and a precipitant agent solution placed one above the other and separated by the gel interface (Fig. 6.1). The generic crystallization cell is a chamber whose length and height are L and H, respectively; the interface is placed at $y \cong h$. At the initial time, both solutions can be assumed at constant concentration.

In practice, the shape of the gel interface cannot be horizontal due to the occurrence of a meniscus that is caused by surface-tension effects. The function

$$\xi(x) = h\left\{1.1 - 0.1\left[\sin\left(\frac{\pi}{4} + \frac{\pi}{2}\frac{x}{L}\right) - \frac{\sqrt{2}}{2}\right] \Big/ \left(1 - \frac{\sqrt{2}}{2}\right)\right\} \tag{6.16}$$

has proven to be a realistic model for the interface curvature (it introduces a minimum protruding in the lower chamber at the mean point along the horizontal length of the chamber: ($\xi(x=0) = \xi(x=L)$ and the minimum at $x = L/2$). The gel interface is supposed to be impermeable to the protein.

The precipitant agent is NaCl. Also in this chapter, hen egg-white (HEW) lysozyme is used as the model protein for the reasons described in Chapter 5. The experimental procedure changes, in principle, according to the subject of the investigation: case (1), crystals locked on gel, case (2), crystal motion allowed. Agarose solution usually is prepared by stirring agarose powder in an acetate buffer solution; then lysozyme (case 1) or salt (case 2) is added to the agarose solution. This solution is injected (before gel solidification occurs) in the lower half of the cell. After gel formation the precipitant agent (case 1, Fig. 6.1a) or lysozyme (case 2, Fig. 6.1b) solution are introduced in the upper part of the cell, like a supernatant

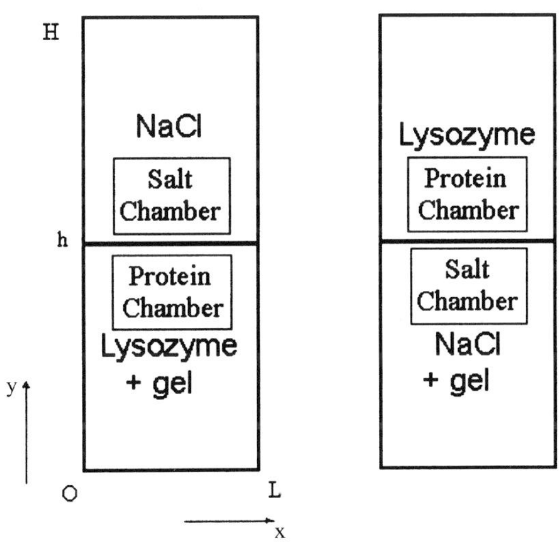

Fig. 6.1 Sketch of the configurations: (a) no sedimentation allowed (b) lysozyme sedimenting on the gel interface.

liquid above the gel. Crystallization occurs by diffusion of salt in the part of the cell containing the protein solution: lower part in case (1) and upper part in case (2).

The next section deals with numerical prototype examples and macroscopic behaviors. In the spirit of the book, a systematic presentation is adopted (processing from simple to complicated), and rational approaches are used to establish physical concepts that can have wide applications in the experimental studies usually carried out by organic crystal growers.

6.9. Protein precipitation in gel

For the case of gellified protein chamber (case (1), diffusive-transport regime), different values of the initial protein concentration are usually considered by organic crystal growers to investigate the effects of supersaturation level on the system dynamics. Accordingly, for the present prototype application three values of the initial protein concentration are considered ($C_{(o)} = 4 \times 10^{-2}$, 6×10^{-2} and 8×10^{-2} g/cm^3 for a fixed initial concentration of salt $C_{NaCl(o)} = 14 \times 10^{-2}$ g/cm^3). The geometrical parameters in Fig. 6.1 are $L = 1$ cm, $H = 4$ cm, width $= 0.1$ cm (the extension along z is very small with respect to L and H), the average height of the gel interface is $h = 1.95$ cm.

According to the computations, some bands of crystals are produced in the gel, i.e. discrete nucleation events separated by certain time (and thus space) intervals occur in the protein chamber. These bands are not spatially uniform (see Figs. 6.2–6.4 for $C_{(o)} = 4 \times 10^{-2}$ g/cm^3; note that in these figures the salt chamber is located on the top of the protein chamber).

6.9.1 *Periodic precipitation and underlying mechanisms*

As expected, and in good agreement with the experimental results (see Fig. 6.2), the simulated phenomenon is characterized by a certain degree of periodicity in time and in space. Lysozyme precipitates to produce deposits spaced in both the x and y directions (see Figs. 6.3a–f and 6.4a–f). The crystal formation results modulated in time and in space, due to the nonlinear interplay among transport, crystal nucleation, and growth. In particular, according to the evolution of the concentration field (Fig. 6.3) nucleating particles deplete their surroundings of protein, which causes a drop in the local level of supersaturation such that the nucleation rate falls in the neighborhood, leading naturally to a spacing between regions of nucleation that, in turn, gives rise to the alternate presence of depleted zones and crystals. As anticipated, this behavior exhibits interesting analogies with the phenomena known as Liesegang patterns (see, e.g., Garcìa Ruiz, 1986a,b; Henisch, 1988; Carotenuto et al., 2002) that have instigated much research also in other fields of physical sciences.

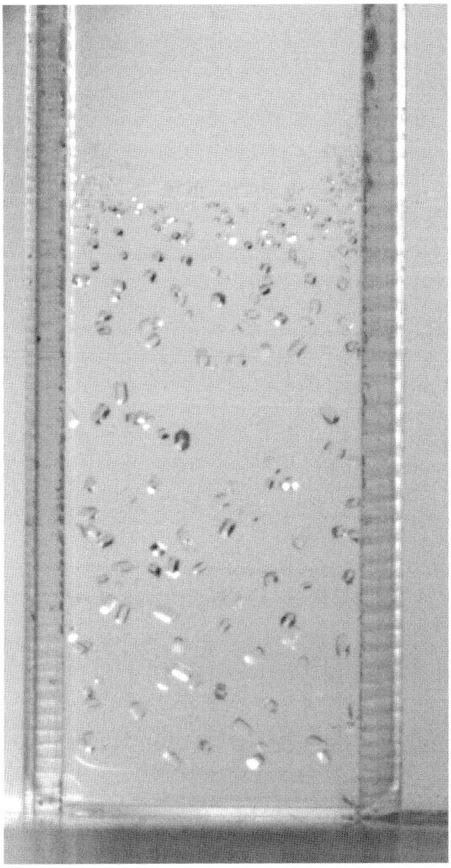

Fig. 6.2 Experimental distribution of the crystals – (gellified configuration, $C_{(o)} = 4 \times 10^{-2}$ g/cm^3) Time $= 1.73 \times 10^5$ s.

Figure 6.4a shows in detail that after 5.25×10^3 s, some crystals appear at the interface separating the protein chamber from the salt chamber. Correspondingly (see Fig. 6.3a), a depletion zone is visible around each solid particle. Owing to the relatively large number of nuclei created in a small time interval, a band, i.e. a region of low protein concentration, having a certain width is created all around the crystals. This band protrudes from the interface in the protein chamber and has a width somewhat proportional to the amount of solid mass since the depletion of the protein concentration distribution is strictly related to the deposits created.

The presence of solid particles near the interface finds its theoretical justification in the fact that salt diffuses in the protein chamber through the interface of gel so that solubility modulation first occurs there. As time passes, the size of these particles

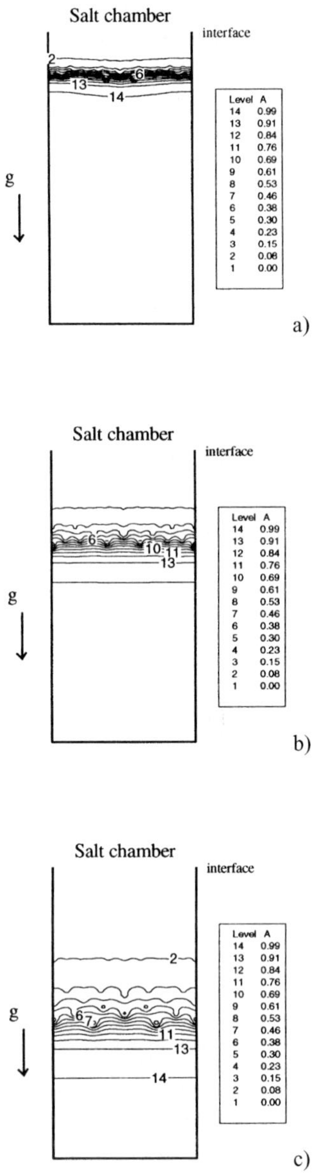

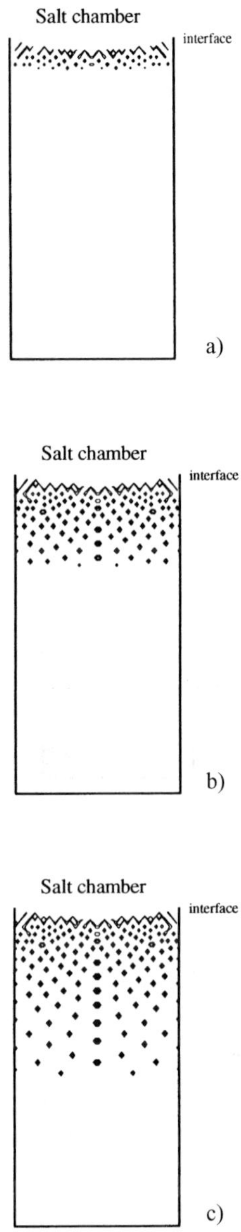

Figs. 6.3–6.4 On the left, Fig. 6.3: Nondimensional protein concentration contour lines ($C_{(o)} = 4 \times 10^{-2}$ g/cm^3); On the right, Fig. 6.4: Corresponding distribution of crystals; (a) $t = 5.25 \times 10^3$ s, (b) $t = 2.625 \times 10^4$ s, (c) $t = 6.825 \times 10^4$ s; (d) $t = 8.925 \times 10^4$ s, (e) $t = 9.975 \times 10^4$ s, (f) $t = 1.73 \times 10^5$ s.

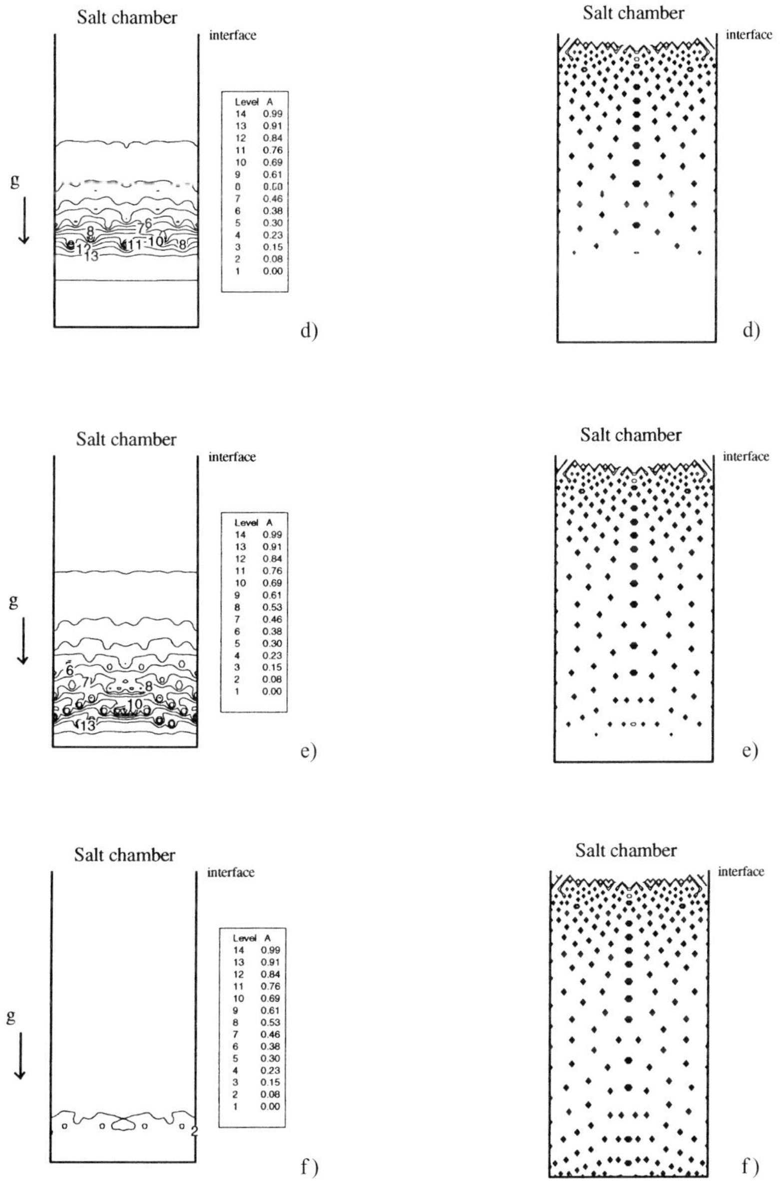

Figs. 6.3–6.4 Continued.

increases due to further precipitation of lysozyme. Figures 6.3b and 6.4b show, in fact, that no further nucleation occurs close to the gel interface and solid deposits present therein continue to grow. This effect increases the extension of the depleted band. The absence of additional nucleation close to the interface is due to the intense depletion of protein solution therein. The formation of the first nuclei locally depletes the protein solution; protein diffuses towards nuclei promoting their growth and enlarging the depletion zone. Therefore C is decreased and, even if salt diffuses downwards, lowering solubility further nucleation is prevented so far as $\sigma < \eta$.

By numerical analysis on the phenomenon, the increase in the solid particles size, can be explained on the basis of a delicate balance between two counteracting effects: (1) protein condenses on any solid crystal already existing, augmenting its size; this depletes the protein concentration and leads to lower values of C; (2) salt continues to diffuse through the interface so that σ is increased due to solubility modulation. If the second effect prevails over the first one, i.e. the protein concentration is larger than the solubility then further protein precipitation in the considered region is possible.

As time passes the extension of the depleted zone increases towards the bottom of the protein chamber (Fig. 6.3a–f). This is due to the combined effect of the nucleation and precipitation phenomena.

The propagation towards the bottom of the supersaturation front (i.e. the front where new nucleation occurs) depletes the protein chamber of lysozyme and further depletion occurs due to the phenomenon of subsequent precipitation. These two effects behave as an advancing wavefront (Figs. 6.3 and 6.4). As time passes new regions are involved in the nucleation and ensuing growth processes, whereas previous regions are no longer characterized by these phenomena.

Insights into the physics provided by the numerical simulation deal, in particular, with the explanation of the macroscopic spatiotemporal evolution of the process: it is mainly driven by the availability of protein in the liquid phase (many crystals compete for growth, the related depletion zones intersect and overlap) and by the rhythmic increase and decrease of supersaturation due to the interplay between protein depletion and salt diffusion through the gel interface.

The distribution of the solid particles not being uniform is one of the reasons why the term "periodic" (the aforementioned denomination of periodic precipitation) should be interpreted with caution and tolerance for the problem under investigation. Strictly speaking, this term demands a period (i.e. a constant time or space interval), something that the phenomena under discussion here fail to show. These phenomena are nevertheless periodic. The periods are not constant, but they are not random either.

Additional details about this aspect are discussed in the following section.

6.9.2 *Spatial distribution of the crystals*

The spacing among different solid particles and the size of the particles seem to vary according to the distance from the origin of the imposed concentration gradient (for the present case the gel interface between the salt and protein chambers). The space distance among near solid particles, in fact, tends to decrease uniformly towards the interface of the protein chamber. Correspondingly, their size is minimum at the interface and increases towards the bottom of the chamber to a constant value in the bulk. Then it decreases again near the wall due to the effect of the solid boundary that prevents solid particles from capturing solute protein from the bottom. In view of these arguments, the density of particles defined as the ratio of the number of particles in a fixed reference volume and the reference volume, is maximum near the interface and almost constant in the bulk. These behaviors are shown in Fig. 6.5.

In addition to the previous case and as anticipated, the precipitation from supersaturated solutions with $C_{(o)} = 6 \times 10^{-2}\,\text{g/cm}^3$ and $C_{(o)} = 8 \times 10^{-2}\,\text{g/cm}^3$, has been investigated in order to discern the effect that the initial amount of protein available in the solute condition may have on the spatiotemporal behavior of the phenomena (see Figs. 6.6 and 6.7 for the case $6 \times 10^{-2}\,\text{g/cm}^3$, the case $C_{(o)} = 8 \times 10^{-2}\,\text{g/cm}^3$ is shown in Figs. 6.10 and 6.11).

For the sake of clarity, attention is first devoted to the spatial distribution of crystals, then the temporal behavior of the phenomena is discussed and analyzed in detail.

The numerical results, in agreement with the experimental ones (Fig. 6.6), show that the spacing among different solid particles is different with respect to the previous case. Particles are closer to each other. The average distance between the particles distributed along x and y is lower if compared with the case $C_{(o)} = 4 \times 10^{-2}\,\text{g/cm}^3$ (compare Figs. 6.2 and 6.7). Moreover, the solid particles are characterized by a smaller diameter.

Regardless of the initial value of protein concentration, the size of the crystals is an increasing function of the distance from the gel interface up to a certain distance; then it becomes almost constant. The values in Fig. 6.6a for the case $C_{(o)} = 6 \times 10^{-2}\,\text{g/cm}^3$, however, are always lower than the corresponding ones for $C_{(o)} = 4 \times 10^{-2}\,\text{g/cm}^3$ (Fig. 6.5a).

On the basis of these results, some general rules can be introduced. In particular, the average size of particles is a decreasing function of the initial amount of protein available in the solute condition (Lappa et al., 2003b).

If one looks at the density-of-particles distribution, it is also evident that the decrease of the average size of the particles is coupled to an increase of the number of crystals (Fig. 6.8 summarizes this effect in a single plot).

6.9.3 *Temporal analysis: propagation of the nucleation and local growth laws*

The temporal analysis of the phenomenon (Fig. 6.9) shows that the supersaturation front in the case $C_{(o)} = 6 \times 10^{-2}\,\text{g/cm}^3$ propagates faster than in the case $C_{(o)} = 4 \times 10^{-2}\,\text{g/cm}^3$. The effect of increase of the initial protein concentration is to accelerate the propagation of the nucleation phenomena towards the bottom (Fig. 6.9d), the faster propagation of the nucleation front being strictly related to the

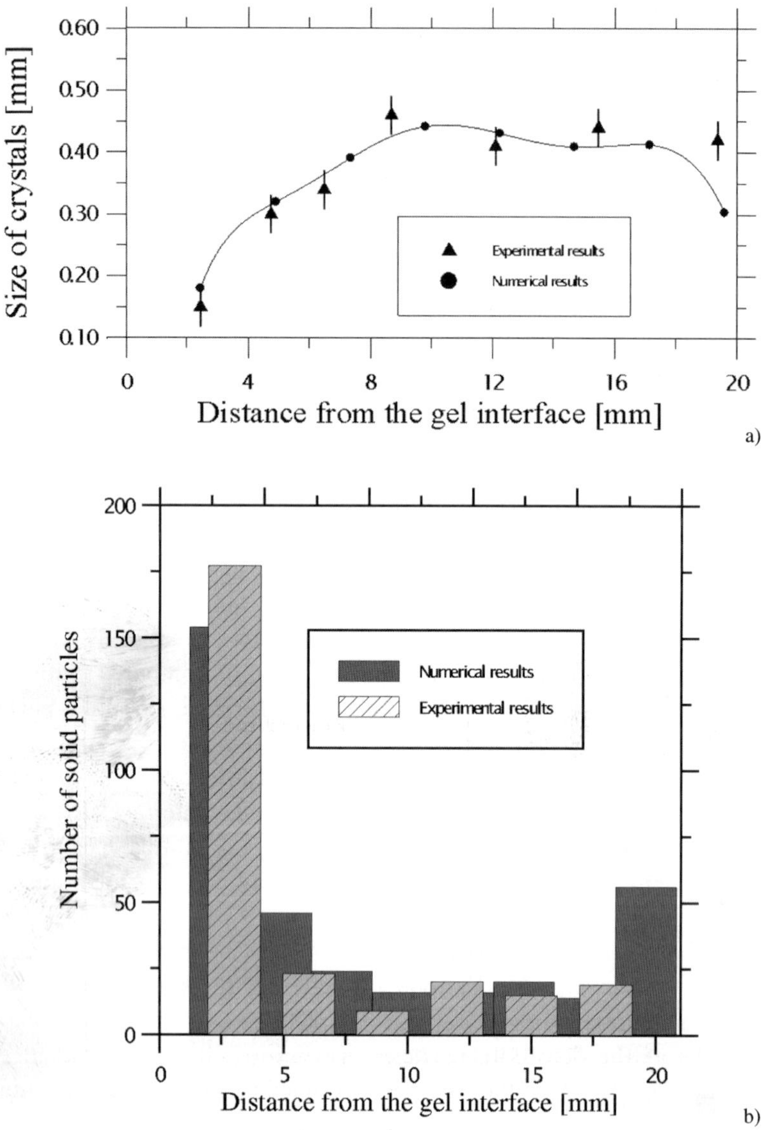

Fig. 6.5 Comparison between numerical and experimental results (gellified configuration, $C_{(o)} = 4 \times 10^{-2}$ g/cm^3): (a) solid-particle-size distribution, (b) density-of-particles distribution.

reduced amount of salt needed to reach the supersaturation limit. This trend is confirmed by the results obtained in the case $C_{(o)} = 8 \times 10^{-2}$ g/cm^3.

The shape of the profiles in Fig. 6.9a–c is also noteworthy since it clearly shows that at the beginning, the propagation of the nucleation front is mainly driven by the

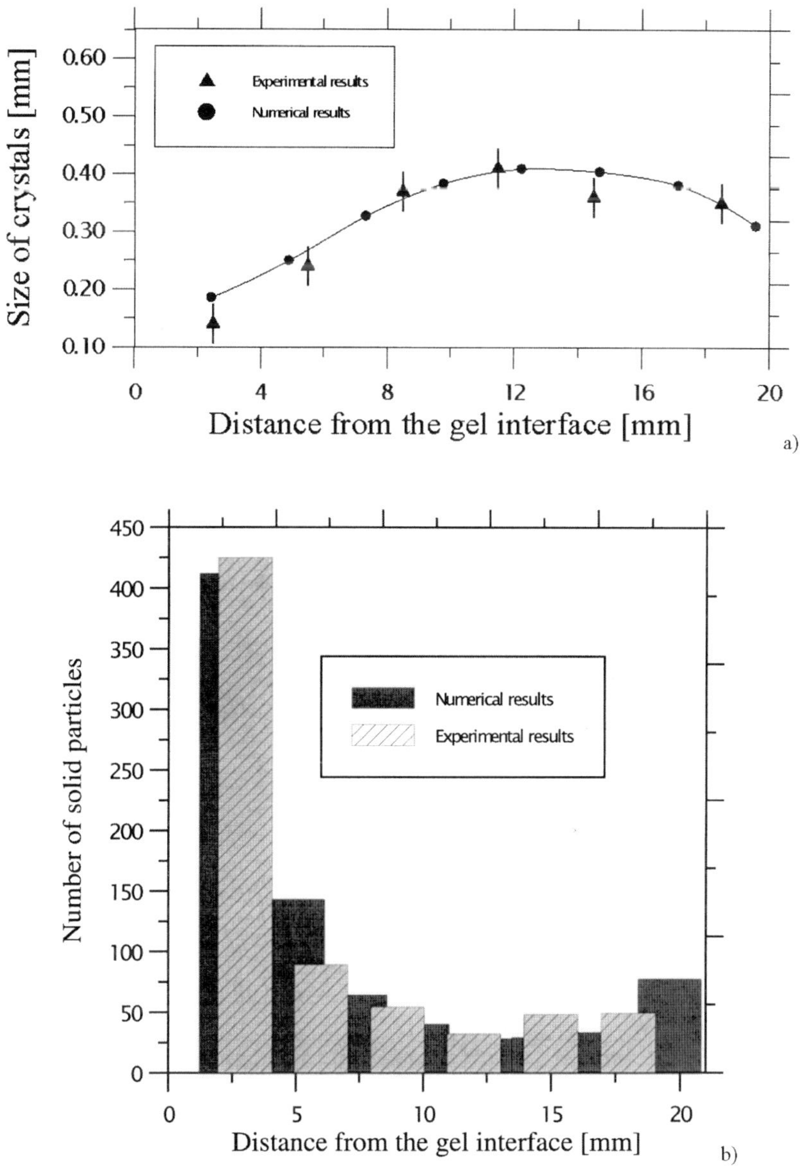

Fig. 6.6 Comparison between numerical and experimental results (gellified configuration, $C_{(o)} = 6 \times 10^{-2}$ g/cm^3): (a) solid-particle-size distribution, (b) density-of-particles distribution.

diffusion of salt (the shape of the profile is concave and according to diffusion law the nucleation front position y propagates as $y \propto (D_{salt} \cdot t)^{1/2}$). However, as time passes the profile departs from such behavior and becomes convex, its deviation being related to the aforementioned nonlinear interplay among transport, crystal nucleation, and growth that deplete the solution of protein.

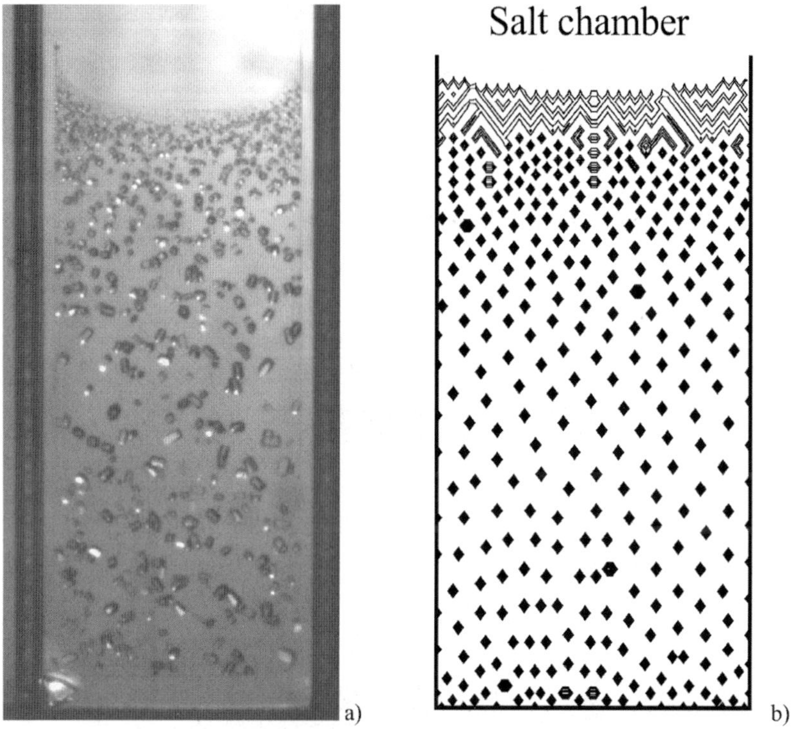

Fig. 6.7 (a) Experimental distribution of the crystals, (b) numerical distribution of the crystals (gellified configuration, $C_{(o)} = 6 \times 10^{-2}\,\mathrm{g/cm^3}$) Time $= 1.7 \times 10^5$ s.

As an example, typical observations are reported in Fig. 6.10 that shows a sequence of interferograms (versus time) for a value of initial protein concentration corresponding to $8 \times 10^{-2}\,\mathrm{g/cm^3}$.

Fringes represent sites of equal refractive index, n, that is dependent on both solute concentrations (lysozyme and NaCl); therefore, fringes may represent salt concentration only where protein concentration could be assumed uniform, i.e. in the absence of nucleation and growth. Initially, before the occurrence of the nucleation, the protein is uniformly distributed inside the gel and the fringes appear at the gel interface as soon as the salt solution is injected. The interferometric pattern is in agreement with the diffusion of salt inside the gel, forming regular horizontal fringes, until the nucleation takes place. Gradually, while the salt diffuses, leading the system to supersaturation conditions, the occurrence of nucleation produces local concentration changes (the aforementioned protein depletion) that affect the regular diffusive fringe pattern. The rectangles in Fig. 6.10 highlight the presence of the nucleation front that propagates through the gel. As the nucleation and growth,

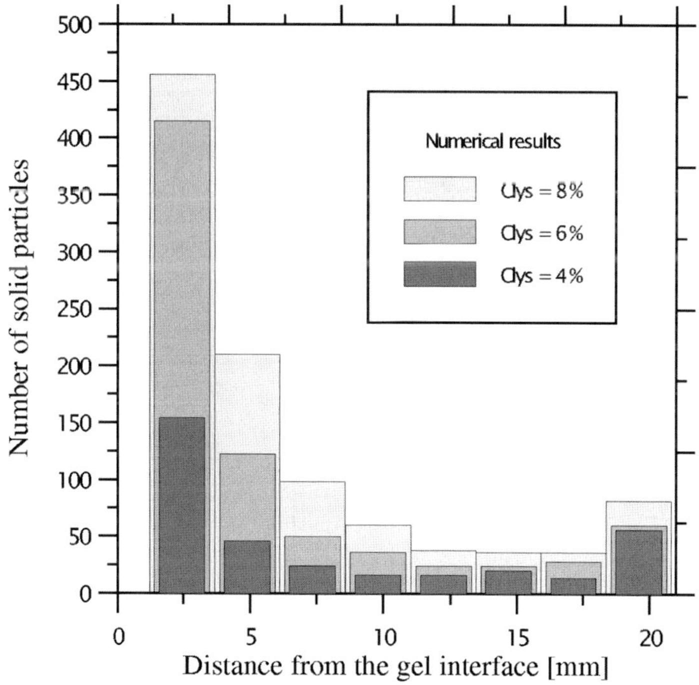

Fig. 6.8 Numerical results – density-of-particle distribution – comparison among $C_{(o)} = 4 \times 10^{-2}\,\text{g/cm}^3$, $C_{(o)} = 6 \times 10^{-2}\,\text{g/cm}^3$ and $C_{(o)} = 8 \times 10^{-2}\,\text{g/cm}^3$.

and therefore the depletion of the protein inside the gel, proceed, the front becomes disturbed and no longer regular. Numerical simulations for the same protein concentration as in the experiments are shown in Fig. 6.11. The grey zones correspond to regions with high density of small-size crystals (see legend), the dimensions of which are nondetectable by the CCD.

In the light of all these arguments, the numerical results are in good quantitative agreement with the experimental ones both for the space properties (particles size and density distributions) and the time evolution of the phenomena (velocity of the nucleation front). It has been illustrated, moreover, how the parametric investigation for different values of the initial supersaturation can provide rich information about the general trends of the system (see, e.g., Figs. 6.8 and 6.9d).

Along these lines, it is also worthwhile to stress that, even if the model has been conceived in order to leave aside local details, it is also somehow able to provide a description of crystal growth as a function of local conditions established in the protein chamber.

As an example, Fig. 6.12 shows the growth law for a reference crystal located in the bulk of the protein chamber ($x = 0.4\,\text{cm}$, $y = 0.9\,\text{cm}$); the growth is compared

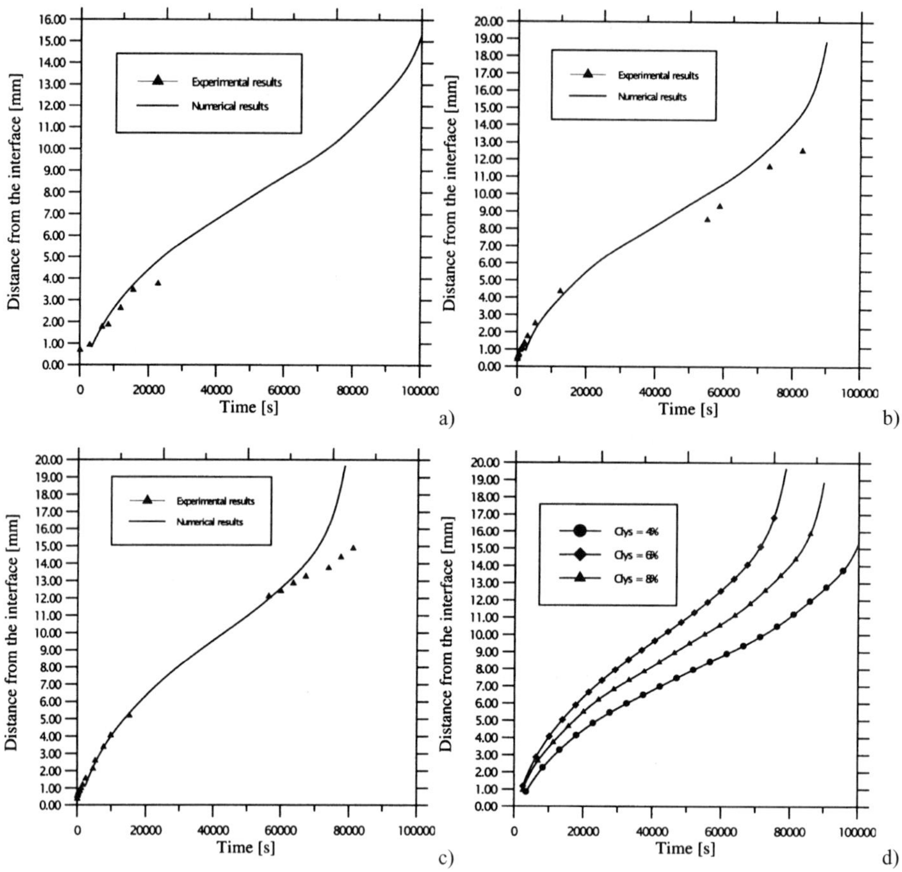

Fig. 6.9 Position of the nucleation front versus time: (a) $C_{(o)} = 4 \times 10^{-2}\,\mathrm{g/cm^3}$, (b) $C_{(o)} = 6 \times 10^{-2}\,\mathrm{g/cm^3}$ and (c) $C_{(o)} = 8 \times 10^{-2}\,\mathrm{g/cm^3}$, (d) comparison among the different cases (numerical results).

with the evolution of local supersaturation. This crystal is formed at $t \cong 7 \times 10^4$ s and reaches its final size after almost 10^5 s. A measure of the average growth rate is also possible. The growth rate (computed over the time interval required to reach 90% of the final mass of the crystal) is 1.27×10^{-9} g/s.

This is a relevant application demonstrating the potentialities of the integral formulation (see Sections 6.3 and 6.4) that, in the light of the aforementioned example, can also go beyond a macroscopic description to provide relevant information about microscopic conditions and growth laws (an approach to the crystal in terms of morphological evolution requires, however, application of the methods described in Chapter 5).

The proposed model is both sufficiently complex to do justice to physical reality and sufficiently simple to focus on the most significant aspects of the given situation.

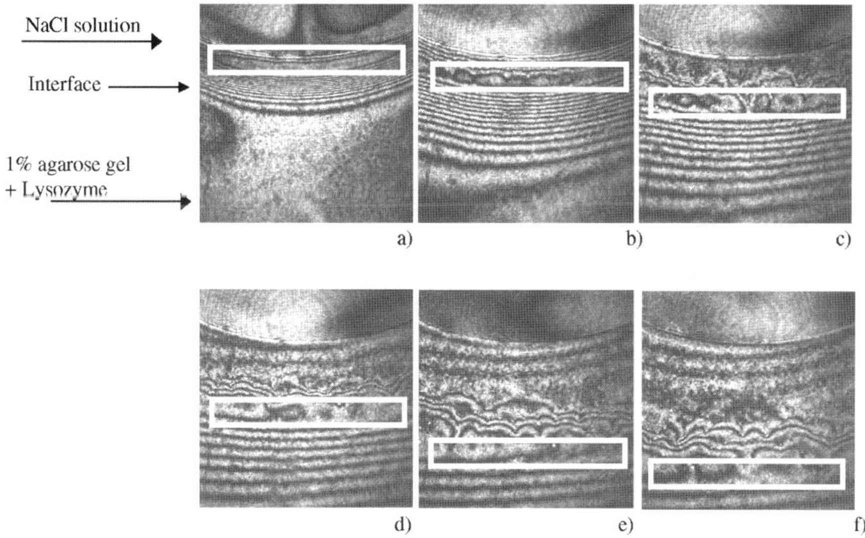

Fig. 6.10 Sequence of interferograms in the case $C_{(o)} = 8 \times 10^{-2}$ g/cm^3 (salt solution is injected at $t = 0$), for times: (a) 240 s, (b) 1.4×10^3 s, (c) 3.1×10^3 s, (d) 5.3×10^3 s, (e) 9.1×10^3 s, (f) 1.26×10^4 s.

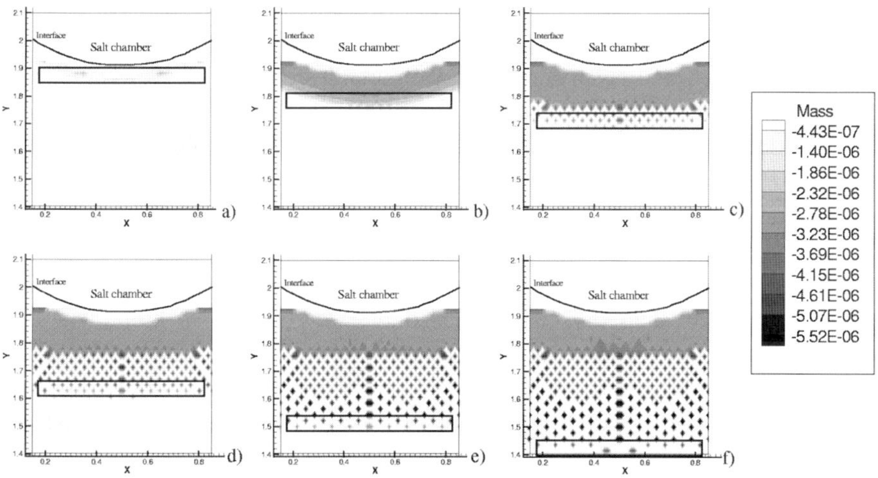

Fig. 6.11 Crystals distribution for the same times as in Fig. 6.10 (numerical results).

Among other things, it can also be used for evaluation of the supersaturation threshold for nucleation, as will be discussed in the next section. This aspect is very important since such a threshold is affected by a large uncertainty in the literature even if well-known proteins are considered and is generally a critical requirement for both experimental and numerical protein studies.

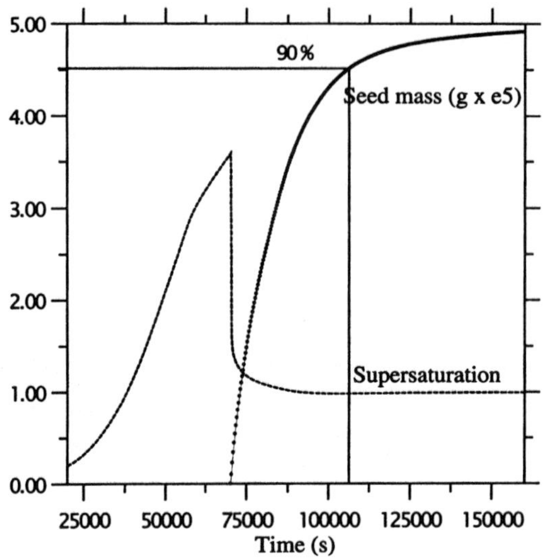

Fig. 6.12 Computed growth law of a reference seed ($x = 0.4$ cm, $y = 0.9$ cm) located in the bulk of the protein chamber ($C_{(o)} = 4 \times 10^{-2}$ g/cm^3).

6.10. Numerical simulation as a useful tool to estimate the nucleation threshold

The present section shows how the uncertainty affecting the aforementioned threshold can be overcome through the comparison between the experimental and numerical results obtained for different values of this parameter. This is an additional demonstration of the insight into the physics that can be obtained by combining numerical simulations and critical analysis of the results.

As an example, Fig. 6.13 show parametric analysis carried out ($3 \leq \eta \leq 5$) for the case $C_{(o)} = 4 \times 10^{-2}$ g/cm^3. According to the computations the problem is quite sensitive to η. It can influence the number of locations where nucleation occurs (i.e. the number of solid crystals produced due to nucleation) and their spatial distribution too, i.e. the density of particles. In Fig. 6.13 the size of particles versus the distance from the gel interface and their density are plotted as a function of the parameter η. As seen from the plots, the value of η should be assumed in the range 3–4 in order to be compatible with experimental observations. Finally, in fact, the value $\eta = 3.6$ has been found for this case and used to fit the experimental data as demonstrated in Section 6.9.

The same parametric analysis carried out for the cases $C_{(o)} = 6 \times 10^{-2}$ g/cm^3 and $C_{(o)} = 8 \times 10^{-2}$ g/cm^3 points out that the best agreement between numerical and experimental results is achieved for $\eta = 2.3$ and $\eta = 1.6$, respectively; the corresponding numerical results are in good quantitative agreement with the experimental ones both for the particles size and density distributions (see, e.g., Figs. 6.5 and 6.6).

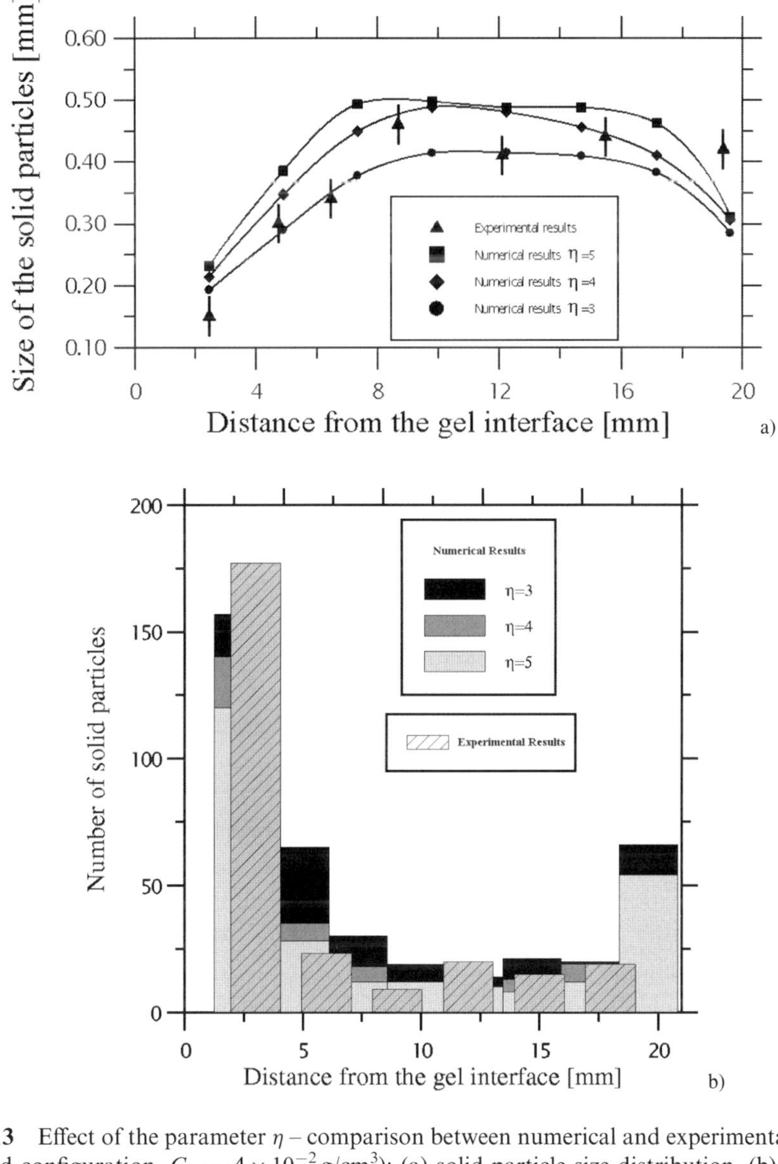

Fig. 6.13 Effect of the parameter η – comparison between numerical and experimental results (gellified configuration, $C_{(o)} = 4 \times 10^{-2}\, g/cm^3$): (a) solid-particle-size distribution, (b) density-of-particles distribution.

According to this study, the value of η behaves as a decreasing function of the initial value of protein concentration, i.e. the larger the amount of protein available in the liquid phase, the lower the supersaturation limit. This trend is not unexpected and explains the different values of η experimentally found by the different investigators. There is no physical reason forcing η to be a constant. On the contrary, one may suspect that a larger initial amount of protein available in the liquid phase can lower the activation energy of the nucleation process.

The present section and the previous ones represent quite an exhaustive attempt to help organic crystal growers to discern the complex interrelations among the various parameters under one's control (that are not independent of one another) and to elaborate rational guidelines relating to physical factors that can influence the probability of success in crystallizing protein substances in gel.

6.11. Crystal sedimentation on the ground

The case where gel matrix is not used in the protein chamber is now investigated (see Fig. 6.1b, $C_{(o)} = 4 \times 10^{-2}\,\mathrm{g/cm^3}$). The buoyancy-driven convection is neglected for the first group of simulations. This artifice is used in order to discern the separate effects of crystal motion and buoyancy forces.

Since protein crystals are denser than the feeding solution, sedimentation occurs. Due to sedimentation, the phenomena under investigation do not show the patterns described in the previous sections. The solid particles, in fact, settle in the growth chamber. The lower boundary of the protein chamber being the interface of gel used to fill the salt chamber, the crystals sediment on this interface. This is clearly shown in Fig. 6.15 where the thickness of the layer of crystals sedimented on the bottom is an increasing function of time due to the continuous mass incoming from the bulk. Figure 6.15a shows that the first nucleation occurs at the free interface. Solid particles accumulate on the interface and then, if the local value of concentration is larger than the solubility, they grow due to further precipitation of protein. As time passes and salt diffuses in the protein chamber, other nucleation occurs far from the interface. Similarly to the configuration investigated in Section 6.9 (i.e. crystals locked on the gel matrix), each nucleation occurring in the protein chamber gives rise to a depletion zone around the location where the solid particle has condensed. For the present case, however, the solid particles are not locked on their initial position and for this reason they leave the position where nucleation has occurred. During the fall to the bottom, the solid particles deplete the surrounding liquid of protein. For this reason, a wake is produced beyond the sedimenting particles and a small depletion spot highlights their initial position (see Fig. 6.14, note that in these figures the salt chamber is located at the bottom of the protein chamber).

As time passes, due to the nucleation process, a depleted band is created starting from the gel interface and protruding in the protein chamber. The amplitude of the

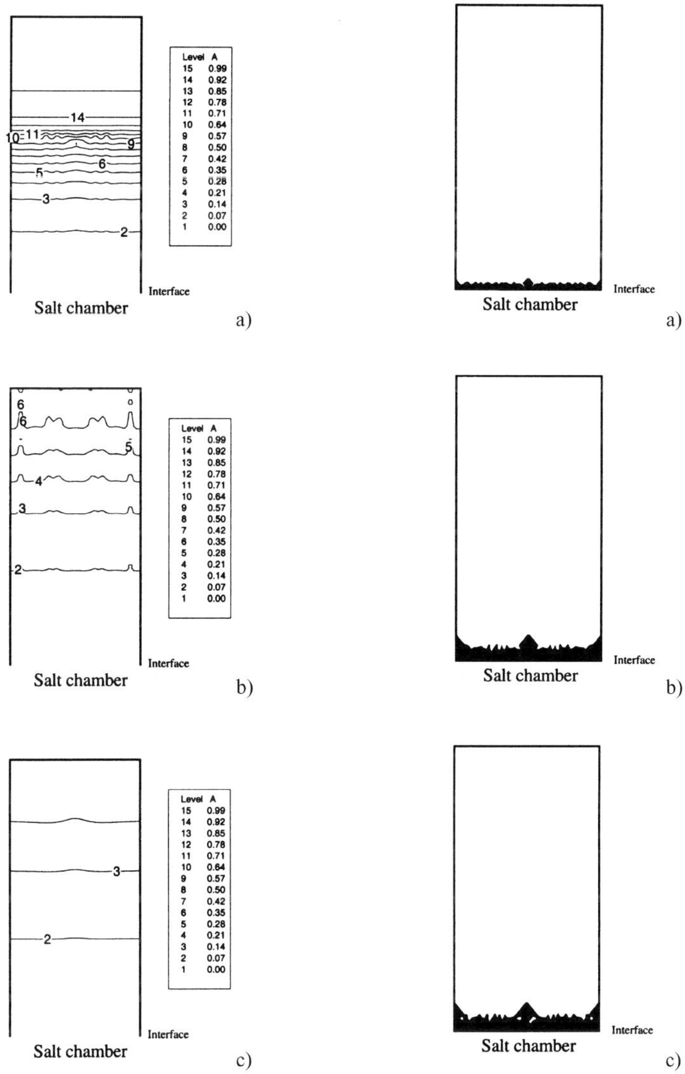

Figs. 6.14–6.15 On the left, Fig. 6.14: Nondimensional protein concentration contour lines ($C_{(o)} = 4 \times 10^{-2}$ g/cm^3); On the right, Fig. 6.15: Corresponding distribution of crystals; (a) $t = 7.35 \times 10^4$ s, (b) $t = 1.365 \times 10^5$ s, (c) $t = 1.68 \times 10^5$ s (sedimentation process on the ground).

band is an increasing function of time. This effect behaves as a disturbance in the concentration field propagating towards the top of the protein chamber (Fig. 6.14).

The propagation towards the top, of the nucleation events depletes the protein solution of lysozyme. This behavior is similar to that described in Section 6.9. For the present case, however, the rate of depletion is lower. In this case, in fact, ensuing depletion due to the phenomenon of subsequent precipitation after nucleation is negligible (it simply applies to the wakes beyond the sedimenting particles). Due to the sedimentation process, solid particles leave the regions of high protein concentration where further precipitation could occur. The crystals reach the bottom of the protein chamber where the concentration is low due to previous protein absorption. Since all the solid particles tend to be clustered and accumulated in the same zone of low protein concentration, the possibility of additional solute absorption and consequent size augmenting is reduced with respect to the case of locked crystals (gel matrix).

For the aforementioned reasons, the value of the protein concentration in the bulk of the chamber tends to be higher in the case of gellified protein solution. This trend is confirmed by the available experimental results.

Sedimentation–convection Figure 6.16 (at each station y the density, computed as $\rho_S = \rho_{H_2O}(1 - \beta_{lys}C - \beta_{NaCl}C_{NaCl})$, has been averaged along x) shows the time evolution of the density profile along y in the protein chamber obtained under the assumption of a protein chamber free of solutal buoyancy convection.

At a given instant, the density has a maximum on the gel interface and a minimum on the top of the protein chamber. This behavior is the sum of two counteracting effects: lysozyme is depleted in regions close to the gel interface (due to the nucleation process and subsequent accumulation of solid particles due to sedimentation) thus reducing the density; salt diffuses in the protein chamber

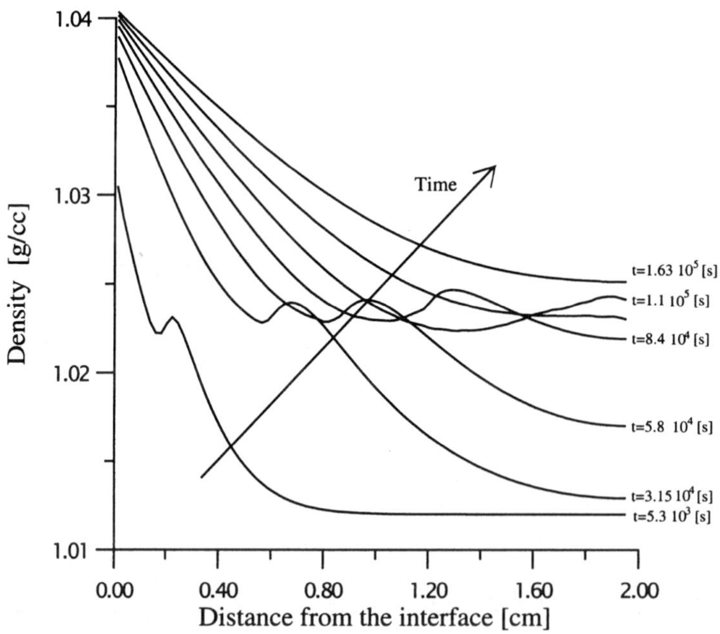

Fig. 6.16 Density distribution in the protein chamber versus time ($C_{(o)} = 4 \times 10^{-2}$ g/cm^3).

through the gel interface thus increasing the density. The latter effect is dominant since salt has a larger diffusion coefficient. For this reason the density has a maximum on the gel interface and then tends to decrease towards the top of the protein chamber. The shape of the density profile, however, is not always decreasing along y. Some local inversions occur in this profile due to nucleation phenomena in the bulk of the protein chamber.

Nucleation phenomena, in fact, absorb protein; this leads to a sink in the density distribution and correspondingly a ripple appears in the shape of the density profile. The position of this ripple is not fixed as time passes. It migrates along y according to the propagation towards the top of the nucleation process. For $t = 1.6 \times 10^5$ s the sink reaches the top of the protein chamber and then disappears on further increase of time.

These results suggest that, for $t < 1.6 \times 10^5$ s, the onset of convection is possible due to buoyancy forces associated with the inversion in the density profile (Rayleigh–Bènard convection). Accordingly, simulations carried out taking into account the solution of the nonlinear and time-dependent Navier–Stokes equations (Fig. 6.17) show that vortex cells arise. These rolls, however, do not extend over the entire bulk of the protein chamber and are confined to the ripple zone of the density profile.

As time passes, the vortex rolls migrate towards the top of the protein chamber (see Fig. 6.17). This behavior is strictly related to the propagation along y of the zone where the inversion occurs. Moreover, the migration of the cells towards the top is coupled to a dynamical change of their number (i.e. the value of the wave number m of the convective instability). This number changes according to the distance from the gel interface. Close to the gel interface m is about 6 (Fig. 6.17a), but this number is reduced when, due to the propagation towards the top, the distance from the gel interface increases. It becomes 5 in Fig. 6.17b and c, and is reduced to $m = 2$ in Fig. 6.17d.

It is known that discrete wave numbers of convective disturbances are selected out of the full spectrum of disturbances because the convection rolls are closed in a special zone geometry (see the discussion in Section 2.2.4, Fig. 2.2 and also the generalized theory for the selection of the azimuthal wave number in liquid bridges introduced in Section 4.10). For the present case, the selection rule is given simply by the constraint that the wavelength must be an aliquot of the length along x of the chamber (L) and by the fact that the convection rolls are confined to the extension along y of the ripple zone (L_i). According to this theory the critical wave number should be related to the length along x of the chamber and to the extension along y of the ripple zone (L_i), i.e. it should scale according to the law

$$m = L/L_i \tag{6.17}$$

Figure 6.16 shows that the extension along y of the ripple zone (L_i) is an increasing function of the distance from the gel interface. This explains why the wave number is

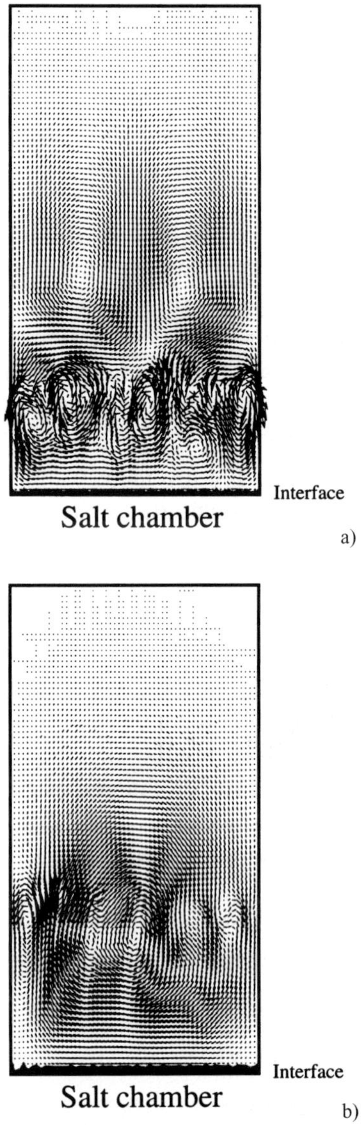

Fig. 6.17 Velocity field – ($C_{(o)} = 4 \times 10^{-2}$ g/cm^3, $Ra = 1.36 \times 10^9$, $Ra_{NaCl} = 9.5 \times 10^8$ sedimentation and convection on the ground): (a) Time $= 1.8 \times 10^4$ s, (b) Time $= 3.9 \times 10^4$ s, (c) Time $= 6.3 \times 10^4$ s, (d) Time $= 9.3 \times 10^4$ s.

reduced as time passes. Since the diameter of the vortex cells is almost equal to L_i (the force driving convection acts along y over a length equal to L_i) and the space available along x is fixed, the number of convective cells decreases if L_i is increased.

The wave number in Fig. 6.17a ($m = 6$) is in good agreement with the experimental results shown in Fig. 6.18. Figure 6.18 shows sinusoidally distorted

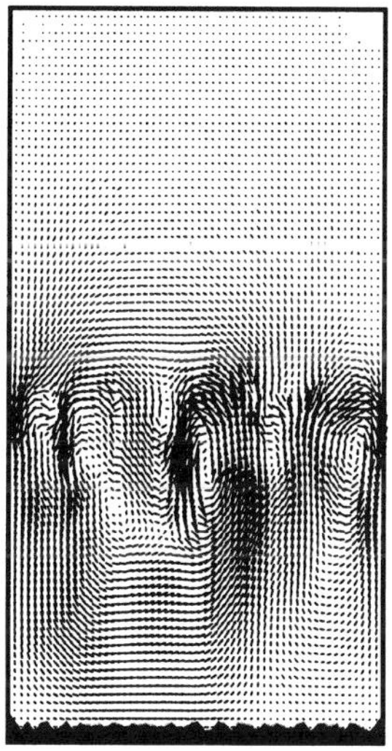

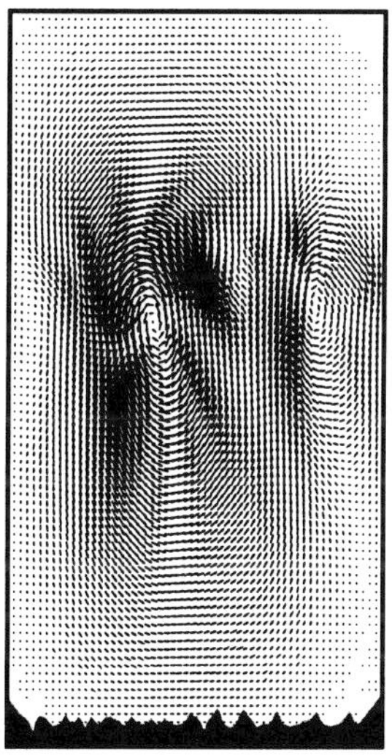

Fig. 6.17 Continued.

fringes (fringes represent sites of equal refractive index (n) that is dependent on protein and salt concentrations). The sinusoidal distortion of the fringes close to the gel interface ($t = 2 \times 10^4$ s) is due to convection and the number of waves is 6–7.

To summarize, if convection due to buoyancy forces is taken into account, a very complex convection pattern arises whose stages of evolution depend on the

Fig. 6.18 Interferogram ($t=2.0\times10^4$ s) showing the number of convective cells close to the gel interface, origin of the imposed concentration gradient (after Lappa et al., 2003b).

nucleation in the bulk of the protein chamber. On the basis of numerical simulation, the convective cells arise owing to local inversions in the density distribution. These vortex cells migrate towards the top of the protein chamber exhibiting a size dependent on the distance from the gel interface and affecting the nucleation dynamics in the bulk. Sedimentation inhibits growth of crystals since they tend to be clustered in a region where the concentration of protein in the liquid phase is small.

6.12. Solutovibrational convection and crystal motion induced by g-jitters

Crystal growth under microgravity proceeds by the same chemical process as on the ground, the only difference between them being fluid dynamics: mass transport is nominally diffusive in the gellified environment, whereas, in the absence of gel (microgravity conditions), concentration (density) gradients existing in the crystallizing solution (as elucidated in the previous section) as an intrinsic consequence of the crystal-growth process may lead to the onset of convection.

Regarding this aspect, it should be pointed out that the benefits of crystal growth in microgravity conditions, can only be obtained if the crystals grow by addition of growth units supplied at a slow rate by mass transport through a stable depletion zone filtering the potentially disturbing protein or impurity concentration changes at the crystal surface (Otàlora et al., 2001). Therefore, for microgravity experiments, the definition of experimental conditions ensuring the development of stable depletion zones is a requirement as important as having a quiet diffusive environment.

The present section deals with the analysis of the phenomenon of lysozyme periodic precipitation (gel absent in the protein chamber, same crystallization cell considered in the earlier sections, $C_{(o)}=4\times10^{-2}$ [g/cm^3]) in the presence of a high-frequency sinusoidal acceleration (corresponding to the US Lab Ergometer conditions, i.e. $f=3$ Hz, $g=b\omega^2\cong1.5\times10^4$ μg, $Ra_V\cong10^4$). The nondimensional frequency and amplitude associated with the Lab Ergometer conditions in Fig. 2.28 are, respectively, $\Omega_{lys}\cong2\times10^7$, $\Lambda_{lys}\cong5\times10^{-4}$ and $\Omega_{NaCl}\cong2\times10^6$, $\Lambda_{NaCl}\cong3.5\times10^{-3}$, i.e. the present case falls with good approximation within the range of validity of the Gershuni formulation (see Chapter 2 and in particular Section 2.6 for additional details on these aspects).

Two different cases are considered for $C_{(o)}=4\times10^{-2}\,\mathrm{g/cm^3}$: g-jitters disturbances perpendicular to the imposed salt and protein concentration gradients (g-jitter parallel to the x axis, worst condition, onset of solutovibrational convection) and parallel to the concentration gradients (g-jitter parallel to the y axis, "ideal" condition).

The comparison between these results and those described in Section 6.9 gives insight into the effect of g-jitter residual accelerations on the phenomena under microgravity conditions. First, attention is focused on the worst case, then the ideal condition is discussed and compared with the worst one.

For the first case, the results show that, due to density gradients associated with salt diffusion and protein depletion occurring in the protein chamber, convection cells arise (velocity is of the order $O(10^{-6}\,\mathrm{cm/s})$). The flow organization is very complex since multicellular structures are formed. Four main vortex cells are visible in the protein chamber. These vortices have the classical Gershuni configuration (see, e.g., Fig. 2.33a for the case $\alpha=90°$ in Chapter 2), i.e. two couples of counter-rotating vortex cells (each couple formed by two convection rolls mirror-symmetric with respect to the position $x=L/2$, see Fig. 6.19a–c). Figure 6.19a–c show that at any instant the line of separation between the two couples of convection cells corresponds (approximately) to the position of the nucleation front (see Fig. 6.20). As time passes the nucleation front and the line of separation migrate, increasing their distance from the gel interface.

According to this behavior, the region where solid particles are present (region between the gel interface and the nucleation front) is affected by the presence of two counter-rotating vortex cells located on the right and on the left of the position $x=L/2$, respectively. Solid particles are transported towards the gel interface near the walls ($x=0$ and $x=L$) and in the opposite direction around $x=L/2$. Crystals located around $x=L/2$ leave regions where protein has been depleted and are transported towards regions where the amount of protein available in the liquid phase is still large. On the contrary, crystals located near the walls are transported towards regions where the protein has been reduced due to previous nucleation and subsequent growth phenomena. For these reasons crystals located around the midsection have a larger size compared to crystals located near the walls (they, in fact, take advantage of the larger amount of protein available with respect to their initial position and vice versa for the crystals located near the walls).

As time passes, the velocity field becomes very complex (Fig. 6.19d–f). According to this behavior the patterns of crystals are no longer characterized by the somewhat repetitive structure and configuration observed in Section 6.9. The distribution of the crystals, in fact, tends to be chaotic and irregular (compare, e.g., Figs. 6.4f and 6.20f).

This behavior shows the high sensitivity of the system to the effect of the g-jitter. The deviation of the crystal distribution from the diffusive one may be regarded as a measure of the effect of the g-jitter disturbances on the complex rhythmic dynamics driven by the availability of protein in the liquid phase and by the rhythmic increase

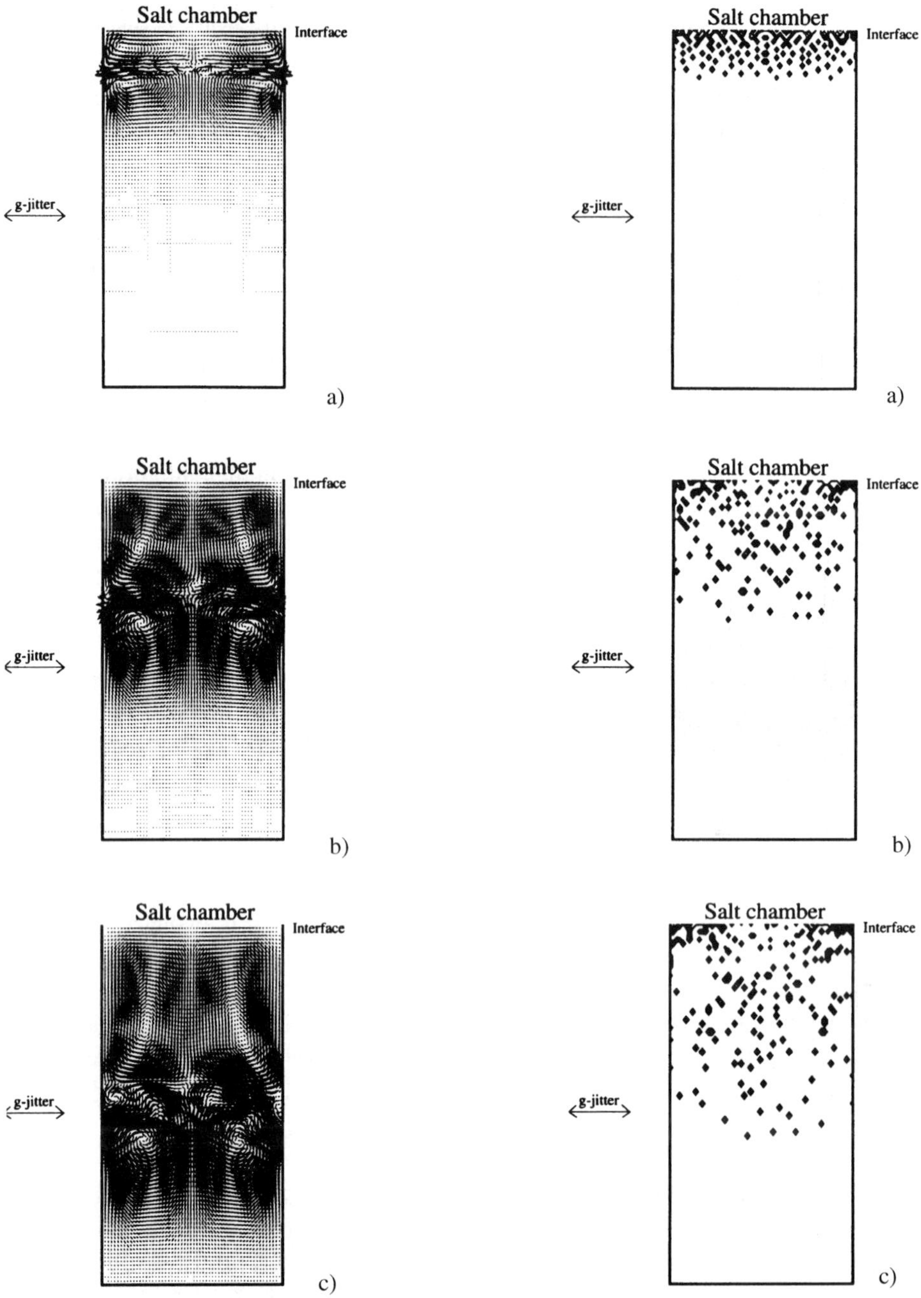

Figs. 6.19–6.20 On the left, Fig. 6.19: Velocity field induced by g-jitter disturbances (microgravity conditions, worst case, $Ra_V=10^4$, $C_{(o)}=4\times10^{-2}$ g/cm^3); On the right, Fig. 6.20: Corresponding distribution of crystals; (a) $t=10^4$ s, (b) $t=4.8\times10^4$ s, (c) $t=9.6\times10^4$ s, (d) $t=9.8\times10^4$ s, (e) $t=1.1\times10^5$ s, (f) $t=1.7\times10^5$ s.

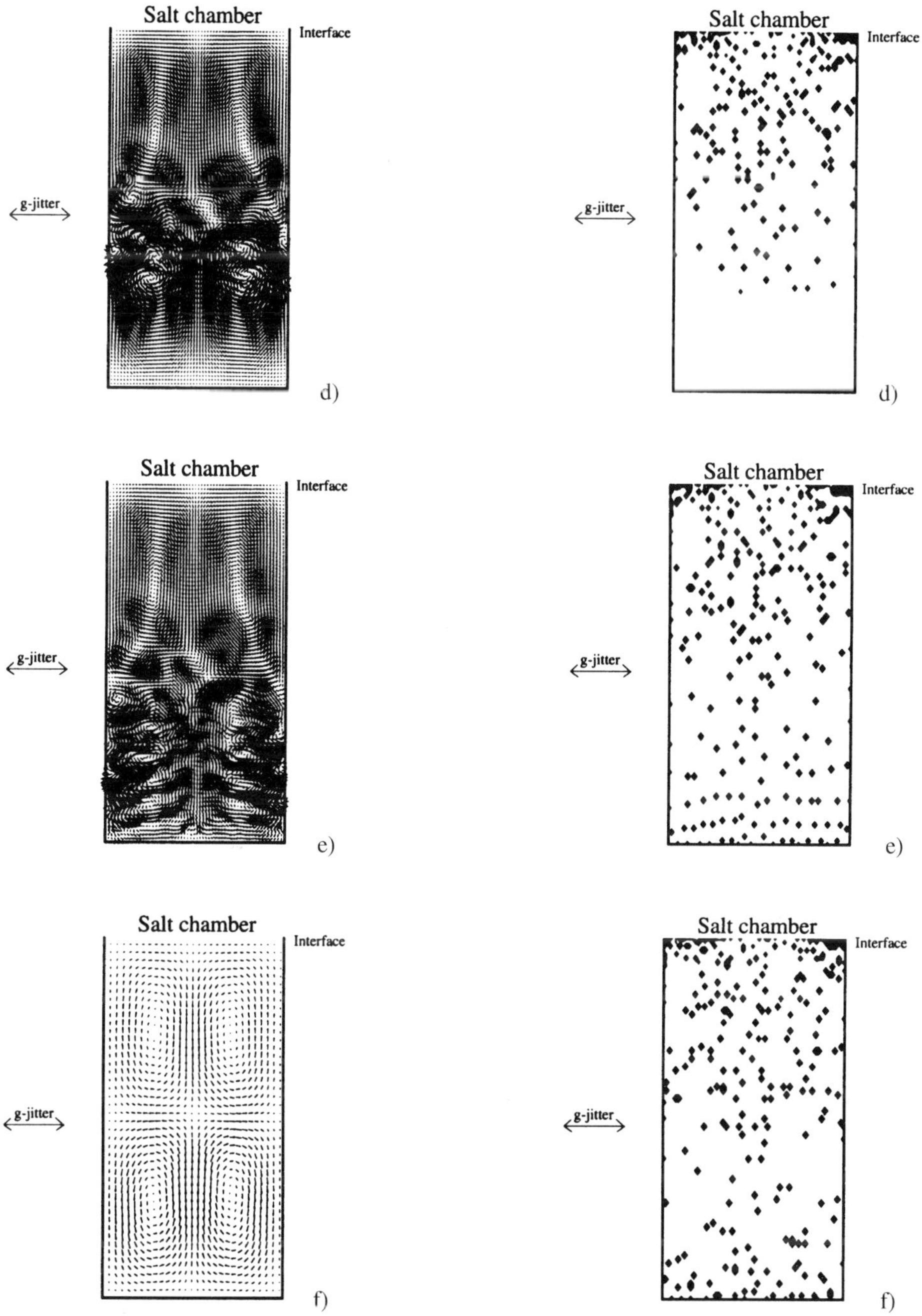

Figs. 6.19–6.20 Continued.

and decrease of supersaturation due to the interplay between protein depletion and salt diffusion through the gel interface.

For the ideal case (g-jitter parallel to the y axis) in the absence of nucleation phenomena (purely diffusive conditions), no vortex cells should arise since it is well known (see, e.g., Section 2.6.5) that concentration (and/or temperature) gradients parallel to the direction of the sinusoidal acceleration do not contribute to the onset of convection. However, due to nucleation phenomena and further growth, concentration gradients perpendicular to the g-jitter direction are created, leading locally to the onset of vortices (velocity is of the order $O(10^{-6}\,\text{cm/s})$ as for the worst case). This is shown in Fig. 6.22a–c. The presence of vortex cells is strictly confined to the zones where new nucleation and/or subsequent precipitation occurs (i.e. around the nucleation front, see Fig. 6.21a–c). The region affected by multicellular convective structures migrates with the nucleation front, whereas the remaining part of the protein chamber is free of convection (Fig. 6.22a–c).

The distribution of crystals at the end of the process (Fig. 6.23) is more regular with respect to the worst condition. However, the effect of g-jitter cannot be considered negligible. The symmetry of the solid pattern is distorted with respect to the gellified conditions. For this reason, this analysis clearly shows that no ideal conditions exist in the case of crystal-pattern formation under the effect of sinusoidal acceleration. While in the case of typical nonisothermal experiments under microgravity conditions, investigators can minimize g-jitter effects orienting their fluid cell up to make the sinusoidal acceleration parallel to the imposed temperature difference (see Fig. 2.35 and the discussion in Section 2.6.5), in the case of experiments dealing with periodic precipitation, this is not possible. In fact, no preferential direction of the concentration gradients can be found or fixed a priori. Even if the crystallization phenomena can be (roughly) schematized for the present case as a wave migrating along a fixed direction (the y axis), however, due to the nucleation phenomena, concentration gradients oriented both in the axial and transverse directions arise (i.e. along x and y) leading locally to the onset of convection.

Note that in the worst case the depletion zone created around each growing crystal is distorted by two different effects: the large-scale flow due to solutal g-jitter and the motion of the crystal itself. In the ideal case, since the presence of vortex cells is strictly confined to the zones around the nucleation front (it is a local phenomenon), more stable depletion zones occur. These aspects are crucial for the quality of the crystals. According to the previously highlighted critical role played by the stability of the depletion zone, better crystals are expected to be obtained in the ideal case (even if the spatial distribution of crystals is not symmetric as for the worst case).

A quantitative investigation of these aspects requires the detailed simulation of the interface attachment kinetics and surface morphology evolution of each crystal. The methods for the treatment of these aspects have been discussed in Chapter 5 dealing with the very detailed investigation of the local evolution.

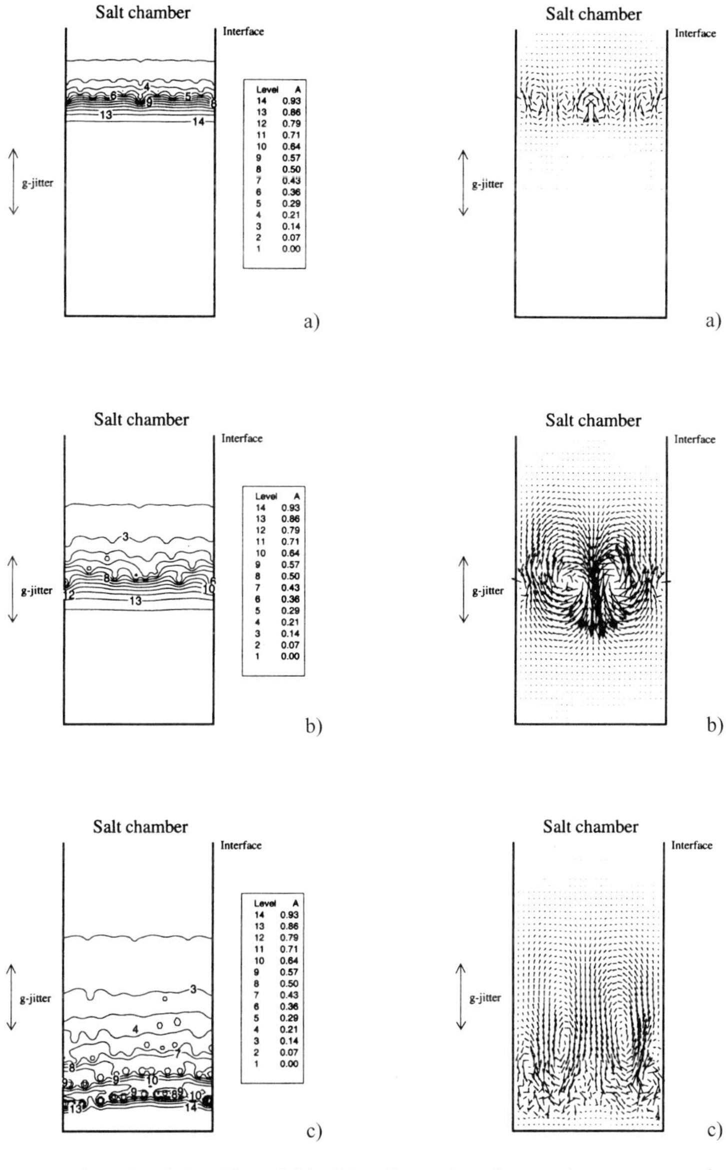

Figs. 6.21–6.22 On the left, Fig. 6.21: Nondimensional protein concentration contour lines ($C_{(o)} = 4 \times 10^{-2}$ g/cm^3); On the right, Fig. 6.22: Corresponding velocity field induced by g-jitter disturbances (microgravity conditions, "ideal" case, $Ra_V = 10^4$) (a) $t = 2.6 \times 10^4$ s, (b) $t = 7.4 \times 10^4$ s, (c) $t = 10.8 \times 10^4$ s.

Finally, Figs. 6.24 and 6.25 compare the solid-particle-size and density-of-particles distributions under the diffusive regime and g-jitter conditions. In the latter environment, the number of crystals tends to be reduced with respect to the case of gellified protein solution. This effect is related to the presence of convection due to

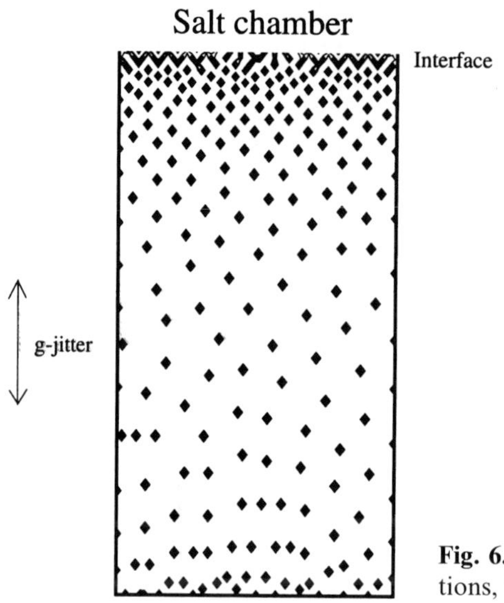

Fig. 6.23 Pattern of crystals – (microgravity conditions, "ideal" case, $Ra_V = 10^4$) Time = 1.7×10^5 s.

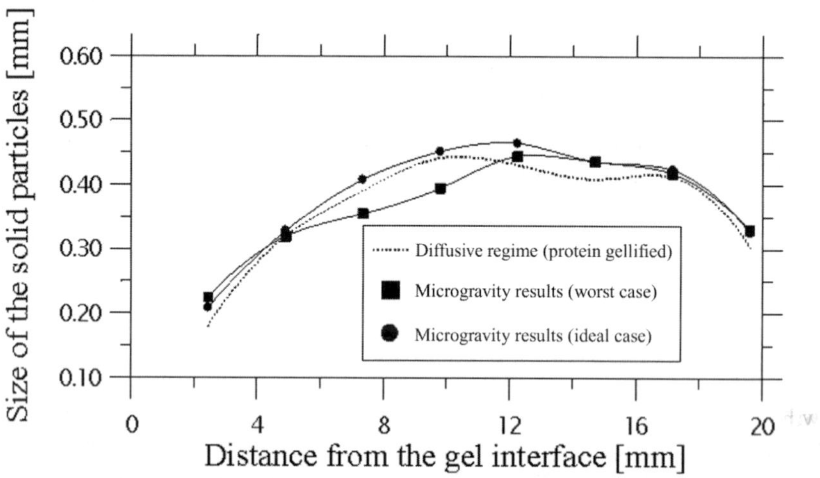

Fig. 6.24 Solid-particle-size distribution: comparison between pure-diffusive regime (gellified configuration) and microgravity conditions (g-jitter disturbance) after Lappa and Carotenuto (2003).

g-jitter that strengthens protein mixing and therefore promotes ensuing growth after nucleation.

The present analysis may support the optimization of growth techniques under the effect of periodic accelerations as it provides information about the relative direction of g-jitter and imposed concentration gradients that should lead to better crystals for diffraction analyses.

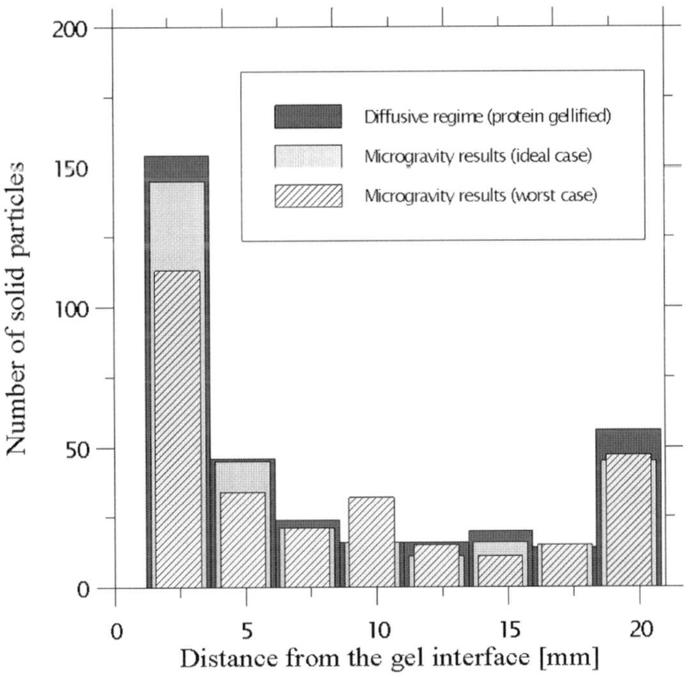

Fig. 6.25 Density-of-particles distribution: Comparison between pure-diffusive regime (gellified configuration) and microgravity conditions (g-jitter disturbance) after Lappa and Carotenuto (2003).

6.13. Solutal Marangoni convection and protein crystallization: the vapor diffusion technique

A final aspect needs clarification with regard to factors that can affect protein growth: The role played by surface-tension-driven convection (see Section 2.3 for additional details about the genesis and the features of this type of flow).

In certain protein crystallization systems, solvent evaporates from the solution in order to create and maintain the supersaturation conditions necessary for crystal growth. Due to differences in surface tension along these evaporating interfaces, the Marangoni effect may introduce flows, even under microgravity conditions.

Since relatively little is known about the possibility of Marangoni convection occurring during protein crystallization (see the discussions in Molenkamp et al., 1994), this topic is treated herein only from a theoretical point of view and in terms of possible numerical strategies applicable for macroscopic analyses.

It is worthwhile to start the discussion noting that Marangoni effects are only possible in the vapor diffusion crystallization methods, as a phase boundary is only present when these methods are used (Marangoni effects are absent in the

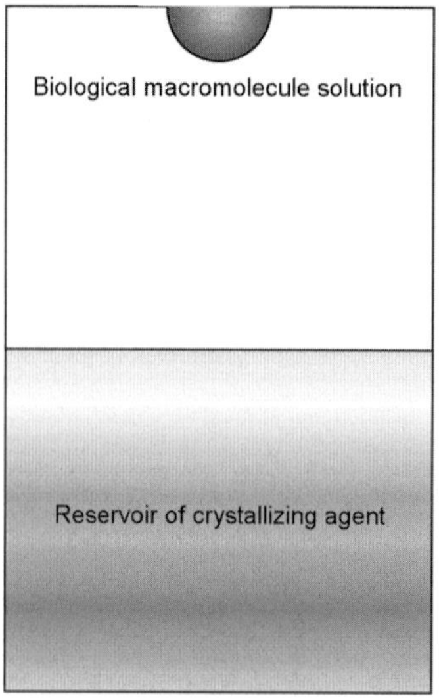

Fig. 6.26 Sketch of the vapor diffusion technique.

gel-based counterdiffusion technique discussed in the previous sections; see also Miller et al., 1992).

Two of the most commonly used ground-based methods for protein crystallization fall under this category. These are known as the "hanging drop" and "sitting drop" methods. Both entail a droplet containing purified protein, buffer, and precipitant being allowed to equilibrate with a larger reservoir containing similar buffers and precipitants in higher concentrations (the hanging drop method differs from the sitting drop method in the vertical orientation of the protein solution drop within the system).

With the hanging drop technique, in turn, two configurations are possible: the half-drop and the full-drop. The differences between the two geometries are that the contact boundary with the drop support is a circular disk of radius R for the half-sphere (Fig. 6.26), and is limited to a small spherical cap, for the full-size drop.

Three kinds of precipitants are used. These are salts, such as sodium chloride and ammonium sulphate, organic solvents, such as ethanol and acetone, and polyethylene glycol (PEG) of various molecular weights. In salt or PEG systems water is transferred from the protein to the precipitant solution, while in organic solvent systems the organic solvent is transferred to the protein solution from the precipitant solution. The use of organic solvents is limited, because organic solvents tend to induce protein denaturation.

Initially, the droplet of protein solution contains an insufficient concentration of precipitant for crystallization, but as water vaporizes from the drop and transfers to the reservoir, the precipitant concentration increases to a level optimal for crystallization.

In such systems, Marangoni convection can be the result of both concentration and temperature gradients along the free interface. Increasing temperature as well as increasing concentration of a surface-tension-lowering solute may locally decrease the surface tension. As a consequence liquid is drawn from places with low surface tension to high surface tension areas.

On the ground, the flow in such configurations is dominated by solutal-buoyancy convection and nuclei move with the fluid until they reach a critical size at which they settle to the bottom of the hanging or sitting drop.

The effect of possible protein concentration gradients under microgravity conditions is discussed in the following:

Suppose the distance of a crystal from the interface of the drop is comparable to the thickness of the depletion zone. The protein concentration at the interface nearest to the crystal is then lower than in the surrounding areas. Therefore, liquid with higher protein concentration is drawn towards this spot, increasing the protein concentration and increasing the speed of growth of the crystal face facing the liquid–gas interface (Fig. 6.27). In this way the Marangoni effect may serve to augment the crystallization growth rate or cause the crystal habit to change (a similar effect has been discussed in Section 5.12 in the case of buoyancy convection).

However, the Marangoni flow may be responsible for an even more important effect similar to that discussed in the previous section in the case of solutovibrational convection. In fact (see Fig. 6.27), for continuity a local radial inward flow is generated that could lead to a pushing of the crystal towards the zones below the free surface where the amount of solute protein is larger.

Since nucleation occurs especially close to the free surface (where supersaturation is very large owing to the aforementioned evaporation process) the Marangoni

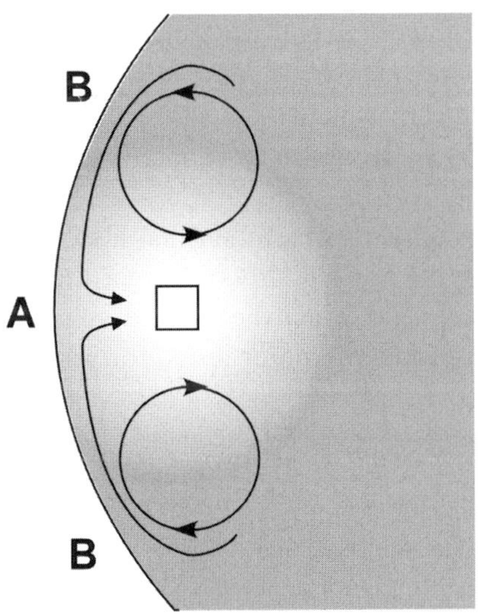

Fig. 6.27 Crystal growing in a drop, surrounded by a protein depletion halo under microgravity conditions (full-drop). When the crystal grows close to the interface, the solutal Marangoni effect may increase the mass-transfer rate of protein towards the crystal face which is closest to the interface and push it towards the interior of the drop.

effects may be responsible for a continuous transport of new crystals towards the interior of the drop.

Such effects are still poorly understood and further investigation is required. In particular possible numerical simulations aimed at the understanding of these behaviors could be carried out within the framework of the numerical techniques discussed in Sections 6.3–6.7 and assuming a polar reference coordinate system.

Concerning the boundary conditions modeling surface tension effects at the free liquid interface, it should be stressed that the gas–liquid interface in a protein crystallization system rarely behaves as a clean Newtonian interface (Molenkamp et al., 1994).

When Newtonian behavior of the liquid is assumed and interfacial properties other than the surface tension and related variations induced by protein solutal effects are neglected, the aforementioned condition could be written as:

$$\mu \frac{\partial \underline{V}}{\partial n} = -\sigma_C \underline{\nabla}_S C \tag{6.18}$$

and in nondimensional form (scaling the lengths by the diameter of the drop $2R$, the velocities by the diffusion velocity and the concentration by a reference ΔC):

$$\frac{\partial \underline{V}}{\partial n} = -Ma \underline{\nabla}_S C \tag{6.19}$$

where $Ma = \sigma_C \Delta C 2R/\mu D$ (μ being the dynamic viscosity) is the so-called solutal Marangoni number. Generally, however, as mentioned above, surface rheology of protein solutions is not Newtonian. The interfacial stress can depend on the history of the interface, resulting in viscoelastic behavior (Graham and Phillips, 1980; Edwards et al., 1991). For instance, it is known (recent experimentation in microgravity) that the flow rates of Marangoni convection decrease exponentially with an increase in the exposure time (time during which the surface has been exposed to the ambient air), and that they can also be decreased by an increase in the protein concentration.

For some interfaces a viscoplastic behavior (Bingham surface fluid) has also been suggested. A Bingham surface only moves when the applied stress is larger than a specific yield stress. Such a surface is completely rigid when the applied stress is smaller. However, no such models have yet been proposed and/or tested for protein solutions.

CHAPTER 7

The growth of biological tissues

7.1. Tissue engineering and microgravity

Given our current expertise in growing organic tissues as model systems for controlled studies of tissue development, and the certain future developments in growing biological tissues as a potential source of functional constructs for organ repair, the concept of tissue engineering is nearing reality.

Tissue engineering is a new field that enables tissue equivalents to be created from isolated cells in combination with biomaterials and bioreactor culture vessels. Potentially it can provide a basis for systematic, controlled in vitro studies of tissue growth and function as well as a detailed knowledge of many important problems, e.g., the chemistry and mechanics of healthy organs and of cancers, infectious diseases, immune-system failures, and so on. Moreover, in future the constructs obtained by these techniques will be used for treatment of damaged human tissues and/or healthy organs.

The key factor in this new field of engineering involves how to define the properties of the artificial environments wherein biological tissues are grown in order to achieve the best possible conditions for growth. Organic tissues are very sensitive to their environment and, like protein crystals, if exposed to sufficiently severe conditions may denature and/or degrade. Often they have to be constantly maintained in a thoroughly hydrated state at or near physiological pH and temperature. For these reasons, very gentle and restricted techniques have been used by the investigators.

A challenge, however, that has plagued tissue science throughout its short history is that most of these gentle cell-culture techniques produce flat, one-cell-thick specimens that offer limited insight into how cells work together. Moreover, without a proper 3D assembly, epithelial cells (the basic cells that differentiate tissue into specific organ functions) lack the proper clues for growing into the variety of cells that make up a particular tissue.

The most promising method to solve these problems seems to be the rotating-wall perfused vessel bioreactor (RWPV), a can-like vessel equipped with a membrane for gas exchange and ports for media exchange and sampling. As the bioreactor turns, the cells continually fall through the medium. Under these quiet conditions, the cells "self-assemble" to form clusters that sometimes grow and differentiate much as they would in the body. It spins a fluid medium to neutralize most of gravity's effects and encourage cells to grow in a natural manner. The rotating bioreactor, in fact, was invented as a model of microgravity effects on cells.

Ground tests of the bioreactor yielded three-dimensional biological specimens approximating natural growth, a striking change from the pancake shapes of traditional cultures. As cells replicate, they self-associate to form a complex matrix of collagens, proteins, fibers, and other chemicals. Eventually, on Earth, the clusters become too large to fall slowly and research has to be continued in the true weightlessness of space.

With regard to this aspect note that unique and landmark experiments were carried out by Freed et al. (1997). They grew specimens of bovine cartilage tissue under both normal (Earth) and microgravity conditions (Mir) and reported that initially disk-like specimens tend to become spherical in space, whereas constructs grown on Earth maintained their initial diskoid shape. Experience aboard Mir turned microgravity bioreactor research into a mature science and proved that the gravity effect may play a crucial role in determining the properties and the growth conditions of organic tissues.

Further progress in creating proper environments for the growth of the tissues requires an understanding of how chemical, mechanical and other environmental factors influence growth. Recent advances in cellular and molecular biology have furthered our knowledge of the biochemical aspects of this problem. Moreover, some theories have been proposed to take into account mechanical factors.

According to many investigators (see, e.g., Rodriguez et al., 1994; Taber, 1998a, b; Taber and Chabert, 2002), growth and remodeling in tissues may be modulated by mechanical factors such as internal stress. A general continuum formulation for finite volumetric growth in soft elastic tissues based on their internal mechanical state therefore has been proposed. This dependence has been introduced through the so-called *growth law*, which is a constitutive equation for the rate of change of the growth tensor, and describes its dependence on mechanical quantities such as stress, strain, and strain energy. The growth tensor is determined from the growth law. Two excellent implementations of this theory have been used to model continuous growth of soft tissues. The first, developed by Taber and coworkers (Taber, 1998a, b; Taber and Chabert, 2002), has been used to investigate the growth of the heart, arteries, and skeletal muscles. In this approach, the growth law was defined on an initial fixed reference configuration for the entire growth process. In the second method, proposed by Hoger and coworkers (Rodriguez et al., 1994), the growth model was defined on the current configuration of the loaded, growing material. This method was used to elucidate how residual stress arises during growth and how, in turn, the growth is affected by stress in the tissue.

Other theories have focused on the transport of gas in the culture medium surrounding the growing tissue, since it is expected that this factor may also be important. The role played by the supply of gas on the rates of synthesis was modeled by Obradovic et al. (2000). They developed a model taking into account the transport of dissolved oxygen in the feeding solution and, in particular, the effect of the gas tension. It was found that more efficient gas transport in both the culture medium and at the construct surfaces stimulated rapid tissue growth, while cultivation at low oxygen tension has the reverse effect.

The basic approach used in these theories was to combine published experimental results with relatively simple theoretical models. First, the trends in available data are used to postulate qualitatively biomechanical principles for growth. Then, explicit growth laws are proposed and tested for their ability to predict known patterns of growth leading to important theoretical results and agreement with the experiments.

It is important to stress that the available mechanical theories have focused on "what happens inside the tissue" whereas the biomechanical laws that govern soft-tissue growth in terms of "surface incorporation/conversion conditions" (interface kinetics of the growth) remain poorly understood. Superimposed on this is the poor state of our current understanding of the effect of fluid-dynamic shear forces. On the contrary, on the basis of experimental results (e.g., the aforementioned experiments of Freed et al., 1997) these effects seem to play a crucial role in determining the final shape and size of the specimens.

This chapter is intended to demonstrate how appropriate models and numerical simulations can be used as useful tools to discern the environmental factors affecting the surface kinetics of growing biological tissues (mechanisms of absorption of nutrients available in the culture medium and subsequent generation of new cells and production of extracellular matrix) as well as their relative importance in determining growth. Of course, the attention is focused on fluid-dynamic aspects (in particular on CFD moving-boundary methods). The potentialities of the proposed models and methods are illustrated by means of a prototype application designed to reproduce the landmark experiments of Freed et al. (1997) as well as to draw information about the intrinsic mechanisms of growth.

Working engineering CFD models and generalized functional correlations are discussed, while providing a framework for deeper understanding and effective treatment of relevant phenomena encountered in the new field of tissue engineering.

7.2. The rotating vessel: how it "simulates" microgravity conditions

This device was developed by NASA within the framework of the ideas and concepts originally pointed out in the pioneering studies of Briegleb (see, e.g., Briegleb, 1988, fast-rotating clinostat-technique). In as early as the 1960s Briegleb recognized the need to study the influence of weightlessness on living cells. As access to real weightlessness always was very limited, Briegleb and coworkers (Briegleb, 1988; Briegleb, 1992; Hemmersbach-Krause et al., 1993) were the first to use a fast-rotating clinostat for research with single mammalian cells and unicellular organisms in functional weightlessness. In his group, two fast rotating clinostat microscopes (CLIMIs) were constructed that were used by many scientists (see also Cogoli, 1992; Todd and Gruener, 1992; Kordyum, 1994).

As anticipated, cells derived from diverse tissue sources lose their specialized features and dedifferentiate when grown under traditional two-dimensional cell culture conditions. Over the years, suspension culture has become the most popular method of preventing this problem. The rich and diverse array of available culture vessels for suspension culture makes selection of specific culture devices astonishing (see, in particular, Hammond and Hammond, 2001 and the references therein).

Several recent reviews have summarized the use of suspension culture (Helmrich and Barnes, 1998), scaleup of suspension culture for industrial use (Mather, 1998), application to anchorage-dependent cells by the use of microcarriers (Gao et al., 1997a; Pollack et al., 2000), and automation of roller-bottle forms of suspension culture (Kunitake et al., 1997).

NASA investigators brought together the various elements of suspension culture (continuous sedimentation of particles through culture medium, suspension of cells and microcarriers without inducing turbulence, a rotating-wall vessel, a coaxial oxygenator to allow gas exchange by dissolved gases without undesired bubbles) for the first time. This led to an efficient (but simple) design for suspension culture vessels embodied in the rotating-wall perfused vessel, a horizontally rotating cylindrical culture vessel with a coaxial tubular oxygenator (Fig. 7.1).

These vessels have characteristic features that determine their utility. First, fluid flow is near solid body at most operating conditions. If the inner and outer cylinders of the rotating-wall vessel rotate at the same angular velocity (rpm), then the (radial) laminar flow fluid velocity gradient is minimized. Secondly, the culture medium is gently mixed by rotation, avoiding the necessity for stirring vanes. The mixing is the result of a secondary flow pattern induced by particle sedimentation through the fluid media (on the ground conditions) or by laminar flows established when differential rotation rates are chosen for the vessel components (microgravity conditions).

The rotating-wall perfused vessel was invented to grow cells in space, with the additional benefit of working in Earth's gravity. It tries to simulate the weightless environment of space on Earth by putting biological constructs in a growth medium that constantly rotates and keeps them in endless free-fall. It has applications on the ground as a simulator of microgravity since it provides some of the physical operating conditions commonly associated with microgravity, such as randomization of the gravity vector and allowing aggregates to fall at terminal velocity.

Of course, the rotating vessel does not really cancel gravity, but ideally maintains continual free-fall conditions similar to that experienced by astronauts in the microgravity of space. Earth-based bioreactor cultures typically maintain cell growth for at least 60 days. On Earth, a sample then becomes so large (about 1 cm) that it is no longer suspended. In long space missions, large-sample growth in bioreactors can be studied in order to understand the extended growth.

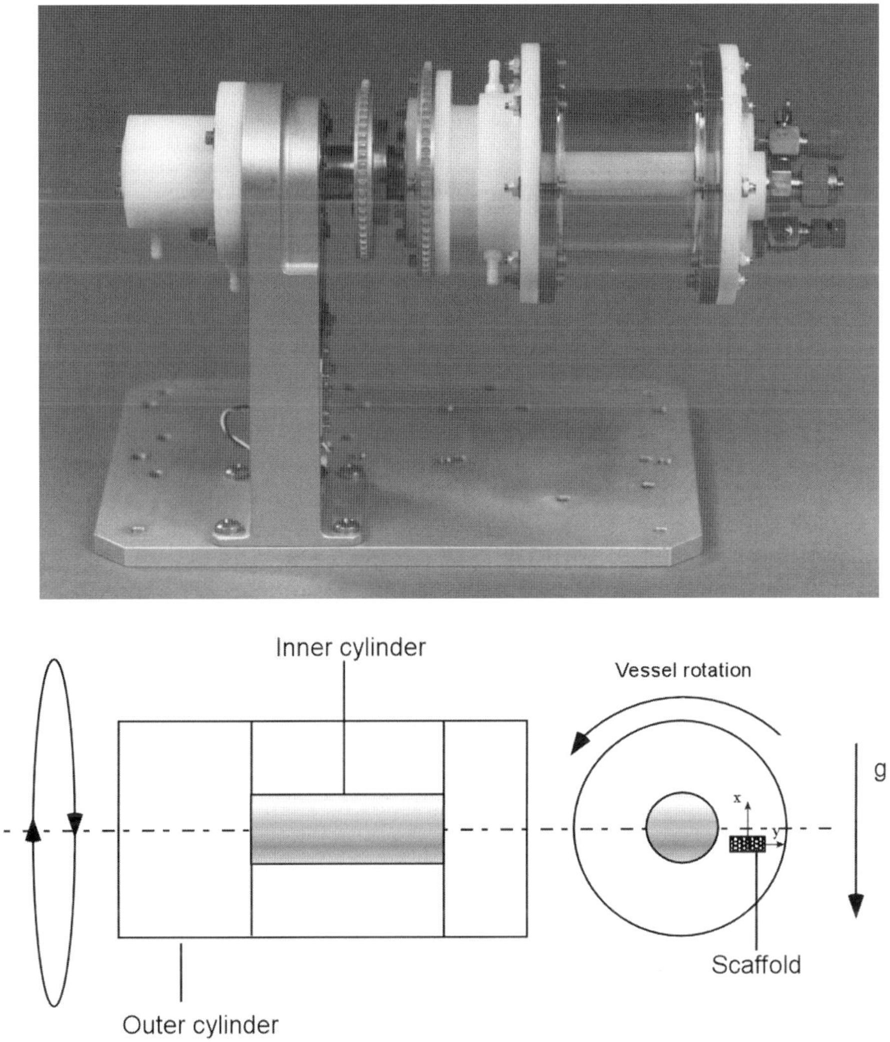

Fig. 7.1 The rotating bioreactor (on the ground conditions).

Owing to the different technique used to guarantee mixing of nutrients in the feeding solution (dynamically seeding of constructs through the liquid posed in solid-body rotation on Earth; differential rotation of the inner and outer vessel walls in space) the structure of the flow field completely changes according to the environment (normal *g* or microgravity conditions).

Within the context of the present chapter attention is focused on the on the ground conditions since most of the available experimental data for tissue engineering come from on the ground experimentation. To optimize suspension culture, the aforementioned aspects need to be expressed in simple mathematical relationships, quantitated, and controlled.

7.3. Scaffolds, microcarriers, and terminal velocity

The RWPV approach involves seeding cells onto 3D polymer scaffolds. The cell/scaffold constructs are placed in the rotating bioreactor that supplies the cells with nutrients and gases and removes wastes. The scaffold induces cell differentiation and degrades at a defined rate (it slowly biodegrades as the cells develop into a full tissue), whereas the bioreactor maintains controlled in vitro culture conditions that permit tissue growth and development.

Concentration gradients within the bioreactor are minimized by the convection associated with gravitational construct settling during solid-body rotation of the bioreactor in unit gravity. The vessel is rotated as a solid body in a horizontal plane around its central axis. The rotation rate can be adjusted so as to suspend each tissue construct about a fixed position within the vessel as viewed by an external observer (Obradovic et al., 2000). This condition guarantees dynamic equilibrium of the operative forces gravity, buoyancy and drag, while maintaining each tissue construct in a state of continuous free-fall. In other words, a tissue construct falling through the culture media in the RWPV is fed as it effectively sieves through the nutrient media.

On Earth the settling of diskoid scaffolds tends to align their flat circular areas perpendicular to the direction of motion (Freed et al., 1997). In order to minimize shear and turbulence, usually the terminal velocity of the constructs is minimized by choosing scaffolds and culture media as close in density as possible.

Many data on the fluid-dynamic environment in rotating-wall perfused vessel bioreactors can be found in the excellent analyses of Kleis et al. (1990), Wolf and Schwartz (1991, 1992), Gao et al. (1997a, b), Meaney et al. (1998), Begley and Kleis (2000), Pollack et al. (2000), Hammond and Hammond (2001), Coimbra and Kobayashi (2002).

These studies mainly concern the case of small aggregates, however, which have motion and dynamics that are quite different with respect to the large constructs cultivated by Freed et al. (1997). Many investigators make cell aggregates rotate around the core of the vessel instead of hovering them at fixed points. For the sake of completeness, in fact, it should be pointed out that particles cultivated in suspension culture generally do not maintain an approximately steady position within the vessel. Rather they follow almost circular trajectories. Also, under these conditions some equilibrium positions would exist within the reactor (these special points behave as attractors for the particles). In this case, however, tangential Coriolis-induced motion is significant and leads to spiraling of particles in the fluid stream with respect to a laboratory reference frame. Owing to this effect small particles tend to fall on large-radius orbits, which periodically strike the vessel wall.

In the case of relatively large cell aggregates and/or large scaffolds the behavior may be different, as demonstrated by Freed et al. (1997). For some set of conditions the rotation rate can be adjusted so as to let each tissue construct "hover" about a stationary position with respect to a laboratory reference system even if small oscillations with respect to this position and impact with walls may occasionally occur. In this case, many (quasistable) equilibrium positions are possible between the central oxygenator and the lowest portion of the wall (see Fig. 7.1). The prefix "quasi" is used since the constructs actually circulate in small loops around the aforementioned stationary fixed positions, due to the characteristic hydrodynamic patterns of disk settling. Gao et al. (1997a) noted that the suspension of the scaffolds has a periodic nature, but, because the oscillations of the relative velocity are slight, (in agreement with the present assumptions) they concluded that this value may be assumed to be almost constant.

For these reasons, for simulating the experiments of Freed et al. (1997) and Obradovic et al. (2000), a simplified model is assumed herein, based on the results of Clift et al. (1978). In the range $1 < Re < 10^3$ the terminal velocity U of a cylinder having diameter d and thickness L (aspect ratio $E = L/d$), free-falling with its axis of symmetry parallel to the direction of motion (and of gravity) through a liquid of kinematic viscosity ν, can be computed according to the equations

$$C_{\mathrm{D}} Re^2 = \frac{2g\Delta\rho E d^3}{\rho\nu^2} \tag{7.1a}$$

$$C_{\mathrm{D}} = \frac{64}{\pi Re}\left(1 + 0.138\, Re^{0.792}\right) \tag{7.1b}$$

where $\Delta\rho$ is the density gradient between the cylindrical body and the surrounding fluid and $Re = Ud/\nu$ is the diameter-based Reynolds number associated with the dynamical construct sedimentation. These equations provide a simple and efficient method for computation of the terminal velocity. Of course, the direction of motion is not exactly parallel to the direction of gravity as assumed by Eqs. (7.1). Generally, the terminal-velocity vector has a component along the direction of gravity and also a perpendicular component. The latter, however, is small and for this reason it is not taken into account in the present computations, even if it could explain some weak deviations with respect to the axisymmetric shape that sometimes characterize the specimen (see Section 7.7 and discussion therein).

Under these assumptions, the mechanism of equilibrium is very simple: the net weight of the tissue is balanced by the viscous resistance and the flow around the construct is axisymmetric (Eqs. (7.1)). According to Galileo's invariance principle, this situation corresponds (see also Section 7.7) to the specimen at rest as seen in a laboratory frame and the fluid moving upward in the direction opposite to the gravity force with velocity equal to U.

Other effects such as Coriolis forces can be considered negligible for the dynamics of the scaffold/construct since the major direct determinants of the terminal velocity in this case are gravity and the viscous resistance (see Hammond and Hammond, 2001).

In the experiments of Freed et al. (1997) and Obradovic et al. (2000) cartilage was used as a model musculoskeletal tissue. Cartilage was selected because of its resilience and low metabolic requirements. Their method was based on cartilage cells (chondrocytes) and biodegradable polyglycolic acid (PGA) scaffolds having a cylindrical shape. Chondrocytes were isolated from bovine calf articular cartilage and seeded onto the fibrous, biodegradable polyglycolic acid scaffolds (these disk-like structures mimic the body's internal environment). PGA scaffolds (the same material used to make Dexon absorbable surgical sutures) were 0.5 cm diameter × 0.2 cm thick discs formed as a 97% porous mesh of 13-μm diameter fibers. Culture medium consisted of DMEM with 4.5 g/l glucose and other substances (see Freed et al. (1997) for further details on the composition of the culture medium).

For this case the density difference between the construct and the feeding solution was $\Delta\rho = 4.9 \times 10^{-2}\ \text{g/cm}^3 \rightarrow 2g\Delta\rho E d^3/\rho\nu^2 = 7.4 \times 10^4$. Equations (7.1), solved using a Newton-Raphson method, give

$$Re = \frac{Ud}{\nu} = 290 \rightarrow U = 4.64\ \text{cm/s};$$

this velocity is in agreement with the average vessel rotation rate (≅30 rpm, see Obradovic et al., 2000) that was used by the investigators to maintain each tissue construct floating at an approximately steady position within the vessel. The bioreactor was configured, in fact, as the annular space between a 5.75-cm diameter polycarbonate outer cylinder and a 2-cm diameter hollow inner cylinder with a porous wall. This corresponds to an average velocity of the fluid ≅6 cm/s that is in reasonable agreement with the terminal velocity computed according to Eqs. (7.1).

7.4. OTGVOF – The Organic Tissue Growth Volume of Fraction Method

It is worthwhile to stress how in the (aforementioned) analyses available in the literature, the morphological evolution of the particle moving in the bioreactor was not taken into account. The macroscopic motion of the particle was studied but how this motion may affect the shape evolution of the growing specimen, and how in turn the changing shape may affect the flow around the construct, was not investigated (on the contrary these aspects are the main goal of the present analysis).

7.4.1 *A general formulation of the surface kinetics*

As widely discussed in Chapter 3, VOF methods have enjoyed a widespread use for the simulation of typical problems associated with gas–liquid or liquid–liquid systems where the surface-tension effects play a critical role in determining the shape of the fluid/fluid interface and/or its motion (see, in particular, Section 3.5). The present section shows how the fundamental principles at the basis of the aforementioned methods can lead to a new class of techniques specifically devoted to tissue engineering.

In practice, the OTGVOF method proposed herein (see also Lappa, 2003d) exhibits notable analogies with the corresponding method introduced in Chapter 5 for the morphological evolution of macromolecular crystals (in this regard see also the discussion in Section 5.7).

It, in fact, accounts for the organic solid mass stored in the generic computational cell by assigning an appropriate value of ϕ to each mesh point ($\phi = 1$ for biological tissue, $\phi = 0$ for feeding solution, and $0 < \phi < 1$ for an interfacial cell). The key element for the OTGVOF method is its technique for adjourning ϕ. Upon changing phase, the ϕ-value of the cell is adjusted to account for mass absorption, this adjustment being reflected in the concentration distribution of nutrients in the liquid phase as a sink. The modeling of these phenomena leads to the introduction of a group of differential equations, strictly related, from a mathematical point of view, to the "kinetic conditions" used to model mass transfer at the tissue surface.

Growth proceeds by the incorporation of growth units (atoms, molecules or small aggregates) from the feeding solution to the biological construct. Nutrients available in the culture medium are incorporated and converted into the main tissue components. This incorporation produces a concentration-depletion zone around the specimen (this mechanism is somewhat similar to that already elucidated in Chapters 5 and 6 for macromolecular crystals growing from solution). The size and shape of the depletion zone are controlled by the coupled transport of the feeding species in solution to the growing surface and the processes allowing these species to be incorporated into the tissue matrix.

In the case of protein crystals (see Chapter 5), in particular, it is well known that the growth rate (i.e. the surface growth kinetics) depends on the steepness of the feeding concentration gradient and that this gradient, in turn, depends on the degree of supersaturation. By analogy with these models it is reasonable to assume that the concentration of nutrient at the tissue surface must satisfy a condition such as:

$$D\left.\frac{\partial C}{\partial n}\right|_i = \lambda C_i \tag{7.2}$$

where C_i is the concentration of the nutrient (in the experiments of Freed et al. (1997) it is glucose) at the construct/liquid interface, D is the related diffusion coefficient,

λ is a kinetic coefficient having the dimensions of a velocity (e.g., cm/s) and n denotes the direction perpendicular to the sample surface.

Equation (7.2) simply follows from Eq. (5.1) by replacing the role played in the case of organic crystals by the degree of supersaturation with the local value of concentration of nutrient in the liquid phase. If $C_i = 0$, in fact, no net increase in the proportion of solid phase can accrue since nutrients are not available (the tissue does not grow). On the contrary, if $C_i > 0$, tissue growth occurs and the rapidity of the phenomena is driven by the value of the kinetic coefficient λ.

The growth rate (see, e.g., Pusey et al. (1986)) can be defined as:

$$\theta = \frac{D}{\rho_S} \frac{\partial C}{\partial n}\bigg|_i \tag{7.3}$$

where ρ_S is the total density of the solution and C satisfies Eq. (7.2).

It is very important to point out how the above model, introduced in the present chapter by analogy with models for organic crystal growth, ignores the fact that, according to biomechanical growth laws proposed in the case of soft tissues (see, e.g., Fung, 1990; Taber, 1998a,b), the rate of growth must depend on the stresses. For instance, the fluid shear stresses generated by blood flow in the vasculature can profoundly influence the phenotype of the endothelium by regulating the activity of certain flow-sensitive proteins (enzymes, for example) as well as by modulating gene expression (see, e.g., Topper and Gimbrone, 1999). This physical force can induce changes in cell metabolism and function. In particular, for the reference case considered in this book (the experiments of Freed et al., 1997) the stress environment can elicit a physiologic response from the cells that are the building blocks of the construct, causing them to undergo internal division and produce (ECM) extracellular matrix (as shown by direct experimental observation).

To model these aspects the kinetic condition should be rewritten as:

$$D \frac{\partial C}{\partial n}\bigg|_i = \lambda C_i f(\tau) \tag{7.4}$$

where f is a function (depending on the type of tissue) of the fluid-dynamic shear stress τ at the tissue/liquid interface (f is always positive since, contrary to the case of protein crystals, the tissue cannot redissolve and consequently the second member of Eq. (7.4) cannot be negative).

This condition finally takes into account (from a theoretical point of view) the main aspects of the growth behavior, i.e. the availability of nutrients (C_i), the slow surface kinetics (λ) and the intriguing effect of surface shear stress (τ). The explicit form of the function f for the case of cartilage construct will be discussed in Section 7.6. In the next sections the generic form of Eq. (7.4) is maintained to obtain a method that is not related to a fixed tissue in particular.

7.4.2 Governing field equations

In the presence of convection, the flow is governed by the continuity, Navier–Stokes and species equations, that in nondimensional conservative form read:

$$\underline{\nabla}\ \underline{V} = 0 \tag{7.5}$$

$$\frac{\partial \underline{V}}{\partial t} = -\underline{\nabla} p - \underline{\nabla} \cdot [\underline{V}\,\underline{V}] + S_c \nabla^2 \underline{V} - S_c \frac{1}{\eta} \underline{V} \tag{7.6}$$

$$\frac{\partial C}{\partial t} = \left[-\underline{\nabla} \cdot (\underline{V} C) + \nabla^2 C\right] \tag{7.7}$$

where $S_c = \nu/D$ is the Schmidt number (ν is the kinematic viscosity of the culture liquid). The nondimensional form of the equations results from scaling the lengths by a reference distance (e.g., $L = 1$ cm), the time by L^2/D, velocity $\underline{V}$ and pressure p by D/L and $\rho_S D^2/L^2$, respectively; the initial value of nutrient is $C_{(o)}$. Note that concentrations are not posed in nondimensional form (g/cm^3).

Like the case of macromolecular crystals, assumptions invoked in the development of equations for this continuum model include: laminar flow, Newtonian behavior of the phases (this implies that the constructs should be treated as highly viscous fluids), constant phase densities.

Moreover, the tissue is assumed to be nondeforming and free of internal stress, while the multiphase region (region where nutrients are absorbed and increase of mass occurs) is viewed as a porous material characterized by an isotropic permeability η. The term $-S_c\underline{V}/\eta$ in Eq. (7.6) is the Darcy term added to the momentum equation to damp convection in the solid phase. In the present analysis permeability is assumed to vary according to the Carman–Kozeny equation (Bennon and Incropera, 1987a,b) $\eta = (1 - \phi + \varepsilon)^3/(\phi + \varepsilon)^2$, $\varepsilon = 10^{-5}$. The effect of η has been exhaustively described in Chapter 4 (Section 4.2) and in Chapter 5 (Section 5.6).

Since the momentum and species equations are valid throughout the entire domain, explicit consideration need not be given to boundaries between solid, multiphase and liquid regions.

7.4.3 Phase-field equation

On the surface of the tissue ($|\underline{\nabla}\phi| \neq 0$, $0 < \phi < 1$), nutrient concentration must satisfy the kinetic condition that in nondimensional form reads:

$$\left.\frac{\partial C}{\partial n}\right|_i = \tilde{\lambda} C_i \tilde{f}(\tilde{\tau}) \tag{7.8}$$

where $\tilde{\lambda} = \lambda L/D$, $\tilde{\tau} = ((\partial u/\partial y) + (\partial v/\partial x))$, u and v being the nondimensional velocity components along x and y, respectively.

In Eq. (7.8) the concentration gradient can be computed as:

$$\frac{\partial C}{\partial n} = \underline{\nabla} C \cdot \underline{\hat{n}} \tag{7.9}$$

where $\underline{\hat{n}}$ is the unit vector perpendicular to the construct/liquid interface pointing into the culture medium:

$$\underline{\hat{n}} = -\frac{\underline{\nabla}\phi}{|\underline{\nabla}\phi|} = (\alpha, \beta) \tag{7.10}$$

$$\alpha = -\frac{\partial \phi}{\partial x} \Big/ \sqrt{\left(\frac{\partial \phi}{\partial x}\right)^2 + \left(\frac{\partial \phi}{\partial y}\right)^2}, \qquad \beta = -\frac{\partial \phi}{\partial y} \Big/ \sqrt{\left(\frac{\partial \phi}{\partial x}\right)^2 + \left(\frac{\partial \phi}{\partial y}\right)^2} \tag{7.11}$$

since $\partial C/\partial n = \alpha \partial C/\partial x + \beta \partial C/\partial y$ (hereafter the subscript "i" is omitted), Eq. (7.8) can be rewritten as

$$\alpha \frac{\partial C}{\partial x} + \beta \frac{\partial C}{\partial y} = \tilde{\lambda} C \tilde{f}(\tilde{\tau}) \tag{7.12}$$

$\tilde{\lambda} C \tilde{f}(\tilde{\tau})$ represents the mass-exchange flux between solid and liquid phase (i.e. tissue and feeding solution). The mass (tissue matrix coming from incorporation and conversion of nutrients) stored in computational cells that are undergoing phase change can be computed according to the equation:

$$\frac{\partial M}{\partial t} = \frac{\lambda L^4}{D} C \tilde{f}(\tilde{\tau})\, \mathrm{d}s \tag{7.13}$$

where ds is the "reconstructed" portion of the tissue surface (by definition perpendicular to the interface normal vector $\underline{\hat{n}}$) "bounded" by the frontier of the control volume (computational cell) located astride the tissue surface.

$$\mathrm{d}v|_{\mathrm{stored}} = \frac{1}{L^3} \frac{M}{\sum_k \rho_k} \tag{7.14}$$

where ρ_j are the partial densities of the main tissue components; correspondingly:

$$\phi = \frac{\mathrm{d}v|_{\mathrm{stored}}}{\mathrm{d}v} \tag{7.15}$$

where dv is the volume of the computational cell.

Therefore the phase-field equation reads

$$\frac{\partial \phi}{\partial t} = 0, \quad \text{if } |\underline{\nabla}\phi| = 0$$

$$\frac{\partial \phi}{\partial t} = \frac{\tilde{\lambda} C \tilde{f}(\tilde{\tau})\, \mathrm{d}s}{\sum_k \rho_k\, \mathrm{d}v}, \quad \text{if } |\underline{\nabla}\phi| \neq 0, \quad 0 < \phi < 1 \tag{7.16}$$

with C satisfying Eq. (7.12).

From mass-balance, the solution for the normal velocity at the interface is:

$$\underline{V} \cdot \hat{\underline{n}} = (\rho_{\mathrm{T}}/\rho_{\mathrm{S}})\theta^* \quad \text{if } |\underline{\nabla}\phi| \neq 0, \quad 0 < \phi < 1 \tag{7.17a}$$

where ρ_{S} and ρ_{T} are the total density of the solution and the total density of the specimen, respectively, and the nondimensional growth rate (by analogy with Eq. (5.17b)) can be computed as:

$$\theta^* = \left(\frac{1}{\rho_{\mathrm{S}}}\right)\frac{\partial C}{\partial n} = \tilde{\lambda} C \tilde{f}(\tilde{\tau})/\rho_{\mathrm{S}} \tag{7.17b}$$

with C (glucose concentration) satisfying Eq. (7.12).

Equations (7.12), (7.16), and (7.17) behave as "moving-boundary conditions," their solution being strictly associated with the computational check on the value of ϕ and its gradient. For instance, in the case of Eqs. (7.16), the phase variable ϕ is updated (i.e. $\partial\phi/\partial t \neq 0$) only where $|\underline{\nabla}\phi| \neq 0$ and $0 < \phi < 1$, i.e. close to the tissue/solution interface where tissue enlargement occurs due to the aforementioned "internal cell division" and "production of extracellular matrix." On the contrary, ϕ does not change (i.e. $\partial\phi/\partial t = 0$) far from the surface. Note that, typically, for organic tissue growth, the normal velocity at the interface is very small with respect to the terminal velocity, and for this reason it could be neglected.

7.4.4 *Discretization*

Equations (7.5–7.7) subjected to the initial and boundary conditions can be solved numerically in primitive variables by a control-volume method. In the following, for educational aims and clarity, forward differences in time are used to discretize the partial differential equations, obtaining (n superscript indicates time step):

$$\underline{V}^{n+1} = \underline{V}^n + \Delta t\left[-\underline{\nabla}\cdot[\underline{V}\,\underline{V}] + S_{\mathrm{c}}\nabla^2\underline{V}\right]^n - \Delta t\frac{S_{\mathrm{c}}}{\eta}\underline{V}^{n+1} - \Delta t\,\underline{\nabla}p^n \tag{7.18}$$

$$C^{n+1} = C^n + \Delta t\left[-\underline{\nabla}\cdot[\underline{V}C] + \nabla^2 C\right]^n \tag{7.19}$$

Like the OCGVOF method, the orientation of the surface of the crystal is used here in determining the face fluxes for the computation of C at the crystal surface (Eq. (7.12)). The interface orientation depends on the direction of the volume-fraction gradient of the phase within the cell, and that of the neighboring cell (or cells) sharing the face in question.

The unit vector $\underline{\hat{n}}$ results from the gradient of a smoothed phase field $\hat{\phi}$, where the transition from one phase to the other takes place continuously over several cells (4 or 5). The smoothed phase field $\hat{\phi}$ is obtained by convolution of the unsmoothed field ϕ with an interpolation function (see, e.g., Eq. (3.22)).

Depending on the interface's orientation, concentration gradients must be discretized by forward or backward schemes. For this reason Eq. (7.12) in discretized form reads (note that the i and j subscripts indicate the relative position along x and y, respectively, of the nodal values of C):

$$\alpha > 0, \quad \beta > 0: \qquad C_{i,j}^{n+1} = \frac{\left[\alpha C_{i+1,j}^{n+1}/\Delta x + \beta C_{i,j+1}^{n+1}/\Delta y\right]}{\left[\tilde{\lambda}\tilde{f}(\tilde{\tau}) + \alpha/\Delta x + \beta/\Delta y\right]} \tag{7.20a}$$

$$\alpha < 0, \quad \beta > 0: \qquad C_{i,j}^{n+1} = \frac{\left[-\alpha C_{i-1,j}^{n+1}/\Delta x + \beta C_{i,j+1}^{n+1}/\Delta y\right]}{\left[\tilde{\lambda}\tilde{f}(\tilde{\tau}) - \alpha/\Delta x + \beta/\Delta y\right]} \tag{7.20b}$$

$$\alpha > 0, \quad \beta < 0: \qquad C_{i,j}^{n+1} = \frac{\left[\alpha C_{i+1,j}^{n+1}/\Delta x - \beta C_{i,j-1}^{n+1}/\Delta y\right]}{\left[\tilde{\lambda}\tilde{f}(\tilde{\tau}) + \alpha/\Delta x - \beta/\Delta y\right]} \tag{7.20c}$$

$$\alpha < 0, \quad \beta < 0: \qquad C_{i,j}^{n+1} = \frac{\left[-\alpha C_{i-1,j}^{n+1}/\Delta x - \beta C_{i,j-1}^{n+1}/\Delta y\right]}{\left[\tilde{\lambda}\tilde{f}(\tilde{\tau}) - \alpha/\Delta x - \beta/\Delta y\right]} \tag{7.20d}$$

$C_{i,j}^{n+1}$ is computed from Eq. (7.20), then the phase variable can be updated using Eq. (7.16):

$$\phi_{i,j}^{n+1} = \phi_{i,j}^{n} + \Delta t \frac{\tilde{\lambda}\tilde{f}(\tilde{\tau})(C_{i,j}^{n+1})\Delta s}{\left(\sum_k \rho_k\right)\Delta v} \tag{7.21}$$

According to Eqs. (7.20) and (7.21), if the nutrient concentration in the culture medium is locally depleted, correspondingly, the solid mass stored in the computational cell grows and the phase variable is increased. These phenomena are driven by the surface-kinetic condition and by the shear-stress distribution, i.e. the existing deposit grows if the nutrient concentration is not zero and the mass exchange is proportional to the kinetic coefficient and to the local value of the fluid shear stress.

In Eq. (7.21) Δs is the "reconstructed" portion of the solid wall. The determination of Δs requires a well-defined "interface-reconstruction" technique (the shape of the construct for a fixed time is not known a priori and must be determined as part of the solution, see, e.g., the nonconnecting straight lines PLIC technique of Gueyffier et al. (1999), also discussed in Chapter 3 and in Section 5.6).

The solution procedure is summarized in the flow scheme below for additional clarity; it proceeds in four major stages:

1. Solution of Eq. (7.18) according to the SMAC method (the necessary details can be found in Section 2.8): $\underline{V}^{n+1}$ is computed as a function of $\underline{V}^{n}$;
2. Solution of the species equation (Eq. (7.19)) in the region where $\phi = 0$: C^{n+1} is computed as a function of the corresponding distribution C^{n} and of $\underline{V}^{n}$;
3. Updating of the local values of C at the tissue/culture-medium interface according to Eqs. (7.20). This stage accounts for glucose depletion (due to incorporation into the tissue) at time $n+1$;
4. Adjournment of the phase-field variable distribution. The exact location of the interface is reconstructed according to the PLIC scheme. Then the phase-field variable is updated according to its distribution at the previous time step (n) and on the basis of the newly computed values of concentration C^{n+1} at the construct interface (see Eq. (7.21)). This final stage accounts for the increase of size of the specimen at time $n+1$.

7.5. The OTGLSET – Organic Tissue Growth Level-Set Method

By analogy with the case already discussed for the macromolecular crystallization, in the case of tissue engineering it is also possible to introduce a level-set technique in a simple and natural manner. As also discussed in Section 5.8, this method, first introduced by Osher and Sethian (1998), is conceptually similar to a phase-field model in that the tissue/liquid interface Γ is represented as the zero contour of a level-set function, $\varphi(r, t)$, which has its own equation of motion. A possible solution procedure is summarized in the flow scheme below; it proceeds in five major stages (Lappa, 2004f): (i) advancing the interface, (ii) reinitializing the level-set function to be a signed distance function, (iii) solving for the new concentration and velocity fields in the culture medium (Eqs. (7.5–7.7)), (iv) adjourning of the local values of C at the tissue/feeding-solution interface (Eq. (7.12), this stage accounts for solute depletion), (v) computing the surface growth-rate distribution. The growth velocity is not directly imposed but it results from internal conditions related to the intrinsic mechanisms of growth of the tissue. From mass balance, the nondimensional velocity (q_n) of the advancing tissue interface reads:

$$q_n = \frac{\rho_{\mathrm{T}}}{(\rho_{\mathrm{S}})^2}\tilde{\lambda}\tilde{f}(\tilde{\tau})C \tag{7.22}$$

The motion of the interface is accomplished by the solution of the equation:

$$\frac{\partial \varphi}{\partial t} + F|\underline{\nabla}\varphi| = 0 \tag{7.23}$$

that advects the contours of φ along the directions normal to the interface according to the velocity field F, which varies in space. F is constructed to be an extension of the interface velocity, q_n, such that $F = q_n$ for points on the interface and the lines of constant F are normal to the interface.

After solving Eq. (7.23) for one time step, the level-set function will no longer be equal to the distance from the interface. It is necessary to *reinitialize* φ to be a signed distance function. This step is accomplished by solving

$$\frac{\partial \psi}{\partial \tau} + \text{sqn}(\varphi)\left[|\underline{\nabla}\psi| - 1\right] = 0 \tag{7.24}$$

$$\psi(\underline{r}, 0) = \varphi(\underline{r}, t)$$

The reader will find additional theoretical background and explanations about the level-set technique in Section 5.8.

7.6. Mathematical formulation of the kinetics of cartilage tissue

For practical applications the details of the specific tissue under consideration have to be introduced in the model (see Eqs. (7.14) and (7.16)).

As anticipated, functional cartilaginous constructs for scientific research and eventual tissue repair were cultivated by Freed et al. (1997) in bioreactors starting from chondrocytes immobilized on polymeric scaffolds and fed by glucose. The scaffolds gradually degraded as the cells regenerated a cartilaginous tissue matrix consisting of water (soft tissues consist, in fact, primarily of various cell types, an extracellular matrix, and abundant water), glycosaminoglycan GAG and type II collagen.

GAG and collagen were found to be the main cartilage components by Freed et al. (1997), i.e.

$$\sum_k \rho_k \cong (\rho_{\text{GAG}} + \rho_{\text{collagen}})$$

with the partial densities being given by $\rho_{\text{GAG}} = 6 \times 10^{-2}\,\text{g/cm}^3$ and $\rho_{\text{collagen}} = 2.7 \times 10^{-2}\,\text{g/cm}^3$, respectively.

As previously pointed out, nutrients available in the culture medium are incorporated and converted into the main tissue components. With regard to this aspect it is important to stress how, though several experimental studies have shown that the growth rates are sensitive to a very large number of parameters, no information has been provided about the functional dependencies for the aforementioned kinetics (the parameter λ and the function f are unknown even if well-known tissues are considered). This problem can be overcome, as briefly illustrated below, by means of a particular iterative procedure (Lappa, 2004g):

Numerical simulations can be carried out on the basis of mathematical models of the growth process similar to those usually employed for macromolecular crystals (that grow due to the slow addition of solute molecules to the surface, e.g., Eq. (7.4)). The fluid dynamics occurring in the surrounding culture medium can be taken into account according to the solution of the Navier–Stokes equations.

Full understanding of the role played by each single factor (availability of nutrients, kinetic coefficients and also fluid-dynamic shear forces) and the relative "importance" with respect to each other in affecting the tissue growth process can be achieved by computational extensive parametric investigation.

During the real experiments the different effects cannot be separated; they act all together and the related mutual interference can be very complex. This limitation, however, can be removed in the virtual environment provided by numerical simulation.

Some parameters can be systematically taken into account while "switching off" the effect of the remaining ones. Computer graphics (animation) allows the effect and "the intensity" of each parameter to be discerned through direct visualization of the morphological evolution of the tissue shape. Finally, in the light of these findings, mathematical models taking into account all the environmental factors and their relative importance can be introduced and used to simulate the "ensemble behavior" of the system (i.e. the proposed "global" growth laws can be tested for their ability to predict the known patterns of growth, e.g., for the present case the experiments of Freed et al. (1997)).

This approach leads to the following relationships for the case of cartilage construct

$$\tilde{f} = b(\tilde{\tau})^{1/2} \rightarrow \frac{\partial C}{\partial n} = \tilde{\lambda} b(\tilde{\tau})^{1/2}(C) \tag{7.25}$$

where b is a nondimensional constant.

According to Eq. (7.25), the determined dependence of the effective flux of nutrients absorbed by the construct on the local feeding concentration at the construct surface is linear, like the case of protein crystals. These crystals, in fact, as anticipated in Section 7.4.1 and widely discussed in Chapter 5, exhibit quite a similar sensitivity on the local value of supersaturation. The aforementioned parametric investigation, however, points out that the presence of the sole term C_i (i.e. $(\partial C/\partial n)|_i = \tilde{\lambda}(C_i)$) is not sufficient to explain the observed evolution of the tissue shape in the experiments of

Obradovic et al. (2000). With such a relationship the tissue would tend to grow equally in all directions maintaining its initial shape, the concentration of nutrients around the specimen provided by the rotating vessel being almost uniform.

The presence of the term $\tilde{f}(\tilde{\tau})$ is necessary. Equation (7.25) is a novel contribution to tissue-engineering models. It provides the required theoretical formalism for the heretofore controversial and poorly understood effect of the fluid-dynamic surface stress; according to Eq. (7.25) the diffusion of momentum towards the tissue interface deeply influences the growth process; in other words there is a "direct" effect of the fluid-dynamic shear stress on the surface incorporations and conversion kinetics. It alters the manner in which nutrients are incorporated into the tissue mass and the tissue grows.

With regard to the explicit dependency of $\tilde{f}$ on $\tilde{\tau}$ (in particular the choice of the exponent 1/2), the analysis has initially shown that $\tilde{f}$ does not behave as a linear function of $\tilde{\tau}$. For this reason the general form $f \propto \tilde{\tau}^m$ has been assumed and a number of simulations have been carried out for different values of the exponent m to match the behavior observed during the experiments as well as the final shape of the tissue after 40 days of growth. Such a strategy leads to $m = 1/n$, $n = 2$ (other values do not reproduce the experiments). The parametric investigation is not shown in detail for the sake of brevity.

The term $\tilde{\tau}^{1/2}$, in particular, accounts for the physiologic response of the biological tissue to the surface stress distribution that acts to modify the internal cell division and addition of extracellular matrix mechanisms responsible for tissue enlargement and therefore the rate of absorption of nutrients.

At this stage it is worthwhile to emphasize how, despite the aforementioned analogies, there is a notable difference with respect to the case of lifeless macromolecular crystals treated in Chapter 5 where only solute transport and crystal intrinsic mechanisms of "slow addition" of molecules to the crystalline structure play a significant role (the term $\tilde{\tau}$ does not appear in the equation for the kinetics). Of course, in the case of protein crystals convective effects can also be present, but they do not enter directly the growth mechanisms. Rather they have an indirect effect on the concentration distribution around the growing crystal modifying the local values of $\partial C/\partial n$ with respect to purely diffusive conditions, as discussed in Section 5.12.

In conclusion, it is noteworthy how, according to Eq. (7.25), the growth phenomena in the case of living materials are not limited to the mere effect of the availability of nutrients. They depend on both transfer of mass (the aforementioned availability of nutrients) and momentum (surface shear stress) to the tissue surface.

7.7. A prototype application

The equations and the initial and boundary conditions are solved numerically in cylindrical coordinates. The specimen (a disk-shaped cartilage construct having an initial size 5 mm × 2 mm) is supposed to be maintained settling at an approximately steady position within the vessel (dynamic equilibrium of the operative forces, gravity, buoyancy and drag, while maintaining a state of continuous free-fall, see Sections 7.2 and 7.3). According to the experimental observations (Freed et al., 1997; Obradovic et al., 2000) the fluid filling the rotating vessel laps against the tissue whose position is held constant in time with the flat circular area perpendicular to the direction of motion.

The specimen is supposed to maintain an axisymmetric shape during the growth process, its symmetry axis being coincident with the symmetry axis of the cylindrical scaffold initially used to seed the chondrocytes. The velocity field is supposed to be uniform and parallel to the above-mentioned axis at a sufficient distance (about 1 cm) from the construct (undisturbed conditions). Its intensity there is assumed to be equal to the terminal velocity computed in Section 7.3.

A 20 mm high and 20 mm wide (radius = 10 mm) computational domain is simulated. Figure 7.2 shows the geometry of the computational domain and the boundary conditions (note that due to the axisymmetry of the model, for the sake of simplicity and brevity, only half of the generic meridian plane is shown): $y=0$ corresponds to the symmetry axis, $y=1$ to the distance from the axis at which undisturbed flow conditions prevail, $x=0$ corresponds to the inflow section (in the laboratory frame the construct holds a fixed position and the fluid moves upward

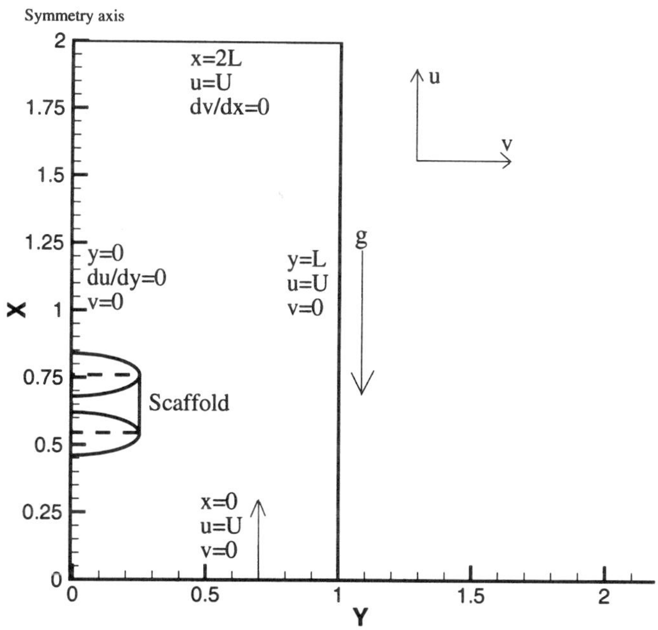

Fig. 7.2 Sketch of the computational domain and boundary conditions (U is the terminal velocity, u and v are the velocity components in axial and radial directions, respectively).

in the direction opposite to the gravity force), correspondingly $x=2$ is the outflow section.

For $y=0$ mirror boundary conditions are imposed, i.e. the radial-velocity component (v) is zero and the axial one (u) does not change across the axis ($du/dy=0$); for $y=1$ undisturbed flow conditions are imposed, i.e. $v=0$ (the undisturbed flow does not have a radial-velocity component) and $u=U$; for $x=0$ again undisturbed flow conditions are imposed; for $x=2$ (outflow section) the axial-velocity component is equal to the terminal velocity as for the inflow section (the volumetric rate of flow has to be constant throughout the system the liquid being an incompressible fluid) but a nonzero radial-velocity component may be present due to the interaction of the moving fluid with the obstruction created by the presence of the organic construct; for this reason the condition $dv/dx=0$ is imposed there (this condition guarantees that the radial velocity at the outflow section is free to change according to the behavior of the wake produced beyond the specimen).

On the basis of a grid-refinement study (not shown for the sake of brevity) a mesh with 200 points in the axial direction and 100 points in the radial direction is used. Growth is obtained from a solution with initial concentration $C_{\text{glucose(o)}}=4.5\times10^{-3}\,\text{g/cm}^3$ (the diffusion coefficient is $D_{\text{glucose}}=6.7\times10^{-6}\,\text{cm}^2/\text{s}$, the kinematic viscosity is $\nu=8\times10^{-3}\,\text{cm}^2/\text{s}$). The frontier of the domain is supposed to be at constant concentration during the growth process (i.e. $C_{\text{glucose}}=C_{\text{glucose(o)}}$).

In the experiments (see Obradovic et al., 2000) cells proliferated over the first 5–7 days. Only after cell growth ceased did the cells begin to separate themselves by synthesizing matrix components (GAG + Collagen). For these reasons the investigators focused on the period ranging between the first ten days of growth and six weeks (42 days). The same period (even if computationally very expensive) is simulated within the framework of the present prototype application.

Note that, of course (the present mathematical model and associated numerical method are original and are applied here for the first time), no data are available in the literature about the value to assign to the kinetic coefficient in Eq. (7.25).

Following the philosophy discussed in Section 7.6, this value can be determined in a very special way: several simulations can be carried out for different values of this parameter, then the effective λ can be selected out on the basis of the experimentally found rate of growth of the specimen size (the size of the construct increased at a constant rate of $70\,\mu\text{m/day}=8.1\times10^{-8}\,\text{cm/s}=8.1\,\text{Å/s}$, see Obradovic et al., 2000). Such an approach gives $\lambda=3\times10^{-6}\,\text{cm/s}$ and $b=5\times10^{-5}$.

Validation of this special procedure, of course, is necessary. In practice, it is provided by the very good agreement of numerical predictions obtained using these values with the experimentally found morphological data (this agreement is illustrated in the next pages). The two parameters λ and b are determined, as explained above, through comparison with the experimental data of Obradovic et al. (2000) dealing with the rate of increase of the tissue size. This rate is an average shape-independent parameter. On the contrary, the criterion to conclude that the model is correct is based on the morphological (shape-change) evolution of the tissue. The numerical results, in fact, are able to reproduce the morphological evolution observed by Obradovic et al. (2000) even if no data are provided by Obradovic et al. (2000) with regard to the laws and kinetics modeling the shape evolution.

The growth process is shown in Fig. 7.3. Each figure of the sequence corresponds to a different snapshot of the growth process.

A toroidal vortex roll is created behind the growing specimen. This behavior is due to the obstruction created in the fluid flow by the presence of the disk-shaped tissue with its circular area perpendicular to the direction of motion. The toroidal convection roll appears in the generic meridian plane in the form of two vortices located behind the body in the downstream direction. The two vortices are embedded in a low-velocity region hereafter referred to as "wake."

Note that the interaction of the flow entering into the computational domain from the lower boundary with the axisymmetric body embedded in the inner space of the reactor leads to two main effects. One is the aforementioned creation of a wake, the second is an effect of deceleration/acceleration of the fluid. The flow is strongly curved since it enters the computational domain directed along x and is then forced to skirt around the construct. Due to this path and its initial conditions, the fluid is initially decelerated along x up to the stagnation conditions on the tissue surface; then it is accelerated in the downstream direction due to the cross-sectional-area reduction associated with the presence of the specimen.

This structure is crucial in determining the distribution of surface shear stress and thus the growth behavior.

The simulations show that the corners and edges of the tissue are more readily supplied with solute (glucose) than the center of sides (this leads to morphological instability and to a macroscopic depression around the center of the faces, see Fig. 7.3). This is due to the pattern of the nutrient concentration field around the growing construct. As previously discussed, absorption of the solute into the tissue causes a local depletion in concentration and a solutal concentration gradient to form between the bulk solution and the growth interface. Like the case of protein crystals, the "steepness" of the gradient determines the rate of solute transport to the growth interface, the steepness being maximum around the corners (see, e.g., Fig. 7.4). Superimposed on this is the fact that a protuberance on the interface sees a higher shear stress and, according to Eqs. (7.12) and (7.16), grows faster than a depression, which sees a lower shear stress.

This behavior is evident in Fig. 7.5a where the growth-habit simulation and the progression of cartilaginous matrix deposition are shown. Note that, as anticipated, the onset of morphological instabilities and the existence of "depressions" around the center of the faces of growing tissues (shown by the present numerical results) is in very good agreement with the experimental results of Freed et al. (1997) and Obradovic et al. (2000) (see Fig. 7.5b). Figures 7.5, in particular, allow the reader to compare the numerical results with the experimental ones (Obradovic et al., 2000) by

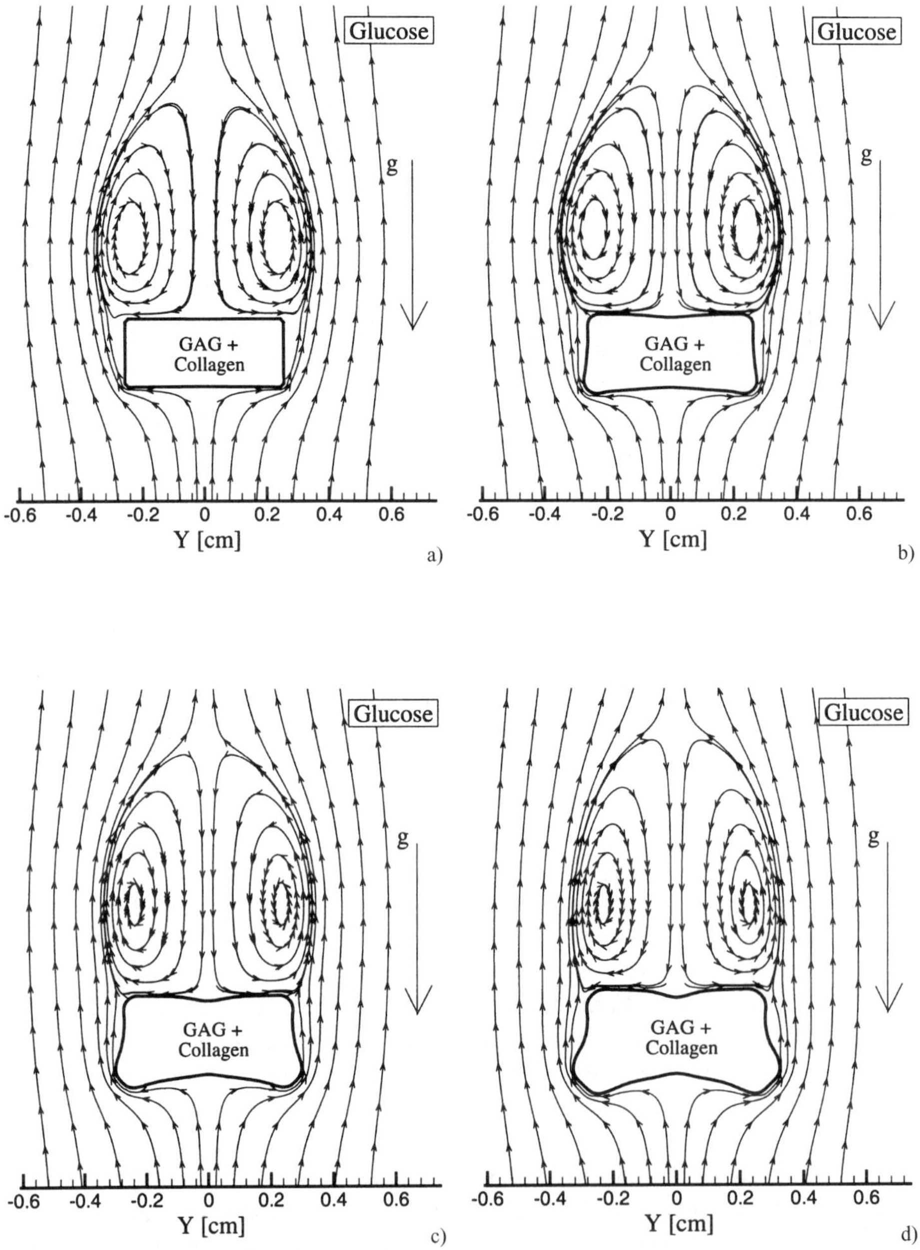

Fig. 7.3 Snapshots of growing tissue and surrounding velocity field: (a) $t = 10$ days, (b) $t = 16$ days, (c) $t = 22$ days, (d) $t = 28$ days, (e) $t = 34$ days, (f) $t = 40$ days. With regard to the experiments of Obradovic et al. (2000), note that frame (a) and frame (f) correspond to their initial condition (scaffold) and to the construct cultured for six weeks, respectively.

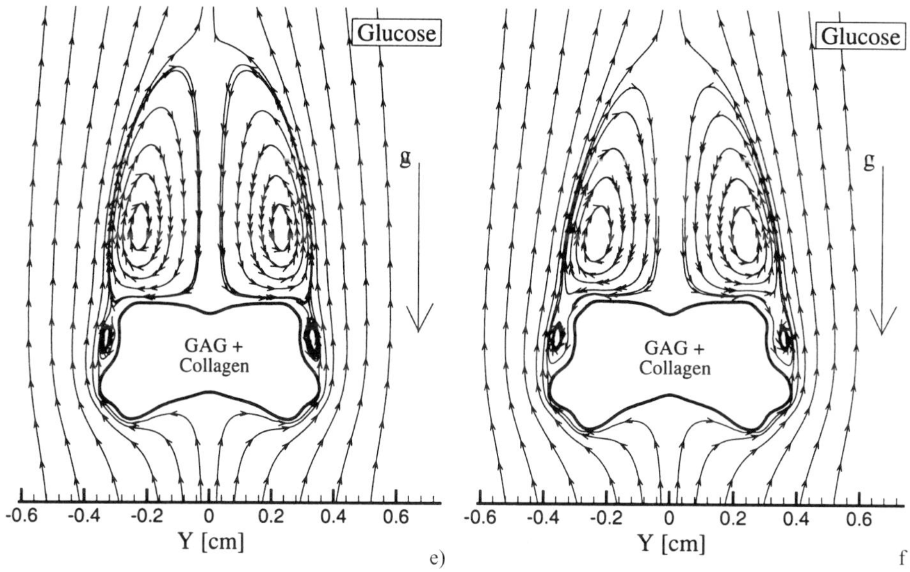

Fig. 7.3 Continued.

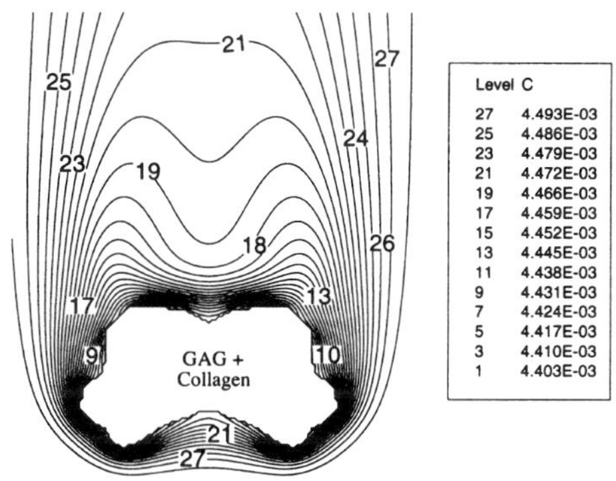

Fig. 7.4 Glucose-concentration distribution ($t = 40$ days).

highlighting some characteristic points along the border of the construct (please consider the correspondence between the points A–F in Figs. 7.5a and b).

It is worthwhile to point out that the "depth" of the face depressions is proportional to the size of the specimen, i.e. it increases during the growth (Fig. 7.5a).

As time passes and the tissue widens, in fact, the disturbance in the flow field produced by the presence of the construct becomes larger. This behavior, in turn, increases the value of the shear stresses responsible for the growth of protuberances on the surface (see Fig. 7.6).

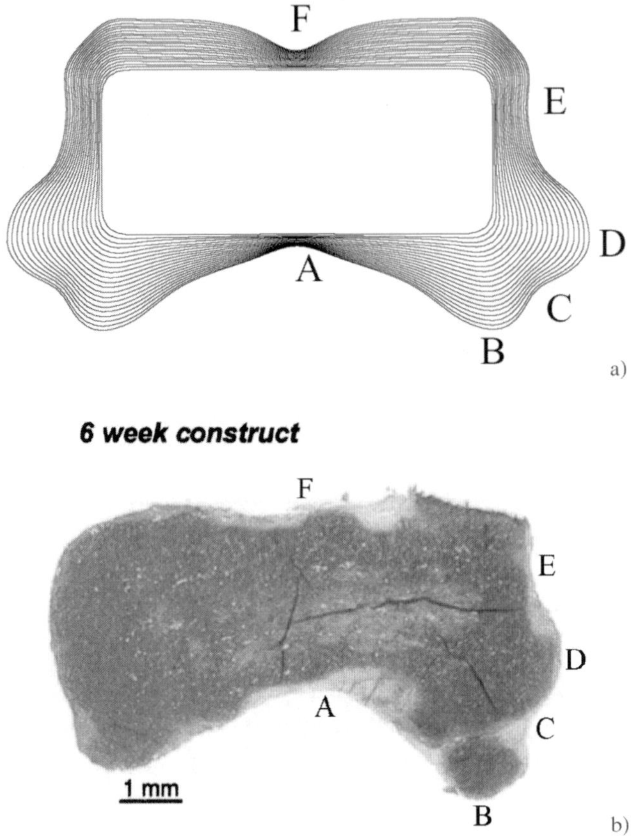

Fig. 7.5 (a) Tissue growth-habit simulation and progression of cartilaginous matrix deposition: snapshots of the tissue shape versus time ($\Delta t = 5.2 \times 10^4$ s), (b) experimental histological cross section of cartilage construct cultured for six weeks (Obradovic et al., 2000. Reproduced with permission of the American Institute of Chemical Engineers. Copyright © 2000 AIChe. All rights reserved).

A detailed description of these phenomena requires a separation of the analysis for the different faces of the construct.

The "face" (average) growth rate exhibits a different value according to the orientation of the face with respect to the main direction of flow, i.e. an important parameter is the relative direction of the different faces, which induces varying conditions for each face.

Figure 7.7 shows in detail how the convection effect results in higher local shear stresses near the surface where the flow is incoming (lower face) and in lower shear stresses near the surface where the flow is outgoing (upper face). This behavior can be explained according to the fact that the upper side faces a region where recirculating flow occurs and therefore the fluid is close to stagnation conditions. For this reason the shear stresses causing growth there are weakened and correspondingly the

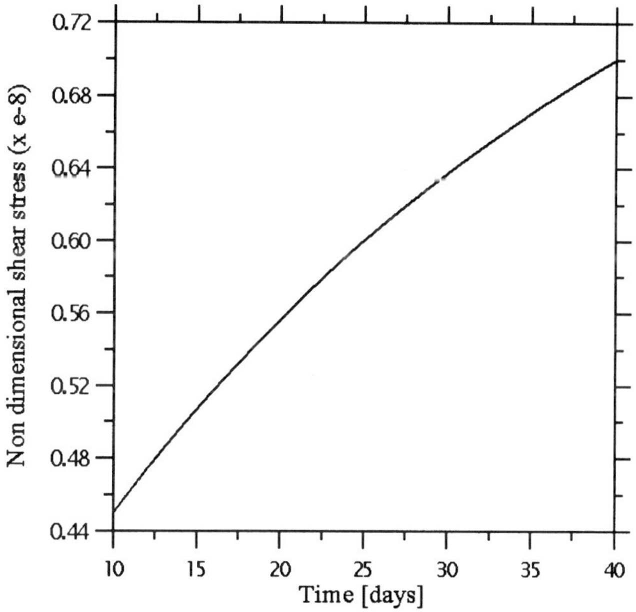

Fig. 7.6 Maximum surface shear stress versus *t*.

growth rate is reduced. For the same reasons onset of morphological instabilities is prevented for the upper side (the tissue surface facing the wake is almost plane and without irregularities).

Note, moreover, that as time passes, the toroidal vortex close to the upper side becomes larger (Fig. 7.3). The expansion of the wake is strictly related to the behavior of the tissue that puffs out due to growth.

At the same time, for the same reason (increasing size of the sample), the shear stresses (and associated growth rates) at the corners of the lower surface increase. This is due to the acceleration of the flow around the corners induced by the progressive reduction of the available cross-sectional area in the direction perpendicular to the flow and provides a theoretical explanation for the trend shown in Fig. 7.6. It is noteworthy how the expansion of the lower edges in the radial direction (see Figs. 7.3e–f) is also responsible for the presence of an additional secondary small vortex ring circumferentially wrapped around the construct.

In Fig. 7.7 the experimental histological cross section of cartilage construct cultured for six weeks (Obradovic et al., 2000) has been superimposed on the independently computed tissue shape and on the velocity and shear-stress distributions at the same time. This artifice is used to highlight how the numerical and experimental results are in very good agreement and how edges and protuberances correspond to the regions of high shear stress. The superposition

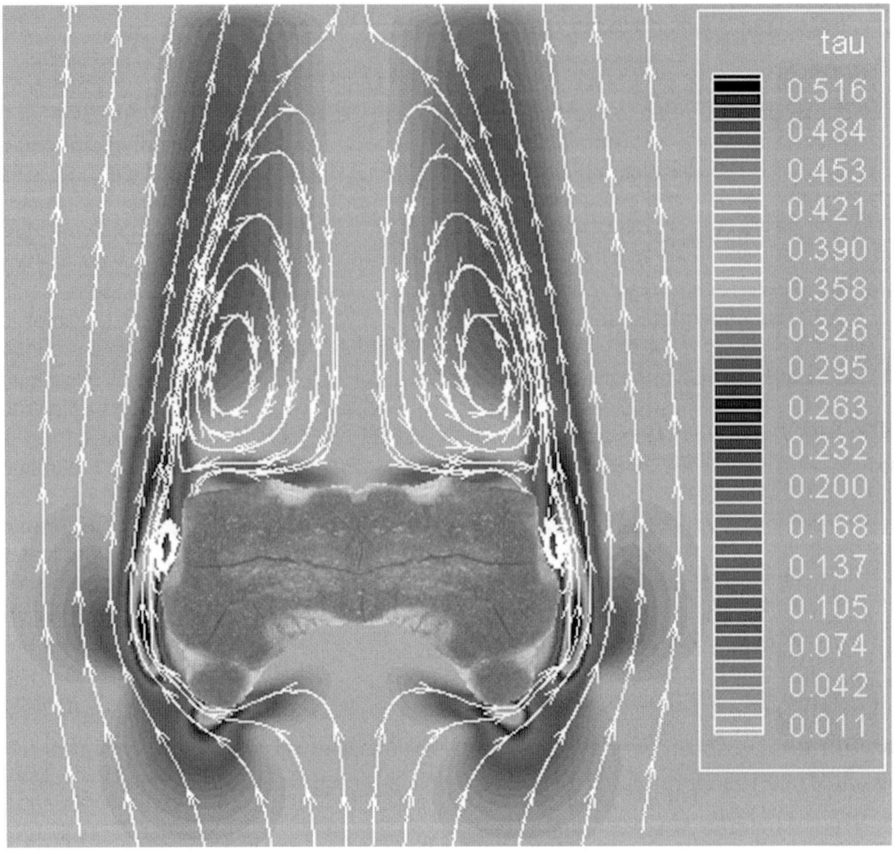

Fig. 7.7 Velocity field and nondimensional shear-stress distribution ($\times 10^{-8}$, $t = 40$ days, the experimental histological cross section of cartilage construct cultured for six weeks obtained by Obradovic et al., 2000, has been superimposed on the independently computed tissue shape and on the velocity and shear-stress distributions at the same time).

provides validation of the present shear-stress-based kinetic model and associated moving-boundary (CFD) computational method. Note that the left side of the experimental image has been replaced by the mirror image of the right side. This artifice has been used since the numerical computations for the prototype example are axisymmetric. For this reason they cannot reproduce the weak asymmetry that may characterize the specimen from an experimental point of view. The prediction of weak asymmetries arising in the final shape of the tissue due to environmental factors such as 3D instabilities of the flow field in the wake behind the specimen, Coriolis forces, terminal velocity nonparallel to the symmetry axis, small asymmetries in the initial shape of the scaffold, curvature of the flow due to the rotation of the vessel, etc. would require 3D computations.

7.8. The controversial effect of the fluid-dynamic shear stress

In the translation of mechanical culture conditions into tissue effects, several lines of evidence seem to support a role for shear stress. The last chapter of this book gives a definitive answer to this question providing heretofore unseen theoretical background supported by numerical simulations. In the light of the arguments pointed out in Section 7.6, the fluid shear distribution turns out to be crucial for the surface-absorption kinetics and for the overall tissue-growth process. It acts by modifying the internal cell division and addition of extracellular matrix mechanisms responsible for tissue enlargement and leads to different growth rates for the different sides of the specimen and to the onset of morphological instability (i.e. depressions and/or protuberances of the tissue/culture-medium interface). According to the present analysis, moreover, the shear-stress distribution is not constant in time but changes due to the increase of size of the organic construct. The interplay between the increasing size of the tissue and the structure of the convective field is essential in determining the time evolution of its shape.

It has been shown herein that the size of the growing specimen plays a "critical role" in determining the intensity of convection and of the shear stresses. Convective effects, in turn, have been found to impact growth rates, particle size and morphology as well as the mechanisms driving growth.

In the simulations, in fact, the growth velocity and/or the "growth law" have not been directly imposed but result from conditions related to solute transport and shear-stress distribution (i.e. mass and momentum transfer) at the biological construct surface.

The present analysis completely supports, by quantitative models and computations, Freed's supposition (Freed et al., 1997) about the effect of fluid shear stress in her experiments: on Mir the constructs were exposed to uniform shear and mass transfer at all surfaces such that the tissue grew equally in all directions, whereas on Earth the particular fluid-dynamic environment increases shear and mass transfer circumferentially such that the tissue tends to grow preferentially in the radial direction (around the lower corners).

For the first time the experiments of Freed et al. (1997) proved the effect of the fluid-dynamic environment on the growth and morphological evolution of the tissue through direct comparison of experiments carried out in the true weightlessness of space and on the ground. On the basis of the lines of evidence supported by the cross-check of these experiments and by the present models and computations, some very important conclusions can be drawn: On the ground the rotating vessel does not really cancel the effects of gravity since the endless free-fall of the specimen through the liquid leads to a convective pattern that, due to shear-stress-dependent surface kinetics, influences the final shape of the biological construct.

The results presented herein fit the processes that control tissue development. The consistency of model predictions with experimental data suggests that rate-controlling steps have been taken into account, and that simplifications do not distort actual behavior. The method has been proven to be able to predict morphological instabilities (i.e. habit/shape change) of the tissue.

Determining the growth laws and models is central to understanding how environmental conditions affect growth. Moreover, this knowledge is essential for growing replacement organic constructs in vitro through the techniques of tissue engineering. The present chapter demonstrates both the potential and challenges of mathematical modeling of in vitro organic tissue growth.

The proposed models and methods exhibit heretofore unseen capabilities to predict and elucidate experimental observations and to identify cause-and-effect relationships, i.e. to give insight into the mechanisms driving the phenomena under investigation. Of course, further investigation is needed to couple the present techniques with other existing ones taking into account the effect of the internal fiber stress of the tissue and the complex 3D fluid dynamics occurring in the rotating-vessel system.

7.9. Comparison with macromolecular crystal growth

In Chapter 5 (Section 5.4) some effort was made to elucidate the difference between inorganic and organic crystal growth; similarly the present section is devoted to a critical comparison between the cases of living tissues and macromolecular crystals.

It forms a preliminary but quite an exhaustive attempt to shed some light on biological-tissue-growth phenomena from an organic crystal grower's point of view. Regardless of the nature of the growing sample (organic protein crystal or biological tissue) some common features are pointed out. This structured and systematic analysis is intended to provide a common source (in terms of original mathematical models, ideas and results) to the scientific community on an optimistic note that the contacts established between the two fields in the last part of the present book will develop into an ongoing, mutually beneficial dialogue (see also Lappa, 2003e).

7.9.1 *Analogies and similarities*

Analogies about the macroscopic behavior of growing macromolecular crystals and growing biological tissues:

- In both cases the specimens are very sensitive to their environment and if exposed to sufficiently severe conditions may denature and/or degrade. They must be constantly maintained in a thoroughly hydrated state at or near physiological pH and temperature (mother liquor for protein crystals, culture medium for biological tissues).
- Hence the protein crystal and the organic tissue grow in "feeding solutions" of known (initial) concentration.

- Protein crystals and organic tissue specimens consist of: (a) organic molecules forming a network; (b) "bound water" that lines the large channels formed by the wide-open organic material network in bond-like arrangements; (c) "bulk water."

Protein crystals, in fact, have widely open structures and incorporate up to 90% by volume of solvent within the network of protein molecules. Macromolecular crystals are mainly composed of solvent. The protein occupies the remaining volume so that the entire crystal is in many ways an ordered gel with extensive interstitial spaces through which solvent and other "small" molecules may freely diffuse (see, e.g., Rosenberger, 1986; McPherson, 1990).

In the case of biological tissues, as cells replicate, they self-associate to form a complex matrix of collagens, proteins, fibers, and other chemicals. Also in this case, water is the major component of the construct. Soft tissues consist, in fact, primarily of various cell types, an extracellular matrix, and abundant water (see, e.g., Humphrey et al., 2001).

- In both cases, on a molecular level, growth follows from the successive addition/incorporation of "growth units" or "building blocks" to a lattice. Growth units can consist of single molecules or possibly clusters molecules. In solution, growth units are typically solvated, i.e. surrounded by highly regularly arranged shells of solvent molecules that interact with the growth unit in a bond-like fashion. In the case of macromolecular crystals the growth units are merely added to the crystal surface (the molecules orient and attach themselves to the growing surface) without changing their initial composition. In the case of biological tissues, of course, the mechanism is more complex. The growth units are incorporated and converted into the main tissue components. Consequently, tissue enlargement occurs due to internal cell division and production of extracellular matrix supported by the aforementioned incorporation of nutrients in the biological cells.
- In both cases, the surface kinetics are considerably lower than in most inorganic systems.
- The presence of a distinct boundary layer about the growing organic material (due to depletion of the feeding substance in the liquid phase) supports (in both the cases of protein crystal and organic tissue growth) the idea of competitive transport and surface growth kinetics in limiting the growth rate of the process.
- In both cases the intrinsic mechanisms of the growth process are influenced by microgravity conditions. Crystallization experiments carried out in space, in fact, yielded protein crystals that resulted in diffraction data of significantly higher resolution than the best crystals of these proteins grown on Earth. On the other hand, more spherical tissue samples have been obtained in orbiting platforms.
- What turns out from the studies carried out under microgravity conditions is that the characteristics of the final samples change depending on the fluid-dynamic

conditions under which they are operated. Hence, an understanding of the transfer of both momentum and mass within the growth reactors as well as of the interplay of such a flow with the intrinsic kinetics of the growth (still poorly understood) is crucial for further improvements in the fields of protein crystal and tissue engineering.

- In both cases the production of larger specimens require true microgravity conditions: the quasidiffusive conditions provided by such an environment have a beneficial effect on the mechanisms with which the solute molecules orient and attach themselves to the growing surface of macromolecular crystals; in the case of tissue specimens the absence of gravity is required for growing large samples (about 1 cm) that are no longer suspended on Earth, even if the RWPV technology is used.
- On Earth, in both cases, plume-like solutal patterns are formed. In the case of cylindrical scaffolds (Freed et al., 1997) the settling process tends to align their flat circular areas perpendicular to the direction of motion (see Fig. 7.3). This leads to the formation of a wake beyond the floating construct and to a depletion zone (Fig. 7.4) for the nutrient distribution whose configuration is very similar to that occurring in the case of protein crystal growth due to gravitational solutal pluming phenomena (see Fig. 5.9). Due to the fluid skirting around the tissue sample, in fact, a plume-like depletion zone is created above (behind) the tissue.
- In both cases, the growth process is associated with the onset of morphological instabilities (i.e. habit/shape change of the growing constructs). Corners and edges of the specimen grow faster than the center of faces and the different faces exhibit different behaviors according to their orientation with respect to the gravity direction.
- In both cases the nonconstant size of the specimen significantly affects the mechanisms underlying the growth process: the "depth" of the face depressions is proportional to the size of the sample, i.e. it increases during the growth (see Fig. 5.7 for the case of macromolecular crystals, Fig. 7.5a for biological tissue).

7.9.2 *Differences*

In the light of the arguments highlighted in Section 5.12 and in the present chapter (Section 7.7) the following differences about cause-and-effect relationships can be pointed out:

- in the case of crystals growing while attached to the walls of the reactor, convective flow is originated by the buoyancy forces (natural convection) associated with solutal effects (liquid with impoverished solute that rises, being lighter than the surrounding);
- in the case of tissue growing in a rotating reactor about a fixed position within the vessel as viewed by an external observer (due to equilibrium of external forces

gravity and drag), convective flow is due to a forced convection; fluid motion is not due to pluming phenomena but follows from a secondary flow pattern induced by particle sedimentation through the fluid media (on the ground conditions) or by laminar flows established when differential rotation rates are chosen for the inner and outer cylinders (true microgravity conditions)

- in the case of macromolecular crystal growth, morphological instability is a consequence of the nonuniform distribution of the nutrients around the crystal due to diffusive/convective effects. Mass transfer is higher close to the corners of the crystals for two reasons: (a) a protuberance on the interface sees a higher supersaturation and grows faster than a depression, which sees a lower supersaturation; (b) convection increases the steepness of the concentration gradient close to protuberances of the surface (see, for instance, Figs. 5.8, 5.22, and 5.23). There is no direct effect of the convective field on the intrinsic mechanism of growth. The crystals grow according to the local level of supersaturation and according to the steepness of the concentration gradient. Both these effects in turn change according to the intensity of the velocity field;
- biological tissues do not behave as macromolecular crystals; the diffusion of momentum towards the tissue interface alters the way in which nutrients are incorporated into the tissue mass and the tissue grows; there is a direct effect of the fluid-dynamic shear stress on the surface incorporations and conversion kinetics and/or on the internal cell division and addition of extracellular matrix mechanisms responsible for tissue enlargement (i.e. the physical forces induce changes in cell metabolism and function); accordingly, corners grow faster since a protuberance on the interface is subject to higher convective shear stress;
- in the case of crystal growth, the increasing size of the sample plays a role through its effect on the buoyancy plume. The intensity of these plumes, in fact, depends on the size of the face upon which they are generated;
- in the case of biological tissues, the increase of size of the construct leads to a progressive reduction of the available cross-sectional area in the direction perpendicular to the flow. This effect accelerates the fluid laterally skirting around the construct and increases the shear stresses at the lower corners.

In conclusion, it is evident how numerical simulations of these aspects carried out by using the proposed sophisticated techniques (in Chapters 5 and 7) can provide useful information allowing, at the same time, to gain important information and data about similarities and common features. The knowledge of the analogies in the macroscopic behavior of growing macromolecular crystals and growing biological tissues as well as of the differences in the cause-and-effect relationships is crucial for determining generalized criteria to obtain controlled growth (i.e. final samples of desired shape and size).

With regard to the mathematical models introduced in this chapter and in Chapter 5, it is worthwhile to remember how the surface conditions have been coupled to the exchange mass flux at the specimen/solution interface and have led to the introduction of a group of differential equations for the nutrient concentration around the sample and for the evolution of the mass displacement. These models take into account the sensitivity of the construct/liquid interface to the level of supersaturation in the case of macromolecular crystal growth and to the direct effect of the fluid-dynamic shear stress in the case of biological tissue growth. They then have been used to show how the proposed surface kinetic laws can predict (through sophisticated numerical simulations) many of the known characteristics of protein crystals and biological tissues produced using well-known and widely used reactors. This procedure has provided validation of the models and associated numerical methods and has given an insight into the (often hidden) mechanisms of the phenomena.

References

A

Agble D. and Mendes Tatsis M.A., (2000), "The effect of surfactants on interfacial mass transfer in binary liquid–liquid systems", Int. J. Heat Mass Transfer, 43: 1025–1034.

Alexander J.I.D., (1990), "Low gravity experiment sensitivity to residual acceleration: a review", Microgravity Sci. Tech., l. 3: 52–68.

Anderson D.M. and Davis S.H., (1995), "The spreading of volatile liquid droplets on heated surfaces", Phys. Fluids, 7: 248–265 .

Arakawa T. and Thimasheff S.N., (1985), "Theory of protein solubility", Meth. Enzymology, 114: 49–77.

Aziz K. and Hellums J.D., (1967), "Numerical solution of the three-dimensional equations of motion for laminar natural convection", Phys. Fluids, 10(2): 314–324.

Azouni M.A., Kalita W., and Yemmou M., (1990), "On the particle behaviour in front of advancing liquid–ice interface", J. Cryst. Growth, 99: 201–205.

B

Balasubramaniam R. and Subramanian R.S., (1999), "Axisymmetric thermal wake interaction of two bubbles in a uniform temperature gradient at large Reynolds and Marangoni numbers", Phys. Fluids, 11(10): 2856–2864.

Balasubramaniam R. and Subramanian S.R., (2000), "The migration of a drop in a uniform temperature gradient at large Marangoni numbers", Phys. Fluids, 12(4): 733–743.

Bassano E., (2003), "Numerical simulation of thermo-solutal-capillary migration of a dissolving drop in a cavity", Int. J. Numer. Meth. Fluids, 41: 765–788.

Baumgartl J. and Muller G., (1992), "Calculation of the effects of magnetic field damping on fluid flow: comparison of magnetohydrodynamic models of different complexity", 2(4): 197–202.

Baumgartl J., Gewald M., Rupp R., Stierlen J., and Muller G., (1990), "The use of magnetic fields and microgravity in melt growth of semiconductors: a comparative study", In Proceedings VIIth European Symposium on Materials and Fluid Sciences in Microgravity, Oxford (United Kingdom), 10–15 September, ESA SP 295.

Bazhlekov I.B., Anderson P.D., and Meijer H.E.H., (2004), "Nonsingular boundary integral method for deformable drops in viscous flows", Phys. Fluids, 16(4): 1064–1081.

Bazzi H., Nguyen C.T., and Galanis N., (2000), "Transient behaviours of float zones in μ-g and 1-g environments", Int. J. Num. Meth. Heat Fluid Flow, 10(3), 307–333.

Beckermann C., Diepers H.-J., Steinbach I., Karma A., and Tong X., (1999), "Modeling melt convection in phase-field simulations of solidification", J. Comput. Phys., 154(2): 468–496.

Begley C.M. and Kleis S.J., (2000), "The fluid dynamic and shear environment in the NASA/JSC rotating-wall perfused vessel bioreactor", Biotechnol. Bioeng., 70: 32–40.

Bejan A. and Tien C.L., (1978), "Fully developed natural counterflow in a long horizontal pipe with different end temperatures", Int. J. Heat Mass Transfer, 21: 701–708.

Ben Hadid H. and Roux B., (1990), "Thermocapillary convection in long horizontal layers of low-Prandtl number melts subject to a horizontal temperature gradient", J. Fluid Mech., 221: 77–103.

Ben Hadid H. and Henry D., (1997b), "Numerical study of convection in the horizontal Bridgman configuration under the action of a constant magnetic field. Part 1. Two dimensional flow", J. Fluid Mech., 333: 57–83.

Ben Hadid H., Henry D., and Kaddeche S., (1997a), "Numerical study of convection in the horizontal Bridgman configuration under the action of a constant magnetic field. Part 2. Three dimensional flow", J. Fluid Mech., 333: 23–56.

Bennon W.D. and Incropera F.P., (1987a), "A continuum model for momentum, heat and species transport in binary solid–liquid phase change systems-I. Model formulation", Int. J. Heat Mass Transfer, 30(10): 2161–2170.

Bennon W.D. and Incropera F.P., (1987b), "A continuum model for momentum, heat and species transport in binary solid–liquid phase change systems-II. Application to solidification in a rectangular cavity", Int. J. Heat Mass Transfer, 30(10): 2171–2187.

Benz K.W., (1990), "Factors controlling crystal perfections during growth under microgravity", In Proceedings VIIth European Symposium on Materials and Fluid Sciences in Microgravity, Oxford (United Kingdom), 10/15 September, ESA SP 295: 59–61.

Benz S., Hintz P., Riley R.J., and Neitzel G.P., (1998), "Instability of thermocapillary-buoyancy convection in shallow layers. Part 2. Suppression of hydrothermal waves", J. Fluid Mech., 359: 165–180.

Bi R.-C., Gui L.-L., Han Q., Shen F.-L., Shi K., Wang Y.-P., Chen S.-Z., Hu Y.-L., Niu X.-T., Dong J., Zhou Y.-C., and Lin N.-Q., (1994), "Protein crystallization in space", Microgravity Sci. Technol., VII/2: 203–206.

Birikh R., Briskman V., Myznikov V., Roux B., and Velarde M.G., (1992), "Vibrational-thermocapillary parallel flows and stability", Microgravity Q., 2(4): 221–226.

Birikh R.V., Briskman V.A., Chernatynski, and Roux B., (1993), "Control of thermocapillary convection in a liquid bridge by high frequency vibrations", Microgravity Q., 3: 23–28.

Boistelle C.W. and Astier J.P., (1988), "Crystallization mechanisms in solution", J. Cryst. Growth, 90: 14–30.

Bontoux P., Roux B., Schiroky G.H., Markham B.L., and Rosenberger F., (1986), "Convection in the vertical midplane of a horizontal cylinder. Comparison of two-dimensional approximations with three-dimensional results", Int. J. Heat Mass Transfer, 29(2): 227–240.

Boussinesq J., (1903), "Théorie analytique de la chaleur, mise en harmonie avec la thermodynamique et avec la théorie mècanique de la lumière", Gauthier-Villars, Paris, 2: 1901–1903.

Brackbill J.U., Kothe D.B., and Zemach C., (1992), "A continuum method for modeling surface tension", J. Comput. Phys., 100(2): 335–354.

Braun R.J., Murray B.T., Boettinger W.J., and McFadden G.B., (1995), "Lubrication theory for reactive spreading of a thin drop", Phys. Fluids, 7: 1797–1810.

Brent A.D., Voller V.R., and Reid J., (1988), "Enthalpy-porosity technique for modelling convection–diffusion phase change: application to the melting of a pure metal", Num. Heat Transf., 13: 297–318.

Briegleb W., (1988), "Ground-born methods in gravitational cell biology", The Physiologist, 31: 44–47.

Briegleb W., (1992), "Some qualitative and quantitative aspects of the fast-rotating clinostat as a research tool", ASGSB Bull., 5(2): 23–30.

Broom M.B.H., Witherow W.K., Snyder R.S., and Carter D.C., (1988), "Preliminary observations of the effect of solutal convection on crystal morphology", J. Cryst. Growth, 90: 130–135.

Bunde A. and Havlin S. (Ed.), (1996), "Fractals and Disordered Systems", Springer-Verlag, New York.

Burguete J., Mukolobwiez N., Daviaud N., Garnier N. and Chiffaudel A., (2001), "Buoyant-thermocapillary instabilities in extended liquid layers subjected to a horizontal temperature gradient", Phys. Fluids, 13(10): 2773–2787.

C

Camel D., Tison P., and Carlberg T., (1995), "Floating zone crystal growth of Germanium", In Proceedings of the Norderney Symposium on Scientific Results of the German Spacelab Mission D2 (Edited by P.R. Sahm, M.H. Keller, and B. Schiewe), 494–501.

Carotenuto L., Piccolo C., Castagnolo D., Lappa M., and Garcìa-Ruiz J.M., (2002), "Experimental observations and numerical modelling of diffusion-driven crystallisation processes", Acta Cryst. D, D58: 1628–1632.

Carruthers J.R., (1977), "Thermal convection instabilities relevant to crystal growth from liquids", In Preparation and Properties of Solid State Materials (Edited by W.R. Wilcox and R.A. Lefever), Vol. 3, Marcel Dekker, New York, 1977.

Cetegen B.M., (1997), "Behaviour of naturally oscillating and periodically forced axisymmetric buoyant plumes of helium and helium–air mixtures", Phys. Fluids, 9(12): 3742–3753.

Cetegen B.M. and Ahmed T.A., (1993), "Experiments on the periodic instability of buoyant plumes and pool fires", Combustion and Flame, 93: 157–184.

Cetegen B.M. and Kasper K.D., (1996), "Experiments on the oscillatory behaviour of buoyant plumes of helium and helium–air", Phys. Fluids, 8(11): 2974–2984.

Chandrasekhar S., (1981), "Hydrodynamic and Hydromagnetic Stability", Dover Publications, New York.

Chen G. and Roux B., (1991), "An analytical study of thermocapillary flow and surface deformations in floating zones", Microgravity Q., 1(2): 73–80.

Chen G. and Roux B., (1992), "Instability of thermocapillary convection in floating zones", In Proceedings VIIIth European Symposium on Materials and fluid Sciences in Microgravity, Brussels (Belgium), 12/16 April 1992, ESA SP-333: 73–78.

Chen H. and Saghir M.Z., (1994), "Three-dimensional Marangoni convection in the asymmetrically heated float zone", Microgravity Q., 4, 39–46.

Chen J.C. and Chin G.H., (1995), "Linear stability analysis of thermocapillary convection in the floating zone", J. Cryst. Growth, 154: 98–107.

Chen J.C. and Chu C.F., (1995), "Numerical computation of fluid flow of floating-zone crystal growth of molybdenum", Int. J. Heat Mass Transfer, 38(10): 1841–1853.

Chen J.C. and Wu H.K., (1996), "Numerical computation of heat flow, fluid flow and interface shapes in the float zone of lithium niobate during a melting process", Int. J. Heat Mass Transfer, 39(17): 3707–3716.

Chen Q.S. and Hu W.R., (1998) , "Influence of liquid bridge volume on instability of floating half-zone convection", Int. J. Heat Mass Transfer, 41: 825–837.

Chen J.C., Chu C.F., and Ueng W.F., (1994), "Thermocapillary convection and melt–solid interface in the floating zone", Int. J. Heat Mass Transfer, 37(12): 1733–1748.

Chen G., Lizee A., and Roux B., (1997a), "Bifurcation analysis of the thermocapillary convection in cylindrical liquid bridges", J. Cryst. Growth, 180: 238–247.

Chen S., Merriman B., Osher S., and Smereka P., (1997b), "A simple level set method for solving Stefan problems", J. Comput. Phys., 135(1): 8–29.

Chen Q.S., Hu W.R., and Prasad V., (1999), "Effect of liquid bridge volume on the instability in small-Prandtl-number half zones", J. Cryst. Growth, 203: 261–268.

Chen X.S., Xiang P.M., (1990), "Review of the roles of facilities with the microgravity of short duration in material science experiments", Microgravity Q., 1(1): 53–56.

Chenier E., Delcarte C., and Labrosse G., (1999), "Stability of the axisymmetric buoyant-capillary flows in a laterally heated liquid bridge", Phys. Fluids, 11(3): 527–541.

Chun C.H., (1980), "Experiments on steady and oscillatory temperature distribution in a floating zone due to the Marangoni convection", Acta Astronautica, 7: 479–488.

Chun C.H. and Wuest W., (1979), "Experiments on the transition from the steady to the oscillatory Marangoni convection of a floating zone under reduced gravity effect", Acta Astronautica, 6: 1073–1082.

Clift R., Grace J.R., and Weber M.E., (1978), "Bubbles, Drops and Particles", Academic Press, New York.

Cogoli M., (1992), "The fast rotating clinostat: a history of its use in gravitational biology and a comparison of ground-based and flight experiment results", ASGSB Bull., 5(2): 59–67.

Coimbra C.F.M. and Kobayashi M.H., (2002), "On the viscous motion of a small particle in a rotating cylinder", J. Fluid Mech., 469: 257–286.

Colinet P., Legros J.C., and Velarde M.G., (2001), "Nonlinear Dynamics of Surface-Tension-Driven Instabilities", Wiley-VCH, Berlin: 1–527.

Coriell S.R., Chernov A.A., Murray B.T., and McFadden G.B., (1998), "Step bunching: generalized kinetics", J. Cryst. Growth, 183: 669–682.

Crochet M.J., Dupret F., Ryckmans Y., Geyling F.T., and Monberg E.M., (1989), "Numerical simulation of crystal growth in a vertical Bridgman furnace", J. Cryst. Growth, 97: 173–185.

Cröll A., Muller-Sebert W., and Nitsche R., (1990), "Transition from steady to time-dependent Marangoni convection in partially coated silicon melt zones", In Proc. 7th Europ Symp. on Materials and Fluid Sciences in Microgravity, ESA SP-295: 263.

Cröll A., Kaiser Th., Schweizer M., Danilewsky A.N., Lauer S., Tegetmeier A., and Benz K.W., (1998a), "Floating-zone and floating-solution-zone growth of GaSb under microgravity", J. Cryst. Growth, 191: 365–376.

Cröll A., Szofran F.R., Dold P., Benz K.W., and Lehoczky S.L., (1998b): "Floating zone growth of silicon in magnetic fields II: strong static axial fields", J. Cryst. Growth, 183: 554–563.

D

Daniels P.G., (1984), "Roll-pattern evolution in finite-amplitude Rayleigh–Bènard convection in a two-dimensional fluid-layer bounded by distant sidewalls", J. Fluid Mech., 143: 125–152.

Davis R.H., (1999), "Buoyancy-driven viscous interaction of a rising drop with a smaller trailing drop", Phys. Fluids, 11(5): 1016–1028.

Davis R.H., Schonberg J.A., and Rallison J.M., (1989), "The lubrication force between two viscous drops", Phys. Fluids, 1(1): 77–81.

Davis R.H., Zhang X., and Wang H., (1993), "Hydrodynamics of dispersions and coagulation", In Immiscible Liquid Metals and Organics (Edited by L. Ratke), Proceedings of an International Workshop Organized by the ESA Expert Working Group, "Immiscible Alloys" held in the Physikzentrum, Bad Honnef 1992, Germany: 163–182.

Day J. and McPherson A., (1992), "Macromolecular crystal growth experiments on International Microgravity Laboratory-1", Protein Science, 1: 1254–1268.

De Luca L. and Costa M., (1997), "Instability of a spatially developing liquid sheet", J. Fluid Mech., 331: 127–144.

Dell'Aversana P. and Neitzel G.P., (1998), "When liquids stay dry", Physics Today, 51(1): 38–41.

Dell'Aversana P. and Neitzel G.P., (2004), "Behavior of noncoalescing and nonwetting drops in stable and marginally stable states", Exp. in Fluids, 36: 299–308.

Dell' Aversana P., Banavar J.R., and Koplik J., (1996), "Suppression of coalescence by shear and temperature gradients", Phys. Fluids, 8: 15–28.

Dell'Aversana P., Tontodonato V., and Carotenuto L., (1997), "Suppression of coalescence and of wetting: the shape of the interstitial film", Phys. Fluids, 9(9): 2475–2485.

De Saedeleer C., Garcimartin A., Chavepeyer G., Platten J.K., and Lebon G., (1996), "The instability of a liquid layer heated from the side when the upper surface is open to air", Phys. Fluids, 8(3): 670–676.

Deville M.O., Mund E.H., and Fischer P.F., (2002), "High-order Methods for Incompressible Fluid Flow", Cambridge University Press: 1–528.

Diefenbach S., Ratke L., Prinz B., and Ahlborn H., (1993), "Marangoni motion during melting of a hypermonotectic alloy – numerical simulations for the D2-experiment IHF04", In Immiscible Liquid Metals and Organics (Edited by L. Ratke), Proceedings of an International Workshop Organized by the ESA Expert Working Group, "Immiscible Alloys" held in the Physikzentrum, Bad Honnef 1992, Germany: 291–298.

Dold, P., Cröll, A., and Benz, K.W. (1998), "Floating zone growth of silicon in magnetic fields I: weak static axial fields", J. Cryst. Growth, 183: 545–553.

Dold, P., Cröll, A., Lichtensteiger, M., Kaiser, Th., and Benz, K.W., (2001), "Floating zone growth of silicon in magnetic fields: IV: rotating magnetic fields", J. Cryst. Growth, 231: 95–106.

Dupret F. and Van der Bogaert N., (1994) "Modelling Bridgman and Czochralski growth", In Handbook of Crystal Growth (Edited by D.T.J. Hurle), North-Holland, Amsterdam, 2: 877–1010.

Durbin S.D. and Feher G., (1986), "Crystal growth studies of lysozyme as a model for protein crystallization", J. Cryst. Growth, 76: 583–592.

E

Eckert K., Bestehorn M., and Thess A., (1998), "Square cells in surface-tension-driven Bènard convection: experiments and theory", J. Fluid Mech., 356: 155–197.

Edwards D.A., Brenner H., and Wasan D.T., (1991), "Interfacial Transport Processes and Rheology", Butterworth-Heinemann, Boston.

Ermakov M.K. and Ermakova M.S., (2004), "Linear-stability analysis of thermocapillary convection in liquid bridges with highly deformed free surface", J. Cryst. Growth, 266: 160–166.

Esmaeeli A. and Tryggvason G., (1998), "Direct numerical simulations of bubbly flows. Part I – low reynolds number arrays", J. Fluid Mech., 377: 313–345.

Esmaeeli A. and Tryggvason G., (1999), "Direct numerical simulations of bubbly flows. Part II – moderate Reynolds number arrays", J. Fluid Mech., 385: 325–358.

Eyer A., Leiste H., and Nitsche R., (1984), In Proc. of the 5th European Symp. on Material Science under Microgravity, ESA SP-222: 173–181.

Eyer, A., Leiste, H., and Nitsche, R., (1985), "Floating-zone growth of silicon in a sounding rocket", J. Cryst. Growth, 71: 173–182.

F

Farouk B. and Fusegi T., (1985), "A coupled solution of the vorticity–velocity formulation of the incompressible Navier–Stokes equations", Int. J. Numer. Meth. Fluids, 5: 1017–1034.

Favier J.J., (1990), "Recent advances in Bridgman growth modelling and fluid flow", J. Cryst. Growth, 99: 18–29.

Fedoseyev A.I. and Alexander J.I.D., (2000), "Investigation of vibrational control of convective flows in Bridgman melt growth configurations", J. Cryst. Growth, 211: 34–42.

Feigelson R.S., (1989), "The relevance of small molecule crystal growth theories and techniques to the growth of biological macromolecules", J. Cryst. Growth, 90(1–3): 1–13.

Feuillebois F., (1993), "A thermo-hydrodynamic mechanism for the repulsion of a particle by a freezing front", In Immiscible Liquid Metals and Organics, Proceedings of an International Workshop Organized by the ESA Expert Working Group, "Immiscible Alloys" held in the Physikzentrum, Bad Honnef 1992 (Edited by L. Ratke), Germany: 231–238.

Fletcher C.A.J., (1991) "Computational Techniques for Fluid-dynamics", Springer Verlag, Berlin.

Frank S. and Schwabe D., (1998), "Temporal and spatial elements of thermocapillary convection in floating zones", Experiments in Fluids, 23: 234–251.

Freed L.E., Langer R., Martin I., Pellis N.R., and Vunjak-Novakovic G., (1997), "Tissue engineering of cartilage in space", Proc. Natl. Acad. Sci. USA, 94: 13885–13890.

Fukai J., Zhao Z., Poulikakos D., Megaridis C.M., and Miyatake O., (1993), "Modelling of the deformation of a liquid droplet impinging upon a flat surface", Phys. Fluids A, 5: 2588–2599.

Fukai J., Shiiba Y., Yamamoto T., Miyatake O., Poulikakos D., Megaridis C.M., and Zhao Z., (1995), "Wetting effects on the spreading of a liquid droplet colliding with a flat surface: experiment and modeling", Phys. Fluids A, 7: 236–247.

Fung Y.C., (1990), "Biomechanics: Motion, Flow, Stress, Growth", Springer, New York.

G

Galkin O. and Vekilov P.G., (1999), "Direct determination of the nucleation rates of protein crystals", J. Phys. Chem. B, 103: 10965–10971.

Gao H., Ayyaswamy P.S., and Ducheyne P., (1997a), "Dynamics of a microcarrier particle in the simulated microgravity environment of a rotating wall vessel", Microgravity Sci. Technol., X: 154–165.

Gao, H., Ayyaswamy P.S., and Ducheyne P., (1997b), "Numerical simulation of global diffusive mass transfer in a rotating wall vessel bioreactor", Advances in Heat and Mass Transfer in Biotechnology, ASME, HTD 335/BED-37: 59–67.

Gelfgat A.Yu., (1999a), "Different modes of Rayleigh–Benard instability in two- and three-dimensional rectangular enclosures", J. Comput. Phys., 156: 300–324.

Gelfgat A.Yu., Bar-Yoseph P.Z., and Yarin A.L., (1999b), "Stability of multiple steady states of convection in laterally heated cavities", J. Fluid Mech., 388: 315–334.

Gelfgat A.Yu., Bar-Yoseph P.Z., and Yarin A.L., (1999c), "Non-symmetric convective flows in laterally heated rectangular cavities", Int. J. Comput. Fluid Dynamics, 11: 261–273.

Gelfgat A.Yu., Bar-Yoseph P.Z., and Solan A., (2000), "Axisymmetry breaking instabilities of natural convection in a vertical Bridgman growth configuration", J. Cryst. Growth, 220: 316–325.

Gelfgat A.Yu., Bar-Yoseph P.Z., and Solan A., (2001), "Effect of axial magnetic field on three-dimensional instability of natural convection in a vertical Bridgman growth configuration", J. Cryst. Growth, 230: 63–72.

Gershuni G.Z. and Lyubimov D.V., (1998), "Thermal Vibrational Convection", Wiley & Sons.

Gershuni G.Z. and Zhukhovitskii E.M., (1986), "Vibrational thermal convection in zero gravity", Fluid. Mech.-Sov. Res., 15: 63–84.

Gershuni G.Z., Zhukhovitskii E.M., and Yurkov Yu.S., (1982), "Vibrational thermal convection in a rectangular cavity", Izv. Akad. Nauk SSSR Mekh. Zhidk. Gaza, 4: 94–99.

Gershuni G.Z., Laure P., Myznikov V.M., Roux B., and Zhukhovitsky E.M., (1992), "On the stability of plane-parallel advective flows in long horizontal layers", Microgravity Q., 2(3): 141–151.

Gershuni G.Z., Lyubimov D.V., Lyubimova T.P., and Roux, B., (1994), "Convective flows in a liquid bridge under the influence of high frequency vibrations", Microgravity Q., 4(2): 113–122.

Giege R. and Ducruix A. (Eds.), (1999), "Crystallization of Nucleic Acids and Proteins: A Practical Approach" (Practical Approach Series, 210), 2nd edn., Oxford University Press, New York.

Graham D.E. and Phillips M.C., (1980), "Proteins at liquid interfaces. IV. Dilatational properties", J. Colloid Interface Sci., 76(1): 227–239.

Gresho P.M., (1991), "Incompressible fluid dynamics: some fundamental formulation issues", Ann. Rev. Fluid Mech., 23: 413–453.

Gresho P.M. and Sani R.T., (1987), "On pressure boundary conditions for the incompressible Navier–Stokes equations", Int. J. Numer. Meth. Fluids, 7: 1111–1145.

Griaznov V., (1994), "CFD simulation of the oscillatory floating zone convection for High Prandtl numbers", In Proceedings of the Microgravity Science and Applications Session, International Aerospace Congress, Moscow, August 16–17.

Gropp W., Lusk E., and Skjellum A., (1994), "Using MPI: Portable Parallel Programming with the Message-Passing Interface", publication TPD-0011, MIT Press.

Gueyffier D., Li J., Nadim A., Scardovelli S., and Zaleski S., (1999), "Volume of fluid interface tracking with smoothed surface stress methods for three-dimensional flows", J. Comput. Phys., 152: 423–456.

H

Haj-Hariri H., Shi Q., and Borhan A., (1997), "Thermocapillary motion of deformable drops at finite Reynolds and Marangoni numbersa", Phys. Fluids, 9(4): 845–855.

Hammond T.G. and Hammond J.M., (2001), "Optimized suspension culture: the rotating-wall vessel", Am. J. Physiol. Renal Physiol., 281. F12–F25.

Han J. and Tryggvason G., (1999), "Secondary breakup of axisymmetric liquid drops. I. Acceleration by a constant body force", Phys. Fluids, 11(12): 3650–3667.

Harlow F.H. and Welch J.E., (1965), "Numerical calculation of time-dependent viscous incompressible flow with free surface", Phys. Fluids, 8: 2182–2189.

Häuser J. and Williams R.D., (1992), "Strategies for parallelizing a Navier–Stokes code on the Intel Touchstone Machines", Int. J. Num. Meth. Fluids, 15: 51–58.

Helmrich A. and Barnes D., (1998), "Animal cell culture equipment and techniques", Methods Cell Biol., 57: 3–17.

Hemmersbach-Krause R., Briegleb W., Häder D.-P., Vogel K., Grothe D., and Meyer I., (1993), "Orientation of Paramecium under the conditions of weightlessness", J. Eur. Microbiol., 40(4): 439–446.

Henisch H.K., (1988), "Crystals in Gels and Liesegang Rings: In Vitro Veritas", Cambridge University Press (May 1988).

Henisch H.K., (1991), "Periodic Precipitation: A Microcomputer Analysis of Transport and Reaction Processes in Diffusion Media with Software Development", Pergamon Press.

Henisch H.K. and Garcìa-Ruiz J.M., (1986a), "Crystal growth in gels and Liesegang ring formation: diffusion relationships", J. Cryst. Growth, 75: 195–202.

Henisch H.K. and Garcìa-Ruiz J.M., (1986b), "Crystal growth in gels and Liesegang ring formation: crystallization criteria and successive precipitation", J. Cryst. Growth, 75: 203–211.

Hibiya T., Nakamura S., Mukai K., Niu Z.-G., Imaishi N., Nishizawa S.-i., Yoda S.-i., and Koyama M., (1998a), "Interfacial phenomena of molten silicon: Marangoni flow and surface tension", Phil. Trans. R. Soc. Lond. A, 356: 899–909.

Hibiya T., Nakamura S., Imaishi N., Mukai K., Onuma K., Dold P., Croll A., Benz K.-W., and Yoda S.-i., (1998b), "Marangoni flow of Si melt: microgravity experiments and perspective", J. Jpn. Soc. Microgravity Appl. Supplement, Proc. Joint 1st Pan-Pacific Basin Workshop and 4th Japan-China Workshop on Microgravity Sciences.

Hirata A., Nishizawa S., and Sakurai M., (1997a), "Experimental results of oscillatory Marangoni convection in a liquid bridge under normal gravity", J. Jpn. Soc. Microgravity Appl., 14: 122–129.

Hirata A., Sakurai M., and Ohishi N., (1997b), "Effect of gravity on Marangoni convection in a liquid bridge", J. Jpn. Soc. Microgravity Appl., 14: 130–136.

Hirata A.A., Sakurai M., Ohishi N., Koyama M., Morita T., and Kawasaki H., (1997c), "Transition process from laminar to oscillatory Marangoni convection in a liquid bridge under normal and microgravity", J. Jpn. Soc. Microgravity Appl., 14: 137–143.

Hirsch C., (1994), "Numerical Computations for Internal and External Flows", John Wiley and Sons.

Hirt C.W. and Nichols B.D., (1981), "Volume of fluid (VOF) method for the dynamics of free boundaries", J. Comput. Phys., 39: 201–225.

Hu W.R. and Tang Z.M., (2003), "Influence of liquid bridge volume on the floating zone convection", Cryst. Res. Technol., 38(7–8): 627–634.

Hu W.R., Tang Z.M., Shu J.Z., and Zhou R., (1995), "Influence of liquid bridge volume on the critical Marangoni number in thermocapillary convection of half floating zone", Microgravity Q., 5(2): 67–74.

Humphrey J.D., Rao I.J., and Rajagopal K.R., (2001), "A mixture model for growth and remodeling in soft tissues", BED-Vol. 50 Bioengineering Conference, ASME 2001: 643–644.

Hurle D.T.J., (1966), "Temperature oscillations in molten metals and their relationship to growth striae in melt-grown crystals", Phil. Mag., 13: 305–310.

Hurle D.T.J. (Ed.), (1994), "Handbook of Crystal Growth", North-Holland, Amsterdam.

Hurle D.T.J. and Cockayne B., (1994), "Czochralski growth", In Handbook of Crystal Growth (Edited by D.T.J. Hurle), North-Holland, Amsterdam.

I

Imaishi N., Yasushiro S., Sato T., and Yoda S., (1999), "Three dimensional numerical simulation of oscillatory Marangoni flow in a half zone of low Pr fluids", Material Research in Low Gravity II (SPIE The International Society for Optical Engineering), Denver, Colorado, 19–21 July 1999, 3792: 344–352.

Imaishi N., Yasushiro S., Sato T., and Yoda S., (2000), "Numerical simulation of three dimensional oscillatory flow in half zone bridges of low Pr fluid", Proceedings 292 of the 4th JSME-KSME Thermal Engineering Conference – October 1–6, 2000, Kobe, Japan: 272–282.

Imaishi N., Yasuhiro S., Akiyama Y., and Yoda S., (2001), "Numerical simulation of oscillatory Marangoni flow in half-zone liquid bridge of low Prandtl number fluid", J. Cryst. Growth, 230: 164–171.

J

Jing C.J., Imaishi N., Yasuhiro S., Sato T., and Miyazawa Y., (2000), "Three-dimensional numerical simulation of rotating spoke pattern in an oxide melt under a magnetic field", Int. J. Heat Mass Transfer, 43: 4347–4359.

Joseph D.D., (1976), "Stability of Fluid Motions I, Springer Tracts in Natural Philosophy", Vol. 27, Springer, Berlin and "Stability of Fluid Motions II, Springer Tracts in Natural Philosophy", Vol. 28, Springer, Berlin.

K

Kasperski G. and Labrosse G., (2000), "On the numerical treatment of viscous singularities in wall-confined thermocapillary convection", Phys. Fluids, 12(11): 2695–2697.

Kasperski G., Batoul A., and Labrosse G., (2000), "Up to the unsteadiness of axisymmetric thermocapillary flows in a laterally heated liquid bridge", Phys. Fluids 12(1): 103–119.

Kassemi M. and Rashidnia N., (1999), "Steady and oscillatory thermocapillary convection generated by a bubble", Phys. Fluids, 12(12): 3133–3146.

Kawamura H. and Harada N., (1998), "Observation of oscillatory Marangoni convection in a liquid column", Proc. of 35th National Heat Transfer Symposium of Japan, III: 749–750 (in Japanese).

Kazarinoff N.D. and Wilkowski J.S., (1990), "Bifurcations of numerically simulated thermocapillary flows in axially symmetric float-zones", Phys. Fluids A, 2: 1797–1806.

Keyes D.E., Chan T.F., Meurant G.A., Scroggs J.S., and Voigt R.G. (Eds.), (1992), Proceedings of "Fifth International Symposium on Domain Decomposition Methods for Partial Differential Equations", SIAM, Philadelphia.

Khine Y.Y. and Walker J.S., (1994), "Thermocapillary convection in a cylinder with a strong non-uniform axisymmetric magnetic field", J. Fluid Mech., 276: 369–388.

Khine Y.Y. and Walker J.S., (1995), "Buoyant convection during Czochralski silicon growth with a strong, non-uniform, axisymmetric magnetic field", J. Cryst. Growth, 147: 313–319.

Kim Y.T., Goldenfeld N., and Dantzig J., (2000), "Computation of dendritic microstructures using a level set method", Phys. Rev. E, 62: 2471–2474.

Kleis S.J., Schreck S., and Nerem R.M., (1990), "A viscous pump bioreactor", Biotech. & Bioeng., 36: 771–777.

Knoll D., Kothe D., and Lally B., (1999), "A new nonlinear solution method for phase change problems", Num. Heat Transfer Part B, 35(4): 439–459.

Knupp P. and Steinberg S., (1994), "Fundamentals of grid generation", CRC Press, London.

Koh C.J. and Leal L.G., (1989), "The stability of drop shapes for translation at zero Reynolds number through a quiescent fluid", Phys. Fluids, 1(8): 1309–1313.

Konkachbaev A. and Morley N.B., (2000), "Stability and contraction of a rectangular liquid metal jet in vacuum environment", Fusion Eng. Design, 51–52: 1109–1114.

Kordyum E.L., (1994), "Effects of altered gravity on plant cell processes: results of recent space and clinostatic experiments", Adv. Space Res., 14: 77–85.

Kothe D.B., Ferrell R.C., Turner J.A., and Mosso S.J., (1997) "A high resolution finite volume method for efficient parallel simulation of casting processes on unstructured meshes", Presented at the 8th SIAM Conference on Parallel Processing for Scientific Computing, Minneapolis, MN (March 14–17, 1997).

Kuhlmann H.C., (1992), "Thermocapillary instabilities in cylindrical liquid bridge", Proceedings VIIIth European Symposium on Materials and Fluid Sciences in Microgravity, Brussels (Belgium), 12/16 April 1992 , ESA SP-333: 79–83.

Kuhlmann H.C., (1995), "Some scaling aspects of thermocapillary flows", Microgravity Q., 5: 29–34.

Kuhlmann H.C. and Rath H.J., (1993a), "Hydrodynamic instabilities in cylindrical thermocapillary liquid bridges", J. Fluid Mech., 247: 247–274.

Kuhlmann H.C. and Rath H.J., (1993b), "On the interpretation of phase measurements of oscillatory thermocapillary convection in liquid bridges", Phys. Fluids A, 5(9): 2117–2120.

Kuhlmann H.C. and Nienhüser Ch., (2002), "Dynamic free-surface deformations in thermocapillary liquid bridges", Fluid Dynamics Res., 31: 103–127.

Kunitake R., Suzuki A., Ichihashi H., Matsuda S., Hirai O., and Morimoto K., (1997), "Fully-automated roller bottle handling system for large scale culture of mammalian cells", J. Biotechnol., 52: 289–294.

Kuo H.P. and Korpela S.A., (1988), "Stability and finite amplitude natural convection in a shallow cavity with insulated top and bottom and heated from the side", Phys. Fluids, 31: 33–42.

Kuznetsov Yu.G., Malkin A.J., Greenwood A., and McPherson A., (1995), "Interferometric studies of growth kinetics and surface morphology in macromolecular crystal growth: canavalin, thaumatin and turnip yellow mosaic virus", J. Structural Biology, 114: 184–196.

L

Lan C.W. and Kou S., (1990), "Thermocapillary flow and melt/solid interfaces in floating-zone crystal growth under microgravity", J. Cryst. Growth, 102: 1043–1058.

Lan C.W. and Liang M.C., (1999), "Multigrid methods for incompressible heat flow problems with an unknown interface", J. Comput. Phys., 152(1): 55–77.

Lan C.W. and Chian C.H., (2001), "Three-dimensional simulation of Marangoni flow and interfaces in floating-zone silicon crystal growth", J. Cryst. Growth, 230: 172–180.

Lan C.W., Liu C.C., et al., (2002), "An adaptive finite volume method for incompressible heat flow problems in solidification", J. Comput. Phys., 178(2), 464–471.

Lappa M., (1997), "Strategies for parallelizing the three-dimensional Navier–Stokes equations on the Cray T3E", Science and Supercomputing at CINECA, 11: 326–340.

Lappa M., (2002a), "Well-posed problems for the Navier–Stokes equations in the microgravity environment", MSSU – Microgravity and Space Station Utilization, 3(4): 23–46.

Lappa M., (2002b) "New mathematical models and CFD numerical tools for the investigation of macromolecular crystal growth at macroscopic and microscopic length scales", MSSU – Microgravity and Space Station Utilization, 3(3): 17–28.

Lappa M., (2003a), "Three-dimensional numerical simulation of Marangoni flow instabilities in floating zones laterally heated by an equatorial ring", Phys. Fluids, 15(3): 776–789.

Lappa M., (2003b), "An 'attachment-kinetics-based' Volume of Fraction Method for organic crystallization: a fluid-dynamic approach to macromolecular crystal engineering", J. Comput. Phys., 191(1): 97–129.

Lappa M., (2003c), "Growth and mutual interference of protein seeds under reduced gravity conditions", Phys. Fluids 15(4): 1046–1057. This article has been also selected by the American Physical Society for the 15 March 2003 issue of the Virtual Journal of Biological Physics Research (Volume 5, Issue 6, section "Physical studies of cell mechanics").

Lappa M., (2003d), "Organic tissues in rotating bioreactors: fluid-mechanical aspects, dynamic growth models and morphological evolution", Biotechnology & Bioengineering, 84(5): 518–532.

Lappa M., (2003e), "The growth and the fluid-dynamics of protein crystals and soft organic tissues: models and simulations, similarities and differences", J. Theor. Biol., 224/2: 225–240.

Lappa M., (2004a), "Combined effect of volume and gravity on the three-dimensional flow instability in non-cylindrical floating zones heated by an equatorial ring", Phys. Fluids, 16(2): 331–343.

Lappa M., (2004b), "Analysis of flow instabilities in convex and concave floating zones heated by an equatorial ring under microgravity conditions", Computers & Fluids (DOI.10.1016/j.compfluids./2004.01.003).

Lappa M., (2004c), "Floating zones heated around the equatorial plane: models and applications", Microgravity Sci. Technol, scheduled for issue XV/3.

Lappa M., (2004d), "An 'attachment kinetics-based' Level-Set Method for protein crystallization under buoyancy-driven convective effects", Int. J. Comput. Fluid Dyn (in press).

Lappa M., (2004e), "Discrete layers of interacting growing protein seeds: Convective and Morphological stages of evolution", Phys. Rev. E (to appear).

Lappa M., (2004f), "A CFD Level-Set method for soft tissue growth: theory and fundamental equations", J. Biomechanics, 38(1): 185–190.

Lappa M., (2004g), "Computational fluid-dynamics and virtual organic tissues", Fluid Dynamics Res. (in press).

Lappa M., (2004h) "Assessment of VOF strategies for the analysis of Marangoni Migration, Collisional Coagulation of Droplets and Thermal Wake effects in Immiscible Metal Alloys under Microgravity conditions", Computers, Materials & Continua (submitted).

Lappa, M., (2004i), "Thermal convection and related instabilities in models of crystal growth from the melt on earth and in microgravity: past history and current status", Cryst. Res. Technol. (in press).

Lappa M. and Savino R. (1999), "Parallel solution of the three-dimensional Marangoni flow instabilities in liquid bridges", Int. J. Num. Meth. Fluids, 31: 911–925.

Lappa M. and Savino R., (2002), "3D analysis of crystal/melt interface shape and Marangoni flow instability in solidifying liquid bridges", J. Comput. Phys., 180(2): 751–774.

Lappa M., Savino R., and Monti R., (2000), "Influence of buoyancy forces on Marangoni flow instabilities in liquid bridges", Int. J. Num. Meth. Heat Fluid Flow, 10(7): 721–749.

Lappa M., Savino R., and Monti R., (2001a), "Three-dimensional numerical simulation of Marangoni instabilities in non-cylindrical liquid bridges in microgravity", Int. J. Heat Mass Transfer, 44(10): 1983–2003.

Lappa M., Savino R., and Monti R., (2001b), "Three-dimensional numerical simulation of Marangoni instabilities in liquid bridges: influence of geometrical aspect ratio", Int. J. Num. Meth. Fluids, 36(1): 53–90.

Lappa M., Castagnolo D., and Carotenuto L., (2002), "Sensitivity of the non-linear dynamics of Lysozyme 'Liesegang Rings' to small asymmetries", Physica A: Statistical Mechanics and its Applications, 314/1–4: 623–635.

Lappa M. and Carotenuto L., (2003), "Effect of convective disturbances induced by g-jitter on the periodic precipitation of lysozyme", Microgravity Sci. Technol., XIV/2: 41–56.

Lappa M. and Castagnolo D., (2003), "Complex dynamics of rhythmic patterns and sedimentation of organic crystals: a new numerical approach", Num. Heat Transfer Part B – Fundamentals, 43(4): 373–401.

Lappa M. and Piccolo C., (2004), "Higher modes of Mixed Buoyant-Marangoni unstable convection originated from a droplet dissolving in a liquid/liquid system with miscibility gap", Phys. Fluids (in press).

Lappa M., Yasushiro S., and Imaishi N., (2003a), "3D numerical simulation of on ground Marangoni flow instabilities in liquid bridges of low Prandtl number fluid", Int. J. Num. Meth. Heat Fluid Flow, 13(3): 309–340.

Lappa M., Piccolo C., and Carotenuto L., (2003b), "Numerical and experimental analysis of periodic patterns and sedimentation of lysozyme", J. Cryst. Growth, 254/3–4: 469–486.

Lappa M., Piccolo C., and Carotenuto L., (2004a), "Mixed buoyant-Marangoni convection due to dissolution of a droplet in a liquid–liquid system with miscibility gap", Eur. J. Mech. B, 23/5: 781–794.

Lappa M., Piccolo C., Carotenuto L., and Savino R., (2004b), "Fluid dynamics and dissolution kinetics in immiscible organic systems with dispersed droplets", Colloids and Surfaces A: Physiochem. and Eng. Asp., Special Issue for the Workshop Bubble and Drop Interfaces (BD2004) held in Genoa, Italy, April 25–28.

Lavrenteva O.M. and Nir A., (2003), "Axisymmetric thermal wake interaction of two drops in a gravity field at low Reynolds and high Peclet numbers", Phys. Fluids, 15(10): 3006–3014.

Lee C.P. and Chernov A.A., (2002), "Solutal convection around growing protein crystals and diffusional purification in space", J. Cryst. Growth, 240: 531–544.

Leib S.J. and Goldstein M.E., (1986a), "The generation of capillary instability on liquid jet", J. Fluid Mech., 168: 479–500.

Leib S.J. and Goldstein M.E., (1986b), "Convective and absolute instability of a viscous liquid jet", Phys. Fluids, 29: 952–954.

Lenski H., (1990), "Advanced facilities for crystal growth", Microgravity Q., 1(1): 47–52.

Levenstam M. and Amberg G., (1995), "Hydrodynamical instabilities of thermocapillary flow in a half-zone", J. Fluid Mech., 297: 357–372.

Levenstam M., Amberg G., Carlberg T., and Andersson M., (1996), "Experimental and numerical studies of thermocapillary convection in a floating zone like configuration", J. Cryst. Growth, 158: 224–230.

Leypoldt J., Kuhlmann H.C., and Rath H.J., (2000), "Three-dimensional numerical simulation of thermocapillary flows in cylindrical liquid bridges", J. Fluid Mech, 414: 285–314.

Li Y.R., Imaishi N., Azami T., and Hibiya T., (2004), "Three-dimensional oscillatory flow in a thin annular pool of silicon melt", J. Cryst. Growth, 260: 28–42.

Lin S.P., (2003), "Breakup of Liquid Sheets and Jets", Cambridge University Press: 1–163.

Lin S.P. and Lian Z.W., (1989), "Absolute instability of a liquid jet in a gas", Phys. Fluids A, 1: 490–493.

Lin S.P. and Lian Z.W., (1993), "Absolute and convective instability of a viscous liquid jet surrounded by a viscous gas in a vertical pipe", Phys. Fluids. A, 5: 771–773.

Lin H., Rosenberger F., Alexander J.I.D., and Nadarajah A., (1995), "Convective–diffusive transport in protein crystal growth", J. Cryst. Growth, 151: 153–162.

Littke W. and John C., (1986), "Protein single crystal growth under microgravity", J. Cryst. Growth, 76: 663–672.

Lizée A. and Alexander J.I.D., (1997), "Chaotic thermovibrational flow in a laterally heated cavity", Phys. Rev., E 56: 4152–4156.

Lopez C.R., Mileham J.R., and Abbaschian R., (1999), "Microgravity growth of GaSb single crystals by the liquid encapsulated melt zone (LEMZ) technique", J. Cryst. Growth, 200: 1–12.

Lou J.Z. and Ferraro R., (1996), "A parallel incompressible flow solver package with a parallel multigrid elliptic kernel", J. Comput. Phys., 125, 225–243.

Lyubimova T.P., Lizee A., Gershuni G.Z., Lyubimov D.V., Chen G., Wadih M., and Roux B., (1994), "High frequency vibrations influence on heat transfer", Microgravity Q., 4(4): 259–268.

Lyubimova T.P., Scuridin R.V., Croell A., and Dold P., (2003), "Influence of high frequency vibrations on fluid flow and heat transfer in a floating zone", Cryst. Res. Technol., 38(7–8): 635–653.

M

Manga M.H. and Stone H.A. (1993), "Buoyancy-driven interactions between two deformable viscous drops", J. Fluid Mech., 256: 647–683.

Mather J.P., (1998), "Laboratory scaleup of cell cultures (0.5–50 liters)", Methods Cell Biol., 57: 219–227.

McPherson A., (1985) "Crystallization of macromolecules: general principles", Meth. Enzymology, 114: 112–120.

McPherson A., (1990), "Current approaches to macromolecular crystallization", Eur. J. Biochem., 189: 1–23.

McPherson A., (1993), "Effects of a microgravity environment on the crystallization of biological macromolecules", Microgravity Sci. Technol., VI/2: 101–109.

McPherson A., (1999), "Procedures for Macromolecular Crystallization, Crystallization of Biological Macromolecules", Cold Spring Harbor Laboratory Press, Cold Spring Harbor, NY.

Meaney D.F., Johnston E.D., Litt M., and Pollack S.R., (1998), "Experimental and numerical investigations of microcarrier motions in simulated microgravity", Adv. Heat Mass Transfer Biotechnol., ASME, HTD 63: 21–27.

Mehdi-Nejad V., Mostaghimi J., and Chandra S., (2003), "Two-fluid heat transfer based on a volume tracking advection algorithm", 9th International Conference on Liquid Atomization and Spray Systems (ILASS – 2003), 13–17 July, 2003, Sorrento, Italy).

Mercier J.F. and Normand C., (1996), "Buoyant-thermocapillary instabilities of differentially heated liquid layers", Phys. Fluids, 8(6): 1433–1445.

"Message Passing Toolkit: MPI programmer's Manual", Cray publication SR-2197 (1996).

Miller T.Y., He X., and Carter D.C., (1992), "A comparison between protein crystals grown with vapor diffusion methods in microgravity and protein crystals using a gel liquid–liquid diffusion ground-based method", J. Cryst. Growth, 122: 306–309.

Molenkamp T., Janssen L.P.B.M., and Drenth J., (1994), "Protein crystallisation and Marangoni convection", In Final Reports of Sounding Rocket Experiments in Fluid Science and Materials Sciences (Edited by O. Minster and B. Kaldeich (ESA SP-1132), Noordwijk, 4: 22–43.

Monaco A. and Rosenberger F., (1993), "Growth and etching kinetics of tetragonal lysozyme", J. Cryst. Growth, 129: 465–484.

Monberg E., (1994), "Bridgman and related growth techniques", In Handbook of Crystal Growth (Edited by D.T.J. Hurle), North-Holland, Amsterdam, 2: 53–97.

Monti R., (1987), "On the onset of the oscillatory regimes in Marangoni flows", Acta Astronautica, 15: 557–560.

Monti R. and Savino R., (1994a), "The basis and the recent developments of OMA and its applications to microgravity", Microgravity Q., 5(1): 13–17.

Monti R. and Savino R., (1994b), "A new approach to g-level tolerability for Fluid and Material Science experiments", Acta Astronautica, 37: 313–331.

Monti R. and Savino R., (1995), "The basis and the recent developments of Napolitano's scaling and order of magnitude analysis", Microgravity Q., 5(1): 13–17.

Monti R. and Savino R., (1996a), "Influence of g-jitter on fluid physics experimentation on-board the International Space Station", ESA SP-385: 215–224.

Monti R. and Savino R., (1996b), "Microgravity experiment acceleration tolerability on space orbiting laboratories", J. Spacecraft Rockets, 33(5): 707–716.

Monti R. and Savino R., (1996c), "Numerical model of non coalescence liquid drops and correlation with experimental results", Microgravity Q., 6(2–3): 102–108.

Monti R. and Savino R., (1997), "Correlation between experimental results and numerical solutions of the Navier–Stokes problem for two non coalescing liquid drops with Marangoni effect", Phys. Fluids, 9: 260–262.

Monti R., Langbein D., and Favier J.J., (1987), "Influence of residual accelerations on fluid physics and material science experiments", In Fluid and Material Science in Space (Edited by H.U. Walter).

Monti R., Albanese C., Carotenuto C., Caotagnolo D., Ceglia E., (1995), "First results from "onset" experiment during spacelab mission D-2", In Proceedings of the Norderney Symposium on Scientific Results of the German Spacelab Mission D2 (Edited by P.R. Sahm, M.H. Keller, and B. Schiewe): 247–258.

Monti R., Savino R., and Lappa M., (1997), "Oscillatory Thermocapillary flows in simulated floating zones with time-dependent boundary conditions", Acta Astronautica, 41: 863–875.

Monti R., Savino R., Lappa M., and Fortezza R., (1998a), "Scientific and technological aspects of a sounding rocket experiment on oscillatory Marangoni flow", Space Forum, 2: 293–318.

Monti R., Savino R., and Lappa M., (1998b), "Microgravity sensitivity of typical fluid physics experiment", Presented at the 17th Microgravity Measurements Group Meeting, Cleveland, Ohio, March 1998, published in the meeting proceedings in NASA CP-1998-208414, 23: 1–15.

Monti R., Savino R., and Tempesta S., (1998c), "Wetting prevention by thermal Marangoni effect. Experimental and numerical simulation", Eur. J. Mechanics B/Fluids, 17: 51–77.

Monti R., Savino R., Lappa M., and Tempesta S., (1998d), "Behaviour of drops in contact with liquid surfaces at non wetting conditions", Phys. Fluids, 10(11): 2786–2796.

Monti R., Savino R., Paterna D., and Lappa M., (2000a), "New features of drops dynamics under Marangoni effect", Recent Research Developments in Fluid Dynamics, Transworld Research Network, Trivandrum, India, 3: 15–32.

Monti R., Savino R., Lappa M., Carotenuto L., Castagnolo D., and Fortezza R., (2000b), "Flight results on Marangoni flow instability in liquid bridges", Acta Astronautica, 47(2–9): 325–334.

Monti R., Savino R., and Lappa M., (2000c), "Influence of geometrical aspect ratio on the oscillatory Marangoni convection in liquid bridges", Acta Astronautica, 47(10): 753–761.

Monti R., Savino R., and Lappa M., (2001), "On the convective disturbances induced by g-jitter on the space station", Acta Astronautica, 48(5–12): 603–615.

Muehlner K.A., Schatz M.F., Petrov V., McCormick W.D., Swift J.B., and Swinney H.L., (1997), "Observation of helical traveling-wave convection in a liquid bridge", Phys. Fluids, 9(6), 1850–1852.

Muller G., (1988), "Convection and inhomogeneities in crystal growth from the melt", In Crystals: Growth, Properties and Applications 12 (H.C. Freyhard), Springer, Berlin.

Mundrane M. and Zebib A., (1994), "Oscillatory buoyant thermocapillary flow", Phys. Fluids, 10: 3294–3305.

Mundrane M. and Zebib A., (1995), "Low Prandtl number Marangoni convection with a deformable interface", J. Thermophys. Heat Transfer, 9(4): 795–797.

N

Nakamura S., Kakimoto K., and Hibiya T., (1995), "Convection visualization and temperature fluctuation measurement in a molten silicon column", Materials and fluids under low gravity (Springer), Proceedings of the IXth European Symposium on Gravity-Dependent Phenomena in Physical Sciences held in Berlin, Germany, 2–5 May 1995.

Nakamura S., Hibiya T., Kakimoto K., Imaishi N., Nishizawa S.-i., Hirata A., Mukai K., Yoda S.-i., and Morita T.S., (1998), "Temperature fluctuations of the Marangoni flow in a liquid bridge of molten silicon under microgravity on board the TR-I4 rocket", J. Cryst. Growth, 186: 85–93.

Nakamura S., Eguchi M., Azami T., and Hibiya T., (1999), "Thermal waves of a non-axisymmetric flow in a Czochralski-type silicon melt", J. Cryst. Growth, 207: 55–61.

Naumann R.J., (1996), "Stabilizing/destabilizing effects of axial accelerations in Bridgman growth", J. Cryst. Growth, 165: 129–136.

Naumann R.J., (2000), "An analytical model for transport from quasi-steady and periodic accelerations on spacecraft", Int. J. Heat Mass Transfer, 43: 2917–2930.

Neitzel G.P. and Dell'Aversana P., (2002), "Noncoalescence and nonwetting behavior of liquids", Annu. Rev. Fluid Mech., 34: 267–289.

Neitzel G.P., Law C.C., Jancowski D.F., and Mittelmann H.D., (1991), "Energy stability of thermocapillary convection in a model of the float-zone crystal-growth process", Phys. Fluids A, 3: 2841–2846.

Neitzel G.P., Chang K.T., Jancowski D.F., and Mittelmann H.D., (1992), "Linear stability of thermocapillary convection in a model of the float-zone crystal-growth process", Phys. Fluids A, 5: 108–114.

Nienhüser Ch. and Kuhlmann H.C., (2002), "Stability of thermocapillary flows in non-cylindrical liquid bridges", J. Fluid Mech., 458: 35–73.

Nobari M.R., Yan Y.J., and Tryggvason G., (1996), "Head-On collision of drops – a numerical investigation", Phys. Fluids, 8: 29–42.

Noh W.F. and Woodward P.R., (1976), "SLIC (simple line interface method)", In Lecture Notes in Phys. (Edited by A.I. van de Vooren and P.J. Zandbergen), Springer-Verlag, Berlin/New York, 59: 330.

Noh D.S., Koh Y., and Kang I.S., (1998), "Numerical solutions for shape evolution of a particle growing in axisymmetric flows of supersaturated solution", J. Cryst. Growth, 183: 427–440.

O

Obradovic B., Meldon J.H., Freed L.E., and Vunjak-Novakovic G., (2000), "Glycosaminoglycan deposition in engineered cartilage: experiments and mathematical model", AIChE J., 46(9): 1860–1871.

Okano Y., Kunikata S.I., Fujioka T., Sakai S., and Koyama M., (1997), "Axisymmetric two-dimensional steady marangoni convection in a floating half-zone under microgravity conditions", Num. Heat Transfer. A, 31: 407–413.

O'Leary D.P. and White R.E., (1985), "Multisplitting of matrices and parallel solution of linear systems", SIAM J. Alg. Disc. Meth., 6: 630–640.

Ory E., Yuan H., Prosperetti A., Popinet S., and Zaleski S., (2000), "Growth and collapse of a vapor bubble in a narrow tube", Phys. Fluids, 12: 1268–1277.

Osher S. and Sethian J.A., (1988), "Fronts propagating with curvature-dependent speed: algorithms based on Hamilton-Jacobi formulations", J. Comput. Phys., 79: 12–49.

Osher S. and Fedkiw R., (2002), "The Level Set Method and Dynamic Implicit Surfaces", Springer-Verlag, New York.

Otàlora F. and Garcìa-Ruiz J.M., (1997), "Crystal growth studies in microgravity with the APCF: computer simulation and transport dynamics", J. Cryst. Growth, 182: 141–154.

Otàlora F., Novella M.L., Gavira J.A., Thomas B.R., and Garcìa-Ruiz J.M., (2001), "Experimental evidence for the stability of the depletion zone around a growing protein crystal under microgravity", Acta Cryst., D57: 412–417.

Otálora F., García-Ruiz J.M., Carotenuto L., Castagnolo D., Novella M.L., and Chernov A.A., (2002), "Lysozyme crystal growth kinetics in microgravity", Acta Cryst. D, 58: 1681–1689.

P

Pallares J., Grau F.X., and Girlt F., (1999), "Flow transitions in laminar Rayleigh–Bènard convection in a cubical cavity at moderate Rayleigh numbers", Int. J. Heat Mass Transfer, 43: 753–769.

Parmentier P., Regnier V., and Lebon G., (1993), "Buoyant-thermocapillary instabilities in medium-Prandtl number fluid layers subject to a horizontal temperature gradient", Int. J. Heat Mass Transf, 36: 2417–2427.

Peltier L. and Biringen S., (1993), "Time-dependent thermocapillary convection in a rectangular cavity: numerical results for a moderate Prandtl number fluid", Phys. Fluids, 257: 339–357.

Perez de Ortiz S.E. and Sawistowski H., (1973), "Interfacial instability of binary liquid–liquid systems, I. Stability analysis", J. Chem. Eng. Sci., 28: 2051–2061.

Petrov V., Haaning A., Muehlner K.A., VanHook S.J., and Swinney H.L., (1998), "Model-independent nonlinear control algorithm with application to a liquid bridge experiment", Phys. Rev. E, 58: 427–433.

Piccolo C., Lappa M., Tortora A., and Carotenuto L., (2002), "Non-linear behaviour of lysozyme crystallization", Physica A: Statistical Mechanics and its Applications, 314/1–4: 636–645.

Pollack S.R., Meaney D.F., Levine E.M., Litt M., and Johnston E.D., (2000), "Numerical model and experimental validation of microcarrier motion in a rotating bioreactor", Tiss. Eng., 6: 519–530.

Prange M., Wanschura M., Kuhlmann H.C., and Rath H.J., (1999), "Linear stability of thermocapillary convection in cylindrical liquid bridges under axial magnetic fields", J. Fluid Mech., 394: 281–302.

Preisser F., Schwabe D., and Scharmann A., (1983), "Steady and oscillatory thermocapillary convection in liquid columns with free cylindrical surface", J. Fluid Mech., 126: 545–567.

Prinz B. and Romero A., (1993), "New casting process for hypermonotectic alloys", In Immiscible Liquid Metals and Organics, Proceedings of an International Workshop Organized by the ESA Expert Working Group, "Immiscible Alloys" held in the Physikzentrum, Bad Honnef, 1992 (Edited by L. Ratke), Germany: 281–289.

Pulicani J.P., Krukowski S., Alexander J.I.D., Quazzani J., and Rosenberger F., (1992), "Convection in an asymmetrically heated cylinder", Int. J. Heat Mass Transfer, 35(9): 2119–2130.

Pusey M.L., Snyder R.S., and Naumann R., (1986), "Protein crystal growth: growth kinetics for tetragonal lysozyme crystals", J. Biol. Chem., 261(14): 6524–6529.

Q

Qi J. and Wakayama N.I., (2000), "Solute convection during the whole process of protein crystal growth", J. Cryst. Growth, 219: 465–476.

R

Ramachandran N., Baugher Ch.R., and Naumann R.J., (1995), "Modeling flows and transport in protein crystal growth", Microgravity Sci. Technol., VIII/3: 170–179.

Rao M.M. and Shyy W., (1997), "Moving boundary computation of the float-zone process", Int. J. Numer. Meth. Eng., 40(7): 1231–1261

Ratke L., (1987), "Simultaneous coarsening of dispersions by growth and coagulation", J. Colloid Interface Sci., 119: 391–397.

Ratke L. (Ed.), (1993), "Decomposition, phase separation, solidification of immiscible liquid alloys", In Immiscible Liquid Metals and Organics, Proceedings of an International Workshop Organized by the ESA Expert Working Group, "Immiscible Alloys" held in the Physikzentrum, Bad Honnef, 1992, Germany: 3–34.

Ratke L., Korekt G., and Drees S., (1996), "Solidification of immiscible alloys", ESA SP 385: 247.

Reddy A.V., Kothe D.B., Beckermann C., Ferrell R.C., and Lam K.L., (1997), "High resolution finite volume parallel simulations of mould filling and binary alloy solidification on unstructured 3-D meshes", In Proceedings of the Fourth Decennial International Conference on Solidification Processing (Edited by J. Beech and H. Jones), The University of Sheffield, UK, July 7–10: 83–87.

Renaut R.A. and Mittelmann H.D., (1995), "Parallel multisplitting for optimization", J. Parallel Alg. Appl., 7: 17–27.

Rider W.J. and Kothe D.B., (1998), "Reconstructing volume tracking", J. Comput. Phys., 141: 112–152.

Riley R.J. and Neitzel G.P., (1998), "Instability of thermocapillary-buoyancy convection in shallow layers. Part 1. Characterization of steady and oscillatory instabilities", J. Fluid Mech., 359: 143–164.

Robert M.C. and Lefaucheux F., (1988), "Crystal growth in gels: principle and applications", J. Cryst. Growth, 90(1–3): 358–367.

Robert M.C., Provost K., and Lefaucheux F., (1992), "Crystallization in gels and related methods", Chapter 6 in "Crystallization of Nucleic Acids and Proteins. A Practical Approach" (Edited by A. Ducruix and R. Giegé), Oxford University Press, Oxford: 127–143.

Rodriguez E.K., Hoger A., and McCulloch A., (1994), "Stress-dependent finite growth in soft elastic tissues", J. Biomechanics, 27(4): 455–467.

Rogers M.J.B., Vogt G.L., and Wargo M.J., (1997), "Microgravity: A Teacher's Guide", EG-1997-08-110-HQ, NASA publication (1997).

Rosenbaum D.F. and Zukoski C.F., (1996), "Protein interactions and crystallization", J. Cryst. Growth, 169: 752–758.

Rosenberger F., (1986), "Inorganic and protein crystal growth: similarities and differences", J. Cryst. Growth, 76: 618–636.

Rother M.A., Zinchenko A.Z., and Davis R.H., (1997) "Buoyancy-driven coalescence of slightly deformable drops", J. Fluid Mech., 346: 117–148.

Rother M.A., Zinchenko A.Z., and Davis R.H., (2002), "Three-dimensional boundary-integral algorithm for thermocapillary motion of deformable drops", J. Colloid Interf. Sci., 245: 358–364.

Roux B. and Hadid H.B. (1989), "Numerical simulation of oscillatory convection in semiconductor melts", J. Cryst. Growth, 97: 201–216.

Rudman M., (1997), "Volume-tracking methods for interfacial flow calculations", Int. J. Numer. Meth. Fluids, 24: 671–691.

Rupp R., Muller G., and Neumann G., (1989), "Three dimensional time dependent modelling of the Marangoni convection in zone melting configurations for GaAs", J. Cryst. Growth, 97: 34–41.

Russo G. and Napolitano L.G., (1984), "Order of Magnitude Analysis of unsteady Marangoni and Buoyancy free convection", 35th Congress of the International Astronautical Federation, October 7–13, 1984, Lausanne, Switzerland.

Rybicki A. and Floryan J.M., (1987), "Thermocapillary effects in liquid bridges. I. Thermocapillary convection", Phys. Fluids, 30: 1956–1972.

S

Saghir M.Z. and Maffei N., (1999), "Float zone of $Bi_{12}GeO_{20}$ STS 77: experimental and numerical results", Microgravity Space Station Utilization Journal, 1(1): 31–35.

Saghir M.Z. and Viviani A., (2001), "Use of magnetic field to reduce convection in low Prandtl number liquid bridge", MSSU, 2(1): 23–27.

Savino R., (1997), "Order of magnitude analysis of g-jitter induced disturbances in fluid-dynamic microgravity experimentation", Microgravity Q., 7(4): 141–154.

Savino R. and Monti R., (1997), "Modelling of non coalescing liquid drops in the presence of thermocapillary convection", Meccanica, 32: 115–133.

Savino R. and Monti R., (1999a), "Convection induced by residual-g and g-jitters in diffusion experiments", Int. J. Heat Mass Transfer, 42: 111–126.

Savino R. and Monti R., (1999b), "Improving diffusion controlled microgravity experiments by facility orientation", J. Aerospace Eng., 212, Part G: 415–426.

Savino R., Monti R., Lappa M., Carotenuto L., Castagnolo D., and Fortezza R., (1999), "Preliminary results of the sounding rocket experiment on oscillatory flows in liquid bridge" (PULSAR I Experiment), ELGRA News, 21: 79 (February 1999).

Savino R., Lappa M., and Paterna D., (2001a), "Experimental and numerical analysis of three-dimensional surface tension and buoyancy-driven flows in cavities", MSSU (Microgravity and Space Station Utilization), 2(1): 13–22.

Savino R., Monti R., and Alterio G., (2001b), "Drops pushing by Marangoni forces", Phys. Fluids, 13(5): 1513–1516.

Savino R. and Lappa M., (2003a), "Assessment of the thermovibrational theory: application to g-jitter on the Space-station", J. Spacecraft Rockets, 40(2), 201–210.

Savino R. and Lappa M., (2003b), "Dissolution and solutal convection in partially miscibile liquid systems", Int. J. Heat Mass Transfer, 47(3), 601–612.

Savino R., Paterna D., and Lappa M., (2003), "Marangoni flotation of liquid droplets", J. Fluid Mech., 479: 307–326.

Scardovelli S. and Zaleski S., (1999), "Direct numerical simulation of free surface and interfacial flow", Ann. Rev. Fluid. Mech., 31: 567–603.

Scriven L.E. and Sternling C.V., (1960), "The Marangoni effects", Nature, 187: 186–188.

Schneider S. and Straub J., (1989), "Influence of the Prandtl number on laminar natural convection in a cylinder caused by g-jitter", J. Cryst. Growth, 46: 125–131.

Schwabe D., (2002), "Buoyant-thermocapillary and pure thermocapillary convective instabilities in Czochralski systems", J. Cryst. Growth, 237–239: 1849–1853.

Schwabe D. and Metzger J., (1989), "Coupling and separation of buoyant and thermocapillary convection", J. Cryst. Growth, 97: 23–33.

Schwabe D., Preisser F., and Scharmann A., (1982), "Verification of the oscillatory state of thermocapillary convection in a floating zone under low gravity", Acta Astronautica, 9(4): 265–273.

Schwabe D., Hintz P., and Frank S., (1996), "New features of thermocapillary convection in floating zones revealed by tracer particle accumulation structures", Microgravity Sci. Technol., 9: 163–168.

Sekerka R.F., (2004), "Morphology: from sharp interface to phase field models", J. Cryst. Growth, 264(4): 530–540.

Sethian J., (1999), "Level Set Methods and Fast Marching Methods", Cambridge University Press.

Shangguan D., Ahuja S. and Stefanescu D.M., (1992), "An analytical model for the interaction between an insoluble particle and an advancing solid/liquid interface", Metall. Trans. A, 23A:669–680.

Shen Y., Neitzel G.P., Jankowsky D.F., and Mittelmann H.D., (1990), "Energy stability of the thermocapillary convection in a model of the float zone crystal-growth process", J. Fluid Mech., 217: 639–660.

Shevtsova V.M. and Legros J.C., (1997), "Thermocapillary stability of strongly deformed liquid bridges", Microgravity Q., 7(2): 53–58.

Shevtsova V.M., Melnikov D.E., and Legros J.C., (2001), "Three-dimensional simulations of hydrodynamical instability in liquid bridges: influence of temperature-dependent viscosity", Phys. Fluids, 13: 2851–2865.

Shevtsova V.M., Melnikov D.E., and Legros J.C., (2003), "Multistability of oscillatory thermocapillary convection in a liquid bridge", Phys. Rev. E, 68(6): 066311–066324.

Shiomi J., Kudo M., Ueno I., Kawamura H., and Amberg G., (2003), "Feedback control of oscillatory thermocapillary convection in a half-zone liquid bridge", J. Fluid Mech., 496: 193–211.

Sim B.-C. and Zebib A., (2002), "Thermocapillary convection in liquid bridges with undeformable curved surfaces", AIAA J. Thermophys. Heat Transfer, 16: 553–561.

Smith M.K., (1986), "Instability mechanism in dynamic thermocapillary liquid layers", Phys. Fluids, 29(10): 3182–3186.

Smith M.K. and Davis S.H. (1983), "Instabilities of dynamic thermocapillary liquid layers. Part 1: convective instabilities", J. Fluid Mech., 132: 119–144.

Smith B., Bjorstad P., and Gropp W., (1996), "Domain Decomposition, Parallel Multilevel Methods for Elliptic Partial Differential Equations", Cambridge University Press.

Soderberg L.D. and Alfredson P.H., (1998), "Experimental and theoretical stability investigations of plane liquid jets", J. Mech. B/Fluids, 17(5): 689–737.

Spenger H.J. and Peters G., (1997), "Promotion of industrial research on the space station", Microgravity Q., 7(1): 15–22.

Stefanescu D.M., Moitra A., Kacar A.S. and Dhindaw B.K., (1990), "The influence of buoyant forces and volume fraction of particles on the particle pushing/entrapment transition during directional solidification of Al/SiC and Al/graphite composites. Metall. Trans. A21, 231–239.

Subramanian R.S. and Balasubramaniam R., (2001), "The Motion of Bubbles and Drops in Reduced Gravity", Cambridge University Press, Cambridge, UK.

Subramanian R.S., Zhang L., and Balasubramaniam R., (2001), "Mass transport from a drop executing thermocapillary motion", Microgravity Sci. Technol., XII(3/4): 107–115.

Sussman M., Smereka P., and Osher S., (1994), "A level set approach for computing solutions to incompressible two-phase flow", J. Comput. Phys., 114: 146–159.

T

Taber L.A., (1998a), "Biomechanical growth laws for muscle tissue", J. Theor. Biol., 193: 201–213.

Taber L.A., (1998b), "A model for aortic growth based on fluid shear and fiber stresses", J. Biochem. Eng., 120: 348–354.

Taber L.A. and Chabert S., (2002), "Theoretical and experimental study of growth and remodeling in the developing heart", Biomechan. Model Mechanbiol., Springer-Verlag: 29–43.

Tang Z.M., Hu W.R., and Zhou R., (1995), "Fractal features of oscillatory convection in the half floating zone", Int. J. Heat Mass Transfer, 38: 3295–3303.

Tang H., Tang Z.M., Hu W.R., Chen G., and Roux B., (1996), "Numerical simulation of g-jitter effects on half floating zone convection under microgravity environment", Microgravity Sci. Technol., IX: 28–34.

Tang Z.M., Hu W.R., Chen G., and Roux B., (1997), "Numerical simulation of 3D thermocapillary oscillatory convection", Microgravity Q., 7(2): 37–42.

Todd P. and Gruener R., (1992), "Clinostats and Centrifuges", ASGSB Bulletin, 5(2): 84–93.

Topper J.N. and Gimbrone M.A., Jr., (1999), "Blood flow and vascular gene expression: fluid shear stress as a modulator of endothelial phenotype", Mol. Med. Today, 5: 40–46.

Touihri R., Ben Hadid H., and Henry D., (1999a), "On the onset of convective instabilities in cylindrical cavities heated from below. I. Pure thermal case", Phys. Fluids, 11(8): 2078–2088.

Touihri R., Ben Hadid H., and Henry D., (1999b), "On the onset of convective instabilities in cylindrical cavities heated from below. II. Effect of a magnetic field", Phys. Fluids, 11(8): 2089–2100.

Tryggvason G., Bunner B., Esmaeeli A., Juric D., Al-Rawahi N., Tauber W., Han J., Nas S., and Jan Y.-J., (2001), "A front tracking method for the computations of multiphase flow", J. Comput. Phys., 169: 708–759.

U

Udaykumar H.S., Mittal R., and Shyy W., (1999), "Computation of solid–liquid phase fronts in the sharp interface limit on fixed grids", J. Comput. Phys., 153(2): 535–574.

Ueno I., Ono, Y., Nagano D., Tanaka S., and Kawamura H., (2000), "Modal oscillatory structure and dynamic particle accumulation in liquid bridge Marangoni convection", In Proc. 4th JSME-KSME Thermal Eng. Conf., Kobe, Japan, Oct. 1–6, Vol. III: 265–270.

Ueno I., Tanaka S., and Kawamura H., (2003), "Oscillatory and chaotic thermocapillary convection in a half-zone liquid bridge", Phys. Fluids, 15(2): 408–416.

Unverdi O.S. and Tryggvason G., (1992), "A front tracking method for viscous incompressible flows", J. Comput. Phys., 100: 25–37.

V

Vekilov P.G. and Chernov A.A., (2002), "The Physics of Protein Crystallization, Solid State Physics" (Edited by H. Ehrenreich and F. Spaepen), Academic Press, Amsterdam, 57: 1–147.

Vekilov P.G., Iwan J., Alexander D., and Rosenberger F., (1996), "Non-linear response of layer growth dynamics in the mixed kinetics-bulk transport regime", Phys. Rev. E, 54(6): 6650–6660.

Velten R., Schwabe D., and Scharmann A., (1991), "The periodic instability of thermocapillary convection in cylindrical liquid bridges", Phys. Fluids A, 3: 267–279.

Vergara A., Lorber B., Zagari A., and Giege R., (2003), "Physical aspects of protein crystal growth investigated with the Advanced Protein Crystallization Facility in reduced-gravity environments", Acta Cryst., D59: 2–15.

Villers D. and Platten J., (1992), "Coupled Buoyancy and Marangoni convection in acetone: experiments and comparison with numerical simulations", J. Fluid Mech., 234: 487–510.

Viviani A., Golia C., and Cioffi M., (1996), "Numerical solution of three-dimensional convection in cylinders", Microgravity Q., 6: 21–32.

Voller V.R. and Prakash C., (1987), "A fixed grid numerical modelling methodology for convection–diffusion mushy region phase-change problems", Int. J. Heat Mass Transfer, 30(8): 1709–1719.

W

Walter H.U., (1984), "Preparation of dispersion alloys-component separation during cooling and solidification of dispersions of immiscible alloys", In Proc. Workshop on Effect of Gravity on Solidification of Immiscible Alloys, Stockholm, January 18–20, 1984, ESA SP-219, March 1984: 47 (post-flight).

Wanschura M., Shevtsova V., Kuhlmann H.C., and Rath H.J., (1995), "Convective instability mechanism in thermocapillary liquid bridges", Phys. Fluids, 5: 912–925.

Wanschura M., Kuhlmann H.C., and Rath H.J., (1996), "Three-dimensional instability of axisymmetric buoyant convection in cylinders heated from below", J. Fluid Mech., 326: 399–415.

Wanschura M., Kuhlmann H.C., and Rath H.J., (1997a), "Instability of thermocapillary flow in symmetrically heated full liquid zones", In Proceedings of the Journal of the Joint Xth European and Vith Russian Symposium on Physical Sciences in Microgravity (Edited by V.S. Avduyevsky and V.I. Polezhaev), 1: 172.

Wanschura M., Kuhlmann H.C., and Rath H.J., (1997b), "Linear stability of two-dimensional combined buoyant-thermocapillary flow in cylindrical liquid bridges", Phys. Rev. E, 55(6): 7036–7042.

Warren C. and Webb W.W., (1969), "Interfacial tension of near-critical cyclohexane–methanol mixtures", J. Chem. Phys., 50(9): 3694–3700.

Wilcox W.R. and Regel L.L., (2001), "Microgravity effects on materials processing: a review", Published in Conference Proceedings of EUROMAT 2001 Rimini, Italy, 10–14 July 2001.

Wolf, D.A. and Schwartz R.P., (1991), "Analysis of gravity-induced particle motion and fluid perfusion flow in the NASA-designed rotating zero-head-space tissue culture vessel", NASA Technical Paper 3143.

Wolf, D.A. and Schwartz R.P., (1992), "Experimental measurement of the orbital paths of particles sedimenting within a rotating viscous fluid as influenced by gravity", NASA Technical Paper 3200.

Wu P.K. and Faeth G.M., (1995), "Onset and end of drop formation along the surface of turbulent liquid jets in still gases", Phys. Fluids, 7(11): 2915–2917.

Wu M., Ludwig A., and Ratke L., (2003), "Modelling the solidification of hypermonotectic alloys", Modelling Simul. Mater. Sci. Eng., 11(5): 755–769.

X

Xu J.J. and Davis S.H., (1984), "Convective thermocapillary instabilities in liquid bridges", Phys. Fluids, 27: 1102–1107.

Xu J. and Zebib A., (1998), "Oscillatory two- and three-dimensional thermocapillary convection", J. Fluid Mech., 364: 187–209.

Y

Yang Y.K. and Kou S., (2001), "Temperature oscillation in a tin liquid bridge and critical Marangoni number dependency on Prandtl number", J. Cryst. Growth, 222: 135–143.

Yasushiro S., Sato T., and Imaishi N., (1997), "Three dimensional oscillatory Marangoni flow in a half zone of Pr=1.02 fluid", Microgravity Sci. Technol., X/3: 144–153.

Yasushiro S., Imaishi N., Kuhlmann H.C., and Yoda S., (1999), "Numerical simulation of three dimensional oscillatory thermocapillary flow in a half zone of Pr=1 fluid", Adv. Space Res., 24: 1385–1390.

Yasushiro S., Imaishi N., Lappa M., and Yoda S., (2001), "Numerical simulation of three dimensional Marangoni convection in non-cylindrical liquid bridges of low Pr fluids", Proc. 17th Jpn. Soc. Microgravity Application Conf., B113, Sendai, Oct. 2001.

Yasushiro S., Sato T., Imaishi N., and Yoda S., (2000), "Three dimensional Marangoni flow in liquid bridge of low Pr fluid", Space Forum, 6: 39–47.

Yost F.G., Hosking F.M., and Frear D.R., (1993), "Introduction: the mechanics of solder alloy wetting and spreading", In The Mechanics of Solder Alloy Wetting and Spreading (Edited by F.G. Yost, F.M. Hosking, and D.R. Frear), The Minerals, Metals and Materials Society, Warrendale, PA.

Young N.O., Goldstein J.S., and Block M.J., (1959), "The motion of bubbles in a vertical temperature gradient", J. Fluid Mech., 6, 350–360.

Z

Zeng Z., Mizuseki H., Higashino K., and Kawazoe Y., (1999), "Direct numerical simulation of oscillatory Marangoni convection in cylindrical liquid bridges", J. Cryst. Growth, 204: 395–404.

Zeng Z., Mizuseki H., Chen J., Ichinoseki K., and Kawazoe Y., (2004), "Oscillatory thermocapillary convection in liquid bridge under microgravity", Materials Transactions, 45(5): 1522–1527.

Zhang Y. and Alexander J.I., (1992), "Surface tension and buoyancy driven flow in a non isothermal liquid bridge", Int. J. Num. Meth. Fluids, 14: 197–215.

Zhou H. and Davis R.H., (1996), "Axisymmetric thermocapillary migration of two deformable viscous drops", J. Colloid Interf. Sci., 181: 60–72.

Zinchenko A.Z., (1994), "An efficient algorithm for calculating multiparticle thermal interaction in a concentrated dispersion of spheres", J. Comput. Phys., 111: 120–134.

Zinchenko A.Z. and Davis R.H., (2003), "Large-scale simulations of concentrated emulsion flows", Phil. Trans. R. Soc. Lond., 361: 813–845.

Zinchenko A.Z., Rother M.A., and Davis R.H., (1997) "A novel boundary-integral algorithm for viscous interactions of deformable drops", Phys. Fluids, 9: 1493–1515.

Zinchenko, A.Z., Rother, M.A., and Davis, R.H., (1999), "Cusping, capture, and breakup of interacting drops by a curvatureless boundary-integral algorithm", J. Fluid Mech., 391: 249–292.

The disciplines treated in this book continuously undergo evolutionary progress.

I will be glad to discuss these aspects at the E-mail addresses marlappa@marscenter.it and marlappa@unina.it.

Any errors are my own fault, and I would greatly appreciate knowing about them along with any suggestion for improving the volume.

Additional material, color figures and animations dealing with the results presented in the book have been placed on the internet: http://staff.marscenter.it/lappa and http://utenti.lycos.it/MarcelloLappa.

Index

Page numbers for illustrations are shown in bold print